Statistical Physics

An Introductory Course

Statistical Physics

An Introductory Course

Daniel J. Amit

Universita di Roma "La Sapienza" and The Hebrew University

Yosef Verbin

The Open University of Israel

Translated from the Hebrew by

Rami Tzafriri

Published by

World Scientific Publishing Co. Pte. Ltd.

P O Box 128, Farrer Road, Singapore 912805

USA office: Suite 1B, 1060 Main Street, River Edge, NJ 07661

UK office: 57 Shelton Street, Covent Garden, London WC2H 9HE

British Library Cataloguing-in-Publication Data
A catalogue record for this book is available from the British Library.

First published in Hebrew in 1995 by The Open University of Israel.

STATISTICAL PHYSICS: AN INTRODUCTORY COURSE

ISBN 981-02-3192-X
ISBN 981-02-3476-7 (pbk)

Printed in Singapore by Uto-Print

Contents

Preface

This book is an introduction to statistical physics aimed at undergraduate students in physics and engineering. It may also serve as a reference for graduate students and researchers. The fact that thermodynamics and statistical physics have a very wide domain of relevance and validity, as well as a very long tradition, often leads to abstract and axiomatic presentations even for beginners. We have chosen the opposite direction, namely to discuss the key ideas and methods through concrete representative systems, and to treat general ideas as casual by-products. This choice is expressed already in the structure of the book, consisting of five parts: (I) *The Kinetic Theory of Gases*; (II) *Statistical Physics with Paramagnets*; (III) *Statistical Physics and Thermodynamics*, which deals with the Einstein solid and monoatomic ideal gases; (IV) *From Ideal Gas to Photon Gas*, which covers also equilibrium of chemical reactions and the Debye model; (V) *Of Fermions and Bosons*.

This approach runs the pedagogical risk that a casual reader may form the impression that he is facing a limited set of special cases. We confront this pitfall technically, by introducing explicit remarks about the generality of results at appropriate places; methodologically, by accumulating enough applications for every major idea to make its validity and generality stand out; and philosophically, observing that physics moves forward most of its ideas by analogies to cleverly chosen simple systems for which profound intuitions have been formed.

Originally this text was the backbone of a course in statistical physics at the Open University of Israel, which is a university for education at a distance. As such, it is vital to provide the student with a text that not only presents the material clearly, but also stimulates him or her to a higher state of active participation, to replace frontal study. This is achieved by inserting a large number of tasks (exercises) into the body of the text. They are aimed at maintaining contact with the experience of the student, either by numerical examples or by rederiving a result in a new way. They are also intended to reduce the amount of inattentive reading, by systematic insertions of break-points. Exercises of a second type serve as corollary applications of newly introduced methods and techniques. A third type fills the gaps left (intentionally) in the process of many derivations. In some places we have preferred not to break the flow of reasoning, introducing first the result and only then the corresponding exercise which calls for the reader to complete the details. Thus, a little patience is required at least at the outset.

In order to raise further the level of active involvement, each part is followed by several "self-assessment" exercises which are generally more extensive and of higher level than the ones in the text. They require frequently an ability to integrate ideas and methods from several parts of the course. The last — and very important — component is the detailed solutions to all exercises of all types, at the end of each part, which should contribute significantly to a successful study.

After the first Hebrew edition was used for about ten years, a second edition was published and is still in use by the Open University. It is this revised and extended version that we now make available to a wider audience. This volume is mainly a translation of the second Hebrew edition, but includes further revisions, additions and updates and should be considered a third edition of the text.

The material of this text corresponds to a semester's course, preferably in a second year. It assumes a prior acquaintance with calculus, basic mechanics, electricity and magnetism and modern physics, as well as some familiarity with thermodynamic concepts. Usually, a statistical physics course is taken after thermodynamics. However, we felt that the text would be more self-contained if a brief compendium of thermodynamic concepts and methods was introduced for coherence with the rest of the text. This led to Chap. 0 of Part II.

A final word concerning notation and units. We have adopted the convention of bold letters for vectors and italics for their absolute value or other scalars. Thus we write, for example, $|\mathbf{v}| = v$. As for units we follow the increasing tendency towards the SI (International System), based on the metric system. However, this convention is used in moderation and we allow, from time to time, other commonly used units, like electron volt (eV) as an energy scale for atomic systems or atmosphere for pressure. The deviations from the SI are particularly pronounced in dealing with magnetic systems. In that case we have chosen to avoid the confusion caused by attributing different units to the magnetic quantities **B**, **H** and **M**, and have adhered to the cgs system.

This book has benefited from the many fruitful suggestions and comments by colleagues, students and reviewers associated with the three editions, including Daniel Bar, David J. Bergman, Rachel Danor, Yossef Dothan, Ofer Eyal, Aharon Kapitulnik, Yoram Kirsh, Ora Maor-Bareket, Guy Sella, Yonathan Shapir, Haim Shinar and Shmuel Weiss. The book would not have taken its present form without their help.

Daniel Amit and Yosef Verbin
Rome and Tel-Aviv, June 1999

Part I

The Kinetic Theory of Gases

Introduction

The theoretical transition from mechanics to the properties of matter is complex as well as technically difficult. Major parts of it are ever-expanding research subjects. In principle, given the laws of dynamics — classical or quantum-mechanical — and a model of the forces acting between the constituent parts of the material, we can determine all that is fit to be determined about the system. Its properties, i.e. material properties, are subject to computation — if the initial conditions are given as well, of course.

However, a moment's reflection is sufficient to make one realize that this is not the right approach. The attempt to characterize initial conditions for a macroscopic system, such as 10^{23} electrons in a cubic centimeter of metal, or to solve 10^{23} differential equations, is bound to convince one that this is technically impossible. A more profound reason for abandoning this approach is the fact that the duration of a typical experiment (or measurement) on a macroscopic system is very long compared to times during which the system changes on a microscopic scale.

For example, let us calculate the average number of collisions of a gas molecule in an experiment that lasts 10^{-3} s. Suppose that the density of the gas is 10^{20} cm^{-3}, and that it is maintained at a relatively low temperature, 100 K. If we assume that the mass of the molecule is about 10^{-27} kg, then its typical speed is about 10^3 m s^{-1}. At the given density the typical distance between molecules is 10^{-9} m, so that a molecule will collide approximately once every 10^{-12} s. Hence, the ratio between the duration of the experiment and the average time between collisions is of order 10^9.

That is to say, during the experiment the system passes over a huge number of states. What is measured in the experiment is, actually, an average of all these states. The detailed dynamical transitions from one state to another are not only technically difficult to follow, but also uninteresting.

Classical thermodynamics demonstrates the above consideration in its extreme form. Not only is the detailed dynamical evolution of the system discarded, but the entire dynamical model. From here stems the vast power of thermodynamics, and also the source of its weakness. Its power lies in its generality, which has withstood to date the upheavals of dynamics — particles, continuum, fields, relativity and quantum mechanics. Its weakness is the lack of detail and its restriction to equilibrium states (or to near equilibrium states; the thermodynamics of states that are far from equilibrium is an important research subject — however, it is still in its infancy).

The most detailed level for the treatment of phenomena in systems with many particles (or degrees of freedom) is statistical mechanics, at which we will arrive in the coming parts of this book. A sort of intermediate level between thermodynamics and statistical mechanics is found in the kinetic theory of gases, which we discuss in detail in this part.

■ The degrees of freedom of a system are a collection of independent variables, required to characterize the system. Thus, for instance, the degrees of freedom of a gas, in a thermodynamic description, are its volume and pressure (or a pair of other variables). The degrees of freedom of a system of particles are the collection of the coordinates and the velocities of all the particles.

This theory, whose development we owe to such giants as Maxwell, Boltzmann and Gibbs, was of crucial importance for the conceptual development of physics in the second half of the 19th century. The successes of the theory, such as the calculation of the dependence of the viscosity upon pressure and temperature and its measurement, which was carried out by Maxwell (see Chapter 3 of this part), contributed to the establishment of the particle description of matter, and thus also to the unifying outlook in physics, i.e. to the strengthening of the relationship between the Newtonian dynamics and the macroscopic properties. On the other hand, kinetic theory has provided the hints which led Willard Gibbs to formulate statistical mechanics, at the beginning of this century, in the form which is still commonly accepted today.

In this and the subsequent parts we will discuss, among other things, topics which may be familiar to the reader from previous study. In most cases such topics constitute tests of the more detailed level of our discussion: each treatment at a detailed level must stand an uncompromising test — it must reproduce the results that are known at the less detailed level. Thus, for instance, results derived from the kinetic theory must be compatible with the laws of thermodynamics; statistical mechanics must agree with the kinetic theory as well as with thermodynamics. In some cases the recurring subjects will appear in a logical order different from before, and from a different standpoint. This does not disqualify, or cast doubt on, the preceding approach. Different approaches make different assumptions and the scope of their results varies: different approaches usually emphasize different aspects — more experimental or more formal. Contemplating and confronting different approaches widens and deepens the understanding of the results and of the deductive process in the physical theory.

Here we approach the kinetic theory of gases in its status as an intermediate level. On the one hand, we will view it as a tool for the calculation of important quantities — such as transport coefficient — and for the development of physical intuition in complex cases. On the other hand, we will try to view it as a corridor for ideas and methods of statistical mechanics, with which we deal in the coming parts.

■ Transport coefficients are quantities that characterize the rate at which a system approaches equilibrium when there is a slight deviation from a state of equilibrium. (See Chapter 3 of this part.)

Chapter 1

Velocity and Position Distributions of Molecules in a Gas

1.1 Avogadro's law, or the equation of state of an ideal gas

The equation of state of ideal gases and its connection with the Boltzmann distribution can be introduced in at least three different ways. One way is to postulate the Boltzmann distribution as the fundamental principle of statistical mechanics, from which the equation of state of an ideal gas can be derived. We opted for the opposite way. We will derive the ideal gas equation of state from a few very simple fundamental considerations. This will serve as a first step towards the introduction of the Boltzmann distribution toward the end of this chapter. You will get to know a third approach for reaching the same point in Part III. Before deriving the equation of state of an ideal gas, which depends on the concept of temperature, we discuss Avogadro's law:

Avogadro's law

At equal pressure and temperature, equal volumes of gases contain an equal number of molecules.

Avogadro deduced this law from an empirical rule in chemistry: if two gases, at the same temperature and pressure, combine in a chemical reaction without leftover constituents, their volumes stand in simple integral proportions. For example, if H_2 and O_2, at the same temperature and pressure, combine to give H_2O, with no leftovers of oxygen or hydrogen, the corresponding volumes must be in a ratio of 2:1. We would say that two H_2 molecules are needed to combine with every O_2 molecule. Here we will show that Avogadro's law is a direct consequence of Newton's laws.

A microscopic description of a dilute gas begins with the clarification of the concept of pressure. In order to keep things simple we will only consider a gas whose molecules all have the same mass. The pressure of a gas is the force which the gas exerts on a unit area. This pressure is measured by the force which must be applied on a piston in order to keep it stationary. The force which the gas exerts on the piston is due to the momentum imparted by the gas molecules which collide with it. For example, if a molecule undergoes an elastic collision with the piston, the magnitude of the component of the momentum that is perpendicular to the piston is conserved, but its sign is reversed; the component parallel to the piston does not change (see Fig. 1.1.1).

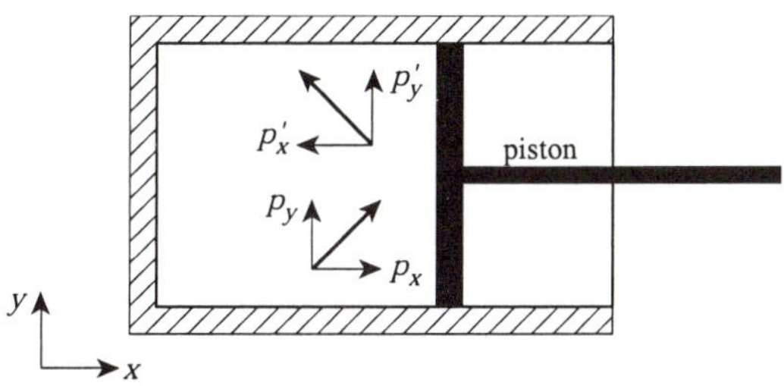

Fig. 1.1.1 A gas molecule colliding with the piston.

Therefore, in such a collision the amount of momentum imparted to the piston is

$$\Delta p_x = 2p_x = 2mv_x\,.$$

The number of molecules with a velocity component v_x, along the x axis, that will hit the piston, of area A, during a very short time interval Δt, is equal to the number of molecules with such a velocity, inside a cylinder of area A and length $v_x\Delta t$. This number is given by the product of the volume $Av_x\Delta t$ and the density. The amount of momentum, transferred to the piston during the time Δt, by this type of molecules (with velocity v_x along x) is

$$\Delta p_x = 2mv_x \cdot v_x \Delta t A \cdot n(v_x)\,, \tag{1.1.1}$$

where $n(v_x)$ is the number of molecules per unit volume with velocity component v_x.

The force exerted on the piston is given by the amount of the momentum transferred to the piston per unit time; this force per unit area is the pressure.

■ Note that this is where Newton's laws enter.

Therefore, the contribution of molecules with a velocity component v_x to the pressure is

$$P(v_x) = 2 \cdot mv_x^2 n(v_x)\,. \tag{1.1.2}$$

Clearly, not all of the gas molecules have the same velocity component along x, and we must sum over all the possible values of v_x. If we carry out this summation in Eq. (1.1.2) we find, for the total pressure exerted on the piston,

$$P = n \cdot \langle mv_x^2 \rangle\,, \tag{1.1.3}$$

where n denotes the total number of molecules per unit volume, and the angular brackets denote an average.

■ Note that the number of molecules per unit volume is

$$n = \sum n(v_x)\,,$$

where the sum is over all values of v_x and the average of v_x^2 is

$$\langle v_x^2 \rangle = \frac{1}{n} \sum n(v_x) v_x^2\,.$$

Exercise 1.1

(a) Prove that in a state of equilibrium the sum of (1.1.2) is indeed equal to (1.1.3). Where did the 2 disappear?

(b) What would the formulation of (1.1.1)–(1.1.3) be if you were to treat the velocity as a continuous variable?

Solution on page 74

In a state of equilibrium the averages of v_x^2, v_y^2 and v_z^2 will be equal, so in this case we can write (1.1.3) as

$$P = \frac{1}{3} nm \langle v^2 \rangle = \frac{2}{3} \frac{N}{V} \left\langle \frac{1}{2} mv^2 \right\rangle = \frac{2}{3} \frac{E}{V}\,. \tag{1.1.4}$$

■ $\langle v^2 \rangle = \langle \mathbf{v}^2 \rangle = \langle v_x^2 \rangle + \langle v_y^2 \rangle + \langle v_z^2 \rangle$.

E is the total kinetic energy of the gas, N is the total number of molecules and V is its volume. Note that the energies used in calculating E are the kinetic energies of the centers of mass of the molecules (or of the atoms, in the case of a monoatomic gas).

Equation (1.1.4) is a very impressive result (even though it was simple to derive!). It states that two gases kept at the same pressure, and whose molecules have the same average kinetic energies (independent of mass, structure and color), will occupy equal volumes, if they contain the same number of molecules. And all this from the direct use of the atomic assumption, Newton's laws and nothing more.

This, however, is not exactly Avogadro's law, but it is not far from it. That is, if we try to identify "temperature" with "average kinetic energy" of a molecule — up to a constant factor, which will take care of the units — we obtain on the one hand Avogadro's law, and on the other the equation of state of an ideal gas. The constant mentioned, k, will be chosen such that

equation of state

$$\left\langle \frac{1}{2} mv^2 \right\rangle = \frac{3}{2} kT \tag{1.1.5}$$

and T is the temperature.

From (1.1.4) we obtain

$$PV = \frac{2}{3}E = NkT\,. \tag{1.1.6}$$

The manner in which we identified the average kinetic energy with the absolute temperature may seem arbitrary. Nevertheless, arbitrary identification is characteristic and necessary when one is passing from one level of description onto another. As we will see later on — for example in the identification of the laws of thermodynamics within the framework of statistical mechanics (Part III) — in transitions of this sort one has to identify within a broader framework (Newtonian dynamics, in the case we discussed) concepts that are defined naturally in the less detailed description (the temperature, in thermodynamics). But the identification is not as arbitrary as it may appear at this stage. It must undergo many consistency tests of additional results derived from it.

Another way of looking at Eq. (1.1.6) is to say: "But is this not the familiar gas law we have always cherished?" Of course it is! And this is an alternative way of validating our interpretation of (1.1.5) as defining temperature.

Exercise 1.2

It is possible to treat electromagnetic radiation in a container, whose walls are mirrors, as a gas of particles (photons) with a constant speed c and whose energy is related to their momentum, p, which is directed parallel to their velocity, by $\epsilon(p) = pc$.

Show that if the container in Fig. 1.1.1 is full of radiation, the equation of state will be

$$PV = \frac{1}{3}E\,. \tag{1.1.7}$$

Solution on page 75

Finally, we note that once the temperature has been identified, it is possible to express the partition of the average kinetic energy as follows: with every direction of motion of a molecule we associate an average energy of $\frac{1}{2}kT$. Instead of "direction of motion" the accepted term is "degree of freedom"; it generalizes the term "direction of motion" also to rotations and internal vibrations. We have seen, therefore, that a molecule in an ideal gas has three degrees of freedom.

degrees of freedom

1.2 Temperature and thermal equilibrium

If we adopt the identification (1.1.5), we obtain the ideal gas equation of state, in which we know how to identify the constant k. This is, of course, the Boltzmann constant:

$$k = 1.38 \times 10^{-23} \text{ J K}^{-1} .$$

If we express the number of molecules in terms of the number of moles ν and Avogadro's number $N_0 : N = N_0\nu$, then Eq. (1.1.6) will take the form

$$PV = \nu RT , \tag{1.1.8}$$

where

$$R = N_0 k = (6.02 \times 10^{23}) \cdot (1.38 \times 10^{-23}) \text{ J K}^{-1} = 8.3 \text{ J K}^{-1} .$$

Exercise 1.3

Calculate the average kinetic energy per gas molecule at room temperature.

Solution on page 76

As mentioned in the previous section, the identification of the temperature requires many additional tests. The temperature has many properties, which are known from our daily experience and from thermodynamics. Central among these is the role the temperature plays in determining the equilibrium between systems, which interact thermally as well as mechanically.

Let us consider a volume in which there is a mixture of two gases. In a state of equilibrium, the number of molecules with a certain velocity $\mathbf{v}$ is independent of direction. This statement deserves additional reflection, since the argument that leads to it is very typical in the framework of the kinetic theory of gases. Indeed, the molecules collide with each other at a high rate. In collisions, even if they are elastic, the directions of the velocities change. Suppose there were a preferred direction, i.e. there were more molecules moving in that direction than in others. Then more molecules moving in the preferred direction would scatter to other (less preferred) directions than the other way around. Therefore, the state of the system would change with time, and this would not be a state of equilibrium. Only if the distribution of the velocity is independent of direction is a state of equilibrium possible. This is a necessary condition. This condition immediately implies that $\langle v_x \rangle = \langle v_y \rangle = \langle v_z \rangle = 0$, and therefore

$$\langle \mathbf{a} \cdot \mathbf{v} \rangle = 0 \tag{1.1.9}$$

for every constant vector $\mathbf{a}$.

We will now prove that in a state of equilibrium, the average kinetic energy per molecule, in a mixture of two gases, is equal. To this end let us consider pairs of molecules that include one molecule of each type of gas. Let $\mathbf{v}_1$ and $\mathbf{v}_2$ denote the velocities of the two molecules of the pair. Each of the gases satisfies Eq. (1.1.9), separately.

In addition, $\mathbf{v}_1$ and $\mathbf{v}_2$ are independent of one another, so the average of their product equals the product of the separate averages, both of which vanish in equilibrium. Hence

$$\langle \mathbf{v}_1 \cdot \mathbf{v}_2 \rangle = 0\,, \tag{1.1.10}$$

where the averaging is over both types of molecules.

We now note that instead of describing the motion of the molecules by the velocities $\mathbf{v}_1$ and $\mathbf{v}_2$, it is possible to describe it using the velocity of the center of mass, $\mathbf{v}_{\text{cm}}$, and the relative velocity, $\mathbf{v}_{\text{rel}}$.

center of mass

■ Reminder:

$$\mathbf{v}_{\text{cm}} = \tfrac{1}{m_1+m_2}(m_1\mathbf{v}_1 + m_2\mathbf{v}_2)\,,$$
$$\mathbf{v}_{\text{rel}} = \mathbf{v}_1 - \mathbf{v}_2\,.$$

If the distributions of $\mathbf{v}_1$ and $\mathbf{v}_2$ are independent so are those of $\mathbf{v}_{\text{cm}}$ and $\mathbf{v}_{\text{rel}}$. In equilibrium, these distributions will also be independent of direction. Consequently, in a state of equilibrium,

$$\langle \mathbf{v}_{\text{cm}} \cdot \mathbf{v}_{\text{rel}} \rangle = 0\,. \tag{1.1.11}$$

From Eqs. (1.1.10) and (1.1.11) we obtain

$$\left\langle \frac{1}{2}m_1 v_1^2 \right\rangle = \left\langle \frac{1}{2}m_2 v_2^2 \right\rangle . \tag{1.1.12}$$

Exercise 1.4

Deduce the equality in (1.1.12) from Eqs. (1.1.10) and (1.1.11).

Solution on page 76

The conclusion is that if the system is in equilibrium, the average kinetic energy per molecule is equal in the two gases. This means that the identification we made [Eq. (1.1.5)] is in accord with the fact that if the two gases are in equilibrium, their temperatures are equal. In the language of mechanics we can say that in a state of equilibrium the distribution of molecules in the heavy gas will tend toward the slower molecules. This means that there will be more slow molecules in the heavy gas than in the lighter gas, and fewer fast molecules (see Fig. 1.1.2).

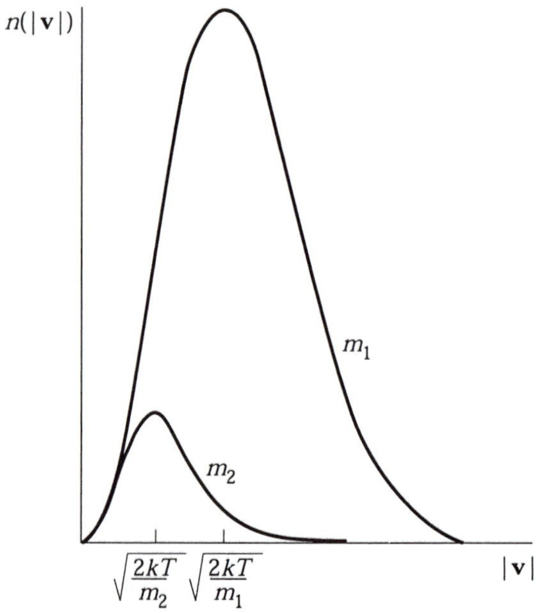

Fig. 1.1.2 A qualitative description of the distribution of the absolute value of the velocity in a gas for two mass values: $m_1 < m_2$. In order to make the picture clearer we have also chosen a state with more light molecules. Concerning the values along the horizontal axis, see self-assessment exercise 2b.

Exercise 1.5

Dalton's law

Dalton's law states that the pressure of a mixture of gases is equal to the sum of pressures that the gases would exert if each of them were separately in the same conditions as the mixture.

Deduce this law from similar reasoning to that we used to obtain (1.1.5).

Solution on page 77

Without going into too many complications, we will treat somewhat superficially, openly so, the case where a piston, of negligible friction, separates two gases inside a cylinder (Fig. 1.1.3). We ask: When will this system be in a state of equilibrium?

First, from mechanics we conclude that the forces (or pressures) exerted on the two sides of the piston must be equal. Equation (1.1.4) implies therefore that

$$\frac{N_1}{V_1}\left\langle\frac{1}{2}m_1v_1^2\right\rangle = \frac{N_2}{V_2}\left\langle\frac{1}{2}m_2v_2^2\right\rangle . \qquad (1.1.13)$$

But this condition is not sufficient, if energy is allowed to pass from side to side, via the piston. That is, if there are very energetic molecules

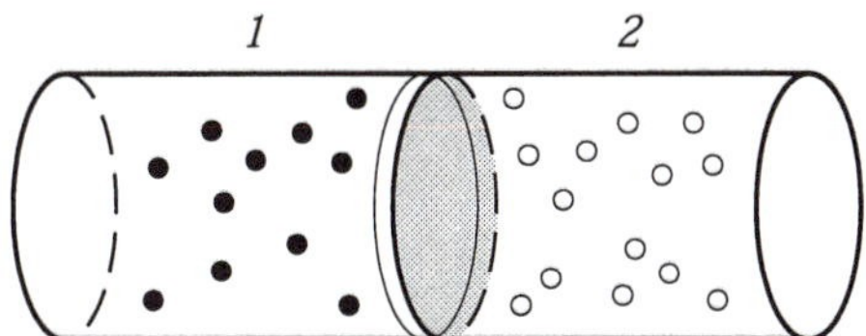

Fig. 1.1.3 Two gases in a container separated by a piston.

on one side, by hitting the piston they will impart the energy to the other side, so that the energy (or velocity) distribution on both sides will change. The balance in (1.1.13) will be violated, and the piston will go over to a new state, in which it will attempt to balance the pressures, and so on and so forth.

If we apply the argument which led to (1.1.12) to a molecule of one of the gases and to the piston (as a single molecule of the other gas), the conclusion will be that in a state of equilibrium the average kinetic energy of the piston, in its small hops left and right, must be equal to that of a gas molecule. The same argument applies to both sides of the piston. Hence, the average kinetic energies on the two sides must be equal in a state of equilibrium. We find again a property that corresponds to a familiar characteristic of temperature.

Exercise 1.6

Show that if in the cylinder of Fig. 1.1.3, the average kinetic energy on the two sides of the piston is equal, and if the total energy in the cylinder is a constant of magnitude E, and the number of molecules (N_1, N_2) of each gas is also constant, then the position of the piston is determined in a state of equilibrium.

What will be the value of the average energy per molecule in this state?

Solution on page 78

1.3 Equipartition of energy per molecule and its constituent parts — a fundamental problem

Till now we have treated molecules as rigid objects devoid of internal structure, which exchange momentum via elastic collisions. We found that the average kinetic energy per molecule in any gas at a given temperature (i.e. in a gas that is in thermal equilibrium with a given system) is constant and is given by Eq. (1.1.5),

$$\left\langle \frac{1}{2}mv^2 \right\rangle = \frac{3}{2}kT \,,$$

and the total energy of the gas is given by the product of the number of molecules and $\frac{3}{2}kT$. This is an adequate description for a monoatomic gas.

But, what happens if the gas molecules are made up of several atoms so that their internal structure comes into play? This question has several components:

(a) What is the average kinetic energy of each atom in a molecule?

(b) What is the average kinetic energy of the center of mass of a molecule?

(c) What is the connection between the answers to (a) and (b)?

(d) How is the gas law affected?

Question (a) arises since inside the molecule there are forces acting between the atoms. Therefore, it seems as if the kinetic energy exchanges in collisions, the same energy exchanges that are responsible for the equilibrium, will not be so free. The answer to (a) is that every atom in the molecule has an average kinetic energy of $\frac{3}{2}kT$ as in Eq. (1.1.5). This is a conclusion from the equipartition theorem, which will be presented in Part III.

equipartition theorem

Although the atoms in the molecule are bound to one another, they move inside it in all directions. When one of the atoms in the molecule collides with another atom of a different molecule, the only important factors are the velocities of the two atoms, immediately prior to the collision. When an atom in the molecule changes its velocity in a collision, the only effect of the forces will be that the internal motion in the molecule will be rearranged. The collision itself takes place as if the atoms were free.

However, if every atom in a diatomic molecule, for example, has an average kinetic energy $\frac{3}{2}kT$, then question (b) arises: What is the kinetic energy of the center of mass of the whole molecule? The answer is surprising: $\frac{3}{2}kT$, as for a monoatomic gas.

One way to obtain this result is to view the system at a lower resolution, so that we see only molecules and not atoms, and to go over the preceding arguments. The molecules will be seen to be moving to and fro at different velocities. They will be seen to collide and as a result to change their velocities, which are the velocities of their centers of mass. At equilibrium we will find that the average kinetic energy is $\frac{3}{2}kT$. But this conclusion leaves question (c) open.

In order to answer this question, we calculate the average kinetic energy of the center of mass, in a diatomic molecule, under the assumption that both parts of the molecule — atoms 1 and 2 — have an average energy, as implied by (a): $\frac{3}{2}kT$. The average kinetic energy of the center of mass will be

$$\begin{aligned}\langle E_k^{\text{cm}}\rangle &= \left\langle \frac{1}{2}(m_1+m_2)\left(\frac{m_1\mathbf{v}_1+m_2\mathbf{v}_2}{m_1+m_2}\right)^2\right\rangle \\ &= \frac{1}{m_1+m_2}\left[\left\langle\frac{1}{2}m_1^2\mathbf{v}_1^2\right\rangle+\left\langle\frac{1}{2}m_2^2\mathbf{v}_2^2\right\rangle+m_1m_2\langle\mathbf{v}_1\cdot\mathbf{v}_2\rangle\right]. \quad (1.1.14)\end{aligned}$$

According to our answer to (a), the first two terms in the sum are equal to $\frac{3}{2}kTm_1$ and $\frac{3}{2}kTm_2$, respectively. Therefore

$$\langle E_k^{\text{cm}}\rangle = \frac{3}{2}kT + \frac{m_1m_2}{m_1+m_2}\langle\mathbf{v}_1\cdot\mathbf{v}_2\rangle\,. \quad (1.1.15)$$

The question is therefore: What is the value of the average $\langle\mathbf{v}_1\cdot\mathbf{v}_2\rangle$ between the two parts of the molecule?

It seems as if there is a correlation between the velocities of the two parts. Nevertheless, it is clear that there is no correlation between the direction of the center of mass velocity $\mathbf{v}_{\text{cm}}$ of the molecule and the relative velocity $\mathbf{v}_{\text{rel}}$ of the two parts. Therefore,

$$\langle\mathbf{v}_{\text{cm}}\cdot\mathbf{v}_{\text{rel}}\rangle = 0\,, \quad (1.1.16)$$

precisely as occurs for two different molecules. But this relation, together with the fact that the average kinetic energy of each part is $\frac{3}{2}kT$, implies that

$$\langle\mathbf{v}_1\cdot\mathbf{v}_2\rangle = 0 \quad (1.1.17)$$

and thus

$$\langle E_k^{\text{cm}}\rangle = \frac{3}{2}kT\,. \quad (1.1.18)$$

Exercise 1.7

Prove that in fact $\langle\mathbf{v}_1\cdot\mathbf{v}_2\rangle = 0$.

Solution on page 79

As to question (d), we must remember that the pressure of the gas is related to the momentum imparted to the piston, and the question is: What is the amount of imparted momentum when the molecule collides elastically with the wall? By elastic collision we mean that there is no change in the internal state of the colliding systems.

■ Of course this is an idealization, but in a state of equilibrium it does not introduce any error.

Therefore, the motion of the center of mass of the molecule is the decisive factor, and the gas law remains the same [Eq. (1.1.6)]:

$$PV = NkT\,,$$

where N is, of course, the number of molecules.

The average kinetic energy per molecule can be calculated from two points of view: as the sum of the kinetic energies of all the atoms of the molecule or as the sum of the energy of the center of mass and the energy of the internal motions of the molecule.

Since the average kinetic energy is shared equally between the three directions of the velocity of the atom (three degrees of freedom per atom), and is also shared equally between all the atoms of the molecule, a molecule of r atoms has an average kinetic energy of

$$\langle E_k^{\text{mol}} \rangle = r \cdot \left(\frac{3}{2} kT \right) = \frac{3r}{2} kT\,. \tag{1.1.19}$$

This means that every molecule has $3r$ degrees of freedom related to its kinetic energy. If we choose the second viewpoint, we should note that $\frac{3}{2}kT$ of this energy is carried by the center of mass, so that $(r-1)\cdot(\frac{3}{2}kT)$ is the average kinetic energy of the internal motions of the molecule.

These motions may be rotations and vibrations of the atoms relative to each other (see Fig. 1.1.4). Here as well we may speak of degrees of freedom that will now be rotational degrees of freedom and degrees of freedom of atomic vibrations. The triatomic molecule depicted in Fig. 1.1.4 has an average kinetic energy of $\frac{9}{2}kT$ (nine degrees of freedom): $\frac{3}{2}kT$ of them related to the center of mass (three degrees of freedom), $\frac{3}{2}kT$ to the three

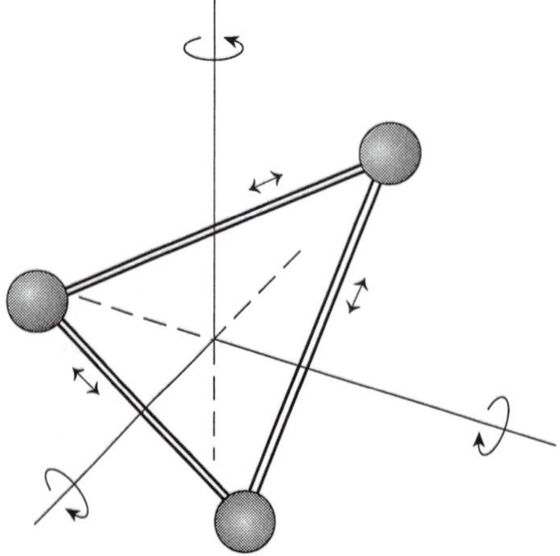

Fig. 1.1.4 Rotations and vibrations of a triatomic molecule.

different vibrations (another three degrees of freedom), and $\frac{3}{2}kT$ to the three possible rotations (another three degrees of freedom). The origin of the internal partition will be clarified in the coming chapters.

■ $\frac{3r}{2}kT$ is not the *total* energy of an r-atomic molecule, since it does not include the *potential* energy of the forces between the atoms. The above discussion has disregarded these forces.

The conclusion is that the right hand side of the gas equation is not $\frac{2}{3}E$, with E the total kinetic energy as in Eq. (1.1.6) for a monoatomic gas, but

$$PV = \frac{2}{3r}E\,, \tag{1.1.20}$$

and if we define

$$\gamma = 1 + \frac{2}{3r} \tag{1.1.21}$$

we obtain

$$PV = (\gamma - 1)E\,. \tag{1.1.22}$$

For the case of a monoatomic gas, for example, $\gamma = 5/3$.

Exercise 1.8

Deduce Eq. (1.1.20) from Eq. (1.1.6).

Solution on page 79

A classical physicist of the nineteenth century could have said, therefore, that the thermodynamic measurement of the curves of adiabatic expansion will provide him with γ, and hence r — which is the answer to the question: What is the number of the fundamental building blocks in familiar molecules? On the other hand, a contemporary physicist would state that r must be huge, as atoms are made up of a large number of electrons, protons and neutrons, and the latter are themselves made up of quarks and antiquarks, and maybe even the latter are not fundamental. This is a formulation from a modern viewpoint of the "heat capacity problem," here discussed in the context of adiabatic expansion. It was one of the most troubling open problems of physics, and was eventually resolved only within the framework of quantum mechanics, as we shall see below.

Classical mechanics allows continuous energy changes, and therefore the momentum exchanges that bring about the equilibrium state can take place separately with each of the molecule's constituents. Every collision with the molecule that is not limited to a change in the motion of the center of mass, causes an internal excitation in the molecule of a vibrational, or rotational, type. Since classical mechanics allows vibrations and rotations of arbitrarily small energies, even a very slow particle can exchange momentum with each part of the molecule separately.

In quantum theory the situation is totally different. In quantum theory internal excitations are allowed only between certain levels which are separated from each other by discrete energy differences. Exciting an internal motion is impossible, for example, if in the collision the energy passed on to the molecule is below a certain threshold, which depends on Planck's constant. Very slow particles, which are the most abundant at low temperatures, will not be able to exchange momentum with molecular parts, since they are unable to excite internal motions. Their collisions will always be elastic as those of rigid bodies.

We can ask, of course: What are low temperatures and what are large energy intervals? The answer is clear: if the average kinetic energy, whose magnitude is about kT, is small compared to the distance between two energy levels of internal excitations, the molecule will behave as if it had less than $3r$ degrees of freedom. In order to make this clearer we redefine the number of degrees of freedom per molecule f, as the number of portions of $\frac{1}{2}kT$ that are contained by the average kinetic energy of the molecule:

$$\langle E_k^{\mathrm{mol}} \rangle = \frac{f}{2} kT \,. \tag{1.1.23}$$

In this sense f is the *effective* number of degrees of freedom. We obtain instead of (1.1.20)

$$PV = \frac{2}{f} E \,, \tag{1.1.24}$$

and if we define

$$\gamma = 1 + \frac{2}{f} \tag{1.1.25}$$

Eq. (1.1.22) remains valid. The quantization of energy finds expression in values of f that are smaller than expected according to classical mechanics, or by values of γ that are too large.

Exercise 1.9

The difference between electronic states in the atom is measured in eV. The difference between nuclear states is measured in MeV = 10^6 eV. Subnuclear energies are are measured in GeV = 10^9 eV.

At what temperature will the electronic, nuclear and subnuclear (quark) degrees of freedom come into play in the calculation of γ?

Solution on page 80

In its original formulation, the "heat capacity problem" focused on trying to predict the value of the constant γ theoretically. For a monoatomic gas $\gamma = 5/3$ [Eq. (1.1.21) with $r = 1$ or (1.1.25) with $f = 3$]. Indeed, the result of the calculation for the rare gases fits the experimental results nicely, even at low temperatures.

The experimental result for He at 100K ($-173°$C) is $\gamma = 1.660$; on the other hand, for diatomic molecules, such as H_2 and O_2, the situation is quite bad. Theory — Eq. (1.1.21) or (1.1.25) — gives $\gamma = 4/3$. The experimental result at 100°C is $\gamma = 1.4$. This result implies that effectively there are fewer than six degrees of freedom. It seems as if the number is five. Attempts to solve this problem by taking into account internal forces were counterproductive, since it was found that internal forces add potential energies and by so doing increase the number of degrees of freedom and decrease the value of γ. In a more quantitative language, we may say that (1.1.23) is replaced by $\langle E^{\text{mol}} \rangle = \frac{f}{2}kT$, where E^{mol} is the total energy of a molecule and f is (therefore) larger than $3r$ and represents all the degrees of freedom of the molecule. This replacement keeps Eqs. (1.1.24) and (1.1.25) intact. In a diatomic gas, for example, the number of degrees of freedom including the internal forces is 7, so that $\gamma = 9/7$ only (see Sec. 1.4, Part IV). Classical mechanics is unable to provide any mechanism that can increase γ (or decrease f) beyond the value that is obtained from calculating the kinetic energies. The solution is found, as already mentioned, in quantum theory, which is able to explain why the internal vibrations of a diatomic molecule do not come into play at a temperature of 100°C (i.e. become "frozen"). The answer is, of course, that the energy difference of adjacent levels of the internal vibrations is too large, and an energy of kT ($T = 100°$C) is not sufficient to excite them, so the number of degrees of freedom does in fact decrease to five. On the other hand, this explanation implies that increasing the temperature will eventually cause an "unfreezing" of the inactive degrees of freedom of the internal vibration, leading to a decrease in γ. Indeed, at high temperatures — 2000°C — the value of γ for the diatomic gases decreases to 1.286, which is very close to 9/7. We revisit all these questions, at a much deeper level, in Chap. 1 of Part IV.

To summarize:

(a) At equilibrium, at a temperature T, each particle has an average kinetic energy of $\frac{1}{2}kT$ for every possible direction of the velocity.

(b) In a molecule of r atoms each atom has an average kinetic energy of $\frac{3}{2}kT$.

(c) The average kinetic energy of the center of mass is also $\frac{3}{2}kT$.

(d) Internal forces increase the number of degrees of freedom, i.e. the number of portions of $\frac{1}{2}kT$ per molecule.

(e) Experiment at moderate temperatures indicates that the number of degrees of freedom is smaller than the number obtained from kinetic energy arguments.

(f) The heat capacity (adiabatic expansion) problem does not have a solution in the framework of classical mechanics. Quantum mechanics provides the solution to this problem.

1.4 The density in an isothermal atmosphere — the Boltzmann factor in the potential energy

How are molecules distributed in a force field? As a special case of this question we will consider the gravitational case. A typical example for this situation is an isothermal atmosphere, namely a volume of gas at a uniform temperature T, in a closed cylinder, as depicted in Fig. 1.1.5. We ask: What is the density of the gas as a function of the height z, at thermal equilibrium? We divide the volume of the cylinder into layers, such that each layer will have a very small thickness compared to the distance that characterizes the rate of change of the force, but is still very large compared to the intermolecular distances. In order to convince yourself that this can really be done, solve the following exercise.

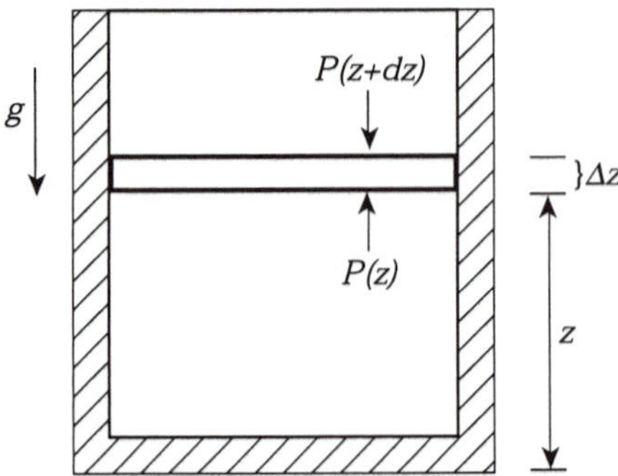

Fig. 1.1.5 Gas inside a container residing in a gravitational field.

Exercise 1.10

Calculate the change in height at the earth's surface that will cause the gravitational field to change by a thousandth of a percent.

What is the ratio between this distance and the average intermolecular distance in a gas at standard conditions?

Solution on page 81

Thermal equilibrium guarantees that the velocity distribution is identical in all the layers; however, due to the gravitational force the pressure differs at each height, since the pressure of the gas is determined by the weight of the gas above it. Hence, the density of the gas changes with height.

The requirement that the thickness of the layers should be small (compared to the earth's radius, for example) allows us to assume that all the molecules of one layer experience an identical gravitational force. Moreover, we can attribute a constant density to the gas contained in a given layer. The requirement that the thickness should be very large compared to the intermolecular distance, allows us to attribute thermodynamic properties to the small volume, as the number of molecules in it will still be very large.

In order to demonstrate the principle of the calculation we will make a further simplifying assumption and neglect the variation of the gravitational force with height. We are discussing, therefore, an isothermal atmosphere in a uniform gravitational field.

If the area of the base of the cylinder is A, the volume of the layer will be $A \cdot \Delta z$ (see Fig. 1.1.5), and the force exerted upon it by gravity is

$$\Delta F = mg \cdot n(z)\Delta z \cdot A\,, \tag{1.1.26a}$$

where m is the mass of a molecule, $n(z)$ is the density at height z, and g is the gravitational free fall acceleration. In a state of equilibrium, this force is balanced by the difference in pressure beneath the layer and above it, and hence

$$[P(z) - P(z + \Delta z)]A = \Delta F\,, \tag{1.1.26b}$$

so that

$$\frac{dP(z)}{dz} = -mg \cdot n(z)\,. \tag{1.1.27}$$

Since our layer is thick enough, we can attribute an equation of state of the type (1.1.6) to it. Namely

$$P(z) = n(z)kT\,. \tag{1.1.28}$$

We assumed that T is independent of z, so we substitute (1.1.28) into (1.1.27), and obtain a differential equation for $n(z)$. The equation is

$$\frac{dn(z)}{dz} = -\frac{mg}{kT}n(z) \tag{1.1.29}$$

and its solution

$$n(z) = n(0)\exp\left(-\frac{mgz}{kT}\right) . \tag{1.1.30}$$

■ Check that this is in fact the solution to the equation.

$n(0)$ is the value of the density at the point $z = 0$. This value is related to the total number of molecules N, since if the total height of the container is h, then

$$\frac{N}{A} = n(0)\int_0^h \exp\left(-\frac{mgz}{kT}\right)dz = \frac{n(0)kT}{mg}\left[1 - \exp\left(-\frac{mgh}{kT}\right)\right] , \tag{1.1.31}$$

so that in a container in which N is constant, the density at its bottom is determined by

$$n(0) = \frac{Nmg}{AkT}\left[1 - \exp\left(-\frac{mgh}{kT}\right)\right]^{-1} .$$

Equation (1.1.30) states that the density of molecules decreases with height; however, the extent of variation depends on the temperature. The higher the temperature, the less significant are the variations in height (see Fig. 1.1.6). The decisive factor in determining the decrease in density between two points whose altitudes differ by Δz is the ratio of the difference in their potential energies $mg\Delta z$ to kT. This is a dimensionless quantity, of course. If the temperature becomes very low, all the particles concentrate at the bottom of the cylinder. The thermal energy is not large enough to overcome the increase in potential.

■ Actually, we use the term potential here instead of *potential energy*. This is common practice in the literature and it is always possible to identify the meaning from the context.

Now we can go back and remove the simplifying assumption about the uniformity of the force field acting on the molecules. This we do in the following exercise:

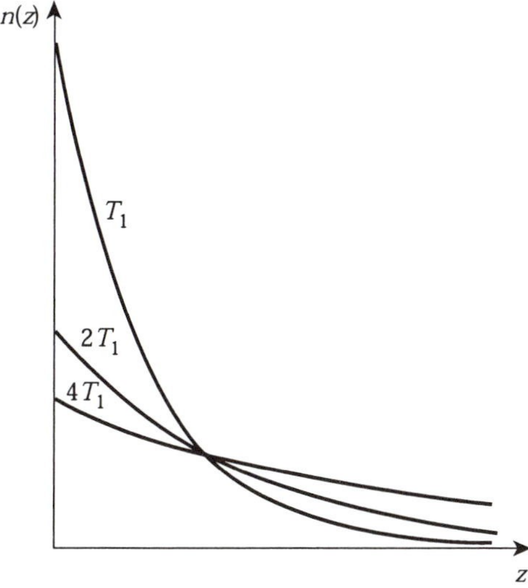

Fig. 1.1.6 The dependence of the density on height in an isothermal atmosphere at three different temperatures. Note: $n(0)$ also depends on T.

Exercise 1.11

Prove that for a gas in a force field that is derived from the potential $U(\mathbf{r})$,

$$n(\mathbf{r}) = n(\mathbf{r}_0) \exp\left[-\frac{U(\mathbf{r})}{kT}\right] . \tag{1.1.32}$$

Solution on page 82

Since our aim is to arrive at statistical mechanics via the kinetic theory, we will describe the result in Eq. (1.1.32) in a slightly different language. Our system contains many molecules. Thus, we can treat $n(\mathbf{r})$ or, more precisely, $n(\mathbf{r})/N$ as the probability for a molecule to be in an infinitesimal volume dV around the point $\mathbf{r}$, where N is the total number of molecules.

▪ Notice that from here to the end of this chapter P denotes the probability and not the pressure.

In other words

$$P(\mathbf{r})dV = \frac{1}{N} n(\mathbf{r})dV \tag{1.1.33}$$

is the probability that a particle will be in the volume dV around $\mathbf{r}$.

It is important to remember that although the volume element dV is "tiny," the number of molecules inside it is very large, and in every such volume there exists a velocity distribution. However, as we mentioned, around each point $\mathbf{r}$ there is exactly the same velocity distribution, since the system is at thermal equilibrium.

Exercise 1.12

Calculate the dimensions of $P(\mathbf{r})$. Is it normalized?

Solution on page 83

Now we want to ask a different question: Not what is the probability of finding *any* particle at a certain place, but what is the probability of a *given configuration* of the system, i.e. what is the probability that N particles will be in the volumes $dV_1, \ldots, dV_N$ around the points $\mathbf{r}_1, \ldots, \mathbf{r}_N$, respectively, in a space with a potential field $U(\mathbf{r})$?

▪ In the present context a configuration is a list of all the coordinates of the particles. Such a collection characterizes a state of the system, and every state of the system is described by an appropriate configuration.

Since the positions of the different particles are independent (ideal gas), the probability is a product of the individual probabilities, i.e.

$$P_N(\mathbf{r}_1, \mathbf{r}_2, \ldots, \mathbf{r}_N) dV_1 \ldots dV_N$$

$$= P(\mathbf{r}_1) dV_1 P(\mathbf{r}_2) dV_2 \ldots P(\mathbf{r}_N) dV_N$$

$$= \left[\frac{n(r_0)}{N}\right]^N \exp\left[-\frac{1}{kT}\sum_{i=1}^{N} U(\mathbf{r}_i)\right] dV_1 \ldots dV_N\,. \tag{1.1.34}$$

▪ Multiparticle probability distributions are discussed at length in the following parts. Note though that if we integrate over $N-1$ of the coordinates, we obtain the probability distribution of the Nth particle.

$\sum U(\mathbf{r}_i)$ is just the total potential energy of the N particles of the gas. Hence, the conclusion is that the probability density for a certain configuration of the system of particles is proportional to the exponential of minus the total potential energy of the configuration, where the potential energy is measured in units of kT.

1.5 The Maxwell–Boltzmann distribution

As already mentioned, $n(\mathbf{r})$, which is the particle density in a small volume dV around the point $\mathbf{r}$, is made up of particles that are moving at different velocities. The distribution of the velocities is independent of $\mathbf{r}$, just as the coordinate distribution $P(\mathbf{r})$ is independent of $\mathbf{v}$. We now enquire how many of the particles in a volume element dV, around the point $\mathbf{r}$, i.e. out of $n(\mathbf{r})dV$, have a velocity inside the volume element $dv_x dv_y dv_z (\equiv d\tau)$ in velocity space around $\mathbf{v}$ (see Fig. 1.1.7). We denote the probability for a particle to have a velocity $\mathbf{v}$ in that velocity volume

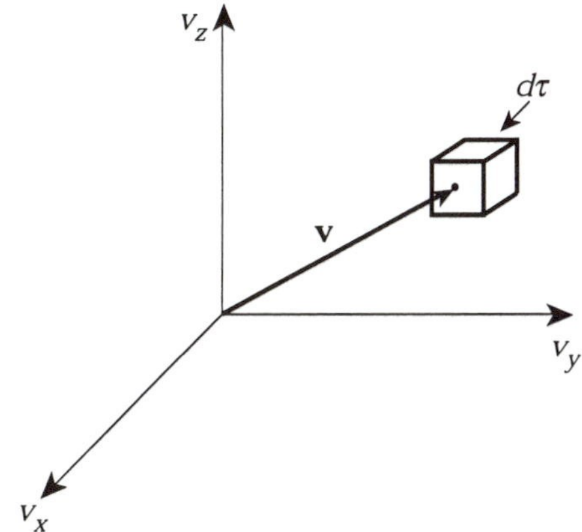

Fig. 1.1.7 A volume element in velocity space.

by $f(\mathbf{v})d\tau$, so that with the help of Eq. (1.1.33),

$$n(\mathbf{r})dVf(\mathbf{v})d\tau = NP(\mathbf{r})f(\mathbf{v})d\tau dV \tag{1.1.35}$$

is the number of molecules with a velocity in a region $d\tau$ around the velocity $\mathbf{v}$, that are located in a volume dV around the location $\mathbf{r}$.

Just as the dimensions of P are length to the minus three, $[L]^{-3}$, so the dimensions of $f(\mathbf{v})$ are velocity to the minus three, i.e. dimensions of length over time, to the minus three $[LT^{-1}]^{-3}$.

As mentioned above, $f(\mathbf{v})$ is the probability per unit velocity volume for a molecule to have a velocity near $\mathbf{v}$. The form of this probability function, named after Maxwell, will be derived next.

■ Many authors obtain the form of $f(\mathbf{v})$ as a special case of the general assertion of the Boltzmann distribution. Here we take the opposite way. Maxwell's result for the velocity distribution is much more fundamental, and provides a clue to the form of the general distribution. Moreover, the simple argument leading to this distribution is so nice that it merits presentation even if one has no bias concerning the direction of the progress of knowledge.

Actually the Maxwell distribution is obtained from two very reasonable and very simple assumptions:

(a) In a state of equilibrium there is no preferred direction.

(b) Orthogonal motions are independent of one another.

The first assumption, whose reasonableness was considered on the way to Eq. (1.1.9), states that $f(\mathbf{v})$ must be a function of v^2 only:

$$f(\mathbf{v}) = h(v^2)\,. \tag{1.1.36}$$

The second assumption implies that $f(\mathbf{v})$ must be a product of the form

$$f(\mathbf{v}) = g(v_x^2)g(v_y^2)g(v_z^2)\,. \tag{1.1.37}$$

The fact that we wrote the functions g as functions of the squares of the components of $\mathbf{v}$ again follows from (a), since a dependence on the sign of the component is impossible. The conclusion from (a) and (b) together is

$$h(v_x^2 + v_y^2 + v_z^2) = g(v_x^2)g(v_y^2)g(v_z^2)\,. \tag{1.1.38}$$

This is a functional equation that determines the forms of h and g.

Before solving Eq. (1.1.38) we note that we already know one possible simple solution: h and g are exponential functions $g(v^2) = e^{\lambda v^2}$. We want to show that this is the only solution and to this end we will solve the equation. In order to do this simply we rename our variables and define

$$v_x^2 = \xi,\; v_y^2 = \eta,\; v_z^2 = \zeta\,, \tag{1.1.39}$$

$$\rho = \xi + \eta + \zeta\,. \tag{1.1.40}$$

We want, therefore, to solve the equation

$$h(\rho) = g(\xi)g(\eta)g(\zeta)\,. \tag{1.1.41}$$

Note that h depends on ξ, η, ζ only through the variable ρ. In order to solve Eq. (1.1.41) we differentiate both sides with respect to ξ. On the left hand side we make use of the chain rule and of the fact that $\partial\rho/\partial\xi = 1$ to obtain

$$\frac{dh}{d\rho} = \frac{dg}{d\xi}g(\eta)g(\zeta)\,. \tag{1.1.42}$$

We now divide both sides of this equation by Eq. (1.1.41), obtaining

$$\frac{1}{h}\frac{dh}{d\rho} = \frac{1}{g}\frac{dg}{d\xi}\,. \tag{1.1.43}$$

Note that the left hand side can in principle depend on all three variables ξ, η, ζ through ρ. However, the right hand side depends only on ξ. Therefore, the function $h^{-1}dh/d\rho$ cannot depend on η and ζ. However, if we differentiate Eq. (1.1.41), this time with respect to η, and repeat the arguments, we will reach the conclusion that the function $h^{-1}dh/d\rho$ cannot depend on ξ or ζ. The conclusion is that $h^{-1}dh/d\rho$ is a constant function. If we denote its value by λ, then we obtain

$$\frac{dh}{d\rho} = \lambda h\,. \tag{1.1.44}$$

The general solution to this equation is of course

$$h = Ce^{\lambda\rho} = Ce^{\lambda v^2}\,, \tag{1.1.45}$$

where C is a constant. From here we obtain, with the help of (1.1.36),

$$f(\mathbf{v}) = h(v^2) = Ce^{\lambda v^2} . \tag{1.1.46}$$

Since $f d\tau$ is a probability, f is positive, and its integral over all possible vectors $\mathbf{v}$ must be 1. Hence, we must have $C > 0$ and $\lambda < 0$. If we denote the negative constant λ by $-A$, we can write

$$f(\mathbf{v}) = Ce^{-Av^2} . \tag{1.1.47}$$

This is the Maxwell velocity distribution. The arguments we have used do not provide the constant A. However, C is determined as a function of A from the normalization condition.

Exercise 1.13

Show that

$$C = \left(\frac{A}{\pi}\right)^{3/2} .$$

Solution on page 83

In order to obtain A some physics must enter our arguments. What is needed is to note that given the distribution $f(\mathbf{v})$ we are able to compute the average of any function of $\mathbf{v}$. In particular we can compute $\langle \frac{1}{2}mv^2 \rangle$, obtaining

$$\left\langle \frac{1}{2}mv^2 \right\rangle = \frac{3m}{4A} ; \tag{1.1.48}$$

however, this average is known to us from the discussion of the ideal gas where it was seen to be equal to $\frac{3}{2}kT$. Hence, we can express A in terms of the temperature and the molecular mass. The result is

$$f(\mathbf{v})d\tau = \left(\frac{m}{2\pi kT}\right)^{3/2} \exp\left(-\frac{mv^2}{2kT}\right) d\tau \tag{1.1.49}$$

for the probability in velocity space.

Exercise 1.14

Prove that (1.1.49) is the only determination of C and A in (1.1.47) which gives a normalized $f(\mathbf{v})$ and the correct average kinetic energy.

Solution on page 84

Now, just as we derived from the coordinate distribution of N particles (1.1.33) the probability for a certain configuration (1.1.34), we want to derive from the velocity distribution (1.1.49) the probability for a given configuration of the coordinates and the velocities of the particles, namely the probability for N particles to be in volumes $dV_1, \ldots, dV_N$ around the points $\mathbf{r}_1, \ldots, \mathbf{r}_N$, respectively, as well as in the velocity region $d\tau_1, \ldots, d\tau_N$ around the respective velocities $\mathbf{v}_1, \ldots, \mathbf{v}_N$. The configuration of the coordinates and velocities characterizes the state of the (classical) system completely. It is possible to think of such a configuration as a point in a space of $6N$ dimensions (six dimensions per molecule).

■ Note that here the term "configuration" signifies the list of the coordinates and *velocities* of all the molecules.

The probability for a molecule to be near $\mathbf{r}$, and for its velocity to be near $\mathbf{v}$, is, as already stated,

$$P(\mathbf{r})dV f(\mathbf{v})d\tau .$$

In an ideal gas with N molecules, the coordinates and velocities of each molecule are independent of the coordinates and velocities of every other molecule. There is, of course, no dependence between the position of a molecule and its velocity. Hence, the probability for a configuration where molecule No. 1 is near location $\mathbf{r}_1$ and has a velocity near $\mathbf{v}_1$; molecule No. 2, near $\mathbf{r}_2$ and $\mathbf{v}_2$; and so on, is

$$\begin{aligned} &P(\mathbf{r}_1)\ldots P(\mathbf{r}_N)dV_1\ldots dV_N f(\mathbf{v}_1)\ldots f(\mathbf{v}_N)d\tau_1\ldots d\tau_N \\ &\quad = C\exp\left\{-\frac{1}{kT}\sum_{i=1}^{N}\left[\frac{1}{2}m_i v_i^2 + U(\mathbf{r}_i)\right]\right\} dV_1\ldots dV_N d\tau_1\ldots d\tau_N . \quad (1.1.50) \end{aligned}$$

Maxwell–Boltzmann distribution

We found, therefore, that given a complete description of the state of the system, the probability of finding such a state is proportional to the exponential of minus the total energy of the system in this state, in units of kT. This distribution is called the Maxwell–Boltzmann distribution.

1.6 Averages and distributions

In the preceding two paragraphs we developed expressions for the density distribution resulting from the presence of a potential — Eq. (1.1.32) — and the distribution of molecules in velocity space — Eq. (1.1.49). Given these distributions we can compute averages of any function that depends on the coordinates and/or the velocities.

We can, for example, calculate the average height of a molecule in the isothermal atmosphere whose density is given by Eq. (1.1.30) with the help of the one-dimensional version of Eq. (1.1.33):

$$\begin{aligned}\langle z\rangle &= \int_0^\infty zP(z)dz = \int_0^\infty zn(z)dz \Big/ \int_0^\infty n(z)dz \\ &= \int_0^\infty z\exp\left(-\frac{mgz}{kT}\right)dz \Big/ \int_0^\infty \exp\left(-\frac{mgz}{kT}\right)dz\,, \qquad (1.1.51)\end{aligned}$$

where zero height denotes the bottom of the cyclinder containing the gas.

In the present case it is possible to evaluate the integrals explicitly. It is, however, unnecessary. $\langle z\rangle$ can be calculated using a trick that is widely used in statistical mechanics: first we note that we can rewrite it as

$$\langle z\rangle = -\frac{d}{d\alpha}\ln Z(\alpha)\,, \qquad (1.1.52)$$

where $Z(\alpha)$ stands for the integral in the denominator of (1.1.51) and α stands for mg/kT. The dimensions of α can be evaluated either directly from the explicit expression or from the knowledge that $z\alpha$, the exponent in (1.1.51), must be dimensionless. The results are, of course, identical:

$$[\alpha] = [\text{length}]^{-1} = [L]^{-1}\,.$$

Since the integral $Z(\alpha)$ has the dimensions of length, and as there is no other variable with dimensions of length in the problem, it must be proportional to $1/\alpha$.

If we write $Z(\alpha) = K/\alpha$, where K is a constant, we immediately obtain the average height of a molecule from (1.1.52):

$$\langle z\rangle = \frac{1}{\alpha} = \frac{kT}{mg}\,. \qquad (1.1.53)$$

Exercise 1.15

Show that the average square deviation (the variance) of the height of the molecules is given by

$$(\Delta z)^2 \equiv \langle (z-\langle z\rangle)^2\rangle = \frac{d^2}{d\alpha^2}\ln Z(\alpha)\,. \qquad (1.1.54)$$

Solution on page 85

Exercise 1.16

Assume that in Eq. (1.1.32) the potential is given by

$$U(z) = U_0 z^n .$$

Using only dimensional arguments calculate $\langle z^n \rangle$ and $\langle U(z) \rangle$.

Solution on page 86

In a similar manner the Maxwell distribution, Eq. (1.1.49), determines the averages of functions of the velocity. Now we can calculate the averages of different powers of the velocity — called moments of the distribution — with the help of dimensional arguments.

First of all it is clear that the average of $\mathbf{v}$, in the distribution (1.1.49), is zero. The average of $\langle |\mathbf{v}|^2 \rangle$ in this distribution, which is also the average square deviation, is given by

$$\langle |\mathbf{v}|^2 \rangle = -\frac{d}{d\alpha} \ln Z(\alpha) , \tag{1.1.55}$$

where this time

$$Z(\alpha) = \int e^{-\alpha |\mathbf{v}|^2} d\tau \tag{1.1.56}$$

and

$$\alpha = \frac{m}{2kT} . \tag{1.1.57}$$

Dimensional analysis gives

$$[\alpha] = [\text{velocity}]^{-2} ,$$

$$[Z(\alpha)] = [\text{velocity}]^3 .$$

Since the only dimensional quantity in $Z(\alpha)$ is α, it must be that

$$Z(\alpha) = C\alpha^{-3/2} , \tag{1.1.58}$$

and substituting this into Eq. (1.1.55) we obtain

$$\langle |\mathbf{v}|^2 \rangle = \frac{3}{2\alpha} = \frac{3kT}{m} , \tag{1.1.59}$$

which is equivalent to (1.1.5). See also (1.1.48).

Exercise 1.17

(a) How does $\langle |\mathbf{v}|^2 \rangle$ vary with the dimensionality of space?

(b) If a gas is placed in a three-dimensional harmonic potential,

$$U(\mathbf{r}) = \frac{1}{2} C |\mathbf{r}|^2 ,$$

what is the average square separation (the average square distance) of a molecule from the center of force?

(c) Under the conditions of (b), show that the average potential energy per molecule in the gas is

$$\langle U(\mathbf{r}) \rangle = \frac{1}{2} DkT , \qquad (1.1.60)$$

where D is the dimensionality of space.

Solution on page 87

Chapter 2

Brownian Motion

2.1 Historical background

The fact that each degree of freedom of a particle which is in equilibrium with gas molecules has an average kinetic energy equal to that of a degree of freedom of the gas molecules, i.e. $\frac{1}{2}kT$, has far-reaching consequences. The possibility of explaining the motion of small specks, hovering in a gas or in a liquid, within the framework of the kinetic theory was, from a historical point of view, the most important of these consequences, due to its contribution to the resolution of the ideological struggle for and against the atomic structure of matter. This titanic struggle went on during the second half of the nineteenth century and involved the prominent physicists of that era. We can learn of its acuteness from the fact that even in the beginning of this century one of the pillars of physics — Ernst Mach — could be heard to say:

"If the belief in the existence of atoms is so crucial in your eyes, I hereby withdraw from the physicist's way of thought... "

An even stronger evidence for the heat of the debate is the fact that the sharp criticism that was aimed at Ludwig Boltzmann seems to have contributed to his suicide in 1906.

The experiment by the botanist Robert Brown, concerning the drifting of specks with radii on the order of micrometers ($1\mu\text{m} = 10^{-6}$ m) in liquids and in gases, had been known since 1827. However, only in 1905 did Einstein explain the phenomenon. Within his explanation, which was based on the kinetic theory, Einstein connected in a quantitative manner the Brownian motion and quantities that appear in the kinetic theory — such as the coefficients of mobility and viscosity — and he brought the debate to a conclusion in a very short time.

But the importance of the subject far transcends the resolution of the heated debate about atomism. This subject is also of great practical importance: understanding the effect of the thermal motions of atoms on sensitive instruments that are not too heavy and are in equilibrium with their surroundings, is very important for understanding the limits on the sensitivity of very accurate measuring instruments. Among these are the galvanometers, which are based on the deviations of a small mirror hanging from a thin thread, and the voltage variations in sensitive electric circuitry containing resistors. The random motion, which is due to the equipartition of thermal energy, is a source of "noise" in many systems. The theoretical treatment of Brownian motion is a workshop for the understanding of such phenomena. Moreover, the treatment of this subject begins to shed some light on the deep and important problem of the connection between thermal fluctuations and the "erosion of energy" (dissipation), or frictional phenomena. **energy dissipation**

2.2 Characteristic scales of Brownian motion

The sort of phenomenon we are about to treat is schematically as follows: a large *body* hovers in a crowd of tiny particles — as a giant bear floats in a huge crowd of bees (see *Winnie the Pooh*, Chapter 1).

The tiny particles are moving to and fro. In this process they make many fast knocks against the large body. The knocks are random and each has a very small effect on the body, since the ratio of the mass of the body to that of the particle is very large.

The hovering particles that Robert Brown was able to see under his microscope have a diameter of a few micrometers. To acquaint ourselves with the data of the problem described in Fig. 1.2.1, we suppose that the body is hovering in a gas under standard conditions. The density of such a gas is $n \approx 10^{26}$ m^{-3}. The thermal energy of a particle at room temperature is

$$\epsilon \approx kT \approx 5 \times 10^{-21} \text{ J}$$

(k is the Boltzmann constant). We are interested only in orders of magnitude, so the factor 3/2 was omitted. We will assume that the density of the body is a tenth of a gram per cm^3. Its mass will therefore be $M \approx 10^{-16}$ kg.

From the assumption that the body is in thermal equilibrium with the gas, we can deduce that its typical velocity is 10^{-2} m s^{-1} (check this). Similarly, if we take a typical value of 10^{-26} kg for the mass of the surrounding molecules, the molecular velocities will be 10^3 m s^{-1}.

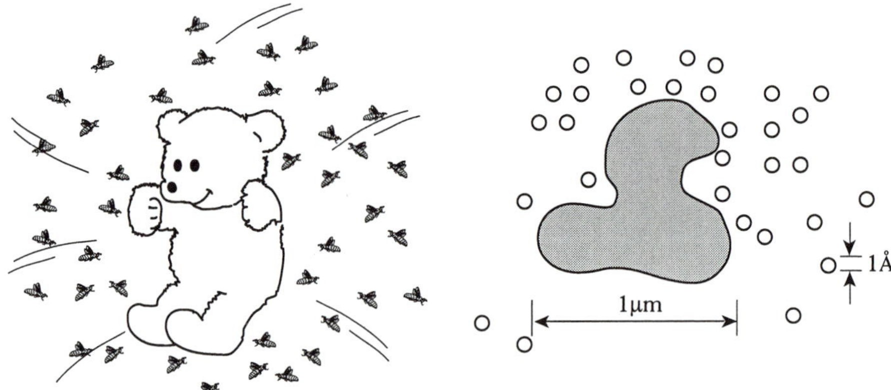

Fig. 1.2.1 A body executing Brownian motion.

Exercise 2.1

Calculate, for the above data, the velocity which a molecule of typical velocity can impart to the body in a head-on elastic collision.

Solution on page 88

---

mean free path

Another scale that characterizes the situation is the distance that the body can traverse between two collisions. This distance is called the *mean free path*, or simply the *free path* (see also Chap. 3). In order to find the orders of magnitude of the free path, we apply the following consideration: if the body were to move, parallel to itself, a distance L (Fig. 1.2.2), it would bump into $N = n \cdot (S \cdot L)$ molecules, where n is the gas density and S is the cross section area of the body. In order to bump into one molecule the body must traverse, typically, a distance L such that $N = 1$. Namely

$$L = \frac{1}{nS} . \tag{1.2.1}$$

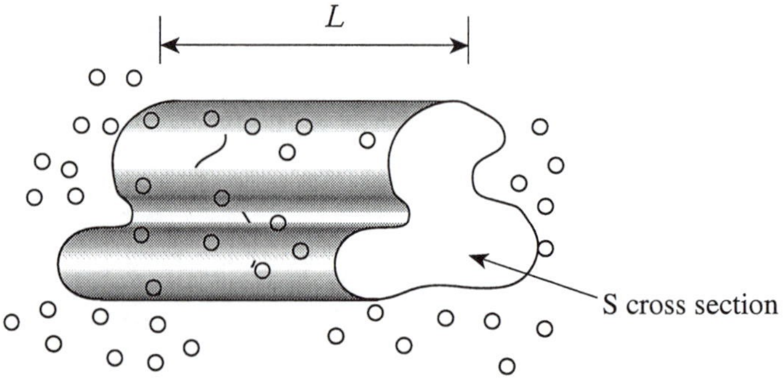

Fig. 1.2.2 Motion of a body in gas.

Under the conditions of our problem $L \approx 10^{-14}$ m (do check), so that even using a very advanced microscope we are unable to discern isolated events: the motion will seem continuous and smooth, but lacking a preferred direction.

Along with the free path, L, it is possible to define the corresponding time scale, *mean free time*, which, as indicated by its name, is the time it takes for a particle to traverse a distance L. In our case $\tau \approx 10^{-12}$ s, which means that the body gets hit by 10^{12} molecules per second!

mean free time

2.3 Random walk

If we follow the motions of different bodies from an initial time $t = 0$, we see that each body follows a different trajectory, reflecting the randomness of the collisions. The average over many trajectories of the displacement, $\mathbf{R}$, of each body from its initial position gives zero. However, the average square distance is not zero. Experiments show that it grows linearly with time. Namely

$$\langle \mathbf{R}^2 \rangle = \alpha t \,. \tag{1.2.2}$$

Thus if, for example, at time $t = 0$ all the bodies were at the same location, they would gradually move from their initial location so that most of them would be found near a sphere of radius proportional to $t^{1/2}$. The major achievement in Einstein's work, as already mentioned, was that the atomic assumption and the use of the kinetic theory enabled him to express α in terms of other quantities of the kinetic theory. How this can be done will be shown in the following paragraphs.

First we inquire whether Eq. (1.2.2) fits a simple intuitive picture of the nature of the process. We assume, for the sake of simplicity, that the collisions of the body with the molecules cause it to move as in a random walk. A two-dimensional example of such a walk is shown in Fig. 1.2.3. The sketched walk consists of 36 steps. The characteristic of such a walk, a drunk's walk from a bar, is a series of steps of more or less constant length, that are randomly directed, in the corresponding space — two-dimensional for the drunk, three-dimensional for the hovering body. Obviously, every drunk will reach a specific point, in a given series of steps. We cannot say where every particular walk will arrive; however, we will have something to say about the average location of a group of drunks who leave the bar together. You may be surprised to find that we do not need to compute the probability of reaching every location. We can get along with a few very general considerations.

Fig. 1.2.3 Thirty-six steps in the walk of a drunk.

We will assume that the length of a step is L, that after N steps the drunk has reached the point $\mathbf{R}_N$, and that his next step will be in the direction $\mathbf{n}$ ($\mathbf{n}$ is the unit vector, which determines the direction of the next step). After $N+1$ steps the position of the drunk will be $\mathbf{R}_{N+1}$:

$$\mathbf{R}_{N+1} = \mathbf{R}_N + L\mathbf{n} \,. \tag{1.2.3a}$$

First, clearly, the average position of the drunks remains at the bar, since each one of them moves in a different direction, $\langle \mathbf{R}_N \rangle = 0$ (see also Exercise 2.2). We can learn about the extent of the scattering of the drunks from the square distance from the initial position, which is given by

$$\mathbf{R}_{N+1}^2 = (\mathbf{R}_N + L\mathbf{n})^2 = \mathbf{R}_N^2 + L^2 + 2L\mathbf{n} \cdot \mathbf{R}_N \,. \tag{1.2.3b}$$

(Note that we used the fact that $\mathbf{n}^2 = 1$.)

We apply the randomness argument to Eq. (1.2.3b) in the following manner: we average both sides of the equation. The fact that the direction of $\mathbf{n}$ is random with respect to $\mathbf{R}_N$ leads to a zero average for the last term on the right hand side [cf. (1.1.9)]. The conclusion is that

$$\langle \mathbf{R}_{N+1}^2 \rangle = \langle \mathbf{R}_N^2 \rangle + L^2 \,, \tag{1.2.4}$$

from which it is immediately inferred that

$$\langle \mathbf{R}_N^2 \rangle = NL^2 \,. \tag{1.2.5}$$

L^2 is constant and N is the number of steps, which is proportional to the duration of the walk, so that the result of Eq. (1.2.5), is not different from that of Eq. (1.2.2).

Finally, we note that exactly the same results can be obtained if we interpret the averages in Eqs. (1.2.4) and (1.2.5) as averages over time for a single drunk, in the following manner:

We observe a certain drunk over a very long period of time — a large number of steps. Every point along the drunk's path can be considered as an initial point (assuming that the effect of the liquor does not diminish). Every such point can be considered as the origin of a coordinate system. For every such choice it is possible to find the point that is reached by the drunk after N additional steps. In a coordinate system whose origin is at that initial point the drunk will be at $\mathbf{R}_N$.

After reaching $\mathbf{R}_N$ the drunk will take his next step in a random fashion (for if it were not so he would not be drunk); this means that for every choice of an initial point $L\mathbf{n}$ will point in a different direction. The averaging over all the choices of the initial point is the averaging from the second point of view.

Exercise 2.2

Prove that for both points of view $\langle \mathbf{R}_N \rangle = 0$.

Solution on page 89

2.4 Brownian motion, random force and friction: the Langevin equation

The treatment we use below is not exactly the same as that formulated by Einstein in 1905, but is similar to the formulation by his close friend, Pierre Langevin, a short while later (1908).

We describe the center of mass motion of a body in a gas as it evolves under an external force $\mathbf{F}_e$ and friction, i.e. a restraining force. The equation of motion is

$$M\ddot{\mathbf{r}} + \mu\dot{\mathbf{r}} = \mathbf{F}_e\,. \tag{1.2.6}$$

■ A simple example of the appearance of a restraining force that is proportional to the velocity can be found in Self-Assessment Exercise 6 of this part.

$\mu\dot{\mathbf{r}}$ is the restraining force, proportional to the velocity — this is a typical description of a frictional force. μ is a *friction coefficient* and is connected, as will be mentioned later, to the viscosity. First, let us elucidate the role of the friction term in Eq. (1.2.6). We do this by way of the following exercises:

Exercise 2.3

(a) Calculate the dimensions of μ.

(b) Show that in the absence of an external force the velocity of the body tends to zero for long times, even if its initial value is different from zero, i.e. the friction restores the system to equilibrium.

Solution on page 89

We will not enter here into an involved discussion of the connection between μ and the viscosity. We will limit ourselves to the following description: The viscosity describes an internal friction between different layers of the fluid (see also Sec. 3.6 below). A body in motion through a liquid drags along nearby liquid layers, giving rise to friction between successive liquid layers. This friction is expressed as a restraining force that acts on the body and is proportional to its velocity. If the body is a ball of radius a then the proportionality coefficient is especially simple, as was found by Stokes: $F = 6\pi a\eta v$. η is called the viscosity coefficient.

viscosity

Its dimensions are $[M][L]^{-1}[T]^{-1}$, i.e. mass divided by length and time, and its units in cgs are poise, P. ($1\text{P} = 10^{-1}$ kg m^{-1} s^{-1}.) For a gas, η is about 10^{-4} P; it is 10^{-2} P for water and 8.5 P for glycerine.

Exercise 2.4

The one-dimensional system (1.2.6) with $F_e = mg$ describes a sky diver in a gravitational field in air. Solve the equation, and verify that for long times the velocity of the sky diver is constant. What are long times?

Solution on page 90

The solution to Exercise 2.4 indicates a way of measuring μ directly. The method is the same as that used for measuring the viscosity. Namely, bodies are dropped in a gravitational field in a certain medium and their final speed is measured. Its absolute value is Mg/μ.

From the fact that the frictional force causes the sky diver of Exercise 2.4 to accelerate at a slower rate than in a free fall, it follows that its total energy decreases with time.

Exercise 2.5

Show that if $\mathbf{F}_e$ is a force derived from a potential, then the rate of change of the total energy of a body whose motion is described by Eq. (1.2.6) is

$$\frac{dE}{dt} = -\mu\dot{\mathbf{r}}^2 .$$

This is the rate of energy dissipation.

■ Reminder: the connection between the force and the potential is given by $\mathbf{F} = -\nabla U$, where U is the potential energy.

Solution on page 91

We have seen, therefore, that due to the friction $\mu\dot{\mathbf{r}}$ the particle can lose energy with time. Where does the friction come from and where does the energy go to?

The friction that restores the body to equilibrium originates of course from the numerous fast collisions with the gas molecules. These collisions also give rise to the Brownian motion of the body. The connection between μ and microscopic factors will be considered in the next chapter. Here we note that $\mu\dot{\mathbf{r}}$ is part of the effect of the momentum exchange between the gas molecules and the body, due to collisions. In these momentum exchanges the body transfers more energy to the gas molecules than it receives from them, due to the fact that in the direction of motion the body makes more frequent and harder collisions than it makes in the opposite direction. In other words, the energy of the body dissipates.

Beyond the damping effect of the collisions with the molecules, these fast momentum exchanges contribute a sort of random force, $\mathbf{F}_e$, that acts on the body in the absence of any external force.

We make here two remarks:

(a) The randomness of the force $\mathbf{F}_e$ is expressed by the fact that, if we average over many particles (averaging over an *ensemble*), or over different initial times, we get **ensemble**

$$\langle \mathbf{F}_e \rangle = 0 \tag{1.2.7}$$

as well as

$$\langle \mathbf{r} \cdot \mathbf{F}_e \rangle = 0\,. \tag{1.2.8}$$

Compare this with the discussion in Sec. 2.3. Figure 1.2.4 shows a series of graphs that describes the behavior with time of a random force. The series can be read as an ensemble of different systems, or as different time intervals in the behavior of the same system along the time axis.

The effect of a random force acting on one system can be substituted by random forces acting on many systems. The assumption that these two approaches lead to the same result, an assumption that seems so natural, is called the ergodic hypothesis. The generality of its validity is still the subject of active research.

(b) It is especially important to note that, if we were to assume that the entire effect of the collisions amounts to the appearance of a random force, i.e. $\mu = 0$ in Eq. (1.2.6), the result would have been that the body could maintain a constant average velocity, without re-equilibrating with its surroundings. The fact that the *average* velocity is conserved, in the absence of friction, is inferred from the observation that the average of a random force is zero — Eq. (1.2.7). Hence, if we substitute $\mu = 0$ in Eq. (1.2.6) and average both sides we obtain

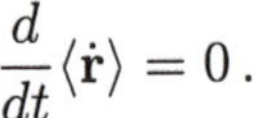

$$\frac{d}{dt}\langle \dot{\mathbf{r}} \rangle = 0 \, .$$

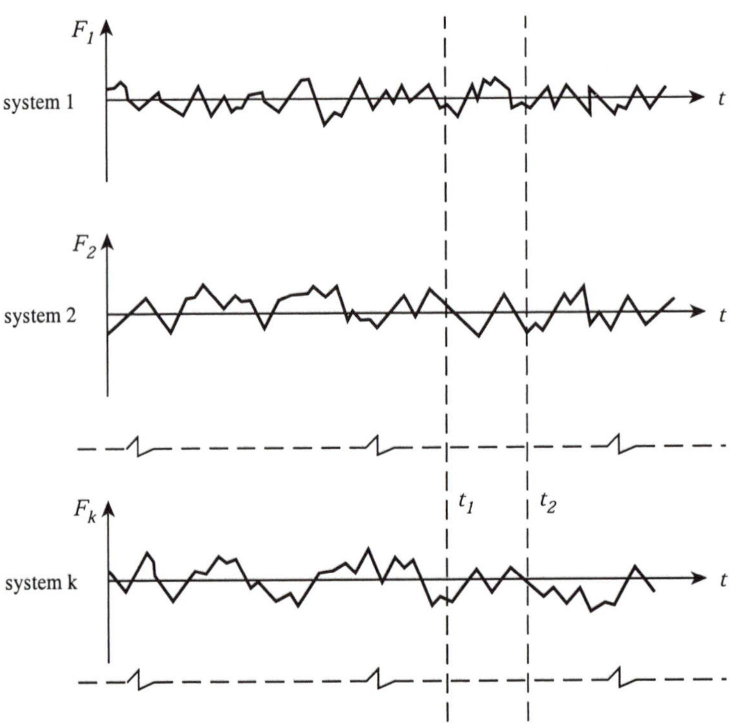

Fig. 1.2.4 Random forces.

Exercise 2.6

Show that the above result is an immediate consequence of Eq. (1.2.6).

Solution on page 92

The practical summary of the discussion above is, therefore, that the effect of the collisions can be written (following Langevin) as a sum of two contributing forces: one gives rise to the friction term and is proportional to the velocity, while the other is the random force that we called $\mathbf{F}_e$. Equation (1.2.6) with a random force $\mathbf{F}_e$ is the Langevin equation.

2.5 Solving the Langevin equation: approximations and orders of magnitude

Let us suppose that the force $\mathbf{F}_e$ in the Langevin equation (1.2.6) is the random force due to collisions with the gas molecules only, and attempt to deduce the time variation of the average square displacement of the body, $\langle r^2 \rangle$. Our aim is to show that this system, described by Eq. (1.2.6), has a solution that describes Brownian motion. We expect, therefore, $\langle r^2 \rangle$ to be proportional to the time.

Since the solution involves analytic arguments along with statistical arguments, we shall discuss it in detail. As we are interested in the change of the magnitude $\langle r^2 \rangle$, we first obtain an equation for r^2:

$$\frac{1}{2} M \frac{d^2 r^2}{dt^2} + \frac{1}{2} \mu \frac{dr^2}{dt} - M \dot{\mathbf{r}}^2 = \mathbf{r} \cdot \mathbf{F}_e \, . \qquad (1.2.9)$$

Note that Eq. (1.2.9) goes beyond the familiar context of differential equations. Beside functions of t and their derivatives, it contains a random element corresponding to some temporal sequence of the force, $\mathbf{F}_e$, as those exemplified in Fig. 1.2.4. It is a *stochastic differential equation.* Here we will restrict ourselves to a few comments concerning such equations:

stochastic differential equation

(a) To any given sequence, $\mathbf{F}_e$, corresponds a particular solution;
(b) The solution corresponding to any particular sequence of the random force is of little interest;
(c) A quantity can be significant only if it is not strongly dependent on the particular sequence;
(d) Such a quantity can be calculated by averaging over all "acceptable" sequences, just because it is insensitive.
(e) The "acceptable" set of sequences, the *ensemble*, has to be specified.

The transition from Eq. (1.2.6) to Eq. (1.2.9) is obtained by taking the scalar product of Eq. (1.2.6) with $\mathbf{r}$ and using the identities

$$\begin{aligned} \frac{d}{dt} r^2 &= 2\mathbf{r} \cdot \frac{d\mathbf{r}}{dt} \, , \\ \frac{d^2}{dt^2} r^2 &= 2\mathbf{r} \cdot \frac{d^2 \mathbf{r}}{dt^2} + 2\dot{\mathbf{r}}^2 \, . \end{aligned} \qquad (1.2.10)$$

Exercise 2.7

Complete the deduction of Eq. (1.2.9).

Solution on page 92

The next step is to average both sides of Eq. (1.2.9) over the ensemble, simplifying the right hand side with the help of Eq. (1.2.8). In this manner we obtain a differential equation for $\langle r^2\rangle$:

$$\frac{1}{2}M\frac{d^2}{dt^2}\langle r^2\rangle + \frac{1}{2}\mu\frac{d}{dt}\langle r^2\rangle - 2\left\langle\frac{1}{2}Mv^2\right\rangle = 0\,,$$

where $\mathbf{v}$ was substituted for $\dot{\mathbf{r}}$.

The last average on the left hand side can be evaluated, at equilibrium, by the equipartition principle. Every degree of freedom, for each dimension of space, is assigned an energy of $\frac{1}{2}kT$. The last term is therefore DkT, where D denotes the number of dimensions of space.

Thus the equation becomes

$$M\ddot{u} + \mu\dot{u} = 2DkT\,, \tag{1.2.11}$$

where we denoted $\langle r^2\rangle$ by u. This equation can be fully solved. The initial conditions are chosen to be $r(t=0)=0$, namely the origin of the coordinate system of each body in the ensemble is chosen as its position at $t=0$. In this case $u(t=0)=\dot{u}(t=0)=0$, and

$$u(t) = \frac{2DkT}{\mu}\left[t + \theta(e^{-t/\theta} - 1)\right]\,, \tag{1.2.12}$$

where

$$\theta = \frac{M}{\mu}\,. \tag{1.2.13}$$

Exercise 2.8

Check that Eq. (1.2.12) is in fact the solution to Eq. (1.2.11), satisfying the initial conditions, and that the parameter θ has the dimensions of time.

Solution on page 93

Let us inquire what happens to the body a very short and a very long time after the initiation of its motion, and compare to our physical intuition. Short and long times must be measured with respect to a characteristic time appearing in the problem. In our case this characteristic time is θ. That is, at short times $t \ll \theta$, it is possible to expand the exponential in Eq. (1.2.12):

$$u(t) = \langle r^2\rangle \approx \frac{DkT}{M}t^2\,. \tag{1.2.14}$$

■ $e^x = 1 + x + \frac{x^2}{2!} + \frac{x^3}{3!} + \cdots$.

Putting it simply, at very short times, relative to times between collisions, the body moves as a free particle with constant velocity. The constant velocity is the thermal velocity as determined at equilibrium, i.e. $v_T = (DkT/M)^{1/2}$. Notice that although we considered short times we have not ignored the body's previous collisions (before $t = 0$), which allowed it to acquire thermal velocity. Otherwise our entire discussion of short times is invalid, since without thermal velocity we cannot use Eq. (1.2.11).

For long times $t \gg \theta$, the exponential decays away, and θ can be neglected, in this limit, compared to t. So

$$u(t) = \langle r^2 \rangle \sim \frac{2DkT}{\mu} t \,. \tag{1.2.15}$$

This result is the same as Eq. (1.2.2), with a bonus of a relation between the coefficient α and the macroscopic characteristics of the problem:

$$\alpha = \frac{2DkT}{\mu} \,. \tag{1.2.16}$$

To obtain an idea of the orders of magnitude of the times for which the two approximations are valid, we compute the magnitude of the time θ. To this end, let us assume that our Brownian particles are spherical. This will allow us to use Stokes' law for a sphere: $\mu = 6\pi\eta a$. We further assume that the small sphere is floating in water, whose viscosity is $\eta = 10^{-2}$ P. The particle's mass, whose radius is about one micrometer and whose specific weight is close to that of water, will be about 5×10^{-12} g, so that

$$\frac{M}{\mu} \approx 2 \times 10^{-7} \text{ s} \,.$$

This means that in any reasonable experiment, lasting more than 10^{-3} s, we will not notice the region $t \ll \theta$ but only the region $t \gg \theta$. Hence, the average square distance of the body from its initial position, will be linear with time.

We further remark on the role of D — the dimensionality of space — in Eq. (1.2.16). In experiment $\langle r^2 \rangle$ is usually measured in a space whose dimensionality is less than that of the space in which the body actually moves. This happens, for example, when we measure the projection of the position of the body on the focal plane of the microscope lens. In this case $D = 2$, though the body's real motion is in three-dimensional space.

As already mentioned, μ and η can be measured directly. With the help of the Brownian motion, i.e. with the help of Eq. (1.2.16) it is possible to measure the Boltzmann constant k. Combining this with the gas equation which gives R, it is possible to obtain Avogadro's number. Indeed, this is how J. Perrin obtained the first precise determination of Avogadro's number in 1908.

2.6 Applications and implications

The Brownian behavior — the fluctuations that are induced in the motion of a fine system as a result of the thermal agitation of the surroundings with which the system is in equilibrium — appears in different contexts. We shall here elaborate on the two cases that were mentioned in Sec. 2.1.

But before doing that let us broaden the discussion in Secs. 2.4 and 2.5, even if in a somewhat artificial manner. Let us assume that the floating body is attached to a spring connected at the origin. In this case Eq. (1.2.6) will take the form

$$M\ddot{\mathbf{r}} + \mu\dot{\mathbf{r}} + C\mathbf{r} = \mathbf{F}_e\,. \tag{1.2.17}$$

The last term on the left hand side originates, of course, from the work required in order to stretch the spring ($\frac{1}{2}Cr^2$). Since it is quadratic in r, the thermal average of $\frac{1}{2}Cr^2$ is given by the equipartition principle, and is identical to that of the kinetic energy, as in Eq. (1.1.60) of Sec. 1.6. That is, at thermal equilibrium (and only then)

$$\left\langle \frac{1}{2}C\mathbf{r}^2 \right\rangle = \left\langle \frac{1}{2}M\dot{\mathbf{r}}^2 \right\rangle = \frac{1}{2}DkT\,. \tag{1.2.18}$$

If we perform on (1.2.17) the same operations that brought us from Eq. (1.2.6) to Eq. (1.2.9) and to Eq. (1.2.11), we obtain for $u = \langle r^2 \rangle$ the equation

$$M\ddot{u} + \mu\dot{u} + 2Cu = 2DkT\,. \tag{1.2.19}$$

Exercise 2.9

Derive the above equation.

Solution on page 93

Equation (1.2.19) appears exactly like the one-dimensional version of Eq. (1.2.17), except that instead of the random external force a constant force is acting. Alternatively, it is possible to transfer the "force" $2DkT$ to the left hand side, and to imagine that u describes the displacement of the spring not from a loose state but from a state in which it is stretched on average according to $2Cu = 2DkT$. Thus, if we define a new variable $v = u - DkT/C$ we obtain an equation that is identical to the equation of a damped harmonic oscillator:

$$M\ddot{v} + \mu\dot{v} + 2Cv = 0\,. \tag{1.2.20}$$

The solution to Eq. (1.2.20) includes an exponentially decaying factor, as for the case $C = 0$, in addition to a restoring force ($2Cv$) producing vibrations around $v = 0$. Thus, even without investigating the solution in detail we can reach the conclusion that at long times v tends to its equilibrium value, i.e. zero, and that u tends to its equilibrium value, DkT/C, as inferred from Eq. (1.2.18).

If we search for a solution of the form $e^{-\gamma t}$, we find that the substitution in Eq. (1.2.19) gives two possible values (both positive) for γ:

$$\gamma_{1,2} = \frac{1}{2\theta}\left(1 \pm \sqrt{1 - \frac{8CM}{\mu^2}}\right), \quad \gamma_2 > \gamma_1 . \tag{1.2.21}$$

And the solution corresponding to the initial conditions $u(t = 0) = \dot{u}(t = 0) = 0$ is

$$\langle r^2 \rangle = u(t) = \frac{DkT}{C}\left(1 - \frac{\gamma_2 e^{-\gamma_1 t} - \gamma_1 e^{-\gamma_2 t}}{\gamma_2 - \gamma_1}\right) . \tag{1.2.22}$$

Indeed, for long times $\langle r^2 \rangle$ tends to its equilibrium value. (See also Fig. 1.2.5.)

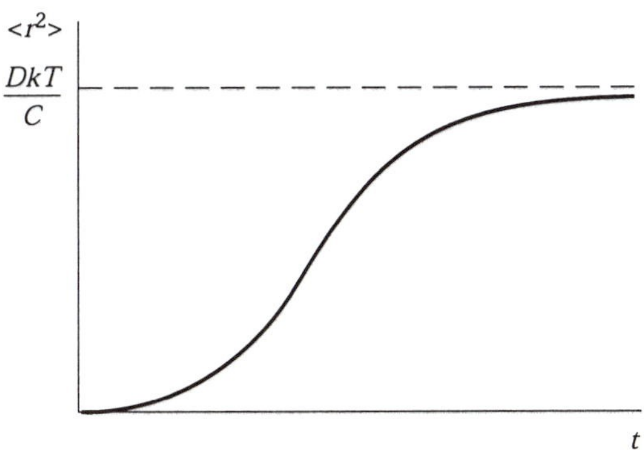

Fig. 1.2.5 A graphic representation of Eq. (1.2.22).

Finally, we note that when the square root in Eq. (1.2.21) becomes imaginary, i.e. when the damping is small, the exponential solution becomes a solution of damped oscillations. This solution also tends to DkT/C at long times.

Exercise 2.10

Obtain Eq. (1.2.22), and check its short time behavior. When is $u(t)$ a linear function of time?

Solution on page 93

Notice that the behavior at times which are not too long resembles that of ordinary Brownian motion without an elastic force. At first u

grows quadratically with time, when the effect of the collisions and of the elastic force are still negligible. Later, when the effect of the collisions starts to be significant, but the elastic force is still negligible, the body does indeed move as in Brownian motion. In contrast, at long times the elastic force dominates and does not allow further separation beyond the limit set by the temperature. This situation does not describe a group of drunk people who move freely from the moment they leave the bar, but rather a group of drunk horses that are tied by flexible straps at the entrance to the bar.

A more physical example for this state of affairs is the galvanometer. This instrument is used to measure very small currents, by means of the very small angles of rotation of a quartz whisker that these currents induce. We shall not halt here to explain how the currents induce the rotations of the whisker, but rather concentrate on the way these rotations are measured. A tiny mirror is connected to the whisker. Light is projected on the mirror and the angle of rotation ϕ of the whisker is measured by registering the angle into which light is reflected from the mirror, on a scale (see Fig. 1.2.6).

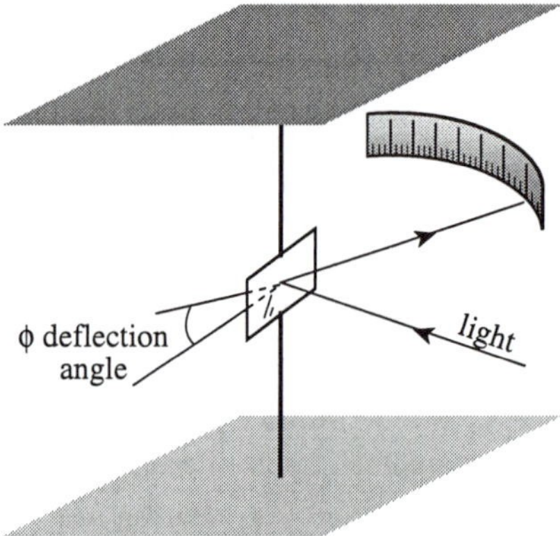

Fig. 1.2.6 Measurement of the angle of reflection in a galvanometer.

Here as well, the mirror and the quartz whisker are in thermal equilibrium with the surrounding gas. The rotations of the whisker have a kinetic energy that depends on the moment of inertia of the system, i.e. $\frac{1}{2}I\dot{\phi}^2$, and a potential energy $U = \frac{1}{2}a\phi^2$, where $-a\phi$ is the restoring force of the whisker. In this case the temporal behavior of $\phi(t)$ is determined by Eq. (1.2.17), with the following substitutions:

$$\text{dimensionality} \quad D = 1 \; , \; C \to a \; , \; M \to I \; , \; \mathbf{r} \to \phi \, .$$

Here, too, there exists a random force or torque, originating from the collisions of the whisker and the mirror with the gas molecules, that gives a friction term $\mu\dot{\phi}$, which restores equilibrium, as well as the force $\mathbf{F}_e$, whose average vanishes.

The result is that even in the absence of a current, deviations will appear due to thermal fluctuations. The average square angular deviations will be given by Eq. (1.2.22) in terms of the parameters of the galvanometer. For long times one finds

$$\langle \phi^2 \rangle = \frac{kT}{a},$$

which is appropriately dimensionless. This is a noise that limits the precision of the instrument. In order to reduce it, the instrument has to be cooled. But where? The answer can be found in Sec. 2.4. The fluctuations that give rise to $\langle \phi^2 \rangle$ are the ones that produce the friction, and therefore the part in which friction is created has to be cooled. This can be, for example, the mirror which suffers collisions with the gas molecules. In this case the gas has to be cooled.

Exercise 2.11

How can we measure the restoring force constant a of the whisker?

Solution on page 94

In Sec. 2.5 we saw that the characteristic time for Brownian motion $\theta = M/\mu$ is very short — 10^{-7} s. Thus, it is impossible to observe in an experiment on Brownian motion the exponential decay to the linear region. Here, on the other hand, $\theta = I/\mu$, but μ can be reduced by reducing the pressure of the gas, making it possible to increase θ up to measurable magnitudes and to check the predictions of the theory in great detail, as was done by Kappler in 1931.

Johnson noise

quality factor

Finally, we mention another analogous instance — the Johnson noise, caused by the fluctuations in a resistor, of a resonating circuit with a high Q factor (see Fig. 1.2.7). The thermal fluctuations, caused by the collisions of electrons, which form the current I, with the atoms in the resistor, produce also here the double effect — resistance R, which damps the system and drags it towards equilibrium at zero current, and a random

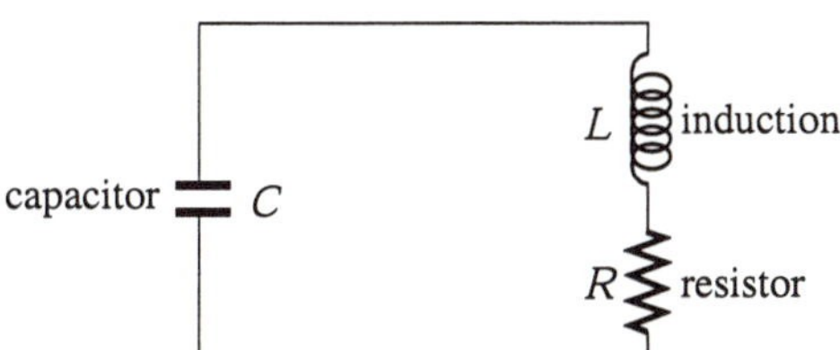

Fig. 1.2.7 A resonating circuit. A high Q value means that the circuit amplifies significantly only in a very narrow band around the resonant frequency $\omega = 1/\sqrt{LC}$.

(electromotive) force. The equation of the circuit determines the charge on the capacitor q:

$$L\ddot{q} + R\dot{q} + \frac{q}{C} = V_e \,. \tag{1.2.23}$$

Here C is the capacitance, L the induction and V_e the random electromotive force originating from the fluctuations in the resistor. This is the Langevin equation, analogous to Eq. (1.2.17), with the substitutions

$$D = 1 \;,\; \mathbf{F}_e \to V_e \;,\; C \to \frac{1}{C} \;,\; \mu \to R \;,\; M \to L \;,\; \mathbf{r} \to q \,.$$

We are of course interested not in the solution to Eq. (1.2.23), but in the solution to the equation that determines the time dependence of $\langle q^2 \rangle$ or of the directly measurable quantity $\langle V^2 \rangle$, where $V = \frac{q}{C}$ is the voltage across the capacitor. The equation obtained for $\langle V^2 \rangle$ has the same form as Eq. (1.2.19), which was solved in detail.

But even without solving the equation we can reach the following conclusions: the first term of Eq. (1.2.23) comes about from the "kinetic energy" term, $\frac{1}{2}LI^2$, while the last term originates in a "potential energy" term, $\frac{q^2}{2C}$. Therefore, at equilibrium we obtain an equation that resembles Eq. (1.2.18):

$$\left\langle \frac{q^2}{2C} \right\rangle = \left\langle \frac{1}{2}LI^2 \right\rangle = \frac{1}{2}kT \,. \tag{1.2.24}$$

Since there are charge fluctuations, there will also be voltage fluctuations (Johnson noise), which are

$$\langle V^2 \rangle = \frac{kT}{C} \tag{1.2.25}$$

or

$$\langle V^2 \rangle = \omega^2 LkT \,, \tag{1.2.26}$$

where ω is the resonant frequency of the circuit.

Chapter 3

Transport Coefficients

3.1 Introduction

One of the strong objections against the kinetic theory of Maxwell and Boltzmann was known as "slow diffusion." In simple words, according to the kinetic theory, the average (thermal) velocities in gases in standard conditions are about 10^3 m s^{-1}. Hence, if a gas is inserted at one end of an empty container, the molecules should appear at its other end within a fraction of a second. However, this is not what is observed in experiment.

diffusion

In order to cope with this observation Clausius introduced (in 1889) the concepts of *mean free path* and *mean free time*. He asserted that it is true that the average distance between molecules in a gas is large compared to the size of the molecules, but because of the high speeds the molecules collide very frequently, changing their direction of motion, so that the gas moves in a given direction at a speed that is much lower than the thermal speed — it moves at the *diffusion speed*.

mean free path

The concept of a mean free path, the average distance traversed by a molecule in the gas between two collisions, is a very central concept of the kinetic theory (it was mentioned in Sec. 2.2 in connection with the Brownian particle). Directly connected with the mean free path is the mean free time, which is the average time between consecutive collisions in the gas. In other words, this is the average time during which the molecule moves as a free particle.

mean free time

The clarification of the concept of a mean free path and its quantitative evaluation opened the way for the calculation of many important quantities, transport coefficients, amongst which we find the mobility, diffusion coefficients of sorts, the viscosity, thermal conductivity and more.

The transport coefficients describe the behavior of the system when there is a slight deviation from an equilibrium state. Such a deviation can be caused by the application of an external force (in the cases of mobility and viscosity) or by creating concentration gradients (in the case of diffusion) or temperature gradients (in the case of thermal conductivity). Such deviations from equilibrium create currents, which drive the system back to equilibrium. The ratios between the currents and the disturbances that create them are the transport coefficients. All of these will be treated in this chapter.

3.2 The mean free path and mean free time

The problem at hand is the calculation of an average distance, ℓ, for which an average gas molecule will collide at least once — this is the mean free path. For the sake of simplicity we will suppose that the molecules are rigid balls, of radius a, so that their trajectories change only when there is direct contact between them. This is a reasonable approximation for the noble gases.

If we will know ℓ, as a function of the characteristic parameters of the gas (temperature, density, molecular radii, etc.), we will be able to obtain the mean free time from

$$\tau \cdot \bar{v} = \ell\,, \tag{1.3.1}$$

where $\bar{v}$ is the average velocity of the molecules.

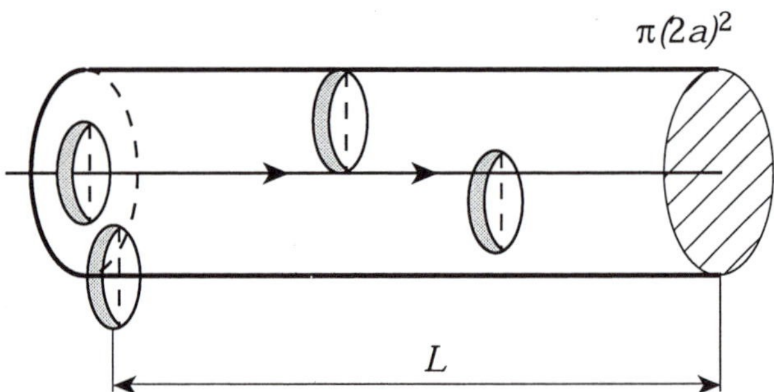

Fig. 1.3.1 Collision cross section of a rigid ball of radius a.

We start with a very rough calculation, in the spirit of Sec. 2.2, and proceed to a more sophisticated and complicated calculation, which exposes some of the approximations made in the rough calculation, and of their precision. Figure 1.3.1 visualizes the fact that a molecule moving in a straight line will collide with every molecule whose center is found in a cylinder along its directon of motion whose radius is twice the radius of the molecule. The number of molecules in a cylinder of this radius and length L is $(4\pi a^2 L)n$, where n is the density. The average distance traversed by a molecule until the first collision is the distance in which the average number of molecules in this cylinder is 1. Hence

$$\ell = \frac{1}{4\pi a^2 n}\,. \qquad (1.3.2)$$

We will make a few remarks on this simple expression. First, clearly, the true meaning of the size $4\pi a^2$ is: the cross section for the collision of two molecules. **cross section**

■ Scattering cross section: the area normal to the direction of motion, such that every molecule that passes through it is bound to collide with the target. In our case this is a circle whose radius is twice the radius of a molecule.

Second, the properties of ℓ, as reflected from Eq. (1.3.2), correspond nicely with intuition. As the density of molecules or their cross section areas increase, ℓ must decrease.

Exercise 3.1

Is the dependence on n and a, together with a dimensional analysis, sufficient to lead to the expression (1.3.2) for ℓ?

Solution on page 95

Third, the arguments we used are very rough, and may be criticized in many respects. Some of the implicit assumptions in this calculation will remain even in the more detailed calculation that will be made below.

Exercise 3.2

What are the implicit assumptions of the above calculation?

Solution on page 95

We now want to calculate the mean free path and the average time between collisions, namely the mean free time, in a more precise fashion, while still retaining two simplifying assumptions:

(a) The gas is dilute so that collisions occur only between pairs of molecules.
(b) The molecules behave as rigid balls of radius a.

Before turning to the calculation of the mean free time we calculate a simpler quantity that will be useful later on, that is the average rate at which the molecules beat against a unit area of the side of the container.

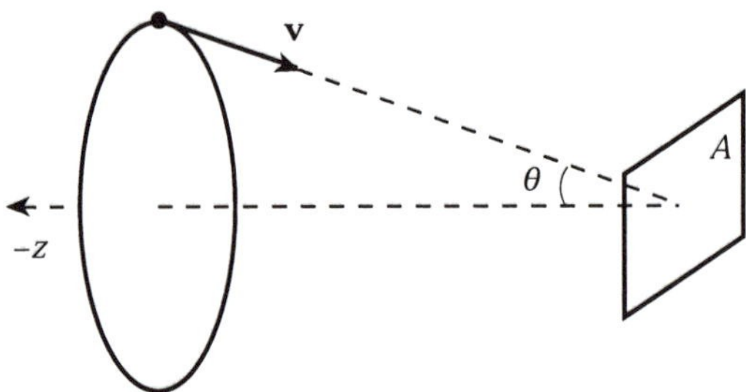

Fig. 1.3.2 A molecule with velocity **v** will strike an area A of the side of the container.

Actually we almost calculated this quantity on our way to Eq. (1.1.3). The number of molecules striking an area A during time Δt and whose velocity component normal to the surface is v_z, was found in Sec. 1.1 to be

$$\Delta N(v_z) = v_z \Delta t A n(v_z)\,, \tag{1.3.3}$$

where $n(v_z)$ is the density of molecules of velocity v_z.

■ Notice that what we called the x direction in Sec. 1.1 we here call z.

In order to find the total number of molecules that strike the surface, we have to sum over all the values of the velocity v_z such that $v_z > 0$, since molecules for which $v_z < 0$ are moving away from the surface and will not strike it. We have

$$\Delta N = A\Delta t \int_0^\infty v_z n(v_z) dv_z\,. \tag{1.3.4}$$

The distribution of the velocity component v_z is given by the one-dimensional Maxwell distribution obtained from the integration of $f(v)$ [see Eq. (1.1.49)] with respect to all values of v_x and v_y:

$$n(v_z) = \frac{N}{V} \int f(\mathbf{v}) dv_x dv_y\,. \tag{1.3.5}$$

Substituting this into Eq. (1.3.4) we obtain

$$\Delta N = A\Delta t \frac{N}{V} \int_{v_z>0} v_z f(\mathbf{v}) dv_x dv_y dv_z\,. \tag{1.3.6}$$

Now, because the velocity distribution is isotropic (there is no preferred direction for the velocities) $f(\mathbf{v})$ depends, actually, only on the absolute value of the velocity, v, as is seen on the right hand side of Eq. (1.1.49). Thus, we can perform the integration over all possible angles separately from the integration over v. Because $v_z = v\cos\theta$ and $dv_x dv_y dv_z = v^2 \sin\theta dv d\theta d\phi$, we find that

$$\Delta N = A\Delta t \frac{N}{V} 2\pi \int_0^\infty v^3 f(v) dv \int_0^{\pi/2} \cos\theta \sin\theta d\theta\,. \tag{1.3.7}$$

Note that θ can vary only up to $\pi/2$ — reflecting the fact that we are only taking into account molecules that are approaching the side ($v_z > 0$). The second integral is simple to evaluate and its value is $1/2$. A closer look at the first integral reveals that it is proportional to the average of the absolute value of the molecular velocity, $\bar{v}$, since

$$\bar{v} = \langle |\mathbf{v}| \rangle = \int f(\mathbf{v})|\mathbf{v}| dv_x dv_y dv_z = 4\pi \int_0^\infty f(v) v^3 dv \,. \tag{1.3.8}$$

We have found, therefore, without making any assumptions about the form of the velocity distribution apart from the assumption of isotropy, that the rate at which molecules strike a unit area of a side, ν, is

$$\nu = \frac{1}{4} n \bar{v} \,. \tag{1.3.9}$$

If the velocity distribution is given explicitly it is possible to express $\bar{v}$ in terms of the characteristic quantities of the gas. For example, if the velocity distribution is Maxwellian we find

$$\nu = n \sqrt{\frac{kT}{2\pi m}} \,. \tag{1.3.10}$$

Exercise 3.3

(a) Derive Eq. (1.3.10).
(b) Calculate the escape rate of the gas molecules from a container with a small hole of area A at its side.

Solution on page 96

Next we calculate the average time between two consecutive collisions of a given molecule. We apply considerations of the type introduced in computing the mean free path. As we said there, for a given molecule to encounter another, it must travel a distance $\frac{1}{4\pi a^2 n}$. If one molecule is stationary and the other is moving, this distance is crossed in a time $\frac{1}{4\pi a^2 n \bar{v}}$, on the average. But since both molecules are moving, $\bar{v}$ should be replaced by the average relative velocity at equilibrium, $\bar{v}_{\rm rel}$. In Exercise 3.4 we show that

$$\bar{v}_{\rm rel} = \sqrt{2}\, \bar{v} \,. \tag{1.3.11}$$

Exercise 3.4

Prove Eq. (1.3.11).

Solution on page 96

Hence we find that the mean free time τ is

$$\tau = \frac{1}{\sqrt{2}\,4\pi a^2 n\bar{v}}\,. \tag{1.3.12}$$

Our first inclination would be to substitute this expression for τ in Eq. (1.3.1), and to calculate the mean free path ℓ. If we do this we obtain a result that is closer to the truth than Eq. (1.3.2), yet not totally accurate:

$$\ell \approx \frac{1}{\sqrt{2}} \cdot \frac{1}{4\pi a^2 n} = 0.707\frac{1}{4\pi a^2 n}\,. \tag{1.3.13}$$

The reason for this inaccuracy is the fact that the mean free time for a given molecule depends, as indicated by Eq. (1.3.12), on its velocity. Thus, the accurate way of calculating ℓ will be to first calculate the mean free path for a molecule that moves at a given velocity, and then to average over all the velocities using the Maxwellian distribution function. The result of this calculation, which will not be performed here, will be the replacement of the factor of 0.707 in Eq. (1.3.13) by a factor of 0.677.

Exercise 3.5

Show that in a gas under standard conditions the ratio between ℓ and the intermolecular distance is 10^3, while the ratio of ℓ and a is 10^4.

Solution on page 97

Exercise 3.6

Oxygen fills a cubic container with a side of 5 cm, at a temperature of 100°C.

At what pressure will the mean free path be equal to the size of the container? What will the mean free time then be?

Solution on page 98

Exercise 3.7

Use the simple argument for calculating the mean free path to obtain the mean free path in a mixture of gases of densities n_1 and n_2 and molecular radii a_1 and a_2, respectively.

Solution on page 99

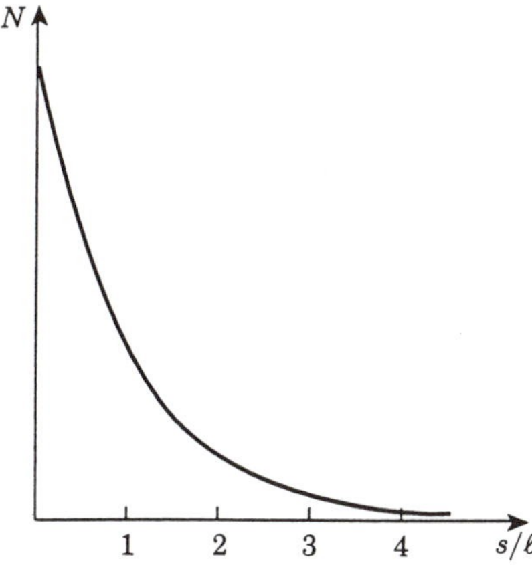

Fig. 1.3.3 The number of molecules that cover a distance s without colliding.

Clearly not all molecules traverse an equal distance between collisions, and not all of them remain a time τ without collisions. What is, therefore, the distribution of the free distances, i.e. the probability of finding molecules that have traversed a distance s without experiencing a collision. Actually it is better to think here in terms of numbers of molecules and to ask how many of the molecules travel a given distance without colliding. Let us suppose that N_0 molecules leave the last collision, and denote by $N(s)$ the number among them that traveled a distance s without colliding. The number of collisions per unit length is $1/\ell$. Hence, the average number of molecules that will collide, when the distance grows from s to $s+ds$, is $N(s)ds/\ell$, i.e. the product of the number of molecules that went a distance s without colliding, and the ratio between the extra distance ds and the distance ℓ required for one more collision.

However, this number is also the reduction in $N(s)$, namely $N(s) - N(s+ds)$, where $N(s+ds)$ is the number of molecules that traveled the distance $s+ds$ without colliding. Hence

$$dN = -\frac{1}{\ell}N(s)ds \tag{1.3.14}$$

and from here

$$N(s) = N_0 e^{-s/\ell}\,. \tag{1.3.15}$$

The actual probability we set out to find is $P(s) = \frac{1}{\ell}e^{-s/\ell}$.

■ Note the difference between:

(a) The probability that the distance traveled by a molecule between two collisions is exactly s.
(b) The probability that the distance traveled by a molecule between two collisions is at least s.

Actually, we mentioned only the distribution corresponding to (b). It is possible to show that distribution (a) is also exponential.

Exercise 3.8

Prove that the average distance traversed by a molecule without experiencing any collisions, with the distribution (1.3.15), is ℓ.

Solution on page 99

Exercise 3.9

Show that the distribution of times between collisions is

$$N(t) = N_0 e^{-t/\tau} . \tag{1.3.16}$$

Solution on page 100

3.3 Self-diffusion

The common diffusion problem deals with a mixture of two materials (gases in our case), whose relative density changes from place to place. Thus, without the application of an external force, currents of the two materials flow — currents that drive the system towards a state of equilibrium, namely to a spatial uniformity.

current density

■ The current J is related to the density of the particles n and to their velocities by $J = nv$. Sometimes J is called the current density, while the term "current" is used to refer to the product of J and a given area perpendicular to the flow, as is done in the definition of the electric current.

We will here treat a simpler problem, yet it contains the main ideas. We will discuss a problem in which the gas is composed of molecules, of equal mass and equal size, that belong to two distinguishable types. An approximation to this situation can be found in a mixture of N_2O and CO_2 (both with molecular mass of 44), or in a mixture of two isotopes. The total density is uniform in the container, but the densities of the constituents vary.

We shall concentrate on one of the constituents and assume, for simplicity, that its density n_1 varies in one direction only, which we choose to be the direction of the z axis. Experiments establish that the current — the number of particles of the type considered, which cross a unit area in the xy plane per unit time — is proportional to the derivative of the density:

$$J_z = -D\frac{\partial n_1}{\partial z} . \tag{1.3.17}$$

The minus sign signifies that the current flows from high density to low density

■ The derivative is partial since n_1 depends on time as well.

D is called the diffusion coefficient. Its dimensions are $[L]^2[T]^{-1}$, and its units are $\mathrm{m}^2\ \mathrm{s}^{-1}$. As will be seen later on, it is closely connected with the coefficient α in Eq. (1.2.2), which can also be expressed in terms of the frictional coefficient and the temperature [see Eq. (1.2.16)].

■ Note that from now on the letter D stands for the diffusion coefficient and not for the dimensionality of the system as in (1.2.16).

Typical values for gases, in these units, are 10^{-5}–10^{-4}. For liquids the typical values are a million times smaller.

The origin of the current and its connection with the density variation are quite obvious. Because the temperature in the gas is uniform, the molecules have equal velocity distributions in all directions at every point. As a result molecules move from place to place at a rate that is equal to the product of the velocity and the density. This rate is the number of molecules that will cross a unit area in unit time. Therefore, more molecules will cross per unit time from a location of high density to a location of low density than in the opposite direction, since the velocity distributions are identical at both locations.

This simple argument will now be turned into a quantitative relationship between D and the characteristics of the gas. This will be done in two ways: the first is simplified, and the second (which appears in the appendix) is more complex and will indicate the sort of arguments that need to be made for a more careful calculation.

Simple calculation. The idea is that the molecules that pass through the plane $z = 0$ will come on average from a distance which is the mean free path, above this plane or below it. On average 1/6 of the molecules have a velocity along $+z$ or $-z$ (1/3 for the choice of a given axis and 1/2 for the choice of direction). The net current density flowing up across $z = 0$ is

$$J_z \approx \frac{1}{6}\bar{v}[n_1(-\ell) - n_1(\ell)]\,, \tag{1.3.18}$$

as is depicted, schematically, in Fig. 1.3.4.

If the density $n_1(z)$ varies slowly — on a scale of the mean free path — we can expand the right hand side of Eq. (1.3.18) in powers of ℓ. If we keep only the linear term, we obtain

$$J_z \approx -\frac{1}{3}\bar{v}\ell\frac{\partial n_1}{\partial z}\,. \tag{1.3.19}$$

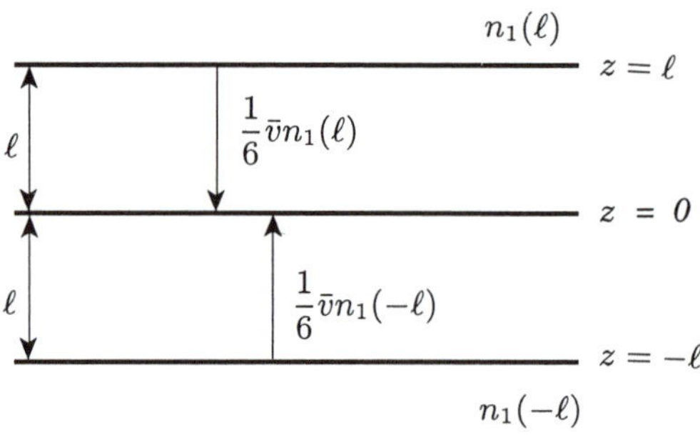

Fig. 1.3.4 Pictorial description of Eq. (1.3.18).

A comparison with Eq. (1.3.17) yields

$$D \approx \frac{1}{3}\bar{v}\ell\,. \tag{1.3.20}$$

Exercise 3.10

Show that at constant temperature the dependence of D on P is given by

$$D \propto \frac{1}{P}\,,$$

and that at a constant pressure

$$D \propto T^{3/2}\,.$$

Solution on page 100

We now draw a few conclusions. First, Eq. (1.3.17) can be generalized to the case where n_1 varies along an arbitrary direction. The generalization is simple and can be deduced from geometrical arguments: since the current is a vector, the right hand side must also be a vector. The vector that can be created from first derivatives is the gradient, so that

$$\mathbf{J} = -D\nabla n_1\,. \tag{1.3.21}$$

Since the gradient of n_1 is directed along the direction of the fastest variation of n_1, it is natural for this to be the direction of the current.

■ The meaning of Eq. (1.3.21) in terms of components is

$$J_x = -D\frac{\partial n_1}{\partial x}\,,\ J_y = -D\frac{\partial n_1}{\partial y}\,,\ J_z = -D\frac{\partial n_1}{\partial z}\,.$$

In order to see that ∇n_1 is directed along the direction of the fastest variation in n_1, we write the change in n_1 between the initial point $\mathbf{r}$ and a nearby point $\mathbf{r} + \Delta\mathbf{r}$:

$$\Delta n_1 \approx \frac{\partial n_1}{\partial x}\Delta x + \frac{\partial n_1}{\partial y}\Delta y + \frac{\partial n_1}{\partial z}\Delta z = (\nabla n_1)\cdot\Delta\mathbf{r}\,,$$

and it is clear that Δn_1 will be maximal when the angle between ∇n_1 and $\Delta\mathbf{r}$ is zero, namely when $\Delta\mathbf{r}$ is in the direction of the gradient.

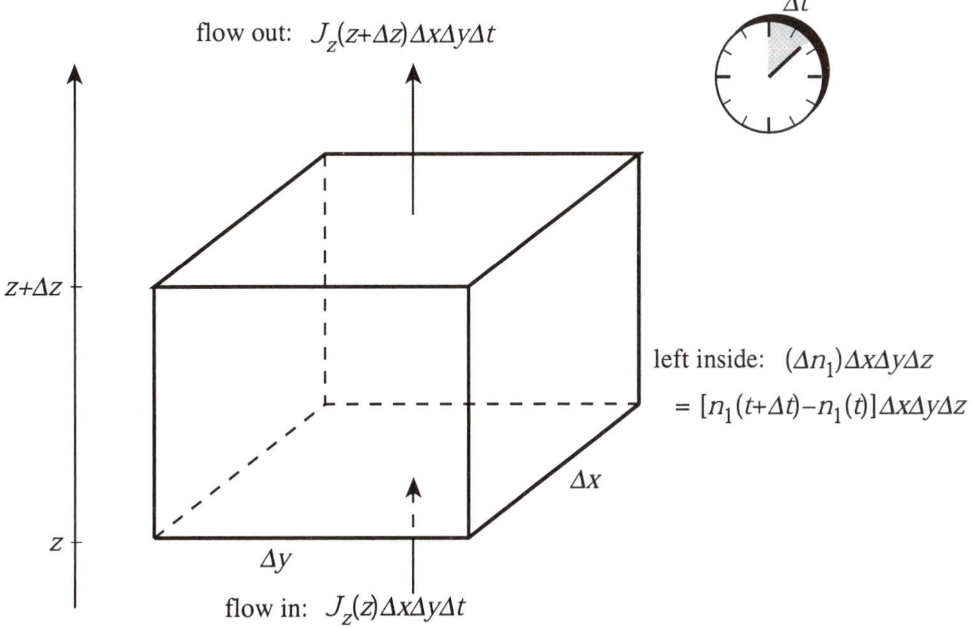

Fig. 1.3.5 The balance of currents in the region around z.

Second, the number of particles of every type is conserved. This means that each type of particle satisfies a continuity equation, expressing the fact that the density changes must come from the balance of the current entering the region around a given point and the current that is leaving it. For particles of type 1, for example, Fig. 1.3.5 is a pictorial derivation of the continuity equation

continuity equation

$$\frac{\partial n_1}{\partial t} = -\frac{\partial J_z}{\partial z}\,. \tag{1.3.22}$$

Exercise 3.11

What is the generalization of Eq. (1.3.22) to the three-dimensional case? Use geometrical arguments.

Solution on page 101

If we substitute Eq. (1.3.17) into Eq. (1.3.22), we obtain the celebrated diffusion equation

diffusion equation

$$\frac{\partial n_1}{\partial t} = D\frac{\partial^2 n_1}{\partial z^2}\,. \tag{1.3.23}$$

For the one-dimensional case, n_1 is a function only of z and t. A typical solution to this equation, with an initial condition that at the time $t = 0$

all the particles of type 1 are concentrated at $z = 0$, is

$$n_1(z,t) = \frac{C}{\sqrt{Dt}} e^{-z^2/4Dt}\,, \qquad (1.3.24)$$

where $C = N_1/\sqrt{4\pi}$ is a normalization factor (see Exercise 3.12). The average square displacement of the particles of type 1 can also be computed in terms of the solution (1.3.24):

$$\langle z^2 \rangle = 2Dt\,. \qquad (1.3.25)$$

Exercise 3.12

(a) Verify that Eq. (1.3.24) is in fact a solution to Eq. (1.3.23).
(b) Show that the total number of particles of type 1 does not depend on time and clarify the meaning of the constant C.
(c) Plot $n_1(z,t)$ as a function of t, at a given value of $z \neq 0$.
(d) Prove Eq. (1.3.25) and find the relation between the time needed for a molecule to traverse a certain distance by diffusion, and the time it needs to traverse the same distance by free motion at the thermal velocity.

Solution on page 101

Finally, we note that in the more realistic case of a three-dimensional system the diffusion equation will take the form

$$\frac{\partial n_1}{\partial t} = D\left(\frac{\partial^2 n_1}{\partial x^2} + \frac{\partial^2 n_1}{\partial y^2} + \frac{\partial^2 n_1}{\partial z^2}\right) \qquad (1.3.26)$$

and the analog of the solution (1.3.24) will be

$$n_1 = \frac{N_1}{(4\pi Dt)^{3/2}} e^{-r^2/4Dt}\,, \qquad (1.3.27)$$

where N_1 is the total number of particles of type 1: $N_1 = \int n_1 dV$. The average square distance will in this case be

$$\langle r^2 \rangle = 6Dt\,. \qquad (1.3.28)$$

A comparison of Eq. (1.3.28) with Eq. (1.2.2) clarifies immediately that α and D are one and the same apart from a numerical factor, i.e. $\alpha = 6$ in this case.

3.4 The mobility coefficient

In our discussion concerning diffusion we saw that a small deviation from a state of equilibrium, in the form of an induced density gradient in the gas, gives rise to a current. Another way to create currents is by applying an external force.

When a particle moves freely, exertion of a force causes it to accelerate and the acceleration, according to Newton, is proportional to the force. As already mentioned, with respect to Brownian motion, a particle that is moving in the gas is exposed to damping and its *velocity* is proportional to the exerted force. That is, if a constant force is acting on a special particle in the gas (different charge, different mass), then its velocity, the drift velocity, will be

$$v_d = KF\,. \tag{1.3.29}$$

K is called the mobility coefficient. Its dimensions are **mobility**

$$[K] = [M]^{-1}[T]\,.$$

It is possible to understand Eq. (1.3.29) and to express K with the help of the parameters of the gas using a very simple argument: between two collisions the particle which feels the force is accelerated according to Newton's law,

$$\mathbf{a} = \frac{\mathbf{F}}{m}\,, \tag{1.3.30}$$

where $\mathbf{a}$ is the acceleration and m is the mass of the particle, so that its velocity at time t after a given collision is

$$\mathbf{v} = \mathbf{v}_0 + \frac{\mathbf{F}}{m}t\,, \tag{1.3.31}$$

where $\mathbf{v}_0$ is the velocity of the particle immediately after the collision. If we now calculate the average of $\mathbf{v}$ over many collisions along its path, the average of $\mathbf{v}_0$ will be zero, since the velocities that particles have immediately after a collision are completely random.

Thus, the average velocity of the accelerated particles will be equal to the acceleration times the average time $\langle t\rangle$ elapsed since the previous collision, which is the same as the average time between collisions τ (see the calculation of $\langle t\rangle$ in Solution 3.13). Namely

$$\mathbf{v}_d = \langle\mathbf{v}\rangle = \frac{\mathbf{F}}{m}\tau\,, \tag{1.3.32}$$

Comparing with Eq. (1.3.29) one obtains the mobility as a function of quantities from the kinetic theory:

$$K = \frac{\tau}{m}\,. \tag{1.3.33}$$

The drift velocity v_d is therefore the "effective" velocity at which the particle advances (along the direction of the force), in spite of the random collisions with the molecules of the gas. The external force "drags" the particle in a specific direction, and in its absence the drift velocity is zero.

■ Note the difference between the drift velocity and the average thermal velocity $\bar{v}$. These are two different averages!

Exercise 3.13

We can try to calculate K in the following way. The distance d that is covered by the particle during the mean free time is

$$d = \frac{1}{2}\frac{F}{m}\tau^2 .$$

The average velocity is the ratio of this distance to the average time. The result is that K is two times smaller than in Eq. (1.3.33). Which result is correct? What happened here?

Hint: calculate v and d with the help of the distribution (1.3.16).

Solution on page 103

The argument given above as a derivation of Eq. (1.3.29) is actually a paraphrase of the description of a body executing Brownian motion in Chap. 2 (Secs. 2.4 and 2.5). In that case the numerous collisions with the molecules of the gas were expressed as a viscosity. The difference with respect to our discussion here is that here an additional constant external force with nonzero average is acting on the body. Hence, a direct averaging of Eq. (1.2.6) yields

$$m\frac{d\langle \mathbf{v}\rangle}{dt} + \mu\langle \mathbf{v}\rangle = \langle \mathbf{F}_e\rangle . \qquad (1.3.34)$$

Since, as we have seen, the characteristic time m/μ is about 10^{-7} s (Sec. 2.5), the average acceleration of the body vanishes almost immediately, and we obtain

$$\mathbf{v}_d = \mu^{-1}\mathbf{F} . \qquad (1.3.35)$$

This means that, in addition to the Brownian motion around its original position, the particle drifts along the direction of the force at a velocity $\mathbf{v}_d$.

If we compare this equation with Eq. (1.3.32), we find that the mobility is none other than

$$K = \frac{1}{\mu} . \qquad (1.3.36)$$

So far we have been talking of the mobility of a special particle in the gas. However, it is possible to repeat the arguments for one of the particles of the gas itself when an external force acts on all or some of them. Examples for this situation are the motion of ions in a gas or in a solution when an electric field is applied, the motion of particles in a centrifuge or in a gravitational field, and so on.

■ In Part V we will apply the same methods to the electric conductivity of metals.

In this situation Eq. (1.3.32) describes the average velocity of one of the particles on which the force is acting. But because there is a macroscopic number of such particles, we can use the ergodic hypothesis to interpret $\mathbf{v}_d$ also as the average velocity of all the particles at a given time. Equation (1.3.32) describes, therefore, a flow of particles in which, in addition to their random motion, there is an ordered (small) velocity component moving them in the direction of the force.

3.5 The connection between the diffusion coefficient and the mobility

Comparison of Eqs. (1.2.15) and (1.3.28) leads to the relation

$$D = \frac{kT}{\mu}, \tag{1.3.37}$$

keeping in mind that in the first D is the number of dimensions, while in (1.3.28) it is the diffusion coefficient. On the other hand, we have established that $K = \mu^{-1}$, so that in three dimensions

$$\frac{D}{K} = kT. \tag{1.3.38}$$

This is a very impressive result: the ratio of two very different quantities depends solely on the temperature, and in a very simple manner. However, the impression is spoiled by the observation that the ratio was obtained by using Langevin's equation, which describes the motion of a Brownian particle. The validity of this equation for the motion of molecules or ions is by no means obvious. We will now see, following Feynman's *Lecture Notes in Physics*, that the relation (1.3.38) is indeed very general.

We assume that n_1 is the number of molecules of type 1 per unit volume, as in Sec. 3.3. These are the labeled molecules. If a force $\mathbf{F}$ is acting on them, their drift velocity is [Eq. (1.3.29)]

$$\mathbf{v} = K\mathbf{F},$$

so that a current of particles of type 1 is created, which is

$$\mathbf{J}_M = n_1\mathbf{v} = n_1 K\mathbf{F}\,. \tag{1.3.39}$$

On the other hand, a diffusion current of molecules of type 1 will develop [Eq. (1.3.21)], which we will denote by $\mathbf{J}_D$:

$$\mathbf{J}_D = -D\nabla n_1\,. \tag{1.3.40}$$

In a state of equilibrium,

$$\mathbf{J}_D + \mathbf{J}_M = 0\,, \tag{1.3.41}$$

and so $\mathbf{F}$ satisfies

$$n_1 K\mathbf{F} = D\nabla n_1\,. \tag{1.3.42}$$

But from the discussion of the isothermal atmosphere (Sec. 1.4) we know that a relationship exists between a force (or potential) field and the density. If the force $\mathbf{F}$ is derived from a potential U, i.e. $\mathbf{F} = -\nabla U$, then the presence of $\mathbf{F}$ corresponds to a spatial distribution or density distribution, Eq. (1.1.32):

$$n_1 = Ce^{-U/kT}\,,$$

and the corresponding density gradient is

$$\nabla n_1 = -\frac{1}{kT}n_1\nabla U = \frac{1}{kT}n_1\mathbf{F}\,. \tag{1.3.43}$$

For this density gradient to satisfy the equilibrium conditions, Eq. (1.3.42), the relation (1.3.38) must hold. That is, the possibility of exerting a force and maintaining a density gradient (even in a thought experiment) requires that the diffusion current be canceled by the drift current, from which $D/K = kT$.

And if the result (1.3.38) seemed impressive at first, then the way it was deduced here is even more impressive, and worth a second and a third reflection.

3.6 Viscosity and thermal conductivity

Viscosity

An experimental situation in which the viscosity will appear under simple conditions is the following:

A container with parallel plane walls is filled with gas. One of the walls, to which the large arrow is attached in Fig. 1.3.6, is moving at a constant velocity u.

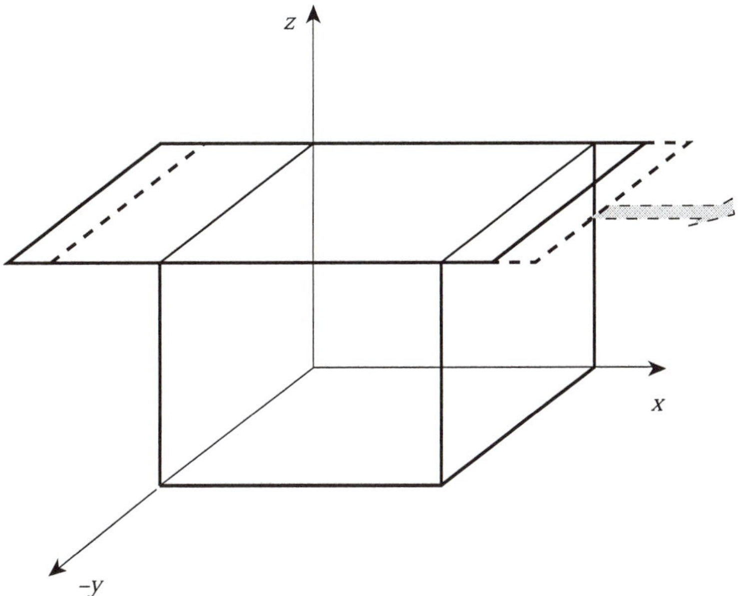

Fig. 1.3.6 A typical situation in which viscosity appears.

A velocity gradient appears in the gas. Each layer in the gas, parallel to the xy plane, advances at a different speed, $u_x(z)$, along the x direction. It turns out that in order to maintain the speed $u_x(z)$, a constant force must be exerted on the moving plane along the x direction, which is proportional to the area in motion and to the speed gradient in the gas du_x/dz. The force, i.e. the viscosity force, acts to stop the motion of the plane. The gas exerts, therefore, a force σ_{xz} per unit area on the moving plane, which is written as

$$\sigma_{xz} = -\eta \frac{du_x}{dz} . \tag{1.3.44}$$

The label xz signifies that a force is exerted in the x direction as a result of the variation of the speed in the z direction.

■ Actually, in more complex situations there can be other components, such as σ_{xy}, σ_{yz}, in which case the derivative in Eq. (1.3.44) is replaced by a partial derivative.

Maxwell was the first to analyze the viscosity in the framework of the kinetic theory. The idea is rather simple. Actually each layer of the gas that is parallel to the xy plane (in the case described in Fig. 1.3.6) exerts a force on every other layer. The reason is that due to the existence of a velocity gradient the molecules in two different layers have different speeds, u_x. This speed exists in addition to the thermal velocities. Thus in two different layers the molecules have different momenta, which on average are different from zero. But the thermal motion transfers molecules from one layer to another. The result is that a certain amount of net momentum is transferred between the layers per unit time per unit area, and this is the force of viscosity. We will perform the calculation of η at the same level that we calculated D, in the previous section. (It is possible, of course, to perform the calculation at the level of the appendix, and it is a worthwhile effort.)

Consider Fig. 1.3.4 in the present context and focus on a layer at $z = 0$. As before, molecules that are crossing the $z = 0$ plane arrive on average from a distance of one mean free path ℓ above or below it. The density of the gas is uniform, and so is the temperature. Again, $1/6$ of the molecules are moving on average at a speed $\bar{v}$ in a direction perpendicular toward the $z = 0$ plane, and the current passing through the plane is a momentum current. Let us stress that this momentum is along x!

In a unit of time the amount of momentum passing through a unit area of the $z = 0$ plane is $\frac{1}{6}n\bar{v}mu_x(\ell)$ from above down, and $\frac{1}{6}n\bar{v}mu_x(-\ell)$ from beneath up. If $u_x(z)$ is an increasing function of z, more momentum is transferred downwards than upwards. The balance is the force that is acting in the x direction per unit area of the moving layer:

$$\sigma_{xz} \approx \frac{1}{6}n\bar{v}m[u_x(-\ell) - u_x(\ell)]\,. \tag{1.3.45}$$

To first order in ℓ this is

$$\sigma_{xz} \approx -\frac{1}{3}n\bar{v}m\ell\frac{du_x}{dz}\,, \tag{1.3.46}$$

which is Eq. (1.3.44) with

$$\eta \approx \frac{1}{3}m\bar{v}\ell n\,. \tag{1.3.47}$$

Exercise 3.14

(a) How will the coefficient of viscosity change with the mass of the gas molecule at a constant temperature?
(b) Show that at a constant temperature η is independent of the pressure.
(c) How does η change with temperature?

Solution on page 104

Thermal conductivity

When the gas is contained between two parallel planes (parallel to the xy plane), kept at different temperatures, heat flows through the gas. If the heat transferred by convection (when parts of the gas move with respect to one another) is negligible, then the amount of heat that has to be supplied per unit time per unit area, in order to maintain the temperature gradient, is experimentally found to be

$$Q_z = -\bar{K}\frac{dT}{dz}\,. \tag{1.3.48}$$

■ Q is called the energy current or the energy current density or the energy flux density, and it is the amount of energy that is transferred per unit time across a unit area.

This time the role of the kinetic theory is to calculate $\bar{K}$ — the thermal conductivity.

The conditions are: the density is uniform in the container, and the average velocity is zero everywhere. But the temperature changes, so that the average energy per molecule $\bar{\epsilon}$ changes from one layer to another along the z direction. The amount of heat passing through a unit area in unit time, along the z direction, is

$$Q_z \approx \frac{1}{6}n\bar{v}\left[\bar{\epsilon}(-\ell) - \bar{\epsilon}(\ell)\right]\,, \tag{1.3.49}$$

and to first order in ℓ, when the temperature gradient is not too large,

$$Q_z \approx -\frac{1}{3}n\bar{v}\ell\frac{d\bar{\epsilon}}{dT}\frac{dT}{dz}\,. \tag{1.3.50}$$

Exercise 3.15

Explain the transition to the last equation.

Solution on page 105

We have therefore obtained Eq. (1.3.48), and found that the thermal conductivity is

$$\bar{K} \approx \frac{1}{3}n\bar{v}\ell c\,, \tag{1.3.51}$$

where

$$c = \frac{d\bar{\epsilon}}{dT}$$

is the specific heat (at constant volume) per molecule.

Exercise 3.16

Show that the thermal conductivity of an ideal gas does not depend on the pressure, and varies as $T^{1/2}$.

Solution on page 105

3.7 Appendix: a more detailed calculation of the diffusion coefficient

Among the prominent flaws in the calculation of the diffusion coefficient, as presented in Sec. 3.3, are the following:

(a) Not all of the molecules have the same velocity — in magnitude or direction.
(b) The molecules do not all start from a distance ℓ.

Let us calculate, therefore, the number of molecules that are crossing an element of area, dA, which was chosen in Fig. 1.3.7 to be in the xy plane surrounding the origin. We are still assuming that the particle density n_1 depends only on z.

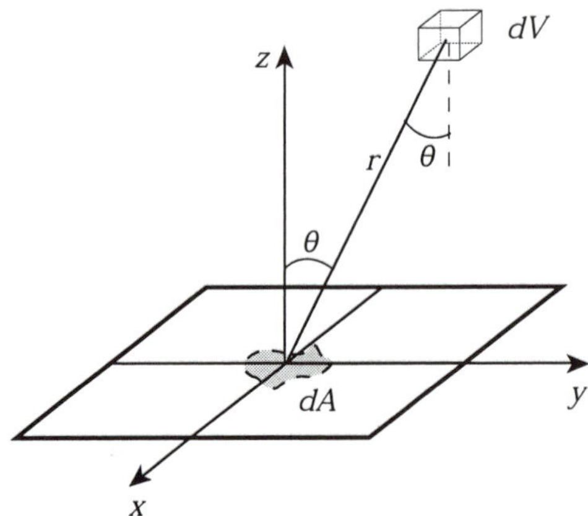

Fig. 1.3.7 The geometric quantities that appear in a diffusion problem.

We start by calculating the number of molecules that are in a volume dV around the point whose coordinates are (x, y, z), and whose velocities are between v and $v + dv$. The number of molecules in the volume is $n_1(z)dV$, which can be approximated, when the gradients are small, by

$$\left(n_1(0) + z\frac{\partial n_1}{\partial z}\right) dV\,. \tag{1.3.52}$$

This number has to be multiplied by the probability for a molecule to have a velocity (in absolute value) between v and $v + dv$, given by the Maxwell–Boltzmann distribution (see Sec. 1.5), namely

$$4\pi v^2 f(v)dv\,. \tag{1.3.53}$$

The directions of the velocities of the molecules are random, so that the probability of finding molecules with velocities in the direction of our unit area is equal to the ratio between the solid angle, spanned by the unit area towards dV, and 4π, or between the effective area of dA and the surface area of the entire sphere of radius r, i.e.

$$\frac{dA\cos\theta}{4\pi r^2}\,. \tag{1.3.54}$$

According to (1.3.15) the fraction of molecules that will arrive from a distance r without colliding is $e^{-r/\ell}$ and the corresponding probability is $e^{-r/\ell}/\ell$. This should be multiplied by v to obtain the fraction per unit time per unit area.

As an intermediate result we find that the number of molecules in dV that will begin to move towards the unit area in unit time, with a velocity between v and $v + dv$, is

$$\left[\left(n_1(0) + z\frac{\partial n_1}{\partial z}\right)dV\right]\cdot\left[4\pi v^2 f(v)dv\right]\cdot\left[\frac{dA\cos\theta}{4\pi r^2}\right]\,. \tag{1.3.55}$$

The total number of molecules passing downwards through the unit area in unit time is obtained inserting the survival probability and dividing by dA. The result is

$$\begin{aligned} J_\downarrow &= \int\left(n_1(0) + z\frac{\partial n_1}{\partial z}\right)\frac{\cos\theta}{4\pi r^2}e^{-r/\ell}dV\int_0^\infty \frac{4\pi}{\ell}v^3 f(v)dv \\ &= \frac{\bar{v}}{\ell}\int_0^{\pi/2}\int_0^\infty\left(n_1 + \frac{\partial n_1}{\partial z}r\cos\theta\right)\sin\theta\cos\theta\frac{e^{-r/\ell}}{2}drd\theta\,. \end{aligned} \tag{1.3.56}$$

In the first line the second integral is proportional to $\bar{v}$ [see Eq. (1.3.8)], while in the first integral, apart from the product of all the factors, we substituted $z = r\cos\theta$ and $dV = r^2\sin\theta drd\theta d\phi$ and integrated over the variable ϕ between 0 and 2π.

The current passing upwards is given by an almost identical expression. The only difference is a minus sign before the factor $\partial n_1/\partial z$, due to the fact that the density entering from below is $n_1(-z)$. The net current is

$$J_z = J_\uparrow - J_\downarrow = -\frac{1}{3}\bar{v}\ell\frac{\partial n_1}{\partial z}\,. \tag{1.3.57}$$

Exercise 3.17

Carry out the integrations in the expression for J_z.

Solution on page 105

We, therefore, rederived Eq. (1.3.19). The result is to be considered still an intermediate result that takes into account only some of the factors which determine the diffusion coefficient.

Self-assessment exercises

Exercise 1 *Solution on page 107*

A gas of molecules of mass m is in thermodynamic equilibrium at temperature T. The velocity of a gas molecule is $\mathbf{v} = (v_x, v_y, v_z)$. Compute the following averages:

(a) $\langle v_x \rangle$
(b) $\langle v_x^2 \rangle$
(c) $\langle v_x^2 v_y \rangle$
(d) $\langle |\mathbf{v}|^2 v_z \rangle$
(e) $\langle (v_x + b v_y)^2 \rangle$
(f) $\langle v_x^2 v_y^2 \rangle$

Exercise 2 *Solution on page 109*

A gas of molecules of mass m is in thermal equilibrium at temperature T.

(a) Calculate the average $\langle 1/|\mathbf{v}| \rangle$, and compare it to the quantity $1/\langle |\mathbf{v}| \rangle$.
(b) Calculate the most probable value for the energy of a molecule in the ensemble, E_m, and compare it to $\frac{1}{2} m v_m^2$, where v_m is the most probable velocity.

Exercise 3 *Solution on page 111*

A gas of molecules is in a central potential, given by

$$U(\mathbf{r}) = C |\mathbf{r}|^n ,$$

where $|\mathbf{r}|$ is the distance from the origin.

Calculate the average potential energy per molecule.

Exercise 4 *Solution on page 112*

Molecules of a monoatomic ideal gas are leaking from a container, which is at a constant temperature T, through a small hole in the container's wall.

(a) From physical considerations (without any calculation!), would you expect that the average energy $\langle E_0 \rangle$ of a molecule in the leaking beam will be larger than, equal to, or smaller than the average energy $\langle E_i \rangle$ of a molecule in the container?
(b) Calculate $\langle E_0 \rangle$. Express your answer terms of $\langle E_i \rangle$.

Exercise 5 *Solution on page 113*

Consider a gas at constant temperature T in a container of volume V. The gas is leaking out slowly through a small hole of cross section area A. The pressure outside the container is low enough, so that it is possible to neglect the leak into the container.

Find the time in which the pressure inside the container will decrease to $1/e$ of its initial value. Express your answer in terms of A, V and the average velocity $\bar{v}$ of a molecule in the gas.

Exercise 6 *Solution on page 114*

A satellite of mass M, in the form of a cube of edge L, is moving in outer space in a direction parallel to one of its edges, at a velocity V. The dilute gas in which the satellite is moving is made up of molecules of mass m and their number per unit volume is n. The temperature of the gas, T, is such that the thermal velocity of the molecules, $\bar{v}$, is much larger than the velocity of the satellite.

(a) Assuming that the collisions of the molecules with the satellite are elastic, calculate the average slowing force exerted on the satellite by the interstellar gas.
(b) No other external force is acting on the satellite. Calculate the time in which the velocity reaches half its initial value.

Exercise 7 *Solution on page 115*

Assume a general situation where the temperature T of a material depends on the time t and the position z. The density of the material is ρ, its specific heat per molecule c and its thermal conductivity $\bar{K}$.

Using considerations similar to those that led to Eq. (1.3.23), obtain the differential equation governing the temperature distribution $T(t, z)$ in a one-dimensional system:

$$\frac{\partial T}{\partial t} = \left(\frac{\bar{K}}{cn}\right)\frac{\partial^2 T}{\partial z^2}\,.$$

This is the Fourier equation.

Exercise 8 *Solution on page 116*

Aluminum grains whose diameter is one micrometer are floating in water at room temperature. The density of aluminum is 3.26 g/cm^3, and the viscosity of water is 1×10^{-2} P.

Find the steady precipitation rate of the aluminum due to gravitation.

Exercise 9 *Solution on page 117*

The viscosity coefficient of gaseous helium at atmospheric pressure and a temperature of 273 K is $\eta_1 = 1.87 \times 10^{-4}$ P. The viscosity coefficient of gaseous argon under these conditions is $\eta_2 = 2.105 \times 10^{-4}$ P. The atomic masses of the two gases are, respectively, $\mu_1 = 4$ amu and $\mu_2 = 40$ amu.

(a) Find the ratio ℓ_2/ℓ_1 of the mean free paths of the two gases, and the approximate values of ℓ_1 and ℓ_2.
(b) Calculate the ratio $\bar{K}_2/\bar{K}_1$ of the thermal conductivities of the two gases.
(c) Calculate the ratio D_2/D_1 of the respective diffusion coefficients.

Solutions to exercises in the text

Solution 1.1

Exercise on page 8

(a) The contribution of the molecules, with a velocity component v_x, to the pressure is

$$P(v_x) = 2 \cdot mv_x^2 \cdot n(v_x) \,. \tag{i}$$

In order to obtain the pressure exerted on the piston, we have to sum over all possible values of v_x. Only molecules with a positive component v_x will hit the piston (molecules with negative v_x will move away from the piston; see Fig. 1.1.1). Thus

$$P = \sum_{v_x > 0} 2 \cdot mv_x^2 \cdot n(v_x) \,. \tag{ii}$$

In a state of equilibrium we expect molecular "chaos," i.e. if there is no external factor forcing the molecules to move in a certain direction, then they will move in all directions with equal probability. Hence, the density of molecules of velocity v_x should be equal to the density of molecules of velocity $-v_x$:

$$n(v_x) = n(-v_x) \,. \tag{iii}$$

We can write

$$v_x^2 n(v_x) = \frac{1}{2}[v_x^2 n(v_x) + (-v_x)^2 n(-v_x)]$$

and replace the summation over $v_x > 0$ in Eq. (ii) by a summation over all the values of v_x:

$$P = \sum_{v_x} mv_x^2 \cdot n(v_x) \,. \tag{iv}$$

The right hand side of Eq. (iv) is almost the average of mv_x^2. The difference is that the average is calculated using the *probability* of finding a molecule with a velocity v_x and not in terms of the *density of molecules* which have a velocity v_x. However, the probability that a molecule will have velocity v_x is $n(v_x)/n$, where n is the number of molecules per unit volume which is constant in a state of equilibrium.

We therefore write Eq. (iv) in the form

$$P = n\sum_{v_x} mv_x^2\frac{n(v_x)}{n} = nm\sum_{v_x} v_x^2\frac{n(v_x)}{n}. \tag{v}$$

The sum on the right hand side of Eq. (v) is the average of v_x^2. Since m is constant, we find that

$$P = n\langle mv_x^2\rangle .$$

(b) When we attribute to v_x a continuous distribution, we must replace the summations of section (a) by an integral. Namely

$$P = \int_0^\infty 2mv_x^2 n(v_x)dv_x . \tag{ii$'$}$$

The conditions (iii) holds here as well, so

$$P = \int_{-\infty}^\infty mv_x^2 n(v_x)dv_x = n\langle mv_x^2\rangle . \tag{iv$'$}$$

Notice that in (a) $n(v_x)$ has the dimensions of number per unit volume, whereas in (b) it has the dimensions of number per unit volume per unit velocity!

Solution 1.2 *Exercise on page 9*

In discussing radiation, it is photons that are colliding with the "piston," which is acting as an ideal mirror. The relation between the energy and the momentum, in the case of photons, is

$$\epsilon(p) = pc , \tag{i}$$

where p is the momentum of the photon, ϵ its energy, and c the speed of light. Since the momentum is parallel to the velocity,

$$\mathbf{p}\cdot\mathbf{v} = pc . \tag{ii}$$

We can repeat the derivation that we made in Solution 1.1. A photon that is reflected from the piston imparts to it a momentum amounting to $2p_x$. All of the photons have the same speed, the speed of light, but there exists a distribution of directions, so that there will again be a distribution of v_x. Equation (1.1.2) can be rewritten as

$$P(v_x) = 2p_x v_x \cdot n(v_x)\,. \tag{iii}$$

We must sum (iii) over all positive v_x, in order to obtain the total pressure:

$$P = \sum_{v_x > 0} 2p_x v_x \cdot n(v_x)\,. \tag{iv}$$

As in Solution 1.1, we get

$$P = n\langle p_x v_x \rangle\,, \tag{v}$$

where n is the photon density in the container.

In a state of equilibrium, the averages of $p_x v_x, p_y v_y$ and $p_z v_z$ will be equal, i.e.

$$\langle p_x v_x \rangle = \langle p_y v_y \rangle = \langle p_z v_z \rangle = \frac{1}{3}\langle \mathbf{p} \cdot \mathbf{v} \rangle = \frac{1}{3}\langle \epsilon \rangle\,, \tag{vi}$$

where in the last equality we have made use of Eqs. (i) and (ii). Substituting in Eq. (v) we obtain the required result:

$$P = n\frac{1}{3}\langle \epsilon \rangle = \frac{1}{3}\frac{N}{V}\langle \epsilon \rangle = \frac{1}{3}\frac{E}{V}\,.$$

Solution 1.3

Exercise on page 10

Room temperature is 300 K. Substitution in Eq. (1.1.5) yields

$$\left\langle \frac{1}{2}mv^2 \right\rangle = 6.21 \times 10^{-21}\ \text{J} = 0.039\ \text{eV}\,.$$

Solution 1.4

Exercise on page 11

Substituting $\mathbf{v}_{\text{cm}}$ and $\mathbf{v}_{\text{rel}}$, by their definitions, in the average $\langle \mathbf{v}_{\text{cm}} \cdot \mathbf{v}_{\text{rel}} \rangle$ we obtain

$$\begin{aligned}
\langle \mathbf{v}_{\text{cm}} \cdot \mathbf{v}_{\text{rel}} \rangle &= \frac{1}{m_1 + m_2}\langle (m_1\mathbf{v}_1 + m_2\mathbf{v}_2) \cdot (\mathbf{v}_1 - \mathbf{v}_2) \rangle \\
&= \frac{1}{m_1 + m_2}\left[\left\langle \frac{m_1\mathbf{v}_1^2}{2} \right\rangle - \left\langle \frac{m_2\mathbf{v}_2^2}{2} \right\rangle\right] + \frac{m_2 - m_1}{m_1 + m_2}\langle \mathbf{v}_1 \cdot \mathbf{v}_2 \rangle\,.
\end{aligned}$$

Equations (1.1.10) and (1.1.11) imply, therefore, that

$$\left\langle \frac{m_1 v_1^2}{2} \right\rangle - \left\langle \frac{m_2 v_2^2}{2} \right\rangle = 0\,,$$

which is (1.1.12).

Solution 1.5 *Exercise on page 12*

The proof of Dalton's law uses mainly the concept of molecular (or atomic) "chaos," and the fact that there is no interaction between the molecules of the gas.

We suppose for simplicity that the mixture is composed of two gases of N_1 and N_2 molecules, respectively (the derivation can easily be generalized to any number of gases). The two gases occupy the same volume V and the total number of molecules $N(= N_1 + N_2)$ is constant. The number of molecules N_1 and N_2 are not necessarily equal.

Since the gases are at thermal equilibrium, their average kinetic energies are equal:

$$\left\langle \frac{1}{2} m_1 v_1^2 \right\rangle = \left\langle \frac{1}{2} m_2 v_2^2 \right\rangle .$$

According to Eq. (1.1.4), if each of the gases were to occupy by itself the same volume V at the same temperature, the partial pressures P_1 and P_2 would have been

$$P_1 = \frac{2}{3}\frac{N_1}{V}\left\langle \frac{1}{2} m_1 v_1^2 \right\rangle ,$$

$$P_2 = \frac{2}{3}\frac{N_2}{V}\left\langle \frac{1}{2} m_2 v_2^2 \right\rangle .$$

■ The partial pressure of a gas in a mixture is the pressure it would have if it were to occupy all the volume by itself.

However, the total pressure of the mixture is

$$P = \frac{2}{3}\frac{N}{V}\begin{pmatrix}\text{average energy}\\ \text{per molecule}\end{pmatrix} = \frac{2}{3}\frac{N_1 + N_2}{V}\begin{pmatrix}\text{average energy}\\ \text{per molecule}\end{pmatrix}$$

$$= \frac{2}{3}\frac{N_1}{V}\left\langle \frac{1}{2} m_1 v_1^2 \right\rangle + \frac{2}{3}\frac{N_2}{V}\left\langle \frac{1}{2} m_2 v_2^2 \right\rangle = P_1 + P_2 .$$

Solution 1.6 *Exercise on page 13*

We are given the total volume V, the number of molecules N_1 and N_2 on both sides of the piston, and the total energy. All these are constants.

The average kinetic energy on both sides of the piston is equal (there are no temperature differences). Hence, from Eq. (1.1.4) or Eq. (1.1.6) it follows that

$$\frac{P_1V_1}{N_1} = \frac{P_2V_2}{N_2} \,. \tag{i}$$

where V_1 and V_2 are the volumes on the two sides of the piston.

At equilibrium, the pressure on the two sides must equalize (otherwise a force will act, and the piston will move). Therefore, Eq. (i) implies that

$$\frac{V_1}{V_2} = \frac{L_1}{L_2} = \frac{N_1}{N_2} \,, \tag{ii}$$

where L_1 and L_2 are the lengths of V_1 and V_2, respectively. (See figure.) However,

$$V_1 + V_2 = V \Rightarrow L_1 + L_2 = L \,, \tag{iii}$$

and we obtain from Eqs. (ii) and (iii)

$$\begin{aligned} V_1 &= \frac{N_1}{N_1+N_2}V \,, \ V_2 = \frac{N_2}{N_1+N_2}V \,, \\ L_1 &= \frac{N_1}{N_1+N_2}L \,, \ L_2 = \frac{N_2}{N_1+N_2}L \,. \end{aligned} \tag{iv}$$

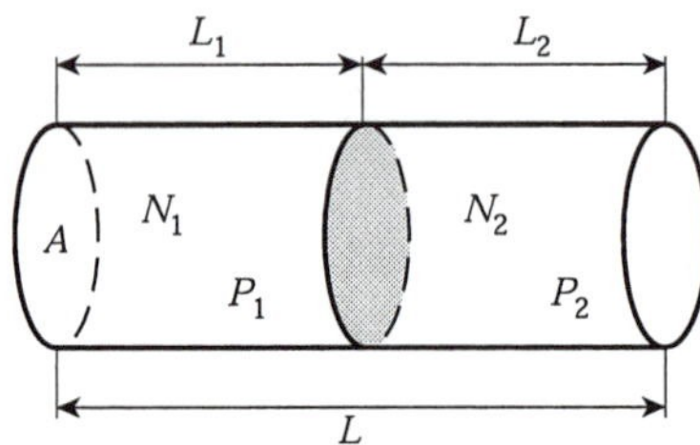

The variables defining the problem.

Now we can calculate the average kinetic energy at equilibrium — as the pressure becomes uniform: The total energy is given by

$$\frac{3}{2}P_1V_1 + \frac{3}{2}P_2V_2 = \frac{3}{2}PV = E \,, \tag{v}$$

and the pressure at equilibrium is consequently

$$P = \frac{2}{3}\frac{E}{V} \,. \tag{vi}$$

The average kinetic energy per molecule to the right of the piston is

$$\left\langle \frac{1}{2}mv^2 \right\rangle = \frac{3}{2}\frac{P_1V_1}{N_1} = \frac{3}{2}P\frac{V_1}{N_1}\,. \tag{vii}$$

Using Eqs. (iv) and (vi) we obtain

$$\left\langle \frac{1}{2}mv^2 \right\rangle = \frac{3}{2}\left(\frac{2}{3}\frac{E}{V}\right)\left(\frac{V}{N_1+N_2}\right) = \frac{E}{N_1+N_2}\,. \tag{viii}$$

Since the result in (viii) is independent of any quantity specific to the gas on the right of the piston, the same result must apply to the average kinetic energy to the left of the piston.

Solution 1.7 *Exercise on page 15*

The assertion

$$\langle \mathbf{v}_{\text{cm}} \cdot \mathbf{v}_{\text{rel}} \rangle = 0$$

implies that

$$\begin{aligned} &\langle (m_1\mathbf{v}_1 + m_2\mathbf{v}_2) \cdot (\mathbf{v}_1 - \mathbf{v}_2) \rangle \\ &\quad = \langle m_1 v_1^2 \rangle - \langle m_2 v_2^2 \rangle + \langle (m_2 - m_1)\mathbf{v}_1 \cdot \mathbf{v}_2 \rangle\,. \end{aligned}$$

But the two parts of the molecule have the same average kinetic energy, i.e.

$$\left\langle \frac{m_1 v_1^2}{2} \right\rangle = \left\langle \frac{m_2 v_2^2}{2} \right\rangle = \frac{3}{2}kT\,,$$

and so

$$\langle (m_2 - m_1)\mathbf{v}_1 \cdot \mathbf{v}_2 \rangle = 0\,.$$

$m_2 - m_1$ can be taken out of the averaging sign, and since in a generic case $m_1 \neq m_2$, we obtain

$$\langle \mathbf{v}_1 \cdot \mathbf{v}_2 \rangle = 0\,.$$

If $m_1 = m_2$, this argument cannot be used. But in this case the velocity distributions of the two atoms in the molecule will also be identical. Thus, calculating $\langle \mathbf{v}_1 \cdot \mathbf{v}_2 \rangle$ is the same as calculating the average of the product of $\mathbf{v}_1$ at time $t = t_1$ and $\mathbf{v}_1$ at time $t = t_2$:

$$\langle \mathbf{v}_1 \cdot \mathbf{v}_2 \rangle = \langle \mathbf{v}_1(t_1) \cdot \mathbf{v}_1(t_2) \rangle\,.$$

Since the collisions are random there is no correlation between $\mathbf{v}_1$ at two different times and the average is zero again.

Solution 1.8 *Exercise on page 17*

(a) The total kinetic energy is given by

$$E = N\frac{3}{2}rkT\,.$$

However, only the center of mass energy produces the pressure in the gas, and so

$$PV = NkT = \frac{2}{3r}E\,.$$

(b) Equation (1.1.22) implies that

$$PdV + VdP = (\gamma - 1)dE\,. \tag{i}$$

For an adiabatic process we have as well

$$dE = -PdV\,, \tag{ii}$$

and substituting this relation into Eq. (i) we obtain

$$\gamma PdV + VdP = 0\,, \tag{iii}$$

and from here

$$\gamma\frac{dV}{V} = -\frac{dP}{P}\,, \tag{iv}$$

Integrating both sides we obtain

$$\ln V^{\gamma} + \ln P = K\,, \tag{v}$$

where K is a constant. From here of course

$$PV^{\gamma} = \text{const}\,.$$

(c) The heat capacity is the quantity of heat required per unit temperature change. At constant volume the heat is equal to the energy change of the gas. Hence

$$C_V = \frac{\partial E}{\partial T}\,. \tag{vi}$$

For a mole of monoatomic ideal gas, $E = \frac{3}{2}N_0kT$ and

$$C_V = \frac{3}{2}N_0k = \frac{3}{2}R = 12.45 \text{ J K}^{-1}\,. \tag{vii}$$

Solution 1.9 *Exercise on page 19*

A degree of freedom will become relevant if its energy intervals, e.g. its excitation energies, are of the order of the equipartition energy kT available at temperature T.

The electronic degrees of freedom in the atom will be excited when

$$kT_e \approx 1 \text{ eV}\,,$$

but

$$1 \text{ eV} = 1.6 \times 10^{-19} \text{ J}$$

so that

$$T_e \approx \frac{1.6 \times 10^{-19}}{1.4 \times 10^{-23}} \approx 10^4 \text{ K}.$$

The nuclear degrees of freedom, whose energies are of the order of MeV, will come into play when

$$kT_N \approx 10^6 \text{ eV} \qquad \text{or} \qquad T_N \approx 10^{10} \text{ K}.$$

The quark degrees of freedom, which signify the intranuclear excitations, will affect the specific heat when

$$kT_Q \approx 10^9 \text{ eV} \qquad \text{or} \qquad T_Q \approx 10^{13} \text{ K}.$$

Solution 1.10 *Exercise on page 21*

The earth's gravitational field E_G at the point $\mathbf{r}$, where r is the distance from the earth's center, is given by

$$E_G = \frac{\alpha}{r^2}, \tag{i}$$

where α is a constant.

The change in the gravitational field, owing to a small change in r, Δr, is

$$\Delta E_G = -\frac{2\alpha}{r^3}\Delta r. \tag{ii}$$

The relative change of the gravitational field is obtained from Eqs. (i) and (ii):

$$\left|\frac{\Delta E_G}{E_G}\right| \approx \frac{\Delta r}{R_E}, \tag{iii}$$

where R_E is the radius of the earth ($R_E \approx 6.4 \times 10^6$ m).

In order to change the gravitational field by a thousandth of a percent, we have to change the height by

$$\Delta r \approx 10^{-5} R_E \approx 100 \text{ m}. \tag{iv}$$

Under standard conditions the intermolecular distance l is given by

$$l \approx \left(\frac{P}{kT}\right)^{-1/3} \approx \left(\frac{10^5}{1.4 \times 10^{-23} \times 300}\right)^{-1/3} \approx 3.5 \times 10^{-9} \text{ m}, \tag{v}$$

where we substituted a pressure of one atmosphere and a temperature of 300 K. From (iv) and (v) we obtain the ratio

$$\frac{l}{\Delta r} \approx \frac{3.5 \times 10^{-9}}{10^2} \approx 10^{-11} \ll 1.$$

That is, the changes in the gravitational field are negligible when one is considering distances on the order of the intermolecular distance, and therefore it is possible to choose a region in space small enough so that the gravitational field within it is constant, and large enough so that we can consider it as a macroscopic system.

Solution 1.11 *Exercise on page 23*

The density of the gas in a force field that is derived from a potential,

$$\mathbf{F}(\mathbf{r}) = -\nabla U(\mathbf{r})\,, \tag{i}$$

can be obtained as in the one-dimensional case, replacing everywhere z with $\mathbf{r}$, and the derivative d/dz with the gradient ∇. Equation (i) gives the force that is acting on a gas molecule at point $\mathbf{r}$.

When U is the gravitational potential, $U = mgz$, Eq. (i) reduces to

$$\mathbf{F} = -\frac{dU}{dz}\hat{\mathbf{z}} = -mg\hat{\mathbf{z}}\,.$$

The equation for the pressure is obtained by generalizing Eq. (1.1.27) to

$$\nabla P(\mathbf{r}) = \mathbf{F}(\mathbf{r})n(\mathbf{r})\,. \tag{ii}$$

Applying the same arguments as in Sec. 1.4, we attribute to the gas a local equation of state [cf. Eq. (1.1.28)] in a small volume around $\mathbf{r}$:

$$P(\mathbf{r}) = n(\mathbf{r})kT\,. \tag{iii}$$

Since T is independent of $\mathbf{r}$, we obtain from Eqs. (i)–(iii) an equation for $n(\mathbf{r})$,

$$kT\nabla n(\mathbf{r}) = -n(\mathbf{r})\nabla U(\mathbf{r})\,, \tag{iv}$$

corresponding to Eq. (1.1.29), with ∇U replacing mg.

Equation (iv) can be written in the form

$$\frac{\nabla n(\mathbf{r})}{n(\mathbf{r})} = -\frac{1}{kT}\nabla U(\mathbf{r}) \qquad \text{or} \qquad \nabla \ln n(\mathbf{r}) = \nabla\left[-\frac{U(\mathbf{r})}{kT}\right]. \tag{v}$$

The solution to this equation is

$$\ln n(\mathbf{r}) = -\frac{U(\mathbf{r})}{kT} + C\,,$$

where C is a constant, or

$$n(\mathbf{r}) = n(\mathbf{r}_0)e^{-U(\mathbf{r})/kT}\,, \tag{vi}$$

where $n(\mathbf{r}_0)$ is the density at the point where the potential $U(\mathbf{r})$ vanishes.

Notice what happens in the case when the potential is the gravitational potential as of a planet. In this case Eq. (iv) implies that very far from the planet the gas density does not vanish but remains constant. This is of course impossible, since the number of molecules, although very large, is finite. It means that a gas cannot be in a state of thermodynamic equilibrium at constant T in a Newtonian gravitational field.

Solution 1.12 *Exercise on page 24*

$P(\mathbf{r})dV$ is a probability, and therefore dimensionless. dV is a volume element, so

$$[dV] = [L]^3 ,$$

and hence

$$[P(\mathbf{r})] = [dV]^{-1} = [L]^{-3} ,$$

which means that the dimensions of $P(\mathbf{r})$ are (volume)$^{-1}$.

$P(\mathbf{r})$ is indeed normalized, since the integration of $n(\mathbf{r})$ yields the total number of molecules, N, which appears in the denominator of Eq. (1.1.33).

Solution 1.13 *Exercise on page 27*

The constant C in the Maxwell–Boltzmann distribution,

$$f(\mathbf{v}) = Ce^{-Av^2} ,$$

is fixed by the normalization condition

$$\int f(\mathbf{v})d\tau = 1 , \tag{i}$$

where the integration is carried out over the entire velocity space. We have to calculate C from

$$C\int_{-\infty}^{\infty}\int_{-\infty}^{\infty}\int_{-\infty}^{\infty} e^{-A(v_x^2+v_y^2+v_z^2)}dv_x dv_y dv_z = 1 . \tag{ii}$$

However, because the integration variables are independent, it is possible to write the triple integral as a product of three integrals:

$$C\int_{-\infty}^{\infty} e^{-Av_x^2}dv_x \int_{-\infty}^{\infty} e^{-Av_y^2}dv_y \int_{-\infty}^{\infty} e^{-Av_z^2}dv_z = C\left(\int_{-\infty}^{\infty} e^{-Av_x^2}dv_x\right)^3 = 1 . \tag{iii}$$

We calculate, therefore, an integral of the type

$$S = \int_{-\infty}^{\infty} e^{-\alpha x^2}dx . \tag{iv}$$

The calculation is not complicated, as we shall see, and the result is $\sqrt{\pi/\alpha}$. It is possible of course to look up the result in a table of integrals.

Substituting the result into (iii) we get

$$C\left(\frac{\pi}{A}\right)^{3/2} = 1 \Rightarrow C = \left(\frac{A}{\pi}\right)^{3/2} .$$

Remark. The integral may be calculated in the following manner: We write

$$S^2 = \int_{-\infty}^{\infty} e^{-\alpha x^2} dx \int_{-\infty}^{\infty} e^{-\alpha y^2} dy = \int_{-\infty}^{\infty} e^{-\alpha(x^2+y^2)} dx dy . \qquad \text{(v)}$$

In the double integral over the xy plane we transform to polar coordinates θ, r. The area element is $rd\theta dr$ and $x^2+y^2 = r^2$. Equation (v) is therefore written as

$$S^2 = \int_0^{\infty} \int_0^{2\pi} e^{-\alpha r^2} r d\theta dr = 2\pi \int_0^{\infty} r e^{-\alpha r^2} dr = 2\pi \left(-\frac{1}{2\alpha} e^{-\alpha r^2}\right)_0^{\infty} = \frac{\pi}{\alpha} , \qquad \text{(vi)}$$

and hence

$$S = \int_{-\infty}^{\infty} e^{-\alpha x^2} dx = \left(\frac{\pi}{\alpha}\right)^{1/2} . \qquad \text{(vii)}$$

Solution 1.14 *Exercise on page 27*

The requirement that $f(\mathbf{v})$ be normalized fixes C as a function of A:

$$C = \left(\frac{A}{\pi}\right)^{3/2} .$$

A is determined from the requirement that the average kinetic energy per molecule be $\frac{3}{2}kT$.

Symmetry considerations give

$$\langle v_x^2 \rangle = \langle v_y^2 \rangle = \langle v_z^2 \rangle ,$$

which means that

$$v^2 = \langle v_x^2 + v_y^2 + v_z^2 \rangle = 3\langle v_x^2 \rangle ,$$

so to calculate $\langle v^2 \rangle$ it is enough to calculate $\langle v_x^2 \rangle$. Thus

$$\begin{aligned}
\frac{3}{2}kT &= \left\langle \frac{1}{2} m v^2 \right\rangle = \int \frac{3}{2}\, m v_x^2 f(\mathbf{v}) d\tau \\
&= \frac{3}{2}\left(\frac{A}{\pi}\right)^{3/2} m \int_{-\infty}^{\infty}\int_{-\infty}^{\infty}\int_{-\infty}^{\infty} v_x^2 e^{-A(v_x^2+v_y^2+v_z^2)} dv_x dv_y dv_z \\
&= \frac{3}{2}\left(\frac{A}{\pi}\right)^{3/2} m \int_{-\infty}^{\infty} v_x^2 e^{-Av_x^2} dv_x \int_{-\infty}^{\infty} e^{-Av_y^2} dv_y \int_{-\infty}^{\infty} e^{-Av_z^2} dv_z \\
&= \frac{3}{2}\left(\frac{A}{\pi}\right)^{1/2} m \int_{-\infty}^{\infty} v_x^2 e^{-Av_x^2} dv_x , \qquad \text{(i)}
\end{aligned}$$

where we have used the result (vii) of the previous exercise for the normalization.

Next we calculate an integral of the form

$$I = \int_{-\infty}^{\infty} x^2 e^{-\alpha x^2} dx .$$

Integration by parts yields

$$I = -\frac{1}{2\alpha} \int_{-\infty}^{\infty} x \left(\frac{d}{dx} e^{-\alpha x^2} \right) dx = \frac{1}{2\alpha} \int_{-\infty}^{\infty} e^{-\alpha x^2} dx = \frac{1}{2} \left(\frac{\pi}{\alpha^3} \right)^{1/2} ,$$

and then [by (i)]

$$\frac{3}{2} kT = \frac{3}{2} \left(\frac{A}{\pi} \right)^{1/2} m \frac{1}{2} \left(\frac{\pi}{A^3} \right)^{1/2} = \frac{3m}{4A} , \tag{ii}$$

which leads to

$$A = \frac{m}{2kT} , \tag{iii}$$

$$C = \left(\frac{A}{\pi} \right)^{3/2} = \left(\frac{m}{2\pi kT} \right)^{3/2} . \tag{iv}$$

Substituting (iv) in $f(\mathbf{v})$ we obtain the distribution function Eq. (1.1.49).

Starting from Eq. (1.1.49) and calculating $\int f(\mathbf{v}) d\tau$ and $\left\langle \frac{1}{2} m v^2 \right\rangle$, we find that Eq. (1.1.49) determines C and A, as required.

Remark. A different method for calculating averages will be introduced in Sec. 1.6. We will repeat the calculation of $\left\langle \frac{1}{2} m v^2 \right\rangle$ using that method.

Solution 1.15 *Exercise on page 29*

We wrote

$$Z(\alpha) = \int_0^{\infty} e^{-\alpha z} dz . \tag{i}$$

Now

$$\frac{d}{dz} \ln Z = \frac{1}{Z} \frac{dZ}{d\alpha} ,$$

$$\frac{d^2}{d\alpha^2} \ln Z = \frac{1}{Z} \frac{d^2 Z}{d\alpha^2} - \frac{1}{Z^2} \left(\frac{dZ}{d\alpha} \right)^2 . \tag{ii}$$

From Eq. (i) we find that

$$\frac{dZ}{d\alpha} = -\int_0^{\infty} z e^{-\alpha z} dz = -Z \langle z \rangle , \tag{iii}$$

$$\frac{d^2 Z}{d\alpha^2} = \int_0^{\infty} z^2 e^{-\alpha z} dz = Z \langle z^2 \rangle , \tag{iv}$$

where we have made use of the definitions of $\langle z \rangle$ and $\langle z^2 \rangle$, Eq. (1.1.51).

Substituting Eqs. (iii) and (iv) into (ii) we obtain

$$\frac{d^2}{d\alpha^2}\ln Z = \langle z^2\rangle - \langle z\rangle^2 . \qquad \text{(v)}$$

But

$$\begin{aligned}\langle (z-\langle z\rangle)^2\rangle &= \langle (z^2 - 2z\langle z\rangle + \langle z\rangle^2)\rangle \\ &= \langle z^2\rangle - 2\langle z\rangle^2 + \langle z\rangle^2 = \langle z^2\rangle - \langle z\rangle^2 ,\end{aligned}$$

so that we find from (v) that the average square deviation is given by

$$(\Delta z)^2 = \langle (z-\langle z\rangle)^2\rangle = \frac{d^2}{d\alpha^2}\ln Z .$$

Solution 1.16 *Exercise on page 30*

The average $\langle z^n\rangle$ can be written as

$$\langle z^n\rangle = \frac{\int_{-\infty}^{\infty} z^n \exp(-U_0 z^n/kT)dz}{Z} , \qquad \text{(i)}$$

where

$$Z = \int_0^{\infty} \exp\left(-\frac{U_0 z^n}{kT}\right) dz . \qquad \text{(ii)}$$

We substitute

$$\alpha = \frac{U_0}{kT} \qquad \text{(iii)}$$

and using $Z(\alpha)$ we can write $\langle z^n\rangle$ as

$$\langle z^n\rangle = -\frac{d}{d\alpha}\ln Z(\alpha) , \qquad \text{(iv)}$$

similar to what we did in Eqs. (1.1.51) and (1.1.52).

The dimensions of $Z(\alpha)$ are the same as those of z, namely dimensions of length. The dimensions of α are determined by the fact that αz^n is dimensionless. Therefore, α has the same dimensions as z^{-n}. Thus

$$Z(\alpha) = K\alpha^{-1/n} . \qquad \text{(v)}$$

Substituting (v) into (iv) we get

$$\langle z^n\rangle = \frac{1}{n\alpha} = \frac{kT}{nU_0} ,$$

where in the last equality we have made use of (iii). Consequently

$$\langle U(z)\rangle = \frac{kT}{n} . \qquad \text{(vi)}$$

Solution 1.17 *Exercise on page 31*

(a) $Z(\alpha) = \int \exp(-\alpha|\mathbf{v}|^2)d\tau$, where $d\tau = dv_1 dv_2 \ldots dv_D$ is an infinitesimal D-dimensional volume.

The only difference between the present calculation and the one performed in Eq. (1.1.56) is that the volume of integration in $Z(\alpha)$ is D-dimensional. The dimensions of α remain as before. From dimensional considerations, identical to the ones we made in Sec. 1.6, we conclude that

$$Z(\alpha) = K\alpha^{-D/2}.$$

Substituting in Eq. (1.1.55),

$$\langle|\mathbf{v}|^2\rangle = -\frac{d}{d\alpha}\ln Z(\alpha),$$

we obtain

$$\langle|\mathbf{v}|^2\rangle = \frac{D}{2\alpha}.$$

In usual, three-dimensional, space ($D = 3$), $\langle|\mathbf{v}|^2\rangle = 3/2\alpha$, as we found in Eq. (1.1.59). In two dimensions, $\langle|\mathbf{v}|^2\rangle = 1/\alpha$; in one dimension, $\langle v^2\rangle = 1/2\alpha$.

(b) For a gas in a three-dimensional harmonic potential

$$U(\mathbf{r}) = \frac{1}{2}C|\mathbf{r}|^2.$$

The potential is centered at $\mathbf{r} = 0$, and therefore we want to find $\langle|\mathbf{r}|^2\rangle$.

In Solution 1.11 we showed that the distribution function is given by

$$n(\mathbf{r}) = n(\mathbf{r}_0)e^{-U(\mathbf{r})/kT},$$

so that in our case

$$n(\mathbf{r}) = n(0)e^{-C|\mathbf{r}|^2/2kT}.$$

There is no need to carry out any further calculations: denoting $\alpha = C/2kT$, we obtain the distribution function we know, where $|\mathbf{r}|$ is written instead of $|\mathbf{v}|$. The result of the calculation of $\langle|\mathbf{r}|^2\rangle$ will not change, because $|\mathbf{r}|$ and $|\mathbf{v}|$ are merely variables of integration, and the integration is over the entire space. The result is therefore

$$\langle|\mathbf{r}|^2\rangle = \frac{3}{2\alpha} = \frac{3kT}{C}.$$

(c) In order to find the average energy per molecule we have to calculate $\langle \frac{1}{2}C|\mathbf{r}|^2 \rangle$.

We use the result of (a) and the arguments of (b) above, to find $\langle |\mathbf{r}|^2 \rangle$ in a D-dimensional space:

$$\langle |\mathbf{r}|^2 \rangle = \frac{D}{2\alpha}, \tag{i}$$

so that

$$\langle U(\mathbf{r}) \rangle = \left\langle \frac{1}{2}C|\mathbf{r}|^2 \right\rangle = \frac{1}{2}C\frac{D}{2\alpha}. \tag{ii}$$

Since $\alpha = C/2kT$,

$$\langle U(\mathbf{r}) \rangle = \frac{1}{2}DkT,$$

according to which each vibrational degree of freedom has an average potential energy of $\frac{1}{2}kT$.

Solution 2.1 *Exercise on page 34*

The problem of a head-on elastic collision between two bodies can, in principle, be solved with the help of the conservation laws for energy and momentum. But we are not interested in an exact solution — an approximate solution is sufficient.

A typical molecule has a mass $m \approx 10^{-26}$ kg and a velocity $v \approx 10^3$ m s^{-1}. The body has a mass $M \approx 10^{-16}$ kg and a velocity $V \approx 10^{-2}$ m s^{-1}, i.e. $m \ll M$ and $v \gg V$.

What is actually taking place in the system is that a small and fast molecule hits the enormous body, which is almost stationary. We therefore expect that the molecule will transfer to the body momentum of order $2mv$ and will be reflected with almost the same speed, v, as it had before the collision. The body will absorb this momentum and change its speed by a very small amount, ΔV. We can therefore write

$$M\Delta V \approx 2mv,$$

so that

$$\Delta V \approx 2\frac{m}{M}v.$$

Substituting the typical values $v \approx 10^3$ m s^{-1}, $m/M \approx 10^{-10}$, we find that in the collision with the molecule the body's velocity changes by about 2×10^{-7} m s^{-1}.

Solution 2.2 *Exercise on page 37*

The drunk's position after $N + 1$ steps is

$$\mathbf{R}_{N+1} = \mathbf{R}_N + L\mathbf{n}\,, \tag{i}$$

where $\mathbf{n}$ is the direction of the $(N+1)$th step, i.e. the direction of the step taken by a drunk located at the point $\mathbf{R}_N$. The origin has been chosen at $\mathbf{R}_0 = 0$.

From the first point of view we say that the drunks of the group that are located at $\mathbf{R}_N$ will advance with equal probability in all the possible directions around $\mathbf{R}_N$, so that

$$\langle \mathbf{n} \rangle = 0\,, \tag{ii}$$

and we obtain

$$\langle \mathbf{R}_{N+1} \rangle = \langle \mathbf{R}_N \rangle\,. \tag{iii}$$

The equality (iii) implies that

$$\langle \mathbf{R}_N \rangle = \langle \mathbf{R}_{N-1} \rangle = \cdots = \langle \mathbf{R}_0 \rangle = 0\,,$$

and so

$$\langle \mathbf{R}_N \rangle = 0 \tag{iv}$$

for all N.

From the second point of view we say that along the drunk's path, each point may be considered as the beginning. If from every point we draw the vector $\mathbf{R}_N$, to which the drunk arrives after N more steps, we find that the number of appearances of $\mathbf{R}_N$ equals the number of appearances of $-\mathbf{R}_N$. Thus on average $\langle \mathbf{R}_N \rangle = 0$.

Solution 2.3 *Exercise on page 38*

(a) From Eq. (1.2.6) we see that $\mu\dot{\mathbf{r}}$ has dimensions of force. Hence

$$[\mu] = \left[\frac{F}{\dot{r}}\right] = \frac{[M][L][T]^{-2}}{[L][T]^{-1}} = [M][T]^{-1}\,.$$

(b) In the absence of an external force, the equation of motion of the body [Eq. (1.2.6)] takes the form

$$M\ddot{\mathbf{r}} + \mu\dot{\mathbf{r}} = 0$$

or

$$M\dot{\mathbf{v}} + \mu\mathbf{v} = 0\,. \tag{i}$$

Equation (i) is a vector equation, so that each component of the velocity satisfies an equation of the form

$$M\dot{v} + \mu v = 0\,. \tag{ii}$$

The solution to (ii) is

$$v = v_0 e^{-(\mu/M)t}\,,$$

so that the solution to (i) is

$$\mathbf{v} = \mathbf{v}_0 e^{-(\mu/M)t}\,, \tag{iii}$$

where $\mathbf{v}_0$ is the initial velocity of the body. From the solution (iii) we immediately see that

$$\lim_{t\to\infty} \mathbf{v} = \lim_{t\to\infty} (\mathbf{v}_0 e^{-(\mu/M)t}) = 0\,.$$

Solution 2.4 *Exercise on page 38*

The motion of the sky diver is one-dimensional along the z axis. We will choose the positive direction to be downwards, so that $F_e = mg$. The equation of motion (1.2.6) then takes the form

$$m\ddot{z} + \mu\dot{z} = mg\,. \tag{i}$$

We choose the origin to be the height from which the sky diver jumped, so that $z(0) = 0$. Furthermore, we assume that the sky diver's initial velocity is zero, namely $\dot{z}(0) = 0$. In order to solve the equation we write (i) in the form

$$m\dot{v} + \mu v = mg\,, \tag{ii}$$

where v is the sky diver's velocity.

The solution to Eq. (ii) with the initial condition $v(0) = 0$ is

$$v = \frac{mg}{\mu}(1 - e^{-(\mu/m)t})\,. \tag{iii}$$

At long times the exponential term in (iii) decays and becomes negligible, so that

$$\lim_{t\to\infty} v = \frac{mg}{\mu}\,.$$

The characteristic time, θ, of the decay is the time after which the term $v_0 e^{-\mu t/m}$ decays to $1/e$ of its initial value, namely $\theta = m/\mu$. Long times are therefore times $t \gg m/\mu$, for which the exponential term is negligible, and then $v \approx mg/\mu$. The solution to the original Eq. (i) is determined by integrating (iii) and choosing the integration constant such that $z(0) = 0$:

$$z = \frac{mg}{\mu}\left[t + \frac{m}{\mu}(e^{-(\mu/m)t} - 1)\right] . \tag{iv}$$

Solution 2.5 *Exercise on page 38*

When $\mathbf{F}_e$ is derived from a potential, Eq. (1.2.6) is written in the form

$$M\ddot{\mathbf{r}} + \mu\dot{\mathbf{r}} = -\nabla U . \tag{i}$$

Taking the scalar product of (i) with $\dot{\mathbf{r}}$,

$$M\dot{\mathbf{r}} \cdot \ddot{\mathbf{r}} + \mu(\dot{\mathbf{r}})^2 = -\dot{\mathbf{r}} \cdot \nabla U . \tag{ii}$$

But

$$\frac{d}{dt}(\dot{\mathbf{r}})^2 = 2\dot{\mathbf{r}} \cdot \ddot{\mathbf{r}} ,$$

and so

$$\frac{1}{2}M\frac{d}{dt}(\dot{\mathbf{r}})^2 + \mu(\dot{\mathbf{r}})^2 = -\dot{\mathbf{r}} \cdot \nabla U . \tag{iv}$$

The right hand side of (iv) is none other than the time derivative of U (with negative sign):

$$\frac{dU}{dt} = \frac{\partial U}{\partial x}\frac{dx}{dt} + \frac{\partial U}{\partial y}\frac{dy}{dt} + \frac{\partial U}{\partial z}\frac{dz}{dt} = (\nabla U) \cdot \dot{\mathbf{r}} ,$$

so that we can write (iv) as

$$\frac{d}{dt}\left[\frac{1}{2}M(\dot{\mathbf{r}})^2 + U\right] = -\mu(\dot{\mathbf{r}})^2 . \tag{v}$$

Observe that the expression in square brackets is the total energy E of the body, and thus

$$\frac{dE}{dt} = -\mu(\dot{\mathbf{r}})^2 . \tag{vi}$$

■ Equation (i) is a vector equation. For each component x_i

$$M\frac{d^2x_i}{dt^2} + \mu\frac{dx_i}{dt} = -\frac{\partial U}{\partial x_i} .$$

Equation (ii) is a scalar equation, and is a sum of three equations,

$$\sum_{i=1}^{3}\left[M\frac{dx_i}{dt}\frac{d^2x_i}{dt^2} + \mu\left(\frac{dx_i}{dt}\right)^2\right] = -\sum_{i=1}^{3}\frac{\partial U}{\partial x_i}\frac{dx_i}{dt} = -\frac{dU}{dt} ,$$

but

$$\frac{d}{dt}\left(\frac{dx_i}{dt}\right)^2 = 2\frac{dx_i}{dt}\frac{d^2x_i}{dt^2},$$

$$\sum_{i=1}^{3}\left(\frac{dx_i}{dt}\right)^2 = \dot{\mathbf{r}}^2,$$

so that

$$\frac{1}{2}M\frac{d}{dt}(\dot{\mathbf{r}}^2) + \mu\dot{\mathbf{r}}^2 = -\dot{U},$$

and this is actually Eq. (v).

Solution 2.6 *Exercise on page 40*

When there is no friction, Eq. (1.2.6) reduces to

$$M\ddot{\mathbf{r}} = \mathbf{F}_e$$

or

$$M\frac{d}{dt}(\dot{\mathbf{r}}) = \mathbf{F}_e\,.$$

Averaging both sides of this equation (the left hand side can be averaged before differentiation), we obtain

$$M\frac{d}{dt}\langle\dot{\mathbf{r}}\rangle = \langle\mathbf{F}_e\rangle = 0\,.$$

Its solution is

$$\langle\dot{\mathbf{r}}\rangle = \text{const}\,,$$

which means motion at constant average velocity $\langle\mathbf{v}\rangle$.

Solution 2.7 *Exercise on page 41*

Taking the scalar product of the equation

$$M\ddot{\mathbf{r}} + \mu\dot{\mathbf{r}} = \mathbf{F}_e \tag{i}$$

with $\mathbf{r}$, we obtain

$$M\mathbf{r}\cdot\ddot{\mathbf{r}} + \mu\mathbf{r}\cdot\dot{\mathbf{r}} = \mathbf{r}\cdot\mathbf{F}_e\,. \tag{ii}$$

Using the identities (1.2.10), we have

$$\mathbf{r}\cdot\ddot{\mathbf{r}} = \frac{1}{2}\frac{d^2r^2}{dt^2} - (\dot{\mathbf{r}})^2,$$

so that Eq. (ii) takes the form (1.2.9):

$$\frac{1}{2}M\frac{d^2r^2}{dt^2} + \frac{1}{2}\mu\frac{dr^2}{dt} - M(\dot{\mathbf{r}})^2 = \mathbf{r}\cdot\mathbf{F}_e\,.$$

Solution 2.8 *Exercise on page 42*

Equation (1.2.11) is identical to the equation of motion of the sky diver in Exercise 2.4, where instead of mg we write $2DkT$, and z is to be replaced here by u. The solution which satisfies the required initial condition is therefore obtained by the appropriate substitutions in Solution 2.4, so (1.2.12) is obtained.

The dimensions of the parameter θ are

$$[\theta] = [M/\mu] = \frac{[M]}{[F/v]} = \frac{[M]}{[M][L][T]^{-2}[T][L]^{-1}} = [T] \, .$$

Solution 2.9 *Exercise on page 44*

Taking the scalar product of Eq. (1.2.17)

$$M\ddot{\mathbf{r}} + \mu\dot{\mathbf{r}} + C\mathbf{r} = \mathbf{F}_e \qquad \text{(i)}$$

with $\mathbf{r}$, and using the identities (1.2.10), we get

$$\frac{1}{2}M\frac{d^2r^2}{dt^2} + \frac{1}{2}\mu\frac{dr^2}{dt} + Cr^2 - M(\dot{\mathbf{r}})^2 = \mathbf{r}\cdot\mathbf{F}_e \, . \qquad \text{(ii)}$$

Averaging Eq. (ii) we obtain

$$\frac{1}{2}M\frac{d^2}{dt^2}\langle r^2\rangle + \frac{1}{2}\mu\frac{d}{dt}\langle r^2\rangle + 2\left\langle\frac{1}{2}Cr^2\right\rangle - 2\left\langle\frac{1}{2}Mv^2\right\rangle = 0 \, , \qquad \text{(iii)}$$

where we wrote $\mathbf{v}$ instead of $\dot{\mathbf{r}}$.

Denoting $u = \langle r^2\rangle$ and using (1.2.18) for the average kinetic energy, we obtain Eq. (1.2.19).

Notice that we have not replaced the average potential energy $\frac{1}{2}Cr^2$ by $\frac{1}{2}DkT$. Had we done so, this would have been equivalent to the assumption that u is a constant independent of time! Since our interest here is in the time dependence of u, we must keep the term $2Cu$ in the equation. We are assuming, therefore, that the numerous collisions have caused the velocity to reach its thermal value before the time $t = 0$.

Solution 2.10 *Exercise on page 45*

Substituting $v = Ke^{-\gamma t}$ into Eq. (1.2.20), we find that γ must satisfy the equation

$$\gamma^2 - \frac{1}{\theta}\gamma + \frac{2C}{M} = 0 \, , \qquad \text{(i)}$$

whose solution is Eq. (1.2.21).

Hence the general solution to (1.2.20) is a sum of two exponentials corresponding to the two solutions for γ:

$$v = Ae^{-\gamma_1 t} + Be^{-\gamma_2 t}. \tag{ii}$$

The initial conditions $u(t=0) = \dot{u}(t=0) = 0$ imply that

$$\begin{aligned} A + B &= -\frac{DkT}{C}, \\ \gamma_1 A + \gamma_2 B &= 0. \end{aligned} \tag{iii}$$

The solution of these equations for A and B leads to Eq. (1.2.22).

To investigate the behavior at short times, we note that the system has two characteristic times: $1/\gamma_1$ and $1/\gamma_2$. Short times are thus short by comparison with $1/\gamma_2$, which is the shorter of the two, and a power expansion yields a quadratic behavior:

$$u(t) = \langle r^2 \rangle \approx \frac{DkT\gamma_1\gamma_2}{2C} t^2. \tag{iv}$$

Notice that $\gamma_1\gamma_2 = 2C/M$, which gives exactly Eq. (1.2.14) again.

The region in which u is approximately linear is around the point at which $\ddot{u} = 0$ (zero "acceleration" means a constant velocity and a distance linear with time). Differentiating Eq. (1.2.22) twice we obtain

$$\gamma_2\gamma_1^2 e^{-\gamma_1 t} - \gamma_1\gamma_2^2 e^{-\gamma_2 t} = 0 \tag{v}$$

or

$$e^{(\gamma_2-\gamma_1)t} = \frac{\gamma_2}{\gamma_1}, \tag{vi}$$

which means that, around a time corresponding to the solution of Eq. (vi), $\langle \mathbf{r}^2 \rangle$ grows linearly with time. It is not the behavior at long times.

Solution 2.11 *Exercise on page 47*

A practical method of measuring the restoring force constant of the whisker is based on the fact that if we shift the mirror by an angular deviation ϕ_0, that is not too large, the mirror will execute harmonic motion at a frequency:

$$\omega^2 = \frac{a}{I}.$$

By measuring the period of the vibrations and the moment of inertia of the mirror, we obtain a.

In order for it to be possible to neglect the influence of the random force on the measurement, the experiment is carried out at a very low pressure.

Solution 3.1 *Exercise on page 51*

It is possible to make the following qualitative argument: the mean free path decreases as the probability for the molecule to collide with other molecules increases. This probability grows with increasing gas density as well as with increasing molecular radius. That is, the mean free path must decrease when the density increases and when the molecular radius increases.

The dimensions of the mean free path are dimensions of length. To obtain length from the density and the radius, by a function that decreases with each of them we note that the dimensions of the density are $(\text{length})^{-3}$, and the radius has dimensions of length, so $1/a^2n$ has the dimensions of length and is a decreasing function of a and n.

However, since there are *two* quantities with dimensions in the problem, n and a, (1.3.2) is not a unique solution. It is possible to create a dimensionless quantity, a^3n, and if $1/a^2n$ is multiplied by an arbitrary function of a^3n, the dimensions of the expression would still be length. For instance, $(1/a^2n)e^{-a^3n}$ satisfies all of our requirements. Thus, in this case, dimensional analysis is not sufficient.

Solution 3.2 *Exercise on page 51*

A few of the implicit assumptions that were made in the calculation are:

(a) We assumed that the distance that a molecule travels between collisions is constant, while this quantity has a distribution and that ℓ is only its average.

(b) The use of the average velocity in Eq. (1.3.2) is also an approximation, because the collision between the two molecules occurs when both are in motion so that $\bar{v}$ is not the average velocity of a gas molecule but the average *relative* velocity.

(c) We assumed that collisions always involve only two molecules. This assumption is good for a dilute gas. When the density of the molecules increases, so does the probability for a simultaneous collision between three molecules, and these collisions have to be taken into account.

(d) We assumed that the cross section for scattering depends only on the geometric dimensions of the molecules. In fact, it also depends on the relative velocity of the colliding molecule and, eventually, also on the force between them.

Solution 3.3 *Exercise on page 53*

(a) In order to prove Eq. (1.3.10) we need only calculate $\bar{v}$ using the Maxwell–Boltzmann distribution

$$\bar{v} = 4\pi \int_0^\infty f(v)v^3 dv = 4\pi \left(\frac{m}{2\pi kT}\right)^{3/2} \int_0^\infty v^3 e^{-mv^2/2kT} dv\,. \qquad \text{(i)}$$

The integral that we obtained, which we denote as I, can be evaluated with the help of integration by parts:

$$I = -\frac{kT}{m}\int_0^\infty \frac{d}{dv}(e^{-mv^2/2kT})v^2 dv = \frac{2kT}{m}\int_0^\infty e^{-mv^2/2kT} v dv$$
$$= 2\left(\frac{kT}{m}\right)^2,$$

so that

$$\bar{v} = \sqrt{8kT/\pi m}\,, \qquad \text{(ii)}$$

and Eq. (1.3.10) follows.

(b) For the calculation of the escape rate of the gas molecules it is possible to repeat the argument that led to Eq. (1.3.9), where this time the area A, which the molecule hits, is the area of the hole at the side of the container. All the molecules that hit it pass through it and leave the container. The escape rate from the container is therefore νA.

Solution 3.4 *Exercise on page 53*

The average relative speed of a pair of molecules is

$$\bar{v}_{\rm rel} = \int f(\mathbf{v}_1)f(\mathbf{v}_2)|\mathbf{v}_1 - \mathbf{v}_2| d\tau_1 d\tau_2\,, \qquad \text{(i)}$$

where $d\tau_1$ and $d\tau_2$ are volume elements in velocity space $d\tau = dv_x dv_y dv_z$, and $f(\mathbf{v})$ is given by Eq. (1.1.49). As was mentioned in Chap. 1, it is also possible to describe the motion of two molecules in terms of the center of mass velocity $\mathbf{v}_{\rm cm}$ and the relative velocity $\mathbf{v}_{\rm rel}$. The product $f(\mathbf{v}_1)f(\mathbf{v}_2)$ is an exponential function of the total kinetic energy of the two molecules, $\exp[-(m_1 v_1^2 + m_2 v_2^2)/kT]$, so we express it in terms of $\mathbf{v}_{\rm rel}$ and $\mathbf{v}_{\rm cm}$. Since

$$\begin{cases} \mathbf{v}_1 = \mathbf{v}_{\rm cm} + \dfrac{m_2}{m_1 + m_2}\mathbf{v}_{\rm rel}\,, \\[2ex] \mathbf{v}_2 = \mathbf{v}_{\rm cm} - \dfrac{m_1}{m_1 + m_2}\mathbf{v}_{\rm rel}\,, \end{cases} \qquad \text{(ii)}$$

we obtain

$$\frac{1}{2}(m_1 v_1^2 + m_2 v_2^2) = \frac{1}{2}(M v_{\rm cm}^2 + \mu v_{\rm rel}^2)\,, \qquad \text{(iii)}$$

where

$$M = m_1 + m_2 \,, \quad \frac{1}{\mu} = \frac{1}{m_1} + \frac{1}{m_2} \,. \tag{iv}$$

This means that the motion of the two molecules is equivalent to the motion of two particles, one with a mass equal to the sum of the molecular masses, M, and velocity $\mathbf{v}_{\rm cm}$, and the other of mass μ (the *reduced mass*) and velocity $\mathbf{v}_{\rm rel}$. All this should be well known from mechanics. Notice that at this stage it is more convenient not to require equal masses and instead to allow m_1 and m_2 to take on arbitrary values.

Returning to the integrand in Eq. (i), it becomes a product of a function that depends only on $\mathbf{v}_{\rm cm}$ and a function that depends only on $\mathbf{v}_{\rm rel}$:

$$f(\mathbf{v}_1)f(\mathbf{v}_2)|\mathbf{v}_1 - \mathbf{v}_2| = \left(\frac{M}{2\pi kT}\right)^{3/2} \exp\left(-\frac{Mv_{\rm cm}^2}{2kT}\right) \cdot \left(\frac{\mu}{2\pi kT}\right)^{3/2} \times \exp\left(-\frac{\mu v_{\rm rel}^2}{2kT}\right) v_{\rm rel} \,. \tag{v}$$

Note that the normalization factors have been changed using the identity $m_1 m_2 = \mu M$. The integration is to be carried out over all possible values of $\mathbf{v}_1$ and $\mathbf{v}_2$ or, alternatively, over all possible values of $\mathbf{v}_{\rm rel}$ and $\mathbf{v}_{\rm cm}$. Integration over $\mathbf{v}_{\rm cm}$ will simply give unity, as the first factor in (v) is precisely the normalized distribution function (1.1.49). Thus, we are left with

$$\bar{v}_{\rm rel} = \left(\frac{\mu}{2\pi kT}\right)^{3/2} \int \exp\left(-\frac{\mu v_{\rm rel}^2}{2kT}\right) v_{\rm rel} d\tau_{\rm rel} \,. \tag{vi}$$

This expression is none other than the familiar expression for the average speed of a particle of mass $m = \mu$, which we calculated in Exercise 3.3. If the two molecules are identical (which is the case for a monocomponent gas), then $\mu = m/2$ and

$$\bar{v}_{\rm rel} = \sqrt{\frac{8kT}{\pi\mu}} = \sqrt{2}\,\bar{v} \,.$$

Solution 3.5 *Exercise on page 54*

The mean free path is approximately [Eq. (1.3.2)]

$$\ell = \frac{1}{4\pi a^2 n} \,,$$

where n is the gas density. Each molecule occupies an average volume of $1/n$, so that the average intermolecular distance is $n^{-1/3}$. Under standard conditions $T \approx 300$ K, $P \approx 10^5$ N m^{-2} and from the equation of state $n \approx 10^{25}$ m^{-3}. For a typical molecule of radius 1 Å we obtain

$$\frac{\ell}{n^{-1/3}} = \ell n^{1/3} = \frac{1}{4\pi a^2 n^{2/3}} \approx 10^3 ,$$

$$\frac{\ell}{a} = \frac{1}{4\pi a^3 n} \approx 10^4 .$$

Solution 3.6 *Exercise on page 54*

We will assume that oxygen is an ideal gas so that

$$P = nkT . \tag{i}$$

Since we have already made an approximation here, there is no point in using the full equations (1.3.12) and (1.3.13), which were themselves obtained under somewhat unrealistic assumptions. Hence we use Eqs. (1.3.1) and (1.3.2). The mean free path is therefore

$$\ell \approx \frac{1}{4\pi a^2 n} . \tag{ii}$$

Equations (i) and (ii) imply that

$$P \approx \frac{kT}{4\pi a^2 \ell} . \tag{iii}$$

We are looking for P such that $\ell = 5$ cm:

$$P = \frac{1.4 \times 10^{-23} \times 373}{4\pi \times 10^{-20} \times 5 \times 10^{-2}} \approx 0.8 \text{ N m}^{-2} \approx 8 \times 10^{-6} \text{ atm} .$$

The mean free time is given by

$$\tau = \frac{\ell}{\bar{v}} .$$

We estimate $\bar{v}$ in terms of the temperature:

$$\frac{3}{2}kT \approx \frac{1}{2}m\bar{v}^2 ,$$

where the mass of an oxygen molecule O_2 is

$$m = 5.3 \times 10^{-26} \text{ kg} ,$$

so finally

$$\tau \approx \ell \left(\frac{m}{3kT}\right)^{1/2} = 0.05 \left(\frac{5.3 \times 10^{-26}}{3 \times 1.4 \times 10^{-23} \times 373}\right)^{1/2} \approx 10^{-4} \text{ s} .$$

Solution 3.7 *Exercise on page 54*

We calculate the mean free path of molecules of type 1, whose radius is a_1 and whose density is n_1. There are also molecules of type 2, whose radius is a_2 and whose density is n_2. We perform the calculation as we did to obtain (1.3.2).

The idea, we recall, is to calculate the number of collisions that the chosen molecule experiences per unit time, and to divide the molecule's thermal speed by this number. The number of collisions experienced by a molecule of type 1 with identical molecules, per unit time, is calculated as in Sec. 3.2.

The result is

$$\nu_{11} = \frac{1}{\tau_{11}} \approx (4\pi a_1^2 \bar{v}) n_1 \,. \tag{i}$$

In order to calculate the number of collisions of the same molecule with type 2 molecules, in the same time unit, we ignore the molecules of type 1. This time the cylinder that the molecule intersects in its collisions, the one that will replace the cylinder of radius $2a$ in Fig. 1.3.1, will be a cylinder of radius $a_1 + a_2$. In other words, a molecule of type 1 will collide with a molecule of type 2 if the distance between their centers is smaller than $a_1 + a_2$. Hence, the number of collisions between 1 and 2 will be

$$\nu_{12} = \frac{1}{\tau_{12}} = \pi(a_1 + a_2)^2 \bar{v} n_2 \,, \tag{ii}$$

and since the total rate for collisions of molecules of type 1 is the sum of (i) and (ii),

$$\nu_1 = \frac{1}{\tau_{11}} + \frac{1}{\tau_{12}} = \frac{\bar{v}}{\ell_{11}} + \frac{\bar{v}}{\ell_{12}} \,,$$

we find that

$$\frac{1}{\ell_1} = \frac{1}{\ell_{11}} + \frac{1}{\ell_{12}}$$

or

$$\ell_1 = \left[4\pi a_1^2 n_1 + \pi(a_1 + a_2)^2 n_2\right]^{-1} . \tag{iii}$$

The mean free path of a type 2 molecule is obtained from Eq. (iii) by interchanging the indices 1 and 2.

Note that a more accurate calculation would require accounting for the fact that the average relative velocity of two molecules of the same type is different from that of two molecules of different types.

Solution 3.8 *Exercise on page 56*

For the distribution in Eq. (1.3.15),

$$N = N_0 e^{-s/\ell} \,,$$

the average distance traversed by a molecule without colliding is

$$\langle s\rangle = \frac{\int_0^\infty sN(s)ds}{\int_0^\infty N(s)ds} = -\frac{d}{d(1/\ell)}\ln\left(\int_0^\infty N_0 e^{-s/\ell}ds\right)$$

$$= -\frac{d}{d(1/\ell)}\ln(N_0\ell) = \frac{d}{d(1/\ell)}\ln\frac{1}{\ell} = \ell\,.$$

Solution 3.9 *Exercise on page 56*

The average distance traversed by a molecule between two successive collisions with other molecules is ℓ, the mean free path. The average time between two collisions is τ. Hence, the average number of collisions experienced by a molecule per unit time is $1/\tau$, and in a time interval dt it will experience dt/τ collisions.

If at time t the number of molecules that did not experience collisions since $t = 0$ is $N(t)$, then $N(t)dt/\tau$ of them will collide between t and $t+dt$. Thus, the change in the number of molecules that did not experience collisions will be

$$dN(t) = -\frac{N(t)dt}{\tau}\,,$$

so that the distribution is

$$N(t) = N_0 e^{-t/\tau}\,,$$

where N_0 is the number of molecules at time $t = 0$.

Solution 3.10 *Exercise on page 58*

The diffusion coefficient is given in Eq. (1.3.20) in the form

$$D = \frac{1}{3}\bar{v}\ell\,, \tag{i}$$

and the mean free path

$$\ell \propto \frac{1}{n}\,. \tag{ii}$$

For an ideal gas

$$P = nkT\,, \tag{iii}$$

and from equipartition

$$\bar{v} \propto T^{1/2}\,, \tag{iv}$$

thus

$$D \propto T^{1/2}\cdot\frac{T}{P} = \frac{T^{3/2}}{P}\,. \tag{v}$$

Therefore, at a constant temperature

$$D \propto \frac{1}{P} \tag{vi}$$

and at constant pressure

$$D \propto T^{3/2}\,. \tag{vii}$$

Solution 3.11 *Exercise on page 59*

To generalize Eq. (1.3.22) to the three-dimensional case, we consider a small box centered around the point $\mathbf{r} = (x, y, z)$, with sides $\Delta x, \Delta y$ and Δz (see Fig. 1.3.5). To obtain the change in the number of particles per unit volume in the box $\frac{\partial n_1}{\partial t}$, we have to sum over the changes due to the flow into and out of its six faces.

The change in the number of particles, due to the flux in the z direction, is $-\partial J_z/\partial z$ [in analogy with Eq. (1.3.22)], where J_z is the component of the flux along the z direction. The two other components of $\mathbf{J}$ do not transport molecules across the faces which are normal to the z axis.

Similarly, the change in the density as a result of the flow in the directions of x and y will be $-\partial J_x/\partial x$ and $-\partial J_y/\partial y$, respectively. Hence we obtain

$$-\frac{\partial n_1}{\partial t} = \frac{\partial J_x}{\partial x} + \frac{\partial J_y}{\partial y} + \frac{\partial J_z}{\partial z} = \nabla \cdot \mathbf{J}\,.$$

■ The divergence operator $\nabla \cdot \mathbf{J}$ can be thought of as a scalar product of the gradient operator $\nabla = \hat{\mathbf{x}}\frac{\partial}{\partial x} + \hat{\mathbf{y}}\frac{\partial}{\partial y} + \hat{\mathbf{z}}\frac{\partial}{\partial z}$ and the vector $\mathbf{J}$.

This equation expresses the conservation of the number of particles and is called the (three-dimensional) continuity equation.

Alternatively, we can make the following argument. n_1 is a scalar and so is $\frac{\partial n_1}{\partial t}$. The flux $\mathbf{J}$ is a vector. The generalization of (1.3.22) must give the change of n_1 in time in terms of the first derivatives of the vector components, and yield a scalar. The only such scalar that can be formed is $\nabla \cdot \mathbf{J}$.

Solution 3.12 *Exercise on page 60*

(a)

$$n_1(z,t) = \frac{C}{\sqrt{Dt}} e^{-z^2/4Dt}$$

$$\Downarrow$$

$$\frac{\partial n_1}{\partial t} = -\frac{C}{\sqrt{Dt}} e^{-z^2/4Dt} \left(\frac{1}{2t} - \frac{z^2}{4Dt^2} \right),$$

$$\frac{\partial n_1}{\partial z} = -\frac{C}{\sqrt{Dt}}e^{-z^2/4Dt}\frac{z}{2Dt},$$

$$\frac{\partial^2 n_1}{\partial z^2} = -\frac{C}{\sqrt{Dt}}e^{-z^2/4Dt}\left[-\left(\frac{z}{2Dt}\right)^2 + \frac{1}{2Dt}\right]$$

$$\Downarrow$$

$$D\frac{\partial^2 n_1}{\partial z^2} = -\frac{C}{\sqrt{Dt}}e^{-z^2/4Dt}\left(-\frac{z^2}{4Dt^2} + \frac{1}{2t}\right) = \frac{\partial n_1}{\partial t},$$

hence (1.3.24) is a solution of (1.3.23).

(b) The number of particles of type 1, N_1, is obtained by integration of $n_1(z,t)$ over all (one-dimensional) "space":

$$N_1 = \int_{-\infty}^{\infty} n_1(z,t)dz = \frac{C}{\sqrt{Dt}}\int_{-\infty}^{\infty} e^{-z^2/4Dt}dz$$

$$= \frac{C}{\sqrt{Dt}}\sqrt{4\pi Dt} = 2C\sqrt{\pi}.$$

Hence N_1 is independent of time. Moreover, we have found an expression for the constant C: $C = N_1/\sqrt{4\pi}$.

(c) The graph of n_1 as a function of time at a given point $z \neq 0$ is drawn in the figure.

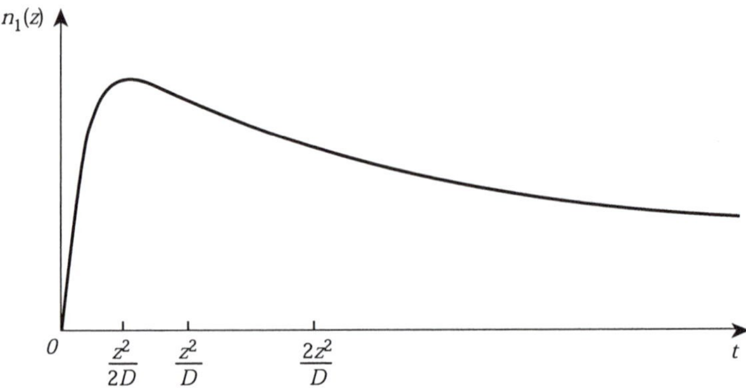

The time dependence of the density at point z.

(d) The average square distance

$$\langle z^2\rangle = \frac{1}{N_1}\int_{-\infty}^{\infty} z^2 n_1(z,t)dz = \frac{1}{2C\sqrt{\pi}}\frac{C}{\sqrt{Dt}}\int_{-\infty}^{\infty} z^2 e^{-z^2/4Dt}dt.$$

Integrating by parts in the last integral we obtain

$$\int_{-\infty}^{\infty} z^2 e^{-z^2/4Dt}dz = (4Dt)^{3/2}\cdot\frac{\sqrt{\pi}}{2},$$

so that

$$\langle z^2 \rangle = 2Dt\,.$$

This means that, on average, the distance of the molecules from their initial position grows as $\sqrt{t}$ (and not as t), which gives a "diffusion time" much longer than the "thermal" time.

The time required to cross a distance L by diffusion is

$$t_1 = \frac{L^2}{2D}\,.$$

The time required for traveling the same distance at the thermal speed $\bar{v}$ is

$$t_2 = \frac{L}{\bar{v}}\,.$$

The ratio of these times is

$$\frac{t_1}{t_2} = \frac{L\bar{v}}{2D}\,.$$

Taking into account Eq. (1.3.20),

$$D = \frac{1}{3}\bar{v}\ell\,,$$

we get

$$\frac{t_1}{t_2} = \frac{3}{2}\frac{L}{\ell}\,,$$

which means that if the distance traveled is very large compared to the mean free path, the diffusion time will be much longer than the time of the respective motion at thermal speed.

Solution 3.13 *Exercise on page 62*

The velocity of the accelerated particle at time t, since its last collision, is

$$\mathbf{v} = \mathbf{v}_0 + \frac{\mathbf{F}}{m}t\,. \tag{i}$$

Calculating its average speed, using the distribution (1.3.16), we obtain

$$v_d = \frac{F}{m}\langle t \rangle = \frac{F}{m}\frac{\int_0^\infty te^{-t/\tau}dt}{\int_0^\infty e^{-t/\tau}dt}\,. \tag{ii}$$

Note that N_0 has canceled out. We have already calculated the ratio of such integrals before. The result is

$$v_d = \frac{F}{m}\tau\,, \tag{iii}$$

namely (1.3.32).

In order to calculate K in terms of the distance traversed by the accelerated particle between its last collision and time t, we write

$$\mathbf{r} = \mathbf{v}_0 t + \frac{\mathbf{F}}{2m} t^2 . \tag{iv}$$

When we average this equation over many intervals along the particle's trajectory (or over an ensemble), the first term on the right hand side of (iv) vanishes so that we obtain

$$\langle \mathbf{r} \rangle = \frac{\mathbf{F}}{2m} \langle t^2 \rangle \tag{v}$$

or

$$d = \frac{1}{2} \frac{F}{m} \langle t^2 \rangle , \tag{vi}$$

but

$$\langle t^2 \rangle = \frac{\int_0^\infty t^2 e^{-t/\tau} dt}{\int_0^\infty e^{-t/\tau} dt} = \frac{2\tau^3}{\tau} = 2\tau^2 , \tag{vii}$$

namely

$$d = \frac{F}{m} \tau^2 , \tag{viii}$$

and consequently

$$\langle v \rangle = \frac{d}{\tau} = \frac{F}{m} \cdot \tau , \tag{ix}$$

as obtained in the other method.

The error in the argument given in the exercise was that we wrote $\langle t^2 \rangle = \tau^2$, but the distribution is an exponential that is linear in t and not in t^2, so $\langle t^2 \rangle = 2\tau^2$.

Solution 3.14 *Exercise on page 66*

The viscosity coefficient: $\eta \propto m \ell \bar{v} n$

The mean free path: $\ell \propto \frac{1}{n}$

The average speed: $\bar{v} \propto \sqrt{\frac{T}{m}}$

Hence

$$\eta \propto m \cdot \frac{1}{n} \cdot \sqrt{\frac{T}{m}} \cdot n = \sqrt{mT} .$$

From here we deduce that:

(a) $\eta \propto m^{1/2}$,
(b) For $T = \text{const}$, $\eta = \text{const}$, so that η is independent of the pressure (or the density).
(c) $\eta \propto T^{1/2}$.

Solution 3.15 *Exercise on page 67*

We have seen that the amount of heat that crosses a unit area per unit time, along z, is

$$Q_z \approx \frac{1}{6} n\bar{v}[\bar{\epsilon}(-\ell) - \bar{\epsilon}(\ell)] .$$

For small enough ℓ and an almost constant temperature, we can use the approximation

$$\bar{\epsilon}(\ell) - \bar{\epsilon}(-\ell) = \left.\frac{d\bar{\epsilon}}{dz}\right|_{z=0} \cdot 2\ell ,$$

$$Q_z \approx -\tfrac{1}{6} n\bar{v}\frac{d\bar{\epsilon}}{dz} 2\ell .$$

The gradient of $\bar{\epsilon}$ is determined in terms of the gradient of T, which is the quantity that varies with z. And we find

$$Q_z \approx -\frac{1}{3} n\bar{v}\ell \frac{d\bar{\epsilon}}{dT}\frac{dT}{dz} .$$

Solution 3.16 *Exercise on page 68*

The thermal conductivity

$$\bar{K} \propto n\ell\bar{v}c . \tag{i}$$

For an ideal gas

$$\ell \propto \frac{1}{n} ,$$

$$\bar{v} \propto \sqrt{T} ,$$

$$\epsilon \propto T \Rightarrow c = \frac{d\varepsilon}{dT} = \text{const} .$$

And substituting in (i) we obtain

$$\bar{K} \propto \sqrt{T} .$$

Solution 3.17 *Exercise on page 70*

When we add $J_\downarrow$ and $J_\uparrow$ the term proportional to $n_1(0)$ cancels out, leaving

$$J_z = -\frac{\bar{v}}{\ell}\frac{\partial n_1}{\partial z}\int_0^{\pi/2} \sin\theta\cos^2\theta d\theta \int_0^\infty r e^{-r/\ell} dr .$$

Note that $\frac{\partial n_1}{\partial z}$ has been treated as a constant in the integration, since it is evaluated at $z = 0$.

The calculation of the two remaining integrals is immediate:

$$\int_0^{\pi/2} \sin\theta \cos^2\theta d\theta = \frac{1}{3},$$

$$\int_0^{\infty} r e^{-r/\ell} dr = \ell^2,$$

and finally

$$J_z = -\frac{1}{3}\bar{v}\ell\frac{\partial n_1}{\partial z}.$$

Solutions to self-assessment exercises

Solution 1

Exercise on page 71

The calculation of the required averages is done by means of the velocity distribution function (1.1.49):

$$f(\mathbf{v}) = \left(\frac{m}{2\pi kT}\right)^{3/2} \exp\left(-\frac{mv^2}{2kT}\right),$$

where $v^2 = v_x^2 + v_y^2 + v_z^2$.

This function is separable, i.e. it can be written as

$$f(\mathbf{v}) = g(v_x)g(v_y)g(v_z),$$

and is symmetric in its velocity components, i.e. $g(x) = g(-x)$.

Before passing on to the calculations themselves, recall that the integral of an antisymmetric function $G(x) = -G(-x)$ from $-\infty$ to ∞ vanishes. This is so because, if we change integration variable according to $y = -x$, both G and dx change sign and the limits of integration switch. Hence

$$\int_{-\infty}^{+\infty} G(x)dx = \int_{+\infty}^{-\infty} G(y)dy\,.$$

When we switch the limits of integration back, we pick up another minus sign and the integral on the right becomes minus the one on the left, but the two integrals are identical. Hence they must vanish.

■ Reminder:

$$\int_a^b G(x)dx = -\int_b^a G(x)dx\,.$$

Also since the distribution function is symmetric, its product with an antisymmetric function is antisymmetric.

If f and A are separable functions of several variables, namely

$$f(r,s,t) = g_1(r)g_2(s)g_3(t)\,,$$

$$A(r,s,t) = h_1(r)h_2(s)h_3(t)\,,$$

we can write the average of A in the form

$$\begin{aligned}\langle A\rangle &= \int h_1(r)g_1(r)dr \int h_2(s)g_2(s)ds \int h_3(t)g_3(t)dt \\ &= \langle h_1\rangle \cdot \langle h_2\rangle \cdot \langle h_3\rangle\,,\end{aligned}$$

where each average is calculated with the one-dimensional distribution. Moreover, if f is symmetric with respect to r, s, t and A is antisymmetric with respect to at least one of these variables, $\langle A\rangle = 0$.

We now pass on to the solution of the problem:

(a) First,

$$\langle v_x\rangle = 0$$

since, as already mentioned, f (or g) is symmetric with respect to v_x, and $A = v_x$ is antisymmetric.

(b) $\langle v_x^2\rangle$ can be calculated directly from (1.1.49):

$$\begin{aligned}\langle v_x^2\rangle &= \left(\frac{m}{2\pi kT}\right)^{3/2} \int_{-\infty}^{\infty} \exp\left(-\frac{mv_x^2}{2kT}\right) v_x^2 dv_x \\ &\times \int_{-\infty}^{\infty} \exp\left(-\frac{mv_y^2}{2kT}\right) dv_y \int_{-\infty}^{\infty} \exp\left(-\frac{mv_z^2}{2kT}\right) dv_z\,.\end{aligned}$$

In fact we have already calculated all these integrals in Exercise 1.14 and obtained

$$\frac{m}{2}\langle v_x^2\rangle = \frac{1}{2}kT\,;$$

see Sec. 1.6 as well.

Hence

$$\langle v_x^2\rangle = \frac{kT}{m}\,.$$

(c) $A = v_x^2 v_y$ is antisymmetric with respect to v_y, and so

$$\langle v_x^2 v_y\rangle = \langle v_x^2\rangle\langle v_y\rangle = 0\,.$$

(d) Here $A = |\mathbf{v}|^2 v_z$ is antisymmetric with respect to v_z, so that

$$\langle |\mathbf{v}|^2 v_z\rangle = \langle v_x^2\rangle\langle v_z\rangle + \langle v_y^2\rangle\langle v_z\rangle + \langle v_z^3\rangle = 0\,.$$

(e) We compute the average of

$$A = (v_x + bv_y)^2 = v_x^2 + b^2 v_y^2 + 2bv_x v_y\,,$$

$$\langle v_x^2\rangle = \frac{kT}{m}\,,$$

$$\langle b^2 v_y^2\rangle = b^2\langle v_y^2\rangle = \frac{b^2 kT}{m}\,.$$

And from considerations of symmetry

$$\langle 2bv_xv_y\rangle = 2b\langle v_xv_y\rangle = 0$$

$$\Downarrow$$

$$\langle (v_x + bv_y)^2\rangle = (1+b^2)\frac{kT}{m}\,.$$

(f) Since f is separable, we have

$$\langle v_x^2v_y^2\rangle = \langle v_x^2\rangle\langle v_y^2\rangle = \left(\frac{kT}{m}\right)^2\,.$$

Solution 2 *Exercise on page 71*

(a)

$$\left\langle\frac{1}{|\mathbf{v}|}\right\rangle = \left(\frac{m}{2\pi kT}\right)^{3/2}\int\frac{1}{|\mathbf{v}|}e^{-mv^2/2kT}d\tau\,.$$

We use spherical coordinates, since $f(\mathbf{v})$ depends only on the magnitude of the velocity.

The volume element of the velocity space in spherical coordinates is

$$d\tau = v^2dv\sin\theta d\theta d\phi\,.$$

The angular integration gives 4π, so that

$$\left\langle\frac{1}{|\mathbf{v}|}\right\rangle = 4\pi\left(\frac{m}{2\pi kT}\right)^{3/2}\int_0^\infty ve^{-mv^2/2kT}dv\,.$$

The last integral is elementary and hence

$$\left\langle\frac{1}{|\mathbf{v}|}\right\rangle = 4\pi\left(\frac{m}{2\pi kT}\right)^{3/2}\left(-\frac{kT}{m}e^{-mv^2/2kT}\right)_0^\infty = \left(\frac{2m}{\pi kT}\right)^{1/2}\,.$$

Using similar considerations we calculate $\langle|\mathbf{v}|\rangle$:

$$\begin{aligned}\langle|\mathbf{v}|\rangle &= \left(\frac{m}{2\pi kT}\right)^{3/2}\int|\mathbf{v}|\exp\left(-\frac{mv^2}{2kT}\right)d\tau\\ &= 4\pi\left(\frac{m}{2\pi kT}\right)^{3/2}\int_0^\infty v^3\exp\left(-\frac{mv^2}{2kT}\right)dv = \left(\frac{8kT}{\pi m}\right)^{1/2},\end{aligned}$$

using the result obtained already in Exercise 3.3 or Eq. (1.3.8). From here

$$\frac{1}{\langle|\mathbf{v}|\rangle} = \left(\frac{\pi m}{8kT}\right)^{1/2}\,.$$

The relation between the results is therefore

$$\left\langle \frac{1}{|\mathbf{v}|} \right\rangle = \frac{4}{\pi} \cdot \frac{1}{\langle |\mathbf{v}| \rangle} \,.$$

That is, the discrepancy is a factor of order 1, as expected.

(b) The probability for the speed of a molecule to be between v and $v+dv$ is obtained by integrating $f(\mathbf{v})$ [Eq. (1.1.49)] over all directions. This yields

$$P(v)dv = \left(\frac{m}{2\pi kT}\right)^{3/2} 4\pi v^2 e^{-mv^2/2kT} dv \,. \tag{i}$$

The most probable molecular speed v_m is the value for which $P(v)$ attains its maximum.

We differentiate $P(v)$, equate its derivative to zero, and obtain

$$2v - \frac{m}{kT}v^3 = 0 \;\Rightarrow\; v_m = \left(\frac{2kT}{m}\right)^{1/2} \,.$$

It is easy to check that this is where $P(v)$ attains its maximum and not its minimum.

Thus

$$\frac{1}{2}mv_m^2 = kT \,.$$

$P(v)dv$ is the probability of finding the molecule with speed between v and $v + dv$. We are looking for a function of E, $\tilde{P}(E)$, such that $\tilde{P}(E)dE$ be the probability of finding the particle with energy in the interval $(E, E + dE)$. v is a function of E, $v(E)$, hence we can write $P(v)$ as $P(v(E))$. But we must take into account that to an interval dv there corresponds an interval in energy $(dv/dE) \cdot dE$. Hence

$$\tilde{P}(E) = P(v(E))\frac{dv}{dE}$$

and since

$$\tilde{P}(E)dE = P(v(E))\frac{dv}{dE} \cdot dE \,, \tag{ii}$$

the integral of $\tilde{P}(E)$ over all E is equal to the integral of $P(v)$ over all v and hence the probability in E is normalized. Since $E = \frac{1}{2}mv^2$, we obtain from (i) and (ii)

$$\begin{aligned} \tilde{P}(E)dE &= \left(\frac{m}{2\pi kT}\right)^{3/2} 4\pi \frac{2}{m} E \frac{dE}{(2mE)^{1/2}} e^{-E/kT} \\ &= \frac{2/\sqrt{\pi}}{(kT)^{3/2}} E^{1/2} e^{-E/kT} dE \,. \end{aligned} \tag{iii}$$

The most probable energy E_m, for which $\tilde{P}(E)$ is maximal, is found here by differentiation with respect to E and equating to zero, which yields

$$\frac{1}{2E^{1/2}} - \frac{E^{1/2}}{kT} = 0 \;\Rightarrow\; E_m = \frac{1}{2}kT\,.$$

Again, it is easy to see that $\tilde{P}(E_m)$ is in fact the maximum of $\tilde{P}(E)$, and not its minimum. Comparing the two results gives

$$\frac{1}{2}mv_m^2 = 2E_m\,,$$

which means that the most probable energy is *not* the energy calculated at the most probable speed! We have another example of this type in section (a) of this exercise.

Solution 3 *Exercise on page 71*

The distribution function that we use is

$$P = Ke^{-U/kT}\,,$$

where K is a normalization constant and

$$U(\mathbf{r}) = C|\mathbf{r}|^n\,.$$

To find the average energy per molecule we have to calculate $\langle|\mathbf{r}|^n\rangle$:

$$\langle|\mathbf{r}|^n\rangle = \frac{\int |\mathbf{r}|^n \exp(-C|\mathbf{r}|^n/kT)dV}{\int \exp(-C|\mathbf{r}|^n/kT)dV}\,, \tag{i}$$

where

$$dV = dxdydz\,.$$

We denote $\alpha = C/kT$, and write (i) in the form

$$\langle|\mathbf{r}|^n\rangle = -\frac{d}{d\alpha}\ln Z(\alpha)\,, \tag{ii}$$

where $Z(\alpha)$ is the normalization factor (actually $1/K$)

$$Z(\alpha) = \int e^{-\alpha|\mathbf{r}|^n}dV\,.$$

With the help of dimensional considerations we can immediately write

$$Z(\alpha) = A\alpha^{-3/n}\,, \tag{iii}$$

where A is a constant.

Substituting (iii) into (ii) we get

$$\langle |\mathbf{r}|^n \rangle = \frac{3}{n\alpha} = \frac{3kT}{Cn} .$$

The average energy of a molecule is therefore

$$\langle U(\mathbf{r}) \rangle = \langle C|\mathbf{r}|^n \rangle = \frac{3kT}{n} .$$

For $n = 2$ (harmonic potential) we obtain the familiar result $\langle U(r) \rangle = \frac{3}{2}kT$.

Solution 4 *Exercise on page 72*

(a) We are assuming a very slow process, so that the equipartition principle applies to the molecules in the container; namely, the average energy per molecule in the container is $\frac{3}{2}kT$.

The situation is a bit different for the molecules of the leaking beam. Here there is clearly a preferred direction, which is the z axis, which is perpendicular to the hole. The degrees of freedom in the x and y directions of the leaking beam will obviously remain unchanged; they will contribute $\frac{1}{2}kT$ per degree of freedom, according to the equipartition principle. The number of molecules crossing the hole per unit time is proportional to the velocity of the molecules in the direction z. Hence more molecules with high v_z will be present in the outgoing beam and the average speed in the beam is higher. Consequently also the average energy will be higher than $\frac{3}{2}kT$, namely higher than that of a molecule inside the container.

(b) Since we are assuming that the process is almost static, the average energy of molecules inside $\langle E_i \rangle$ is given in a state of equilibrium by

$$\langle E_i \rangle = \int \frac{1}{2}mv^2 \left(\frac{m}{2\pi kT}\right)^{3/2} e^{-mv^2/2kT} d\tau = \frac{3}{2}kT . \qquad \text{(i)}$$

The average energy in the leaking beam $\langle E_0 \rangle$, moving out, is calculated in the following manner:

The number of molecules of velocity $\mathbf{v}$ that pass through the hole per unit time per unit area can be found using Eq. (1.3.3):

$$J(\mathbf{v}) = \frac{\Delta N(\mathbf{v})}{A\Delta t} = nf(\mathbf{v})v_z . \qquad \text{(ii)}$$

This is of course the current of particles of velocity $\mathbf{v}$. Hence the energy that is carried out through the hole per unit time per unit area by molecules of velocity $\mathbf{v}$ is

$$S(\mathbf{v}) = nf(\mathbf{v})v_z \cdot \frac{m}{2}|\mathbf{v}|^2 \qquad \text{(iii)}$$

and the total energy flux will be

$$S = n\frac{m}{2}\int_{v_z \geq 0} v_z v^2 f(\mathbf{v}) dv_x dv_y dv_z \,. \tag{iv}$$

The average energy of a molecule that leaves through the hole can be defined by

$$S = J \cdot \langle E_0 \rangle \,, \tag{v}$$

where J is the particle flux [called ν in Eq. (1.3.9)]. It is the integral of (ii) over all velocities,

$$J = n\int_{v_z \geq 0} v_z f(\mathbf{v}) dv_x dv_y dv_z \,, \tag{vi}$$

so that

$$\langle E_0 \rangle = \frac{S}{J} = \frac{m}{2} \cdot \frac{\int_{v_z \geq 0} v_z v^2 f(\mathbf{v}) dv_x dv_y dv_z}{\int_{v_z \geq 0} v_z f(\mathbf{v}) dv_x dv_y dv_z} \,. \tag{vii}$$

Passing to spherical coordinates for the evaluation of the integrals in (vii), we find that

$$\begin{aligned} \langle E_0 \rangle &= \frac{1}{2}m\frac{\int_0^{2\pi} d\phi \int_0^{\pi/2} \sin\theta\cos\theta d\theta \int_0^\infty v^5 e^{-mv^2/2kT} dv}{\int_0^{2\pi} d\phi \int_0^{\pi/2} \sin\theta\cos\theta d\theta \int_0^\infty v^3 e^{-mv^2/2kT} dv} \\ &= \frac{1}{2}m\left[-\frac{d}{d(m/2kT)}\ln\int_0^\infty v^3 e^{-mv^2/2kT} dv\right] \\ &= \frac{1}{2}m \cdot \frac{4kT}{m} = 2kT \,. \end{aligned} \tag{viii}$$

A comparison of the results (i) and (viii) shows that indeed we have $\langle E_i \rangle < \langle E_0 \rangle$.

Solution 5 *Exercise on page 72*

We assume that at each moment t the gas in the container is at equilibrium; namely, it satisfies the ideal gas equation of state:

$$P(t)V = N(t)kT \,, \tag{i}$$

where the temperature and the volume are constant.

The number of molecules that leak through the hole in the short time interval Δt is given by Eq. (1.3.9), so that the number of molecules will change at the rate

$$\frac{dN}{dt} = -\frac{1}{4}\frac{A}{V}\bar{v}N \,, \tag{ii}$$

where the negative sign means that the number of molecules decreases with time. The solution of Eq. (ii) is

$$N(t) = N_0 e^{-t/\tau},$$

where N_0 is the number of molecules at the moment the hole is opened, which is defined as $t = 0$; τ is the time constant for the *escape* of the molecules (τ is not the mean free time):

$$\tau = \frac{4V}{A\bar{v}}.$$

During an interval τ the number of particles decreases by a factor $1/e$.

Since the pressure in the container is proportional to the number of particles inside it, τ is also the time during which the pressure decreases by a factor $1/e$.

Solution 6 *Exercise on page 72*

(a) First we calculate the rate at which the molecules collide with the satellite. Since $V \ll \bar{v}$, it is possible to consider the satellite stationary, so that Eq. (1.3.9) applies. The rate at which the molecules collide with each side of the satellite is

$$\nu = \frac{1}{4} n\bar{v}L^2 \tag{i}$$

and the average time between two collisions with each side is

$$\tau = \frac{4}{n\bar{v}L^2}. \tag{ii}$$

To understand the collision process note that the satellite is much heavier than the molecules and its speed is many times smaller. Therefore, to a good approximation, the molecule will reverse its direction of motion, without changing its speed. Hence, in a collision with the satellite an average molecule will transfer momentum of order $2m(\bar{v} + V)$ if the molecule and the satellite are moving towards one another, and $2m(\bar{v} - V)$ if the molecule is moving in the same direction as the satellite. Thus in an interval Δt the satellite will lose momentum on order

$$\Delta p = -[2m(\bar{v} + V) - 2m(\bar{v} - V)]\frac{\Delta t}{\tau},$$

where $\Delta t/\tau$ is the number of collisions with each side during time Δt. Substituting (ii) into (iii) we find that the restraining force $\Delta p/\Delta t$ will be

$$F = -(m\bar{v}nL^2)V. \tag{iv}$$

Notice the direct proportionality between the force and the velocity which is characteristic of a friction force. (See Sec. 2.4).

(b) We obtain the satellite's equation of motion from Newton's second law:

$$M\frac{dV}{dt} = -\mu V\,, \qquad \mu = m\bar{v}nL^2\,,$$

whose solution is

$$V(t) = V_0 e^{-\mu t/M}\,,$$

where V_0 is the satellite's velocity at time $t = 0$.

The velocity decreases to half its initial value after a time t^*, which is determined by

$$e^{-\mu t^*/M} = \frac{1}{2}\,,$$

namely

$$t^* = \frac{M}{\mu}\ln 2 = \frac{M\ln 2}{m\bar{v}nL^2}\,.$$

We assume that $n \approx 10^9$ m^{-3}, and substitute characteristic orders of magnitude: $M \approx 10^2$ kg, $m \approx 10^{-26}$ kg, $\bar{v} \approx 10^2$ m s^{-1}, $L \approx 10$ m, and we obtain

$$t^* \approx \frac{10^2 \ln 2}{10^{-26} \times 10^2 \times 10^9 \times 10^2} \approx 10^{15}\ \mathrm{s} \approx 10^{10}\ \mathrm{days}\,.$$

Solution 7 *Exercise on page 72*

Equation (1.3.48),

$$Q_z = -\bar{K}\frac{dT}{dz}\,, \tag{i}$$

was obtained from the balance of molecules crossing a given plane $z =$ const, at fixed temperature gradient. The thermal energy current density along the z axis, Q_z, is proportional to the temperature change per unit length along this axis. The proportionality constant is $-\bar{K}$, where $\bar{K}$ is the thermal conductivity coefficient of the material, given in (1.3.51).

Next, we want to obtain a continuity equation as in (1.3.22), relating the energy current Q_z and the energy density, which we denote by u.

Consider a surface element of material, with area $\Delta x \Delta y$, located between the planes z and $z + dz$ as in Fig. 1.3.5. In a short time interval Δt *hot* molecules will enter the layer and *cold* molecules will leave it. If $\partial T/\partial z > 0$, i.e. if the $z+dz$ is warmer than the z plane, then hot molecules will enter through the top plane and cold molecules will exit through the bottom plane. The change of the internal energy in the layer is the difference between the thermal energy brought in by the hot molecules and the thermal energy carried out by the cold molecules. Namely

$$[u(t+\Delta t)-u(t)]\Delta x\Delta y\Delta z=-[Q_z(z+\Delta z)-Q_z(z)]\Delta x\Delta y\Delta t\,. \qquad \text{(ii)}$$

As Δt and Δz tend to zero, one obtains from (ii) the (one-dimensional) continuity equation for the energy:

$$\frac{\partial u}{\partial t}=-\frac{\partial Q_z}{\partial z}\,. \qquad \text{(iii)}$$

We now apply the chain rule to the derivative of the energy density,

$$\frac{\partial u}{\partial t}=\frac{\partial u}{\partial T}\frac{\partial T}{\partial t}\,,$$

and since

$$\frac{\partial u}{\partial T}=cn\,,$$

where c is the specific heat per molecule and n is the density of the molecules, we obtain from Eq. (iii)

$$cn\frac{\partial T}{\partial t}=-\frac{\partial Q_z}{\partial z}\,. \qquad \text{(iv)}$$

Substituting (i) into (iv) we obtain

$$cn\frac{\partial T}{\partial t}=-\frac{\partial}{\partial z}\left(-\bar{K}\frac{\partial T}{\partial z}\right),$$

and since the thermal conductivity is constant in space, we find the required equation:

$$\frac{\partial T}{\partial t}=\left(\frac{\bar{K}}{cn}\right)\frac{\partial^2 T}{\partial z^2}\,.$$

Solution 8 *Exercise on page 73*

The equation of motion for the height of an aluminum grain, z, with respect to the bottom of the container is

$$m\ddot{z}+\mu\dot{z}=-mg \qquad \text{(i)}$$

where μ is given by Stokes' law, $\mu=6\pi a\eta$.

The rate of precipitation of the grain is $\dot{z}$. The steady precipitation rate is the solution of Eq. (i) for long times:

$$\dot{z}=-\frac{mg}{\mu}=-\frac{mg}{6\pi a\eta}=-\frac{2\rho a^2 g}{9\eta}\,. \qquad \text{(ii)}$$

Substituting the numerical data we obtain

$$\dot{z} = -\frac{2 \times 3.26 \times (0.5 \times 10^{-4})^2 \times 980}{9 \times 10^{-2}} \approx -1.8 \times 10^{-4} \text{ cm/s}.$$

■ The solution of this exercise is similar to that of Exercise 2.4 in the text.

Solution 9 *Exercise on page 73*

(a) According to Eq. (1.3.47), η is proportional to $m\bar{v}\ell n$, where m is the mass of a molecule. The ratio between the viscosity coefficients of the two gases is therefore

$$\frac{\eta_2}{\eta_1} = \frac{m_2\bar{v}_2\ell_2 n_2}{m_1\bar{v}_1\ell_1 n_1}.$$

At a constant temperature and pressure, the densities of two ideal gases are equal. The relationship between the average speed of a molecule in the gas and its mass is given by $\bar{v} \propto m^{-1/2}$. Thus

$$\frac{\eta_2}{\eta_1} = \frac{m_2 \cdot m_2^{-1/2} \cdot \ell_2}{m_1 \cdot m_1^{-1/2} \cdot \ell_1} = \left(\frac{m_2}{m_1}\right)^{1/2} \frac{\ell_2}{\ell_1}.$$

Since the ratio of the masses is equal to the ratio of the atomic masses,

$$\frac{\ell_2}{\ell_1} = \left(\frac{\mu_1}{\mu_2}\right)^{1/2} \frac{\eta_2}{\eta_1} = 0.356.$$

It is also possible to obtain an estimate of ℓ_1 and ℓ_2 from Eq. (1.3.47). If we substitute for $\bar{v}$

$$\bar{v} \approx \sqrt{\frac{3kT}{m}}$$

and for n

$$n = \frac{P}{kT},$$

we obtain

$$\ell \approx \frac{n}{P}\sqrt{\frac{3kT}{m}}.$$

At a pressure of one atmosphere and a temperature of 273 K we obtain for helium

$$\ell_1 \approx 2.4 \times 10^{-7} \text{ m}$$

and for argon

$$\ell_2 \approx 8.5 \times 10^{-8} \text{ m}.$$

(b) According to Eq. (1.3.51), $\bar{K}$ is proportional to $n\bar{v}\ell c$, where c is the specific heat per molecule. Since the average energy per atom is independent of the type of atom and depends only on the temperature, the specific heat per atom is also independent of the type of atom.

Moreover, the density of the two gases is equal at equal temperature and pressure. Thus

$$\frac{\bar{K}_2}{\bar{K}_1} = \frac{\bar{v}_2\ell_2}{\bar{v}_1\ell_1} = \frac{m_2^{-1/2}\ell_2}{m_1^{-1/2}\ell_1} = \left(\frac{\mu_2}{\mu_1}\right)^{-1/2}\left(\frac{\mu_1}{\mu_2}\right)^{1/2}\frac{\eta_2}{\eta_1} = \frac{\mu_1\eta_2}{\mu_2\eta_1},$$

where we have used the result of (a) above. Hence

$$\frac{\bar{K}_2}{\bar{K}_1} = 0.113\,.$$

(c) According to Eq. (1.3.20), D is proportional to $\bar{v}\ell$. Thus

$$\frac{D_2}{D_1} = \frac{\bar{v}_2\ell_2}{\bar{v}_1\ell_1} = \frac{\bar{K}_2}{\bar{K}_1} = \frac{\mu_1\eta_2}{\mu_2\eta_1} = 0.113\,.$$

Part II

Statistical Physics with Paramagnets

Introduction

As promised in the Introduction to Part I, we now turn to the treatment of material systems from a more detailed point of view. This means that the system, be it a system of molecules, a system of magnetic moments, or a system of electrons and protons, is described by a dynamical model. The dynamical model has two components: (a) the rules according to which the system evolves in time (the laws of classical mechanics (Newton), or the laws of relativistic mechanics (Einstein), or the laws of quantum mechanics (Schrödinger); (b) the forces governing this evolution (electric, magnetic, gravitational, etc.). Moreover, in such a model we must also define the relevant degrees of freedom (these may be electrons and protons); it may suffice to describe the system as a collection of atoms, or perhaps molecules are sufficiently stable and can serve as building blocks. For instance, the ideal gas discussed in Part I is described by the coordinates and velocities (or momenta) of each of its molecules — a total of $6N$ variables. Its evolution in time is determined by the laws of classical mechanics (Newton), and to describe its microscopic evolution in time we would have to solve $3N$ coupled, second order, differential equations.

We noted already in the Introduction to Part I that we are not interested in a specific trajectory, i.e. a detailed solution of the given problem, and in following each of its degrees of freedom in time. In typical experiments (or measurements) we are not interested in the detailed evolution, but in averages over long times and over large spatial regions. When speaking of long times we mean that the durations of the experiments, or the averaging times, are very long compared to the times during which the system changes its microscopic state. For instance, the microscopic time in the gas will be the time between successive collisions — the mean free time, which is the characteristic time after which the particles change their velocities.

What we need, therefore, is a method of calculating averages over long times, along the evolution trajectory of the system. The system may change its state for different reasons: because of the forces that are acting between its constituents, or because of the effects of its surroundings (the system may be in contact with a source of heat, with a piston on which an external force is acting, or with an electric or magnetic field, etc.).

The calculation of averages over long times along the system's trajectories is almost impossible. Hence the time averaging is replaced by ensemble averaging. The idea is similar to the one we met in the discussion of Brownian motion (Chap. 2 of Part I): instead of following the system's motion from state to state, we assume, following Gibbs, that we are given many similar systems each in a different allowed state of the evolving system. The probability, or the recurrence, of a certain state in the ensemble is equal to that which would have been obtained along the path. The averages are then performed over the ensemble. The next two parts of the book are devoted to the clarification of this method.

ensemble

We mention here two examples of ensembles, two types of assumptions, which will be discussed in detail later on:

(a) In insulated systems the energy is conserved. Therefore, during its evolution, the system will only pass through equienergetic states (there may be other conservation laws or constraints, which will not be violated. The particles will remain, for instance, in their designated volume, etc.). A natural assumption in this case is that the system, evolving in time, will pass through all the allowed states at the same reccurence rate. That is, after a sufficiently long time every state of the system, having the same given energy, will have appeared the same number of times (this is the ergodic hypothesis). The average over long times will equal, therefore, the average over the ensemble of all the equienergetic states, all with the same probability. Such an ensemble is called a microcanonical ensemble.

microcanonical ensemble

(b) A system coupled to a heat reservoir has a given temperature, and is able to exchange energy. Hence if we want to replace the trajectory by an ensemble, we must determine the relative occurrence of systems in states with different energies.

In Part I we saw that the temperature is related to the average energy per degree of freedom, and that states whose energies are much larger than the average energy are less probable. It is not unreasonable to assume, extrapolating from the

> Maxwell–Boltzmann distribution, that equienergetic states have the same probability, whereas the relative recurrence rate of states with different energies is given by their Boltzmann factor, namely by the quantity $e^{-E/kT}$. Such an ensemble is called a canonical ensemble.

canonical ensemble

As was emphasized in the introduction to the previous part, all results from the more detailed approaches must face a consistency test with the consequences of the less detailed theories, such as the kinetic theory, and ultimately with thermodynamics. Part I will serve as a reference for results of the kinetic theory. Thermodynamics is not developed in this course and for reference we have introduced Chap. 0 in this part, to recapitulate essential ideas from thermodynamics. The rest of this part illustrates the application of the idea of ensembles in the simple example of the paramagnet rather than in gases or liquids. This system is attractive because it can be described in terms of discrete degrees of freedom, which makes probabilistic considerations less abstract. Despite the simplicity of the model, it describes a few experimental systems in a surprisingly precise manner. In Part III we will formulate these concepts in a more general context.

Chapter 1 concentrates on clarifying the microscopic and thermodynamic concepts for magnetic variables. The magnetic moments in our paramagnet have discrete states. This would be typical of a quantum theory, and not in the classical one. Here we take this fact as given and concentrate on the statistical technique, which is simpler for discrete variables. Chapter 2 deals with the identification of the paramagnet's microscopic states and the following two chapters treat the insulated paramagnet, namely a system at constant energy. In Chap. 5 we turn to the discussion of a paramagnet in contact with a "heat bath", namely a system at constant temperature. Chapter 6 is dedicated to deepening the understanding of the concept of entropy based on the example of the paramagnet, and the last chapter presents experimental results.

Chapter 0

Essential Background in Thermodynamics

0.1 The first law

This chapter is intended as a condensed review of basic notions in thermodynamics. Its role is to compile ideas that are presupposed to be known from previous study and are either underlying certain parts of this course or derived in it. It can serve as a condensed reference and it includes several bibliographical suggestions for wider expositions of the various topics. Though it is basically a compendium, several exercises have been included along the way to maintain the style of this course. The presentation is based, for concreteness and simplicity, on a specific physical system: the ideal gas.

The first law of thermodynamics is the law of conservation of energy. It is written as

$$dE = \delta Q - \delta W\,. \tag{2.0.1}$$

dE is the increase in the energy of the system, δW is the work done by the system and δQ is the heat transferred to the system. In fact, δQ is defined as the sum of the increase in the system's mechanical energy and the mechanical work done by it. In mechanics this sum is strictly zero. For a gas in a container whose volume expands by dV against a pressure P, $\delta W = PdV$ and Eq. (2.0.1) becomes

$$dE = \delta Q - PdV\,. \tag{2.0.2}$$

A state of a system in thermodynamic equilibrium is specified by a reduced set of *macroscopic variables.* The minimal set of variables defines the *state space* of the system. For an ideal gas of N molecules confined in a volume V at temperature T, any two of the variables (P, V, T) define the state space. Any other quantity at equilibrium is determined when any two of these are specified. This is the case because the ideal gas obeys the *equation of state*:

equation of state

$$PV = NkT \tag{2.0.3}$$

[compare (1.1.6)], which gives any one of the three in terms of the other two. The two that are chosen may be thought of as coordinates in a two-dimensional state space. Thus, any state of the system corresponds to a point in the state space. Other dependent thermodynamic variables which are fully determined by the point in state space are called *functions of state*. For the ideal gas, for any choice of two of the three variables, the third one is a function of state. As we shall see below, the internal energy of the ideal gas is also a function of state [see Eq. (2.0.4), below].

extensive and intensive variables

Thermodynamic variables are classified into two types: *extensive variables* and *intensive variables*. Extensive variables have a magnitude proportional to the size of the system. If two identical systems are combined into one, each extensive variable is doubled in value. The volume and number of particles of the gas are examples. The energy, Eq. (2.0.4), is another. Intensive variables are independent of the size of the system. T and P are such variables.

While the energy is a function of state, work or heat are not. Specifying a thermodynamic state of the system does not fix the values of these quantities. When the state of a system changes, i.e. it undergoes a process of some kind, the amount of work done by the system depends upon the process, i.e. upon the path in the space of states which corresponds to the process and not only upon the initial and final points. The same is true for the amount of heat transferred to the system. In mathematical terms: δQ and δW are not exact differentials and one uses δ instead of d for these infinitesimal quantities. By analogy, in classical mechanics, if a force is not conservative, the initial and final values of the coordinates and the velocities of its particles do not determine the work done by the system in the transition between these two states; the entire path must be specified.

■ Reminder: A differential in two variables x, y is written as $\delta a = A_x(x,y)dx + A_y(x,y)dy$. It is an exact differential if it is the difference between the values of some function $z(x,y)$ at two neighboring points, i.e. if $\delta a = z(x+dx, y+dy) - z(x,y)$. A necessary and sufficient condition for this is that $\frac{\partial A_x}{\partial y} - \frac{\partial A_y}{\partial x} = 0$. If this condition holds, then $A_x(x,y) = \frac{\partial z}{\partial x}$ and $A_y(x,y) = \frac{\partial z}{\partial y}$ for some function $z(x,y)$ and δa is the difference of the values of this function. The above can be simply generalized to any number of variables.

As an example, consider again the ideal gas. The relation between energy and (absolute) temperature is

$$E = \frac{f}{2}NkT\,, \tag{2.0.4}$$

where f is the number of degrees of freedom per molecule. See also Part I, Chap. 1. Keeping N fixed, we write δQ for a process where T, V and P are varied. The state space is two-dimensional. Taking T and V as the independent variables (coordinates), Eq. (2.0.2) gives for δQ

$$\delta Q = Nk\left(\frac{f}{2}dT + \frac{T}{V}dV\right). \tag{2.0.5}$$

The right hand side of (2.0.5) does not give equal cross derivatives with respect to V and T and hence is not an exact differential. Consequently, the amount of heat transferred to the system in a given process, which is obtained by integrating δQ along the appropriate path, is not the difference of the values of some "heat" function between the final and initial points, but depends on the particular process (the path between the two points in state space.)

Several processes are given special names: *isothermal process* — the temperature is kept constant; *isochoric process* — the volume is kept constant; *isobaric process* — the pressure is kept constant; *adiabatic process* — there is no heat exchange: $dE = -\delta W$.

Exercise 0.1

Calculate the amount of heat transferred to an ideal gas of N molecules with f degrees of freedom in:

(a) An isothermal process.
(b) An isochoric process.
(c) An isobaric process.

Solution on page 183

Exercise 0.2

An ideal gas is at temperature T_1 and volume V_1. The gas is taken through an isobaric process to a state of higher temperature T_2. It is then taken via an isochoric process to a state of temperature T_1 and finally back to the initial state in an isothermal process.

(a) Calculate the amount of heat transferred to the gas in the cycle.
(b) Same as (a) but in reverse cycle.
(c) What would be the result if δQ were an exact differential?

Solution on page 183

Exercise 0.3

(a) Calculate the work done by the gas during the cycle described in the previous exercise.

(b) Is it equal to Q?
(c) Calculate the work done by the gas during an adiabatic process.
(d) Show that if the ideal gas undergoes an adiabatic process T and V satisfy

$$VT^{f/2} = \text{const}$$

along the path.

Solution on page 184

The *heat capacity* is the quantity of heat required to change the temperature of the system by δT. Since the heat capacity is defined in terms of the heat, it depends on the process. For example, there is the heat capacity at constant volume, C_V, and the heat capacity at constant pressure, C_P. The heat capacity is not a derivative of Q. One can schematically write

$$C_V = \left(\frac{\delta Q}{\delta T}\right)_V . \tag{2.0.6}$$

It represents the coefficient of dT in the expression for δQ when it is written in terms of the variables T and V. This expression already appears in Eq. (2.0.5) and one has

$$C_V = \frac{f}{2} Nk . \tag{2.0.7}$$

The heat capacity at constant pressure is defined analogously as

$$C_P = \left(\frac{\delta Q}{\delta T}\right)_P . \tag{2.0.8}$$

To calculate it one identifies the coefficient of dT in the expression for δQ when it is written in terms of T and P.

Exercise 0.4

Show that the heat capacity at constant pressure for an ideal gas is

$$C_P = \left(1 + \frac{f}{2}\right) Nk . \tag{2.0.9}$$

Solution on page 185

The ratio between the two heat capacities is simply the adiabatic parameter $\gamma = 1 + \frac{2}{f}$. See also Eq. (1.1.25).

Exercise 0.5

Show that γ is the same γ as in the adiabatic equation, $PV^\gamma = \text{const}$.

Solution on page 185

0.2 The second law and the entropy

Clausius formulation: It is impossible for any engine working continuously in a cycle to transfer heat from a colder to a hotter body and to produce no other effect.

Kelvin formulation: It is impossible for an engine working in a cycle to extract heat from a single reservoir, produce an equal amount of work and have no other effect.

The analysis of the second law leads to the conclusion that $\frac{\delta Q}{T}$ is an exact differential, i.e. a differential of a new function of state. This is the entropy S.

$$dS = \frac{\delta Q}{T} . \tag{2.0.10}$$

It is an extensive variable. In mathematical terms $\frac{1}{T}$ is an integration factor of the differential δQ. It converts it into an exact differential.

The first law, Eq. (2.0.2), may thus be written in terms of state functions only:

$$dE = TdS - PdV . \tag{2.0.11}$$

This relation may be used to calculate the entropy differences between states of a gas. Given that at this stage the entropy is defined only up to an additive constant this is the most one can do. More about this in Part III, Chap. 5.

Exercise 0.6

(a) Starting from (2.0.5) show that the entropy of an ideal gas of N molecules with f degrees of freedom is

$$S = Nk \ln \left(\frac{cVT^{f/2}}{N} \right) . \tag{2.0.12}$$

(b) Calculate the entropy increase of an ideal gas in an isothermal process.
(c) Calculate the entropy increase of an ideal gas in an isochoric process.
(d) What is the process in which the entropy remains constant?

Solution on page 186

An alternative expression for the heat capacity is obtained in terms of the entropy, Eq. (2.0.10). One can write δQ as

$$\delta Q = T \left(\frac{\partial S}{\partial T} dT + \frac{\partial S}{\partial V} dV \right) . \tag{2.0.13}$$

The coefficient of dT is C_V. Hence

$$C_V = T \left(\frac{\partial S}{\partial T} \right)_V . \tag{2.0.14}$$

Analogously

$$C_P = T\left(\frac{\partial S}{\partial T}\right)_P . \tag{2.0.15}$$

C_V and C_P can also be obtained from other functions of state: The first law is $\delta Q = dE + PdV$, hence at constant volume $\delta Q = dE$ and

$$C_V = \left(\frac{\delta Q}{\delta T}\right)_V = \left(\frac{\partial E}{\partial T}\right)_V . \tag{2.0.16}$$

To calculate C_P we define the *enthalpy* H as **enthalpy**

$$\mathsf{H} = E + PV . \tag{2.0.17}$$

It is one member of a family of thermodynamic potentials discussed in the next section. It is a function of state, since it depends only on functions of state.

■ We reserve the ordinary H for the magnetic field.

We write δQ with dE replaced by $d\mathsf{H}$:

$$\delta Q = dE + PdV = d\mathsf{H} - VdP = \left(\frac{\partial \mathsf{H}}{\partial T}\right)_P dT + \left[\left(\frac{\partial \mathsf{H}}{\partial P}\right)_T - V\right] dP . \tag{2.0.18}$$

If T and P are the independent variables, the heat transfer due to a change in the temperature at constant *pressure* equals the change in the *enthalpy* at constant pressure:

$$C_P = \left(\frac{\delta Q}{\delta T}\right)_P = \left(\frac{\partial \mathsf{H}}{\partial T}\right)_P . \tag{2.0.19}$$

Exercise 0.7

Calculate the enthalpy of an ideal gas and from it C_P.

Solution on page 186

0.3 Thermodynamic potentials

The energy is a thermodynamic potential when expressed in terms of S and V: Eq. (2.0.11) gives the variation of the energy in terms of the variations of the independent variables S and V. It is a function of state. From Eq. (2.0.11) one obtains two relations:

$$T = \left(\frac{\partial E}{\partial S}\right)_V , \qquad P = -\left(\frac{\partial E}{\partial V}\right)_S . \tag{2.0.20}$$

If one rewrites Eq. (2.0.11) isolating dS and regards the entropy as a function of E and V, then the entropy S is identified as another thermodynamic potential, $S(E, V)$. It is a function of state and two new relations follow:

$$\frac{1}{T} = \left(\frac{\partial S}{\partial E}\right)_V, \qquad \frac{P}{T} = \left(\frac{\partial S}{\partial V}\right)_E. \tag{2.0.21}$$

thermodynamic potential

This structure is rather general: The basic function of state, or the *thermodynamic potential*, is an extensive quantity which is a function of extensive variables. Its differential is a linear combination of differentials of extensive variables with intensive variables as coefficients. The partial derivative of a thermodynamic potential with respect to an extensive variable yields its *conjugate* intensive variable. Once a thermodynamic potential is given, all the thermodynamic properties can be derived from it.

Exercise 0.8

Given the following expression for the entropy of a system:

$$S = Nk \ln\left(\frac{aVE^{3/2}}{N^{5/2}}\right), \tag{2.0.22}$$

where a is a constant, obtain all the thermodynamic information about this system, P, T etc.

Solution on page 187

When the number of particles N is reconsidered as a thermodynamic variable that may vary from state to state as a result of an exchange of particles with the surrounding, or a chemical reaction, one adds in the expression for the first law the corresponding *chemical work.*

$$dE = TdS - PdV + \mu dN. \tag{2.0.23}$$

chemical potential

The thermodynamic potential E thus becomes a function of S, V and N. The coefficient μ is the *chemical potential.* It represents the change in the energy of the system associated with a unit increase in the number of particles. μ is an intensive variable.

Exercise 0.9

Show that the chemical potential for a monoatomic ideal gas is given by

$$\mu = -kT \ln\left(\frac{bVT^{3/2}}{N}\right), \tag{2.0.24}$$

where b is a constant.

Solution on page 187

Since energy and entropy are not easy to control experimentally, one develops alternative representations of the thermodynamic information, expressed in terms of other thermodynamic potentials, which depend on the temperature instead of the energy or the entropy. This is carried out by a *Legendre transformation.* Starting from an $E(S,V,N)$, Eq. (2.0.20) gives an expression for T in terms of S, V (and N). Hence it provides S as a function of T, V (and N). One then considers the function

$$F = E - TS\,, \tag{2.0.25}$$

called the *Helmholtz free energy* (or the *free energy*). It is a Legendre transform of the energy. To identify the independent variables, one writes the differential of F [using (2.0.23)]: **free energy**

$$dF = dE - TdS - SdT = -SdT - PdV + \mu dN\,. \tag{2.0.26}$$

Hence, it is a thermodynamic potential with T, V and N as its natural variables. Starting from an expression for the free energy one derives the pressure as a function of T, V and N (which is the equation of state) as well the entropy and the chemical potential by

$$S = -\left(\frac{\partial F}{\partial T}\right)_{V,N}, \quad P = -\left(\frac{\partial F}{\partial V}\right)_{T,N}, \quad \mu = \left(\frac{\partial F}{\partial N}\right)_{T,V}. \tag{2.0.27}$$

In contrast to S or E, F is a potential that depends on an intensive variable, T. Since F is extensive, its derivative with respect to T, i.e. S, is extensive as well.

These expressions lead to the *Maxwell relations* between various derivatives of the entropy, pressure and chemical potential. They all express the fact that the mixed second partial derivatives of $F(T,V,N)$ are independent of the order of derivation. One finds that **Maxwell relations**

$$-\left(\frac{\partial P}{\partial N}\right)_{T,V} = \left(\frac{\partial \mu}{\partial V}\right)_{T,N},$$

$$-\left(\frac{\partial S}{\partial N}\right)_{T,V} = \left(\frac{\partial \mu}{\partial T}\right)_{V,N},$$

$$\left(\frac{\partial S}{\partial V}\right)_{T,N} = \left(\frac{\partial P}{\partial T}\right)_{V,N}.$$

Exercise 0.10

Calculate the free energy of a monoatomic ideal gas. Give the result in terms of the variables T, V and N and choose the arbitrary constant to be consistent with (2.0.24).

Solution on page 187

Exercise 0.11

The free energy for a photon gas is given by

$$F = -\frac{a}{3}VT^4\,, \tag{2.0.28}$$

where a is a constant. The origin of this result and the meaning of a will be discussed in Part IV.

(a) Calculate the entropy of the photon gas.
(b) Calculate the pressure of the photon gas and its equation of state.
(c) Calculate the energy of the photon gas.
(d) What is the chemical potential of the photon gas?

Solution on page 188

Exercise 0.12

Calculate the equation of the adiabatics of a photon gas.

Solution on page 189

Exercise 0.13

Verify the Maxwell relations for:

(a) Ideal gas.
(b) Photon gas.

Solution on page 189

To consider systems which may exchange not only energy but also particles with a reservoir, the number of particles N is replaced by its conjugate variable, the chemical potential μ. Mathematically, another Legendre transformation is performed, leading from the free energy to a new thermodynamic potential whose natural variables are T, V and μ. This is done by defining the *grand potential*, or just the thermodynamic potential, Ω:

$$\Omega = F - \mu N = E - TS - \mu N\,, \tag{2.0.29}$$

where the three N's (the explicit one, the one inside S and the one inside E) are expressed in terms of T, V and μ, using the equation for $\mu(T, V, N)$, in Eq. (2.0.27). Then using Eq. (2.0.23) the differential, in terms of the independent variables T, V and μ, becomes

$$d\Omega = -SdT - PdV - Nd\mu. \tag{2.0.30}$$

One obtains a set of relations analogous to (2.0.27):

$$S = -\left(\frac{\partial \Omega}{\partial T}\right)_{V,\mu}, \quad P = -\left(\frac{\partial \Omega}{\partial V}\right)_{T,\mu}, \quad N = -\left(\frac{\partial \Omega}{\partial \mu}\right)_{T,V}, \tag{2.0.31}$$

with the associated Maxwell relations, which are left as an exercise.

Exercise 0.14

(a) Derive the Maxwell relations associated with the thermodynamic potential Ω.

(b) Calculate Ω for a monoatomic ideal gas.

Solution on page 190

Finally, we express the heat capacity at constant volume (and number of particles) in terms of F. C_V is a derivative of the energy with respect to T at constant V and N [Eq. (2.0.16)]. But E is a function of S, V and N [Eq. (2.0.23)], so first one needs the dependence of E on T. This is done using Eq. (2.0.25), substituting $S(T, V, N)$ from (2.0.27). One can write

$$E(T, V, N) = F(T, V, N) - T\left(\frac{\partial F}{\partial T}\right)_{V,N}, \tag{2.0.32}$$

and the heat capacity at constant volume becomes

$$C_V = \left(\frac{\partial E}{\partial T}\right)_{V,N} = -T\left(\frac{\partial^2 F}{\partial T^2}\right)_{V,N}. \tag{2.0.33}$$

This is just a rewriting of Eq. (2.0.14). Correspondingly, C_P should be calculated from the enthalpy which is written as a function of T, P and N. These variables call for a new thermodynamic potential.

0.4 The third law

The experimental failure to reach absolute zero together with the theoretical failure to derive the impossibility of reaching 0 K from the first two laws, led to the formulation of the third law:
Formulation 1: By no finite series of processes is absolute zero attainable.
Formulation 2: As the temperature tends to zero, the magnitude of the entropy change in any reversible process tends to zero. We will return to this issue and its consequences in Part III.

Suggested reading

(1) E. Fermi, *Thermodynamics* (Dover, New York).
(2) A. B. Pippard, *Classical Thermodynamics* (Cambridge University Press, London).
(3) C. J. Adkins, *Equilibrium Thermodynamics* (McGraw-Hill, London).

Chapter 1

Thermodynamics with Magnetic Variables

1.1 Introduction

Since part of the testing of statistical mechanics is its ability to reproduce the laws of thermodynamics, and since the magnetic system is a convenient example, we open the discussion with a thermodynamic description of systems which respond to a magnetic field. We will use these results in Chaps. 4 and 5.

Before passing on to the discussion of the thermodynamic variables, we recall the relevant concepts from the electromagnetic theory. The fundamental concept here is the dipole moment (electric or magnetic), which has two important characteristics. First, a dipole moment in an external field is acted upon by a torque which tends to align it along the direction of the field as described in Fig. 2.1.1. A simple calculation, which will not be given here, shows that the magnitude of the electric dipole moment, p, of a pair of charges $\pm q$ is proportional to the distance between them, d, i.e. $p = qd$, and that the magnetic dipole moment of a planar loop, μ (not necessarily rectangular!), carrying a current I is proportional to its area, $a : \mu = Ia/c$ (c is the speed of light).

electric/ magnetic dipole moment

Since the dipole prefers to align itself along the direction of the field, it is clear that it has a potential energy that depends on the angle between its direction and the direction of the field or, in other words, on their scalar product:

$$E_{\text{el}} = -\mathbf{p} \cdot \mathbf{E}\,, \tag{2.1.1a}$$

$$E_{\text{mag}} = -\boldsymbol{\mu} \cdot \mathbf{B}\,. \tag{2.1.1b}$$

This means that the potential energy decreases as the angle between the direction of the dipole and the direction of the field decreases.

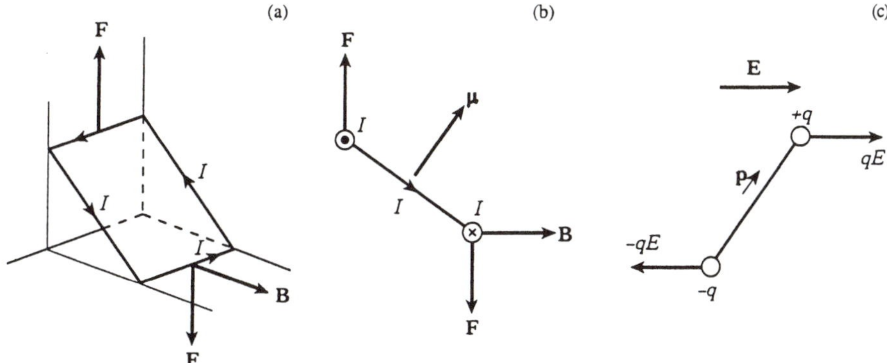

Fig. 2.1.1 The torque acting on a current loop (a) and a side view of the magnetic moment of the loop (b). For comparison the electric dipole is depicted (c).

The second role is that each dipole moment is also a source of an electric or magnetic field. An important fact for us is that the presence of a macroscopic number of dipoles, for example in a bulk of dielectric material, changes considerably the external field that would exist in that region in the absence of the material. The intensity of the field originating from these dipoles is determined by the polarization density, which is the dipole moment per unit volume of the material and is denoted by $\mathbf{P}$ in the electric case and M in the magnetic case. M is also called magnetization density or magnetization per unit volume. The relation between the induced fields and the polarization density is

$$\mathbf{E} = -4\pi \mathbf{P}\,, \tag{2.1.2a}$$

$$\mathbf{B} = 4\pi \mathsf{M}\,. \tag{2.1.2b}$$

■ Notice the difference in sign between the electric and the magnetic equations. It stems from the opposite behavior of the fields "inside" the dipoles.

Consider, for example, a single dipole moment in a medium of many other dipoles: It will feel (via the torque) the external field $\mathbf{H}$ and also the additional field $4\pi\mathsf{M}$, originating from the neighboring dipoles. The sum of these two effects is the magnetic induction $\mathbf{B}$, and the relationship between them is

$$\mathbf{B} = \mathbf{H} + 4\pi \mathsf{M}\,. \tag{2.1.3}$$

A similar relationship can be formulated for the electric case, but we will restrict ourselves to magnetic systems.

To conclude the introduction, we further remark that there is a practical reason for concentrating on the field $\mathbf{H}$, and not the magnetic induction $\mathbf{B}$, which is that $\mathbf{H}$ is the external variable which is under control: We are able to change the currents, thereby affecting what would have happened in the absence of the material. The appearance of the magnetic induction depends on M, the magnetization density, which is the system's response to **H**. Hence, even though **B** is the natural variable in electromagnetic theory, in thermodynamics we prefer to express quantities in terms of **H**.

1.2 The first law in magnetic variables

The thermodynamic description is of a much more general validity than would seem so when dealing with a mechanical set of variables — pressure P and volume V, characteristic of gases and liquids. Analogously, we can describe a system responding to a magnetic field, using the *mechanical-like* variables: the magnetic field **H** and the magnetization (i.e. the total magnetic moment) **M**. The relationship between the magnetization **M** and the magnetization density M is, of course, $\mathbf{M} = \mathsf{M}V$, where V is the volume of the system. The field **H** is an intensive variable, while **M**, the response to **H**, is an extensive variable.

An increase in P leads to a decrease in V, while the increase in **H** increases **M**; however, this is only a difference in sign [see Eq. (2.1.10)]. We can similarly characterize the system by an electric field **E** (intensive) and an electric polarization (extensive), etc.

We saw in the previous chapter that the first law of thermodynamics is written in the form

$$dE = \delta Q - \delta W \,. \tag{2.1.4}$$

dE is the increase in the energy of the system, δQ is the heat transferred to the system and δW is the work done by the system.

When the system expands by dV against a pressure P, $\delta W = PdV$. In order to formulate the first law in magnetic variables, we must first inquire how the work is expressed in these variables, namely what is the work done by the system when the external field changes by $d\mathbf{H}$.

For the sake of simplicity we assume that our system is a long, thin cylindrical bar, with its axis along the magnetic field. The external magnetic field orders the microscopic dipoles in the bar, giving rise to an induced magnetic moment, M. This situation is described in Fig. 2.1.2, where the external magnetic field along the bar is depicted — before the change and after it. Suppose the magnetic system at a given instant is in a microscopic state r in which it has an induced magnetic moment M_r and the field is H. The change of H to $H + dH$ lowers the energy of

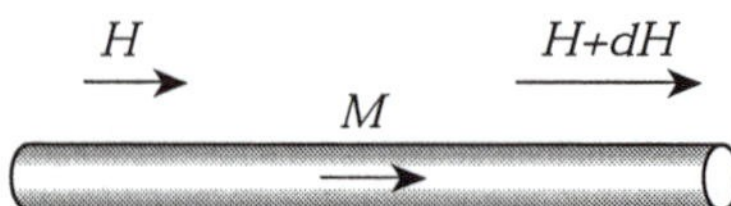

Fig. 2.1.2 A long, thin bar having an induced magnetic moment in an external field.

the state. The change in energy is obtained by multiplying the magnetic moment by the change in the field:

$$dE_r = -M_r dH\,, \tag{2.1.5}$$

which means that in such a change the system will perform work of $M_r dH$.

In thermodynamics we are not interested in a given state of the system, but in the average over long times. Therefore the work performed by the system when H changes will be

$$\delta W = M dH\,, \tag{2.1.6}$$

where M is the average of M_r.

We now write (2.1.4) in the form

$$dE = \delta Q - M dH\,. \tag{2.1.7}$$

And the entropy [Eq. (2.0.10)] is given by

$$dS = \frac{dE}{T} + \frac{M dH}{T}\,. \tag{2.1.8}$$

It is of course possible to define a new energy:

$$E^* = E + MH\,, \tag{2.1.9}$$

which is the energy of the state, together with the energy of the electromagnetic field.

■ This is a standard Legendre transformation which produces a new thermodynamic potential. See e.g. Sec. 0.3.

In terms of E^* we can express the entropy change in the form

$$T dS = dE^* - H dM\,, \tag{2.1.10}$$

which is more reminiscent of the corresponding equation for gases, in that both differentials are of extensive quantities, except for the sign difference:

$$P dV \to -H dM\,. \tag{2.1.11}$$

Exercise 1.1

What is the physical meaning of the different derivatives of S, as inferred from (2.1.8) and (2.1.10), and the relationships between them?

Solution on page 190

Chapter 2

Microscopic States and Averages

2.1 Magnetic states, angular momentum and paramagnetism

Different materials respond differently when a magnetic field is applied to them. The simplest response is that as a result of the application of the field a magnetic moment is induced, which vanishes with the vanishing of the field. If in addition the induced moment is directed along the field, the response is called *paramagnetic.* The magnetic moment (the magnetization) may also be directed opposite to the field. In this case the material is called *diamagnetic.* There is a large diversity of other behaviors, but here we will mention only the *ferromagnet* (*ferro* for iron), a substance which at moderate temperatures retains its magnetization, even after the removal of the magnetic field.

Before going on to the thermodynamic discussion of paramagnetism, we recall that the source of the magnetization that appears in materials is microscopic and each atom or ion possesses an intrinsic magnetic dipole moment. Actually, this intrinsic dipole moment originates from the electrons of the atom or ion, as hinted in Fig. 2.1.1(a). An electron revolving around the nucleus at radius r and velocity v behaves as a circular current loop of area πr^2 and current $ev/2\pi r$. The magnetic moment of the loop is

$$\mu = \frac{1}{2c} evr\,. \tag{2.2.1}$$

Equation (2.2.1) can be written in the form $\mu = e\ell/2mc$ or in its vector version:

$$\boldsymbol{\mu} = -\frac{e}{2mc}\boldsymbol{\ell}\,, \tag{2.2.2}$$

where m is the mass of the electron and $\boldsymbol{\ell}$ is its angular momentum. However, despite the oversimplified assumptions upon which it is based, this connection between the magnetic moment and the mechanical quantities related to the motion of the electron is of general validity. Equation (2.2.2) also holds for noncircular orbits and remains valid even within the framework of quantum mechanics, which is the appropriate theoretical framework for the discussion of electrons in atoms. In quantum theory the angular momentum of the electron (or, in fact, of any other particle) is quantized and can take on values which are integer multiples of the fundamental quantum of angular momentum, $\hbar$.

Planck's constant

$$h = 6.626 \times 10^{-34}\ \mathrm{J \cdot s} = 6.626 \times 10^{-27}\ \mathrm{erg \cdot s}\,,$$

$$\hbar = \frac{h}{2\pi}\,.$$

■ Both h and $\hbar$ are referred to as Planck's constant in the literature. Concerning the quantization of angular momentum, see any text on modern physics.

It is therefore convenient to define a dimensionless vector $\mathbf{L}$ which measures the angular momentum in units of $\hbar$:

$$\mathbf{L} = \frac{\boldsymbol{\ell}}{\hbar}\,. \tag{2.2.3}$$

$\boldsymbol{\ell}$ and $\mathbf{L}$ are both called *angular momentum*; we will leave no room for confusion between them.

Since the angular momentum of the electron is quantized, the magnetic moment related to it is also quantized, and the fundamental quantum of the magnetic moment of the electron is called the Bohr magneton,

Bohr's magneton

$$\mu_{\mathrm{B}} = \frac{e\hbar}{2mc} = 9.273 \times 10^{-21}\ \mathrm{erg/gauss}\,,$$

so that we can write

$$\boldsymbol{\mu} = -\mu_{\mathrm{B}}\mathbf{L}\,. \tag{2.2.4a}$$

But this is not the end of the story, because in addition to the orbital angular momentum $\boldsymbol{\ell}$ the electron also possesses an internal angular momentum or spin $\mathbf{S}$. Like all angular momenta, also the spin assumes only discrete values but, in contrast to the orbital angular momentum, which can only take on integral multiples of $\hbar$, the spin can also take on half-integral values. The spin of the electron can take on the values $\pm\frac{1}{2}\hbar$ only.

spin

■ This fact is conventionally summarized in the literature by the words "the electron has spin $\frac{1}{2}$."

On the basis of Eq. (2.2.2) we could expect the spin itself to create an additional magnetic moment of $\frac{1}{2}\mu_{\mathrm{B}}$, but this is not so: a magnitude twice as large is found experimentally. Hence the relation between the magnetic moment and the spin of the electron is

$$\boldsymbol{\mu} = -\frac{e}{m}\mathbf{s} = -2\mu_{\mathrm{B}}\mathbf{S}\,, \tag{2.2.4b}$$

where again we have defined a dimensionless spin by $\mathbf{S} = \mathbf{s}/\hbar$.

■ The reason for the factor-of-2 discrepancy with the expected result for the electron's magnetic moment is explained by taking into account relativistic effects in the quantum treatment of the electron, as was first done by Dirac at the end of the twenties. Actually, the true value is slightly larger than 2 (by two parts in a thousand). This fact can be explained within the framework of quantum electrodynamics.

The total magnetic moment of a single electron is thus the sum of (2.2.4a) and (2.2.4b). The magnetic moment of the whole atom or ion, which is in fact our main interest, is the sum of the contributions of all its electrons. The electrons in the atom are arranged in shells, so that each electron in every shell has a specific angular momentum. The total angular momentum of electrons in a full shell or subshell is zero. If one electron is added to or removed from a full shell the ion remains with an angular momentum equal in size to the angular momentum of the additional or missing electron. Since a charged particle, possessing angular momentum, has a magnetic moment, we can assign the ion that magnetic moment.

In light of Eq. (2.2.4) the magnetic moment of the ion is proportional to its total (dimensionless) angular momentum:

$$\mathbf{J} = \sum_i (\mathbf{L}_i + \mathbf{S}_i)\,. \tag{2.2.5}$$

Moreover, it is possible to show that in an external magnetic field, the ion increases its energy by an amount which is proportional to $\mathbf{J}$:

$$\Delta E = g\mu_{\mathrm{B}}\mathbf{J}\cdot\mathbf{H}\,, \tag{2.2.6}$$

gyromagnetic ratio

where g is a numerical factor of order 1 which is determined by the structure of the atom. g is called the *gyromagnetic factor* and it actually describes the ratio between the magnetic moment of the ion and its total angular momentum. For a free electron we obtain $g = 2$, as already mentioned.

As an example we consider doubly ionized copper, Cu^{++}. In the external shell, the fourth shell, Cu contains a single electron. The double ionization removes this last electron as well as one from the full, highest subshell of the third shell, 3d. The ion has a single hole in the 3d shell with angular momentum $\frac{5}{2}\hbar$, as explained above.

There are different kinds of paramagnets. Here we discuss a type for which the ions possessing angular momenta are set in fixed positions in a crystal. For example, in ionic crystals of copper salts, manganese or gadolinium, the ions Cu^{++}, Mn^{++}, Gd^{++} give rise to the paramagnetism of the salt. If there are many nonmagnetic atoms in the salt per magnetic ion, then it is reasonable to assume that due to the large separation of the magnetic ions they do not affect one another. This is the case, for example, in copper potassium sulphate: $CuSO_4 \cdot K_2SO_4 \cdot 6H_2O$.

Remark. Equations (2.2.1) and (2.2.2) are valid only in cgs units. In the SI system the factor of c (the speed of light) in the denominator should be dropped. Hence the Bohr magneton is in SI units:

$$\mu_B = \frac{e\hbar}{2m} = 9.273 \times 10^{-24} \text{ J T}^{-1} .$$

All other equations have the same form in both systems.

2.2 Microscopic states, observables

The simplest case that we begin with is a paramagnet in which N magnetic ions have "spin $\frac{1}{2}$." This means that any projection (component) of $\mathbf{S}$ may take on the values $\pm\frac{1}{2}$ only.

■ From here on the total angular momentum of the ion will be referred to as the spin. This is so since we are not interested in the details of such an ion and are considering only its general characteristics, which are its spin and magnetic moment. The spin will serve for the enumeration of states and the magnetic moment for the calculation of its energy in a magnetic field.

If a field $\mathbf{H}$ is applied to a single moment $\boldsymbol{\mu}_i$, the energy is given by

$$\epsilon_i = -\boldsymbol{\mu}_i \cdot \mathbf{H} . \tag{2.2.7}$$

However, in our case, the magnetic moment can have only two projections along the field (μ_B or $-\mu_B$), so that we can write

$$\epsilon_i = -\mu_B H \sigma_i , \tag{2.2.8}$$

where σ_i is a variable that has two possible values: ± 1.

■ σ describes the projection of the spin along the direction of the field. In the case of spin $\frac{1}{2}$, its projection along the field is $\hbar\sigma/2$, where $\sigma = \pm 1$.

Since we are assuming that the magnetic moments are far from one another (which means that there is no interaction between the magnetic ions), we can write the energy of a configuration or of a microscopic state of the system of moments in the form

$$E(\sigma_1, \ldots, \sigma_N) = -\mu_B H \sum_{i=1}^{N} \sigma_i . \tag{2.2.9}$$

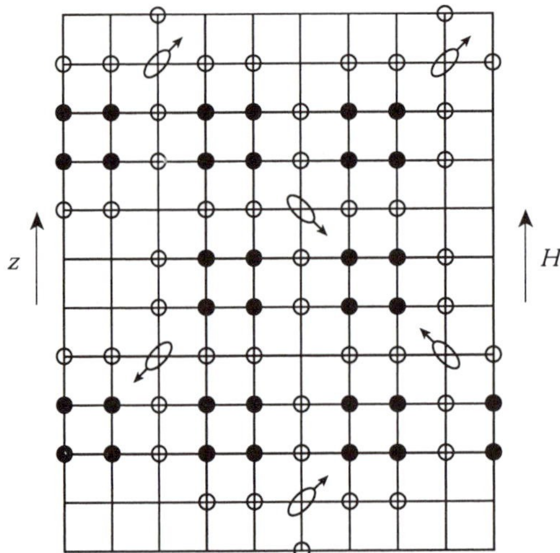

Fig. 2.2.1 A model of a paramagnet in an external field.

By the term *configuration* or *microscopic state* we mean a state in which all the degrees of freedom of the system are specified (see also Sec. 1.5 of Part I).

In the present case the system has N degrees of freedom corresponding to N magnetic ions, which we denote by σ_i. Each degree of freedom can assume one of two possible values, $\sigma_i = \pm 1$ (because we chose ions with spin $\frac{1}{2}$). If each σ_i is assigned a value, a specific configuration, or a specific microscopic state, is determined. For instance, the configuration $\sigma_i = +1$ for every i is the state in which all the magnetic moments point along the field, while $\sigma_i = -1$ for every i is the state in which all the magnetic moments are opposite to the field. Figure 2.2.1 depicts a lattice on which there are "spherical" atoms, black and white, which have zero magnetic moments. There are also "elongated" atoms, with arrowheads representing their magnetic moments. Their serial number is i and the projection of their magnetic moment along the z axis, which is chosen along the field, is $\pm\mu_B$. There is no interaction between them. This is our model system.

This system has 2^N microscopic states. Each of them is characterized by N numbers $(\sigma_1, \ldots, \sigma_N)$, which determine the projection of each spin. In each microscopic state the properties of the system are completely defined. With each microscopic state one can associate observables, $A(\sigma_1, \sigma_2, \ldots, \sigma_N)$, which are numbers associated with the particular state. For instance, the total magnetic moment along the z axis is an observable given by

observables

$$M(\sigma_1, \ldots, \sigma_N) = \mu_B \sum_{i=1}^{N} \sigma_i\,, \tag{2.2.10}$$

as is the total energy given by (2.2.9).

The magnetic moment of a subsystem with N_a of the N magnetic moments is an observable of the full system

$$M_a(\sigma_1, \ldots, \sigma_N) = \mu_B \sum_{i=1}^{N_a} \sigma_i \,, \qquad (2.2.11)$$

despite the fact that it depends only on part of the variables. So is the energy of the subsystem $E_a = -HM_a$. The value of a particular spin σ_i is also an observable, etc.

With a little extra effort it is possible to generalize the present model to the case in which the ions have an angular momentum $J > \frac{1}{2}$, where $\mathbf{J}$ is the sum of the orbital angular momentum and the spin [Eq. (2.2.5)].

In this case the possible values of the projection of the angular momentum along H are

$$J_z = -J, -J+1, -J+2, \ldots, J \,.$$

Thus, there are $2J + 1$ possibile projections. The energy levels of such an ion are given again by multiplying $\mu_B J_z H$ by the gyromagnetic factor. All the expressions that we have written in this section still hold, with the proviso that every degree of freedom will now have $2J+1$ states, instead of just two. Consequently, there will now be $(2J+1)^N$ microscopic states.

2.3 Probabilities and averages

As was mentioned in the Introduction to this part, the role of statistical mechanics is to create a bridge from microscopic states to averages. To this end we want to know the probability for a certain microscopic state to appear along the system's trajectory in configuration space. Such a trajectory is described by giving the N-tuples $(\sigma_1, \ldots, \sigma_N)$ in a certain time interval. This approach is replaced, of necessity, by the ensemble approach, in which the time evolution is ignored, and probabilities are assigned to all allowed states.

The ensemble is defined by the collection of probabilities, or recurrence rates, of every microscopic state. We denote the probability of the state $(\sigma_1, \ldots, \sigma_N)$ by $P(\sigma_1, \ldots, \sigma_N)$. Once P is given, the problem of evaluating averages becomes a mere technicality. Since, if $A(\sigma_1, \ldots, \sigma_N)$ is an observable of the paramagnetic system, in the microscopic state $(\sigma_1, \ldots, \sigma_N)$, the required average is just

$$\langle A \rangle = \sum_{\{\sigma\}} P(\sigma_1, \ldots, \sigma_N) A(\sigma_1, \ldots, \sigma_N) \,, \qquad (2.2.12)$$

where $\sum_{\{\sigma\}}$ denotes the summation over all microscopic states. Thus, the sum is over 2^N states, wherein each spin takes on both possible values. In the case of a spin J, each spin takes on $2J+1$ values, and the sum is over $(2J+1)^N$ microscopic states of the system.

Thus, for example, the average magnetic moment of the system, in an ensemble defined by P, is determined by the average of $A = M = \mu_B(\sigma_1 + \sigma_2 + \cdots + \sigma_N)$, so that

$$\langle M \rangle = \mu_B \sum_{\{\sigma\}} \left(\sum_i \sigma_i \right) P(\sigma_1, \ldots, \sigma_N) . \tag{2.2.13}$$

■ The fog of mathematical symbols can be reduced by writing Eq. (2.2.13) in the simple cases of small N. For $N = 3$ we obtain

$$\langle M \rangle = \mu_B \sum_{\sigma_1, \sigma_2, \sigma_3} (\sigma_1 + \sigma_2 + \sigma_3) P(\sigma_1, \sigma_2, \sigma_3) ,$$

where the summation is performed independently over σ_1, σ_2 and σ_3, each assuming the values ± 1. There are a total of eight possibilities.

Next, we show that $\langle M \rangle$ is proportional to the average magnetic moment of a single spin. To this end we define the probability for a given spin — with $i = 1$ for example — to have the moment $\sigma_1(+1$ or $-1)$:

$$P(\sigma_1) = \sum_{\{\sigma\}}{}' P(\sigma_1, \ldots, \sigma_N) , \tag{2.2.14}$$

where $\sum'$ denotes that we are summing over microscopic states with fixed σ_1, or in other words we are summing over all possible values of $\sigma_2, \ldots, \sigma_N$.

Of course, all the moments are equivalent so there is no special significance to σ_1. The prime can be attributed to any other spin i, and then the sum in Eq. (2.2.14) represents the probability $P(\sigma_i)$ for that spin to have the value $+1$ or -1. Thus we can denote all of them by $P(\sigma)$.

We can therefore rewrite Eq. (2.2.13) in the form

$$\langle M \rangle = \mu_B N \sum_{\sigma = \pm 1} \sigma P(\sigma) = \mu_B N \langle \sigma \rangle , \tag{2.2.15}$$

where $\langle \sigma \rangle$ is the average of the observable $A = \sigma_i$.

Exercise 2.1

Does the result (2.2.15) seem reasonable to you, without a detailed calculation? What would you expect to be, without calculating, the average magnetic moment of the subsystem of the N_a spins [whose moment is given by Eq. (2.2.11)]?

Solution on page 191

The transition from Eq. (2.2.13) to Eq. (2.2.15) was done in the following manner. We change the order of summations in Eq. (2.2.13) and write

$$\langle M \rangle = \mu_{\mathrm{B}} \sum_{i=1}^{N} \sum_{\{\sigma\}} \sigma_i P(\sigma_1, \ldots, \sigma_N). \tag{2.2.16}$$

The summation over the states may be carried out in two stages. First, we hold σ_i constant for a certain i, and carry out the inner summation over all possible σ_j with $j \neq i$. After that we sum over σ_i. The inner sum is

$$\sum_{\{\sigma\}} \sigma_i P(\sigma_1, \ldots, \sigma_N) = \sum_{\sigma_i = \pm 1} \sigma_i \sum_{\{\sigma\}}{}' P(\sigma_1, \ldots, \sigma_N), \tag{2.2.17}$$

where, as before, $\sum'$ denotes a sum in which each spin takes the values ± 1, except for σ_i, whose value is fixed. Using Eq. (2.2.14) we note that the primed sum is $P(\sigma_i)$, so that we may write the right hand side of Eq. (2.2.17) in the form

$$\sum_{\sigma_i = \pm 1} \sigma_i P(\sigma_i) = \langle \sigma_i \rangle. \tag{2.2.18}$$

Now we are left with the outer summation in Eq. (2.2.16), after we substitute Eq. (2.2.18) into Eq. (2.2.16). Since $\langle \sigma_i \rangle$ is independent of i [as $P(\sigma_i)$ are also independent of i], we are summing N equal terms $\langle \sigma \rangle$, and we obtain Eq. (2.2.15).

▪ The independence of i means that we obtain the same result for each of the magnetic moments of the system, the reason for which is of course that they are all identical.

Exercise 2.2

Show that it is possible to write the probability that two spins i and j will have components σ_i and σ_j, respectively, as

$$P(\sigma_i, \sigma_j) = \sum_{\{\sigma\}}{}'' P(\sigma_1, \ldots, \sigma_N), \tag{2.2.19}$$

where σ_i and σ_j are held fixed in the summation, and the summation is carried out over all other $N - 2$ spins.

Solution on page 191

With the help of Eq. (2.2.19) we can write the covariance between two spins, which measures the dependence of one spin (say, $i = 1$) on the state of another (say, $j = 2$). We denote the covariance by $C(1, 2)$ or generally $C(i, j)$. Its definition is

$$C(1, 2) = \langle \sigma_1 \sigma_2 \rangle - \langle \sigma_1 \rangle \langle \sigma_2 \rangle. \tag{2.2.20}$$

Clearly, if the two spins are independent of each other, C vanishes. C is positive if a given value of σ_1 encourages spin 2 to have the same value. C will be negative if σ_1 encourages spin 2 to have the opposite value.

An explicit example for the calculation of $C(1,2)$ will be given in the next chapter.

Chapter 3

Isolated Paramagnet — Microcanonical Ensemble

3.1 Number of states and probabilities

When one is trying to determine the ensemble, to decide on the occurrence rates of the distinct microscopic states of a system, it is vital to take into account the strict constraints. It is clear, for example, that if a gas is in a sealed vessel, the time evolution of the system will not result in the appearance of states in which molecules are found outside the vessel. Similarly, all conservation laws must apply to the states of the ensemble.

The first conservation law which comes to mind is the conservation of energy of an isolated system. Clearly, the trajectory of an isolated system will contain only states whose energy is equal to the initial energy. Taking into account the conservation laws and the external constraints applied to the system, it is natural to assume, in the absence of other information, that all the states are equally probable.

One may ask, of course, how it is possible for the paramagnetic system to pass from one state to another, and not remain in a single microscopic state if the spins do not affect one another. The answer is that there are, necessarily, interactions between the different spins, but that these involve energies that are very small compared to the energy of the moments in the magnetic field. Nevertheless, since we are dealing with many states of the same energy, even a small perturbation can (in the quantum description) transfer the system from state to state, with finite probability. Thus it is reasonable to treat the problem as if the energy were given by (2.2.9) and the system's trajectory, in the space of microscopic states, passes through all of the allowed states.

■ A similar question may be asked for the case of an ideal gas of pointlike molecules, and the answer is, of course, that the collisions of the molecules with one another and with the walls of the container cause the gas to pass from one microscopic state to another.

Moreover, we shall interpret the state of thermal equilibrium as a state to which the system has evolved after a long enough time, so that by now all the microscopic states appear at the same occurrence rate during its time evolution. This collection of states corresponds to a macroscopic state. The ensemble we defined, of constant energy, is called a microcanonical ensemble (there is no point in trying to find a deep meaning to this name).

microcanonical ensemble

In such an ensemble, if the energy is E, the probability for a certain microscopic state is, simply, $1/\Gamma(E)$, where $\Gamma(E)$ is the number of states with energy E. Namely for our paramagnet

$$P(\sigma_1, \ldots, \sigma_N) = \frac{1}{\Gamma(E)}. \tag{2.3.1}$$

Up to this point the considerations are of universal generality. To say more about $\Gamma(E)$ we turn to the simple model of the paramagnet.

The model is so simple that it is possible to answer this question in detail, using combinatorial considerations. The energy of the state depends only on the difference between the number of spins that point along the field, N_+, and those that point in the opposite direction, N_-. Denoting this difference by q, namely

$$q = N_+ - N_-, \tag{2.3.2}$$

we obtain from (2.2.9)

$$E = -\mu_{\text{B}} H q, \tag{2.3.3}$$

which means that constant E is equivalent to constant q.

However, since the sum of N_+ and N_- is also constant, and equal to the total number of spins, constant q means that both N_- and N_+ are constant. But we can immediately write down the number of states with given N_- and N_+, since this is simply the number of ways of dividing N objects into two groups, N_+ in one and N_- in the other. That is,

$$\Gamma(E) = \frac{N!}{N_+! N_-!}. \tag{2.3.4}$$

Exercise 3.1

Show that

$$\Gamma(E) = \frac{N!}{\left(\frac{N}{2} - \frac{E}{2\mu_{\text{B}}H}\right)! \left(\frac{N}{2} + \frac{E}{2\mu_{\text{B}}H}\right)!}. \tag{2.3.5}$$

Solution on page 192

3.2 Calculating averages and correlations

The first calculation we can perform is of the average magnetization. This is an especially simple calculation, since in all the states of equal energy the magnetization is identical. Hence

$$\langle M \rangle = \mu_{\mathrm{B}} q = -\frac{E}{H} \tag{2.3.6}$$

and the average magnetization per degree of freedom (spin) is simply $\mu_{\mathrm{B}} q/N$, or $-E/(NH)$.

Next we calculate $P(\sigma_1)$, which was defined in Eq. (2.2.14) as the probability for spin number 1 to have a projection σ_1 along the field. We start with $\sigma_1 = 1$. That is, we calculate $P(\sigma_1 = 1)$ (as already mentioned and will be verified by the calculation, the result is independent of our choice of spin $i = 1$). We carry out the summation in (2.2.14) with $P(\sigma_1, \ldots, \sigma_N)$ given by $1/\Gamma(E)$ in (2.3.5).

Since the probabilities are all equal, the problem reduces to finding the number of microscopic states with given q, for which $\sigma_1 = +1$.

We have to calculate the number of states of the remaining $N - 1$ spins, with an excess $(N_+ - N_-)$ along $+z$, equal to $q - 1$. In other words, these are the states with N_- spins along $-z$, as in the calculation of $\Gamma(E)$ but with only $N_+ - 1$ along z. This number is

$$\frac{(N-1)!}{(N_+ - 1)!N_-!}\,,$$

so that

$$P(\sigma_1 = +1) = \frac{(N-1)!}{(N_+ - 1)!N_-!} \cdot \frac{1}{\Gamma(E)} = \frac{N_+}{N} = \frac{1}{2} - \frac{E/N}{2\mu_{\mathrm{B}} H}\,. \tag{2.3.7}$$

The answer indicates that we could have calculated $P(\sigma_1 = 1)$ in a simpler way: the probability for spin 1 to have projection $+1$ is equal to the probability for any one of other spins to have projection $+1$. Thus the required probability is also equal to the probability for an arbitrary spin to have projection $+1$, and is of course equal to the ratio of the number of spins with projection $+1$ and their total number: N_+/N.

Exercise 3.2

Show that

$$P(\sigma_1 = -1) = \frac{1}{2} + \frac{E/N}{2\mu_{\mathrm{B}} H}\,. \tag{2.3.8}$$

Solution on page 193

Exercise 3.3

Show that it is possible to deduce $P(\sigma_1)$ from Eq. (2.2.15).

Solution on page 193

In other words, we can say that if the energy is negative, the spins tend to align themselves along the magnetic field. Namely, the probability for a spin to point along the field is larger than $\frac{1}{2}$. If the energy is positive, the spins tend to be opposite to the field. Let us consider a few representative cases:

(a) The maximal energy of the system is $N\mu_{\text{B}}H$: all the spins point opposite to the field. If this is the given energy it is obvious that we will not find any spin that is pointing along the field and so

$$P(\sigma_1 = 1) = 0\ ,\ P(\sigma_1 = -1) = 1\,.$$

The same result will of course be obtained from a direct substitution into (2.3.7) and (2.3.8).

(b) The minimal energy is $-N\mu_{\text{B}}H$: all the spins point along the field —

$$P(\sigma_1 = 1) = 1\ ,\ P(\sigma_1 = -1) = 0\,.$$

(c) $E = 0$: $P(\sigma_1 = 1) = P(\sigma_1 = -1) = \frac{1}{2}$.

The behavior of $P(+1)$ and $P(-1)$ as functions of the energy is depicted in Fig. 2.3.1.

The average moment per spin $\langle\sigma\rangle$ is calculated as follows:

$$\langle\sigma\rangle = \sum_{\sigma} \sigma P(\sigma) = 1 \cdot P(1) + (-1)P(-1) = \frac{N_+}{N} - \frac{N_-}{N} = \frac{q}{N}\,. \qquad (2.3.9)$$

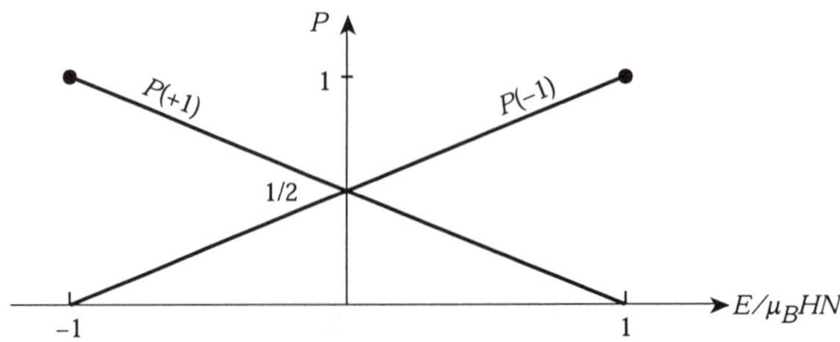

Fig. 2.3.1 A graphical representation of Eqs. (2.3.7) and (2.3.8).

A more interesting question with a somewhat surprising answer, maybe, is the question of the covariance between the spins in the isolated system. $P(\sigma_1, \sigma_2)$ and $C(1,2)$ are defined by (2.2.19) and (2.2.20). An argument similar to the one made concerning $P(\sigma_1)$ will hold here as well, and we can write down

$$\begin{cases} P(+1,+1) = \dfrac{(N_+ - 1)N_+}{N(N-1)}\,, \\ P(+1,-1) = P(-1,+1) = \dfrac{N_+ N_-}{N(N-1)}\,, \\ P(-1,-1) = \dfrac{(N_- - 1)N_-}{N(N-1)}\,. \end{cases} \tag{2.3.10}$$

Note that $P(-1,+1)$ and $P(+1,-1)$ are equal but correspond to different microscopic states. Hence the normalization is

$$P(+1,+1) + P(+1,-1) + P(-1,+1) + P(-1,-1) = 1\,.$$

Exercise 3.4

Produce the detailed argument that leads to (2.3.10).

Solution on page 194

The covariance of the spins number 1 and 2, Eq. (2.2.20), is

$$\begin{aligned} C(1,2) &= \sum_{\sigma_1,\sigma_2} \sigma_1\sigma_2 P(\sigma_1\sigma_2) - \left[\sum_{\sigma_1} \sigma_1 P(\sigma_1)\right]^2 \\ &= \frac{(N_+ - 1)N_+}{(N-1)N} + \frac{(N_- - 1)N_-}{(N-1)N} - 2\frac{N_+N_-}{(N-1)N} - \frac{(N_+ - N_-)^2}{N^2}\,, \end{aligned}$$

in which for the last term we have used Eq. (2.3.9). A bit of algebra leads to

$$C(1,2) = \frac{1}{N-1}\left(\frac{q^2}{N^2} - 1\right) < 0\,. \tag{2.3.11}$$

That is, although there are no interactions between the spins it seems as if they "disturb" one another from aligning in the same direction. This is implied by the fact that $C(1,2) < 0$. However, it is important to note that the correlations are independent of the distance between spin number 1 and spin number 2. This fact points to the origin of the correlations. They appear since, when $N_+ - N_- = q$ is constant, the combinatorics of $N-2$ spins is different from that of $N-1$ spins. In other words, choosing the value of σ_1 to be, for example, $+1$, reduces the number of possibilities for the rest of the spins to have the value $+1$ so that it also reduces the probability of finding $\sigma_2 = +1$. The origin of these correlations, therefore, is not dynamical but combinatorial. In order to learn from the correlations about the dynamics of the system, we must remove from the system the combinatorial effect, which can be done by taking the thermodynamic limit.

The thermodynamic limit is the limit in which N tends to infinity and is the limit that generally interests us in statistical physics. The covariance in Eq. (2.3.11) is of order $1/N$, since q/N is of order 1. So it tends to zero as N tends to infinity, as indeed we would expect to occur. The probabilities (2.3.7) and (2.3.8) etc. depend on the ratio E/N, and do not tend to zero in the thermodynamic limit, since E/N remains finite. This is also the rule for treating all the other extensive quantities: in the thermodynamic limit the extensive quantities are to be evaluated per degree of freedom. For instance, $\langle E\rangle$ and $\langle M\rangle$ are extensive quantities, so that they yield meaningful values for $\langle E\rangle/N$ and $\langle M\rangle/N$ in the thermodynamic limit.

▪ This means that when calculating an extensive quantity we are interested only in the largest term, which is proportional to the number of particles (or to the volume). Since the number of particles is very large, the successive terms are negligible.

Exercise 3.5

Calculate the average square deviation of M from its average $(\Delta M)^2$.

Solution on page 195

Exercise 3.6

What is the probability that spin number 1, 2 and 3 will all be $+1$, in the thermodynamic limit?

Express the result in terms of the number of spins and the energy.

Solution on page 197

3.3 Numerical examples and Stirling's formula

In order to obtain a better sense for the orders of magnitude of the number of states we are dealing with, we shall study $\Gamma(E)$ of Eq. (2.3.5) for different values of N. For $N = 1$, i.e. for a single spin, there are two possible values for E, each of them corresponding to a single state. For the case $N = 2$ there are already three possibilities for E. Two states correspond to $E = \pm 2\mu_{\mathrm{B}}H$, one state to each sign and two more states correspond to $E = 0$, $\Gamma(0) = 2$. For the not-too-simple case $N = 10$ the results are summarized in Table 3.1.

Table 3.1

$E/\mu_B H$	−10	−8	−6	−4	−2	0	2	4	6	8	10
Γ	1	10	45	120	210	252	210	120	45	10	1

Notice the emerging trend that the largest number of states corresponds to $E = 0$. This trend becomes more pronounced as the number of degrees of freedom (spins) increases.

This fact hints that, if our system were not isolated, but could exchange energy with its surroundings, it would have had to compromise between its tendency towards the state with minimal energy and the most probable (microscopic) state with zero energy.

In order to effectively analyze the cases with macroscopic N ($\sim 10^{22}$), we have to find a good method of calculating, at least approximately, factorials of numbers of this order. This is achieved by using Stirling's formula:

$$n! \simeq n^n e^{-n}\sqrt{2\pi n}\,. \tag{2.3.12}$$

■ Stirling's formula gives an approximation that improves as n increases. Already for $n = 10$ substitution into (2.3.12) gives 3.60×10^6 compared to the true value, which is $10! = 3628800$.

Instead of directly calculating $\Gamma(E)$ for large values of N, we define a function S/k as the logarithm of $\Gamma(E)$ in (2.3.5), and calculate it. That is,

$$\frac{1}{k}S \equiv \ln\Gamma \simeq -\left(\frac{N}{2} - \frac{E}{2\mu_B H}\right)\ln\left(\frac{1}{2} - \frac{E/N}{2\mu_B H}\right)$$
$$-\left(\frac{N}{2} + \frac{E}{2\mu_B H}\right)\ln\left(\frac{1}{2} + \frac{E/N}{2\mu_B H}\right). \tag{2.3.13}$$

The first part of (2.3.13) is a definition and is referred to as Boltzmann's formula (it should not come as a great surprise to the reader if we leak at this stage the fact, to be discussed in the next chapter, that S is the entropy). The second part is an approximation, using (2.3.12), in which we keep the first term, which is the largest, in the expansion in powers of N. **entropy**

The term we have kept is proportional to N, since E is proportional to N. The first term that we neglected is independent of N (in the limit $N \to \infty$, the neglected term divided by N tends to 0), so that when N is very large the approximation is justified. We have found, therefore, that S is an extensive quantity.

Exercise 3.7

(a) Use Stirling's formula to deduce Eq. (2.3.13).
(b) Check if (2.3.13) gives values that are close to the ones that appear in Table 3.1.

Solution on page 198

The following fact is of interest: Stirling's approximation is justified for all three factorials in (2.3.5), except for a few states at the edges of the spectrum. The energies of our paramagnets start at $-N\mu_{\mathrm{B}}H$ and end at $+N\mu_{\mathrm{B}}H$. At these energies the number of states is one. Thus S must vanish at the edges of the spectrum. Even though Stirling's approximation does not apply near these points, the leading term in S, (2.3.13), has the property that it vanishes at the edges of the spectrum, as it should.

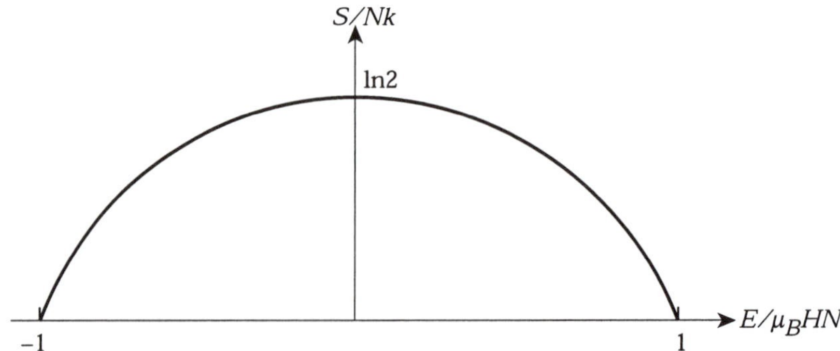

Fig. 2.3.2 The "entropy" per spin for the paramagnet.

The shape of the function S is shown in Fig. 2.3.2. Note that the vertical axis corresponds to S/N, not S. With the help of S we can now calculate the number of states $\Gamma = e^{S/k}$ as a function of E or, better yet, as a function of $E/(N\mu_{\mathrm{B}}H)$, which measures the energy per degree of freedom in units of $\mu_{\mathrm{B}}H$. The shape of $\Gamma(E)$ is described in Fig. 2.3.3 for $N = 10, 100, 1000$. Note the huge increase in the number of states as a function of N and the concentration near $E = 0$.

A study of Fig. 2.3.3 reveals a great similarity to the shape of the Gaussian (normal) distribution. And indeed it is possible to show that when $N \gg 1$ the number of states with energy E depends on E in a Gaussian manner:

$$\Gamma(E) = C \exp\left(\frac{-E^2}{2N\mu_{\mathrm{B}}^2 H^2}\right), \tag{2.3.14}$$

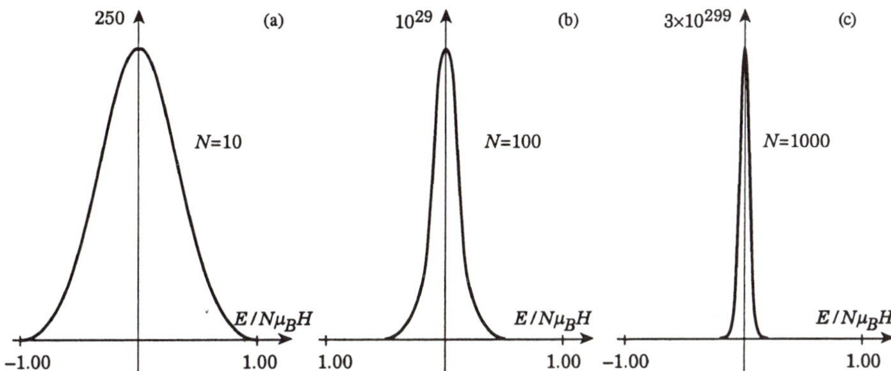

Fig. 2.3.3 Graphs of the number of states as a function of the dimensionless variable $E/(N\mu_B H)$, for (a) $N = 10$, (b) $N = 100$, (c) $N = 1000$. The vertical scale is different in each of the graphs.

where C is a constant that depends only on N (and not on E), and can be calculated, for instance, from the normalization requirement, according to which the integral of $\Gamma(E)$ over all values of E should be 2^N.

Exercise 3.8

Prove (2.3.14), and calculate the normalization constant.

Solution on page 200

Finally, it is worth mentioning that S of Eq. (2.3.13) can be expressed in terms of $P(\pm 1)$, from Eqs. (2.3.7) and (2.3.8), and has the form

$$S = -Nk[P(+1)\ln P(+1) + P(-1)\ln P(-1)]\,. \qquad (2.3.15)$$

Chapter 4

Isolated Paramagnet — Subsystems and Temperature

4.1 Microscopic states and thermodynamic equilibrium

So far only one new concept has been introduced beyond the dynamics of the system, which is the probability within a set of states, or an ensemble. It is time to try and connect this new concept with thermodynamic quantities. The first among these is the temperature. We turn therefore to the identification of the relative temperature of two systems.

Obviously, in order to discuss temperature we need at least two systems, since the temperature is precisely the intensive variable, whose equality characterizes the equilibrium between them when there is no mechanical interaction. We choose, therefore, two paramagnetic systems:

System a with N_a spins and a magnetic field H_a,
System b with N_b spins and a magnetic field H_b.

We isolate both systems from the rest of the universe, but allow them to interact thermally. This means that the total energy of the two systems, E, will be constant but the energies of the two systems E_a and E_b are unconstrained provided $E_a+E_b = E$. No forces will act between the spins, except for the tiny forces we mentioned earlier which drive the system towards thermal equilibrium — namely, to uniform occurrence rates of all microscopic states of the combined isolated system, as explained in Chap. 3.

The energy of a given state of system a is

$$E_a = -\mu_{\mathrm{B}} H_a \sum_{i=1}^{N_a} \sigma_i \qquad (2.4.1)$$

and that of system b is

$$E_b = E - E_a \,. \qquad (2.4.2)$$

If the number of states of system a with energy E_a is $\Gamma(E_a)$, then system b has $\Gamma(E - E_a)$ states, and the total number of states of the composite system, for which system a has energy E_a, is

$$\begin{aligned}\Gamma_T &= \Gamma(E_a, H_a, N_a) \cdot \Gamma(E_b, H_b, N_b) \\ &= \Gamma(E_a, H_a, N_a) \cdot \Gamma(E - E_a, H_b, N_b)\,,\end{aligned} \tag{2.4.3}$$

where we have emphasized the dependence of both factors on the number of spins of the subsystems and on the magnetic field of each of them.

The rest of the argument has the following structure:

When the numbers N_a and N_b are very large, there exists a value of E_a, which we denote by $\bar{E}_a$, for which Γ_T is maximal. Moreover, the maximum is extremely sharp, and the number of states in which E_a differs from $\bar{E}_a$ is, relatively, very small. This will be the equilibrium state, because if the combined system "visits" all the states with total energy E at the same frequency, it will almost always be in a state for which $E_a = \bar{E}_a$. In this case we can identify the intensive quantity that becomes equal in the two subsystems. This quantity will be called the temperature.

In order to proceed, we write (2.4.3) in the form

$$\Gamma_T = \exp\left(\frac{S(E_a, H_a, N_a) + S(E - E_a, H_b, N_b)}{k}\right), \tag{2.4.4}$$

where for each system we have defined separately $S/k = \ln \Gamma$.

The number of states attains its maximum at $\bar{E}_a$, which may be determined by the requirement that the derivative of the exponent with respect to E_a should vanish, or

$$\frac{1}{k}\frac{\partial S(E_a, H_a, N_a)}{\partial E_a}\bigg|_{\bar{E}_a} = \frac{1}{k}\frac{\partial S(E_b, H_b, N_b)}{\partial E_b}\bigg|_{\bar{E}_b} \equiv \beta\,, \tag{2.4.5a}$$

where E_b is not an independent variable but satisfies $E_b = E - E_a$ and $\bar{E}_b = E - \bar{E}_a$. β has the dimensions of energy to the power of -1.

■ Since Γ is a monotonic function of S, if S has a maximum Γ is maximal as well.

In conclusion, we found an intensive quantity which characterizes the maximum of Γ_T and has the same value in the two subsystems. We called this quantity β. Second, if we also find that almost all the states of the composite system satisfy $E_a = \bar{E}_a$, then this will be the system's thermal equilibrium state.

4.2 β and the temperature

Before we proceed to prove the sharpness of the maximum, let us identify the quantity β in our model of a paramagnet. β came about from equilibrium considerations between two systems a and b, but it is possible to define it in general for a single paramagnetic system with a given energy E:

$$\beta = \frac{1}{k}\frac{\partial S(E,H,N)}{\partial E}\,. \tag{2.4.5b}$$

The "entropy" S of an isolated paramagnet is given by Eq. (2.3.13) and by differentiating it we obtain

$$\beta = \frac{1}{2\mu_{\mathrm{B}}H}\ln\left[\left(\frac{1}{2}-\frac{E/N}{2\mu_{\mathrm{B}}H}\right)\Big/\left(\frac{1}{2}+\frac{E/N}{2\mu_{\mathrm{B}}H}\right)\right]\,. \tag{2.4.6}$$

Recalling the expressions (2.3.7) and (2.3.8), for the probabilities for a spin to point up or down, we can write

$$2\mu_{\mathrm{B}}H\beta = \ln\left[\frac{P(\sigma=+1)}{P(\sigma=-1)}\right]\,, \tag{2.4.7}$$

from which we obtain the interesting result

$$\frac{P(+1)}{P(-1)} = e^{2\mu_{\mathrm{B}}H\beta}\,. \tag{2.4.8}$$

Exercise 4.1

Use Eq. (2.4.8) to calculate $P(+1)$ and $P(-1)$. Compare to (2.3.7) and (2.3.8).

Solution on page 201

Namely, if we knew that $\beta = 1/kT$, then (2.4.8) would be the expression for the Boltzmann distribution, as in Part I. That is, the ratio of the probabilities of the two states is $e^{-\Delta E/kT}$, where ΔE is their energy difference.

But at present we cannot conclude that $1/\beta$ is proportional to an absolute temperature, only that it is an increasing function of the relative temperature — since even after demonstrating the sharpness of the maximum, we will only know that β is identical for systems at thermal equilibrium with each other. In order to identify β as defined by Eq. (2.4.5) as an absolute temperature, we have to show that it connects the entropy change with the heat increase, or that it may be identified from the ideal gas law. However, we may note that if S is indeed the entropy, then (2.4.5b) is the connection between the entropy and the absolute temperature.

4.3 Sharpness of the maximum

In order to find the behavior of the number of states of the combined system as a function of E_a, near its maximum, we will use the expression for $\Gamma(E)$ when $N \gg 1$, i.e. Eq. (2.3.14). Inserting it into Eq. (2.4.3) and taking for simplicity $H_a = H_b = H$, we obtain

$$\Gamma_T = C_a C_b \exp\left(-\frac{E_a^2}{2N_a\mu_B^2 H^2}\right) \exp\left[-\frac{(E-E_a)^2}{2N_b\mu_B^2 H^2}\right], \tag{2.4.9}$$

where C_a and C_b are normalization constants that depend on N_a and N_b. In terms of the "entropy" we obtain

$$\frac{1}{k}S_T = \frac{1}{k}(S_a + S_b) = \ln\Gamma_T = \ln(C_a C_b) - \frac{1}{2(\mu_B H)^2}\left[\frac{E_a^2}{N_a} + \frac{(E-E_a)^2}{N_b}\right]. \tag{2.4.10}$$

Next we find the maximum of the entropy. Since (2.4.10) is a quadratic function of E_a, there is no need to differentiate with respect to E_a; it is enough to complete the expression in brackets to a square:

$$\begin{aligned}\frac{E_a^2}{N_a} + \frac{(E-E_a)^2}{N_b} &= \frac{(N_a+N_b)E_a^2}{N_a N_b} - \frac{2EE_a}{N_b} + \frac{E^2}{N_b}\\ &= \frac{N_a+N_b}{N_a N_b}\left(E_a - \frac{N_a E}{N_a+N_b}\right)^2 + \frac{E^2}{N_a+N_b},\end{aligned} \tag{2.4.11}$$

so that

$$\frac{1}{k}S_T = \ln(C_a C_b) - \frac{1}{2(\mu_B H)^2}\left[\frac{N_a+N_b}{N_a N_b}\left(E_a - \frac{N_a E}{N_a+N_b}\right)^2 + \frac{E^2}{N_a+N_b}\right]. \tag{2.4.12}$$

This is a quadratic function of E_a whose maximum is attained at

$$\bar{E}_a = \frac{N_a E}{N_a + N_b}. \tag{2.4.13}$$

As will immediately be shown, Γ_T has a sharp maximum at this energy, so that this will also be the average of E_a. We see, therefore, that the energy is distributed between the two systems in direct proportionality to their size. In order to show that the overwhelming majority of states are concentrated around $E_a = \bar{E}_a$, we return to Γ_T, and rewrite it using (2.4.12) in the form

$$\Gamma_T(E_a) = C_T \exp\left[-\frac{N_a+N_b}{2(\mu_B H)^2 N_a N_b}(E_a - \bar{E}_a)^2\right], \tag{2.4.14}$$

where the constant C_T includes all the factors which do not depend on E_a.

Equation (2.4.14) describes a Gaussian distribution whose width (standard deviation), ΔE_a, is $\mu_B H\sqrt{N_a N_b/(N_a+N_b)}$. If both systems are macroscopic, N_a and N_b will both be of order N ($= N_a + N_b$), and the width of the distribution will be of order $\mu_B H N^{1/2}$. Apparently, this is not the result we anticipated, since instead of decreasing with N the width of the distribution increases with N. However, we must remember that what is relevant is the *relative width* of the distribution, which is $\Delta E_a/\bar{E}_a$. since $\bar{E}_a$ is proportional to N, the relative width tends to zero as $N^{-1/2}$:

$$\frac{\Delta E_a}{\bar{E}_a} \sim N^{-1/2} \xrightarrow[N\to\infty]{} 0\,, \tag{2.4.15}$$

which means that in the thermodynamic limit the states of the combined system for which $E_a = \bar{E}_a$ exhaust the states of the isolated system. Hence, $E_a = \bar{E}_a$ describes thermal equilibrium between the two systems.

It is possible to understand the state of thermodynamic equilibrium between two systems from a slightly different perspective on Eq. (2.4.14). Since in the thermodynamic limit E_a diverges, it is more natural to consider the distribution of the quantity E_a/N_a, which is the energy per degree of freedom. To this end we need only rewrite Eq. (2.4.14) in the slightly different form

$$\Gamma_T = C_T \exp\left[-\left(\frac{E_a}{N_a} - \frac{\bar{E}_a}{N_a}\right)^2 \cdot \frac{N_a/N_b}{2(\mu_B H)^2} N\right]\,. \tag{2.4.16}$$

And it is immediately clear that the width of this (Gaussian) distribution of E_a/N_a behaves as $N^{-1/2}$ and vanishes in the thermodynamic limit.

Exercise 4.2

Verify that (2.4.14) actually implies (2.4.15).

Solution on page 202

Returning to (2.4.10), we find that at thermal equilibrium ($E_a = \bar{E}_a$)

$$S_T = \bar{S}_T = S(\bar{E}_a) + S(\bar{E}_b)\,. \tag{2.4.17}$$

This means that the "entropy" is extensive: It grows linearly with the size of the system.

The distribution of the number of states of the composite system, as a function of E_a/N_a, is depicted in Fig. 2.4.1. In the thermodynamic limit the width of the peak tends to zero. An instructive exercise is to sketch Γ_T from (2.4.3) with Γ from (2.3.5) for small values of N_a and N_b (even 12), and a value of E around $6\mu_B H$.

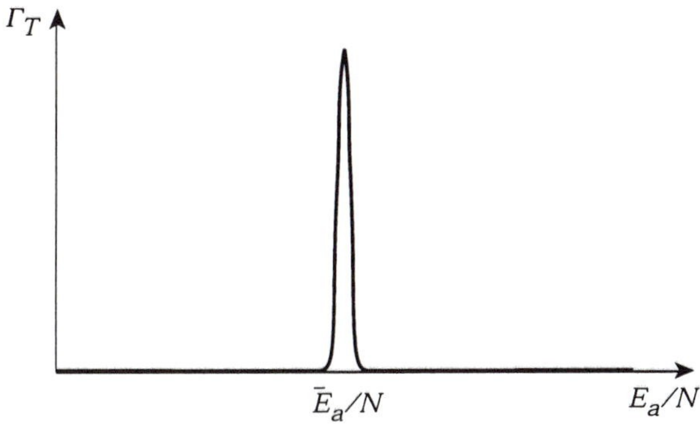

Fig. 2.4.1 The number of states of the combined system (2.4.16).

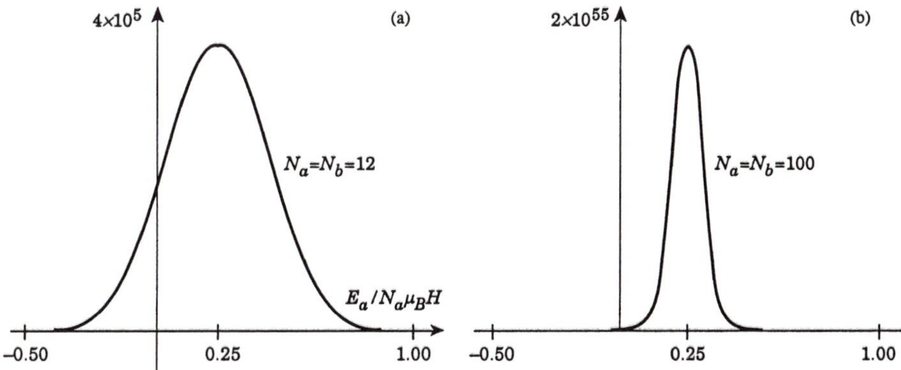

Fig. 2.4.2 The number of states in Eq. (2.4.3) as a function of E_a with $E = N_a\mu_B H/2$ for (a) $N_a = N_b = 12$, (b) $N_a = N_b = 100$. Notice that we have not made use of the approximation (2.4.14).

The calculation of Γ can be performed by a direct calculation of the factorials in (2.3.5) or using the approximation (2.3.13). The result is depicted in Fig. 2.4.2(a). To demonstrate the behavior of the width of the distribution as a function of N, we give in Fig. 2.4.2(b) a sketch of Γ_T for $N_a = N_b = 100$ and $E = 50\ \mu_B H$.

4.4 Identification of temperature and entropy

After finding that the equality of β is the condition for thermodynamic equilibrium, we return to discuss an isolated paramagnet and to identify β. To this end we note that S, as given in Eq. (2.3.13), is a function of M ($M = -E/H$) alone. Namely,

$$S = -\frac{1}{2}k[(N + M/\mu_B)\ln(1 + M/N\mu_B) \\ +(N - M/\mu_B)\ln(1 - M/N\mu_B) - 2N\ln 2]\,. \qquad (2.4.18)$$

Hence changing the magnetization of the system will lead to a change in the "entropy" given by

$$dS = \frac{\partial S}{\partial M} dM = -\frac{k}{2\mu_B} \ln\left(\frac{1 + M/N\mu_B}{1 - M/N\mu_B}\right) dM\,. \qquad (2.4.19)$$

Using (2.4.6) with $M = -E/H$ we obtain

$$dS = -k\beta H dM\,. \qquad (2.4.20)$$

We compare this expression with the expression for the entropy, as it appears in thermodynamics. From Eq. (2.1.10) we can obtain the entropy change due to a change of M or E^*.

But the paramagnet is a special system for which $E = -MH$, so that $E^* = 0$, and the entropy cannot depend on E^*. Thus we obtain from (2.1.10)

$$dS = -\frac{H}{T} dM\,. \qquad (2.4.21)$$

This means that if we identify β as

$$k\beta = \frac{1}{T}\,, \qquad (2.4.22)$$

we obtain a full correspondence between the statistical mechanics of a paramagnet and its thermodynamics. T, which appears on the right hand side of (2.4.22), is of course the absolute temperature. And if k is the Boltzmann constant, then the unit of temperature is identical in the two scales. The identification (2.4.22) also states that S, which thus far has been referred to as "entropy," is really the entropy, since if we insert (2.4.22) into (2.4.20) we immediately obtain (2.4.21).

Now, after having identified the meaning of β and S, we can use the equations we derived not only for a paramagnet at a given energy (isolated) but also for other situations, such as when the system is at a given temperature. This we do in the following exercise.

Exercise 4.3

(a) Show that for a paramagnet at a given temperature, the average value of the spin's projection along the field is

$$\langle \sigma \rangle = \tanh(\beta \mu_B H)\,. \qquad (2.4.23)$$

What is the average magnetic moment M of the whole system? Sketch the average moment as a function of H.

(b) Repeat the calculation of the magnetization $M(\beta, H)$ using Eq. (2.4.6).

(c) Repeat the calculation of the magnetization using Eq. (2.4.21) and the explicit expression for $S(M)$ (2.4.18).

(d) Show that for a paramagnet at a given temperature, the average energy is given by

$$E = -N\mu_B H \tanh(\beta\mu_B H) . \tag{2.4.24}$$

Solution on page 203

4.5 Negative temperature

It is worth noting that the definition of the inverse temperature $1/kT$ via the energy derivative of the entropy, Eq. (2.4.5), gives the paramagnet a range of energies with negative temperatures (see Fig. 2.4.3). This is the range $E > 0$, where the entropy decreases with increasing energy. At $E = 0$, where the slope of S as a function of E vanishes, the temperature is infinite. Systems with negative temperatures are warmer than the systems with positive temperature: if two systems are brought together, one with negative temperature and the other with positive temperature, heat will flow from the first to the second. This phenomenon is typical of systems whose energy spectrum is bounded from above. Here we will not enter into more detail. We mention only that in realistic systems there are also kinetic energies, which we did not account for, thus having an energy spectrum that is not bounded from above. In fact, it is impossible for a state of real thermodynamic equilibrium to exist at negative temperatures.

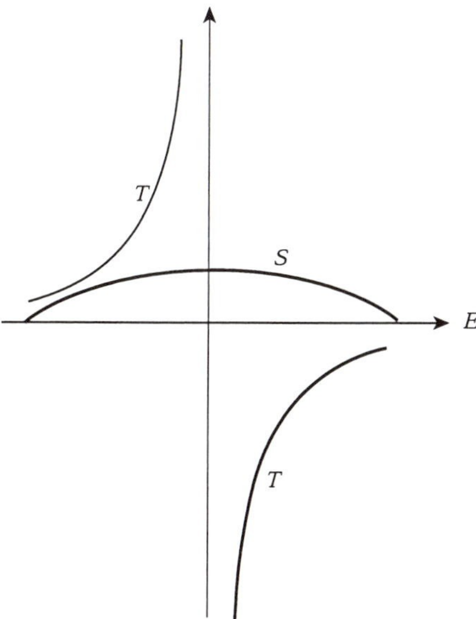

Fig. 2.4.3 The entropy and temperature of a paramagnet vs energy. Since $1/T = \partial S/\partial E$, $T > 0$ for increasing S, "$T = \infty$" for maximal S, and $T < 0$ for decreasing S.

4.6 Summary

The discussion of the paramagnet suggests that in the framework of the microcanonical ensemble, in which each microscopic state of the isolated system has the same probability, it is possible to define entropy and temperature.

The entropy is given by the Boltzmann formula:

$$S(E) = k \ln \Gamma(E)\,, \tag{2.4.25}$$

where $\Gamma(E)$ is the number of states with energy E.

The absolute temperature is determined by studying the equilibrium of subsystems and is defined as

$$\beta = \frac{1}{kT} = \frac{1}{k}\frac{\partial S}{\partial E}\,, \tag{2.4.26}$$

where on the right hand side the external parameters are constant.

Given β, the probability for a single spin to have projection σ along the magnetic field is, according to (2.4.8) and Exercise 4.1,

$$P(\sigma) = Ce^{\beta\mu_B H\sigma} = Ce^{-\beta\epsilon(\sigma)}\,, \tag{2.4.27}$$

where $\epsilon(\sigma)$ is the energy per spin with projection σ and C is the normalization constant of the probability.

Chapter 5

Paramagnet at a Given Temperature

5.1 The canonical ensemble

A system is kept at a fixed temperature when it is in thermal equilibrium with a much larger system, referred to as a heat bath. Clearly, the energy of the system under discussion is not constant, since the system is in thermal contact with the heat bath, so that it may exchange energy with it. We thus have to determine probabilities for states with different energies. In the Introduction (to Part II) we proposed that in such a case, the recurrence of systems in the ensemble, describing the smaller system, will be proportional to the Boltzmann factor $e^{-E/kT}$, where E is the energy of the small system.

The Boltzmannian proposal is not arbitrary. First, the kinetic theory hinted in this direction; second, studying Eq. (2.4.8) or (2.4.27) in the preceding chapter we find that in the isolated paramagnetic system, the single spin satisfies the Boltzmann distribution in a state of equilibrium.

We pick from the isolated system a subsystem a of spins whose number n is very small compared to the total number of spins N ($n \ll N$ but still $n \gg 1$; for example, $N = 10^{23}$, $n = 10^{20}$) — so that we may consider the rest as a heat bath. The probability for these spins to have values $\sigma_1, \ldots, \sigma_n$ will be the product of the probabilities for the single spins. Since as long as $n \gg 1$ there are no correlations, as we have learned in Sec. 3.2 (the artificial correlations, which enter owing to the constraint of a constant energy, vanish when n is very large). Hence

$$P(\sigma_1, \ldots, \sigma_n) = C \exp\left(\beta \mu_{\mathrm{B}} H \sum_{i=1}^{n} \sigma_i\right) = C e^{-\beta(\epsilon_1 + \ldots + \epsilon_n)} . \tag{2.5.1a}$$

But

$$E = \sum_{i=1}^{n} \epsilon_i \tag{2.5.1b}$$

is the energy of a microscopic state with n spins — the microscopic state, given by $(\sigma_1, \ldots, \sigma_n)$.

We have seen therefore that the probability that *in thermodynamic equilibrium* our n-spin system will be found in a specific state $(\sigma_1, \ldots, \sigma_n)$ is indeed proportional to $e^{-E/kT}$, where E is the total energy in this state. Notice that there may exist many states with the same value of E and hence also the same probability. As we will see later on, we will have to take all of them into account.

Exercise 5.1

Show that the constant C in (2.5.1a) is given by

$$C^{-1} = \sum_{\sigma_1=\pm 1} \cdots \sum_{\sigma_n=\pm 1} e^{-\beta E(\sigma_1,\ldots,\sigma_n)} . \tag{2.5.2}$$

Solution on page 204

---

canonical ensemble

In other words, the Boltzmann distribution corresponding to the *canonical ensemble*, describing the system at a given temperature, is derived from the first ensemble we discussed — the microcanonical ensemble, describing an isolated system. It is true that here we have only shown this for the special case of a paramagnet, but the assertion holds quite generally.

The derivation of the canonical distribution, from the microcanonical one, in the general case is not more complicated: The number of states of the isolated system in which the subsystem a is in a specific microscopic state with energy E_a, is given by the second exponential in Eq. (2.4.4) (considered as a product) with $H_a = H_b$. The first one counts the number of states of the subsystem with energy E_a. Since $E_a \ll E$, the second term in the exponent can be expanded to first order in E_a, giving

$$\Gamma_T = \exp\left(\frac{S(E,H,N)}{k}\right) \exp(-\beta E_a) = Ce^{-\beta E_a} , \tag{2.5.3}$$

where use has been made of (2.4.5).

The fact that the two ensembles are equivalent is of double importance. First, the temperature is a much more natural variable than the energy. Second, calculations with constant energy are several times more complicated than those at constant temperature. It implies that the results we obtain for the thermodynamic functions from each of the ensembles will be identical. The difference is that each of the ensembles will give the state functions in terms of different controllable quantities, so that we will have to "translate" from one language to the other. For instance, in the case of the paramagnet, the microcanonical ensemble is described by E, N, H as controlled variables, while the canonical ensemble is described by T, N, H. We re-emphasize the differences and the similarities between the two types of ensembles from a microscopic standpoint:

In a microcanonical ensemble all states have the same energy. All of these states are assigned the same probability. In a canonical ensemble the system may exchange energy with its surroundings, but it is in thermal equilibrium with a heat bath, so that it has a well-defined average energy. The magnitude of the heat exchanged with the heat bath is small compared to the average energy.

In the canonical ensemble the system is assigned a temperature, and every state can appear in it. However, the probability of a state is proportional to $e^{-E/kT}$, if the energy of the state is E.

The canonical ensemble is related to the microcanonical ensemble in that a large system can be subdivided into a relatively small system and a heat bath. The system as a whole is isolated and is described by a microcanonical ensemble; however, the subsystem does not have a constant energy. The distribution of subsystems must be derived from the assumptions of equilibrium and the equality of the probabilities of all states of the large isolated systems (subsystem + heat bath) having the same energy. And, indeed, the result is the Boltzmann result as formulated above.

5.2 The partition function and thermodynamic quantities

We pass on, therefore, to the calculation of the properties of the paramagnet, whose temperature is $1/k\beta = T$. To this end we first define an all-important concept, which is the partition function:

partition function

$$Z = \sum_{\substack{\text{all microscopic}\\ \text{states}}} e^{-\beta E\,(\text{microscopic state})} . \qquad (2.5.4)$$

▪ The common notation for the partition function, Z, originates from its German name *Zustandsumme*, meaning "sum over states."

Actually, we have already seen the partition function as the normalization factor $1/C$ in Eqs. (2.5.2) and (2.5.3). This function, known in probability theory as the "generating function," is very useful in the calculation of averages. The idea is that we can replace the computation of the average of many observables by the computation of derivatives of Z with respect to the appropriate controlled variables.

For instance, the average energy at temperature T is an ordinary average [Eq. (2.2.12)] with probabilities

$$P = Z^{-1} e^{-\beta E} . \qquad (2.5.5)$$

Namely,

$$\langle E \rangle = Z^{-1} \sum_{\{\sigma\}} E(\sigma_1, \ldots, \sigma_n) e^{-\beta E(\sigma_1, \ldots, \sigma_n)} , \qquad (2.5.6)$$

where the summation is over all possible states $(\sigma_1, \ldots, \sigma_n)$.

A brief study of Eq. (2.5.6) reveals that the sum appearing in it can be written as $-\partial Z/\partial\beta$. Thus

$$\langle E\rangle = -\frac{1}{Z}\frac{\partial Z}{\partial\beta} = -\frac{\partial \ln Z}{\partial\beta}\,. \tag{2.5.7}$$

■ Note the similarity to the method of calculating averages which was presented in Sec. 1.6.

In a similar manner it is possible to write the average magnetization in the form

$$\langle M\rangle = \frac{1}{\beta}\frac{\partial \ln Z}{\partial H}\,. \tag{2.5.8}$$

Exercise 5.2

Prove Eq. (2.5.8).

Solution on page 204

Exercise 5.3

Show that if Z is a function of βH only, then we can immediately deduce that $\langle E\rangle = -H\langle M\rangle$ from Eqs. (2.5.7) and (2.5.8).

Solution on page 205

For the paramagnetic system it is possible to directly evaluate Z, since in this case each term of the sum in (2.5.4) is a product, making it possible to write Z as

$$Z(\beta,H) = \sum_{\sigma_1,\sigma_2,\ldots,\sigma_n} e^{-\beta\epsilon_1(\sigma_1)}\cdot e^{-\beta\epsilon_2(\sigma_2)}\cdot\ldots\cdot e^{-\beta\epsilon_n(\sigma_n)}\,. \tag{2.5.9}$$

Now, it is possible to sum over each of the variables $(\sigma_1,\ldots,\sigma_n)$ independently, and to write the sum of products as a product of sums:

$$Z(\beta,H) = \left(\sum_{\sigma_1=\pm1} e^{-\beta\epsilon_1}\right)\cdots\left(\sum_{\sigma_n=\pm1} e^{-\beta\epsilon_n}\right)\,. \tag{2.5.10}$$

It is easy to see that every term of the product we wrote appears in (2.5.9), and every term of (2.5.9) appears in (2.5.10).

■ For example, for $n = 2$ Eq. (2.5.9) takes the form

$$\begin{aligned} Z &= e^{-\beta\epsilon_1(-1)}e^{-\beta\epsilon_2(-1)} + e^{-\beta\epsilon_1(-1)}e^{-\beta\epsilon_2(+1)} \\ &\quad + e^{-\beta\epsilon_1(+1)}e^{-\beta\epsilon_2(-1)} + e^{-\beta\epsilon_1(+1)}e^{-\beta\epsilon_2(+1)}\,, \end{aligned}$$

while [Eq. (2.5.10)]

$$Z = \left(e^{-\beta\epsilon_1(-1)} + e^{-\beta\epsilon_1(+1)}\right)\left(e^{-\beta\epsilon_2(-1)} + e^{-\beta\epsilon_2(+1)}\right)\,,$$

which is exactly equivalent.

All the terms in the product (2.5.10) are identical, and equal to

$$z(\beta, H) = e^{-\beta\mu_{\mathrm{B}}H} + e^{\beta\mu_{\mathrm{B}}H} = 2\cosh(\beta\mu_{\mathrm{B}}H)\,. \tag{2.5.11}$$

Hence

$$Z(\beta, H) = [z(\beta, H)]^n\,. \tag{2.5.12}$$

■ It is possible to think of z as the partition function of a single spin.

Substituting (2.5.12) into (2.5.7), we obtain

$$\langle E\rangle = -n\frac{\partial \ln z}{\partial\beta} = -n\mu_{\mathrm{B}}H\tanh(\beta\mu_{\mathrm{B}}H)\,. \tag{2.5.13}$$

This expression is identical to (2.4.24).

Obviously the magnetization, obtained from substituting (2.5.12) in (2.5.8), is

$$\langle M\rangle = n\mu_{\mathrm{B}}\tanh(\beta\mu_{\mathrm{B}}H)\,, \tag{2.5.14}$$

as was already obtained in Exercise 4.3.

■ It is of course possible to use the relationship

$$\langle E\rangle = -H\langle M\rangle\,.$$

As a final note in this section the reader is warned, if he has not noticed it himself already, of the notational ambiguity, characteristic of statistical mechanics, expressed by the fact that no special notation is used to distinguish between the thermodynamic random variables and their averages. For example, one meets Eqs. (2.5.7) and (2.5.8) repeatedly in the literature (in this text as well!) in the form

$$E = -\frac{\partial \ln Z}{\partial\beta}\,,\quad M = \frac{1}{\beta}\frac{\partial \ln Z}{\partial H}\,,$$

and so on. The reason for this is of course that due to the huge number of degrees of freedom, in a thermodynamic system, each observable takes its average value with negligible relative deviations. Moreover, there is no room for such a distinction within thermodynamics since the concept of probability does not enter its framework at all. This fact requires the reader to be alert and to notice always, especially in the calculation of averages and probabilities, which are the quantities that have already been averaged and which have not. The reader will be warned where necessary.

5.3 Susceptibility and specific heat of a paramagnet

The behavior of the average energy and of the magnetization, as a function of the external variables H and β, is determined by the behavior of $\langle\sigma\rangle$. In order to study it, it is convenient to define the dimensionless variable, $x = \beta\mu_B H$. Now, we can draw $\langle\sigma\rangle$ as a function of x, as depicted in Fig. 2.5.1.

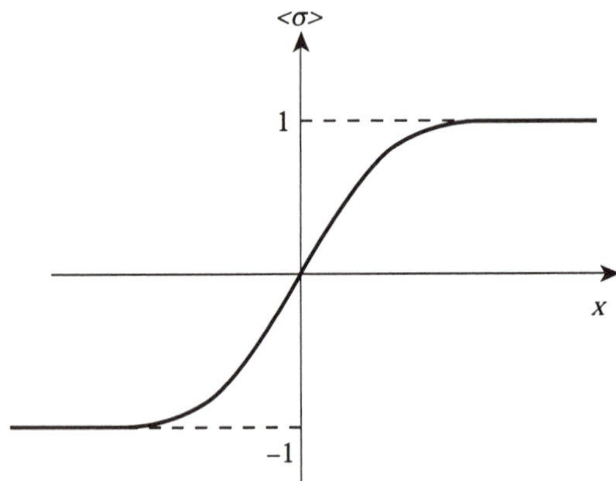

Fig. 2.5.1 $\langle\sigma\rangle$ as a function of $x = \beta\mu_B H$.

This graph may be interpreted in two ways:

(a) As the description of the average projection of the spin along the direction of the field as a function of H, at a constant temperature.

We see that at zero field ($x = 0$) there is no preferred direction, and the average projection is zero. It starts to grow linearly with H (see below) and finally, at a very large field, the average of the projection attains the full value of the spin, $\langle\sigma\rangle \to 1$, namely the polarization saturates. The region $x < 0$ describes the case in which the direction of H is reversed, so that at saturation the direction of the magnetization will also reverse, $\langle\sigma\rangle \to -1$.

(b) As the description of the behavior of the average spin projection at a decreasing temperature, when the external field is held constant. At very high temperatures,

$$x = \beta\mu_B H \to 0 ,$$

there is no difference between the probability for the spin to have a projection along the field and the probability for its projection opposite to the field. Hence $\langle\sigma\rangle$ tends to 0.

When H is constant and $T \to 0$, the thermal energy finds it hard to flip spins, and again the average spin projection tends to saturate the spin, along the field. More and more spins "freeze" along the direction of the field.

Low and high temperatures are defined relative to the characteristic temperature $\Theta = \mu_B H/k$.

■ At very high temperatures the energy of a single moment, $\mu_B H$, is negligible compared to the thermal energy kT. At very low temperatures the reverse is true.

As mentioned above, at low fields the magnetization grows linearly with the field. This property is the main characteristic of the paramagnet. The coefficient of H is called the magnetic susceptibility and it measures the size of the magnetic response of the system to changes in the external magnetic field. If we write **susceptibility**

$$\langle M \rangle \simeq n\chi H\,, \tag{2.5.15}$$

then χ is the magnetic susceptibility per spin. Its value is obtained from (2.5.14) with the help of the fact that $\tanh x \simeq x$ for small x.

The result is

$$\chi = \frac{\mu_B^2}{kT}\,. \tag{2.5.16}$$

The existence of an inverse relationship between the susceptibility and the temperature was found experimentally by P. Curie in 1895, and has since been referred to as the Curie law.

From (2.5.16) we learn that as the temperature decreases the system magnetizes more easily — the slope at the origin, in Fig. 2.5.1, grows. As to the correspondence of this result with experiments, we will return to discuss this point in Chap. 7.

Finally, we calculate the specific heat per degree of freedom at a constant field. Since at a constant field $\delta Q = \delta E$ [see Eq. (2.1.7)], clearly

specific heat

$$c_H = \frac{1}{n}\left(\frac{\partial E}{\partial T}\right)_H\,. \tag{2.5.17}$$

Exercise 5.4

Prove that

$$c_H = \frac{\mu_B^2 H^2}{kT^2\cosh^2(\beta\mu_B H)}\,. \tag{2.5.18}$$

Solution on page 205

In the specific heat the characteristic temperature of the paramagnet appears in the most dramatic manner. c_H is drawn as a function of $1/x = kT/\mu_B H$ in Fig. 2.5.2. The specific heat attains its maximum at

$$kT \simeq 0.8\,\mu_B H\,. \tag{2.5.19}$$

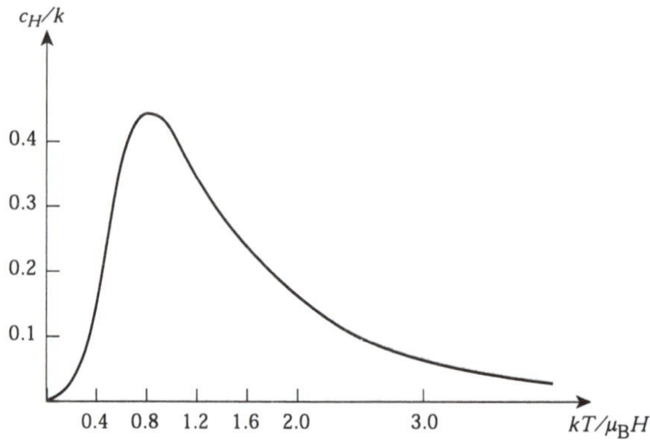

Fig. 2.5.2 The specific heat (2.5.18).

The general appearance of the graph could have been guessed in advance. At low temperatures ($kT \ll \mu_B H$), changing the temperature does not change the energy, since each spin has thermal energy of order kT but requires energy (which it lacks) of order $\mu_B H$ to reverse its direction. Stated differently, the system is saturated and the energy is minimal. At high temperatures ($kT \gg \mu_B H$), disorder is total and cannot be increased by raising the temperature. Thus $c_H \to 0$ when $T \to 0$ and when $T \to \infty$. Since there are in this problem only two characteristic energies, $\mu_B H$ and kT, clearly everything that happens depends on their ratio, $y \equiv kT/\mu_B H$; otherwise, changing the units would affect the behavior of the system. The two limits that we studied are $y \to 0$ and $y \to \infty$. Thus, somewhere between them we expect to attain a maximum, since $c_H(y)$ is nonnegative, and must therefore grow near $y = 0$ and decrease towards $y \to \infty$. The location of the maximum would be around $y = 1$, as both the thermal energy and the magnetic energy are significant in this region.

Exercise 5.5

Sketch the magnetization per spin at a constant field as a function of temperature.

Solution on page 206

Exercise 5.6

(a) Calculate the entropy of the paramagnet as a function of T and H.
(b) Sketch the entropy per spin at a constant field as a function of temperature.
(c) Sketch the entropy per spin at a constant temperature as a function of the field (sketch in one drawing two graphs whose temperatures have a ratio of 5).

Solution on page 206

5.4 Paramagnet with $J > 1/2$

The results we have obtained in this chapter can easily be generalized to the case in which the spin is not $\frac{1}{2}$. Such a spin would have more than two states. If the magnitude of the spin is J, then there are $2J+1$ possible values for the projection of the spin along the direction of the field, taking the values from $-J$ to $+J$ in unit steps.

Since the ion acquires in a magnetic field an additional energy given by Eq. (2.2.6), the energy of ion number i in a magnetic field $\mathbf{H}$ will be given by a generalization of (2.2.8):

$$\epsilon_i = -\frac{1}{2} g\mu_{\mathrm{B}} H \sigma_i \,, \tag{2.5.20}$$

where

$$\sigma_i = -2J, -2(J-1), \ldots, 2J \,. \tag{2.5.21}$$

The number of microscopic states of a system of N spins is not 2^N any more but $(2J+1)^N$. The properties of a paramagnet with general J will be revealed by solving the following exercise.

Exercise 5.7

(a) Calculate the partition function (2.5.4) for a paramagnet with general J.
(b) Calculate the average magnetization per spin, and sketch it as a function of H.
(c) Calculate the susceptibility and compare it with that we calculated for $J = \frac{1}{2}$.
(d) Calculate the specific heat c_H, and compare it with the result for $J = \frac{1}{2}$.

Solution on page 207

Chapter 6

Order, Disorder and Entropy

It is commonly said that the magnitude of the entropy measures the disorder in a system, and that the tendency of the entropy to increase is the same as the tendency for increasing disorder. These concepts and relationships take on a more definite and quantitative character within information theory, which is related in this way to thermodynamics. Here we exemplify only a few of these concepts in connection with the simple paramagnetic model, without giving detailed definitions and without entering into a formal discussion.

If the system is ordered, one glance at the system is enough to determine its state. There are not many states to confuse us. Thus, for example, if a small collection of cubes is ordered into a large cube, it is easy to discern the state of the system. A very small number of parameters will describe it, there are very few states similar to the ordered one. Conversely, if the cubes are scattered, the number of states which create a similar impression is huge, and the system appears to be disordered.

The same applies to our spins. Here each state of the system corresponds to a certain ordering of the spins, some of which are along the field while others are in the opposite direction. If all are pointing along the field ($E = -N\mu_B H$), or in the opposite direction ($E = +N\mu_B H$), we will naturally say that the system is in an ordered state. Notice that there is only one state (corresponding to a single ordering) with each of the energies we mentioned. Thus for these energies

$$\Gamma(E = -N\mu_B H) = \Gamma(E = N\mu_B H) = 1 \qquad (2.6.1)$$

and the entropy for these energy values vanishes [see (2.3.13) or (2.4.18)].

If we know that the system has energy $-(N-2)\mu_B H$, then obviously one spin is pointing in a direction opposite to the field. If this fact is all that is known about the system (this is the information we obtained in our glance), the system can be in one of N different states, in each of which a different spin is the exception. Namely,

$$\Gamma[E = -(N-2)\mu_B H] = N \tag{2.6.2a}$$

and the corresponding entropy is

$$S[E = -(N-2)\mu_B H] = k \ln N\,. \tag{2.6.2b}$$

All these states are equally probable, and the fact that the same amount of information, namely the knowledge of E, is much less informative about the state of the system, indicates the increase in disorder.

We recall a few expressions for the entropy, for example (2.3.13):

$$\begin{aligned}\frac{S}{k} = &-\left(\frac{N}{2} - \frac{E}{2\mu_B H}\right) \ln\left(\frac{1}{2} - \frac{E/N}{2\mu_B H}\right) \\ &-\left(\frac{N}{2} + \frac{E}{2\mu_B H}\right) \ln\left(\frac{1}{2} + \frac{E/N}{2\mu_B H}\right)\,.\end{aligned} \tag{2.6.3}$$

This is the expression which was depicted in Fig. 2.3.2. It has a maximum at $E = 0$, where

$$S(E=0) = kN \ln 2\,. \tag{2.6.4}$$

Since the number of states is maximal, we cannot say anything about the system if only the energy $E = 0$ is given. Disorder dominates. As we have already seen in the solutions of Exercises 3.7 and 3.8, the number of states with $E = 0$ is not given simply by $e^{S/k} = 2^N$ but by

$$\Gamma(E=0) = \frac{2^N}{\sqrt{2\pi N}}\,. \tag{2.6.5}$$

A different expression for the entropy, which we have seen earlier, is (2.3.15), is given in terms of *single* spin probabilities:

$$S = -Nk[P(+1)\ln P(+1) + P(-1)\ln P(-1)]\,. \tag{2.6.6}$$

Exercise 6.1

What assumptions will transfer us from (2.6.6) to (2.6.4)?

Solution on page 211

We will try to view this expression as a particular case of a situation in which there are m states, which we will denote by α, $\alpha = 1, \ldots, m$, whose probabilities are P_α, and S is given by

$$S = -k \sum_{\alpha=1}^{m} P_\alpha \ln P_\alpha\,. \tag{2.6.7}$$

■ Note that $S = -k\langle \ln P \rangle$.

Along with (2.6.7) there exists of course the relationship

$$\sum_{\alpha=1}^{m} P_\alpha = 1 . \tag{2.6.8}$$

Without entering into detailed proofs, which are a mathematical exercise of interest in itself, we note that:

(a) S has a minimum if the system is in a state — say, $\alpha = 1$ — and hence $P_1 = 1$ and $P_\alpha = 0$ for all $\alpha \neq 1$. In this case we infer from (2.6.7) that $S = 0$. We can again identify this probability distribution with order, since among the m states only one is possible, with the information at hand.

(b) S has a maximum if $P_\alpha = 1/m$ for all α, and then

$$S = k \ln m . \tag{2.6.9}$$

In this condition we are unable to distinguish between the different possibilities: disorder dominates.

For example, a paramagnet of N spins can be in one of $m = 2^N$ possible states. If we denote them according to ascending order in energy, then $\alpha = 1$ will be the unique state for which $E = -N\mu_B H$ and $\alpha = 2^N$ will be the unique state for which $E = +N\mu_B H$. Thus, for example, if it is known that $E = -N\mu_B H$, then clearly $P_1 = 1$, and all other P_α vanish. If it is known that $E = 0$, we find from (2.6.5) that the number of states is $2^N/\sqrt{2\pi N}$, and the probability for each state is $\sqrt{2\pi N}/2^N$. Hence a substitution in (2.6.7) or (2.6.9) will give again, in the limit $N \to \infty$, the already familiar result $S(E = 0) = kN \ln 2$, which describes maximal disorder.

Finally, note that Eq. (2.6.7) implies that for the canonical ensemble

$$S = k(\ln Z + \beta\langle E \rangle) . \tag{2.6.10}$$

Exercise 6.2

Prove Eq. (2.6.10).

Solution on page 211

Chapter 7

Comparison with Experiment

The magnetization of several paramagnetic ionic salts has been measured, from very small fields up to saturation. Figure 2.7.1 describes the experimental results for the ions of chromium, iron and gadolinium which are all triply ionized, composed in salts, whose other components are non-magnetic. The correspondence between theory and experiment is very impressive. This includes the fact that measurements performed at different temperatures, for the same salt, when plotted as a function of H/T, fall on the same curve.

Let us consider some characteristics of the results. The first is the saturation value of the magnetization which differs from ion to ion. The reason for this is, of course, the dependence of the magnetization upon the spin, which was obtained in Exercise 5.7(b). At saturation ($H/T \to \infty$) we find that

$$\langle M \rangle / N \to g\mu_B J \,, \tag{2.7.1}$$

so that there is a linear relationship between the magnetization at saturation and the spin. Thus, had we not known the value of J by other means, we could have determined it from these experiments.

The second characteristic of the results is the different slopes of the graphs at the origin, i.e. for $H \to 0$. The relevant feature is the dependence of the susceptibility — described by the slope of the graph at the origin — on the spin, a dependence which was studied in Exercise 5.7(c). It reads

$$\chi = (g\mu_B)^2 \frac{J(J+1)}{3kT} \,. \tag{2.7.2}$$

Actually, we could have used these two equations to find in addition to J the value of g, which is equal in all three ions to 2. In parenthesis we note that the reason for this is that their atomic structure is such that only the spin, and not the orbital angular momentum, contributes to the total angular momentum J [see Eq. (2.2.5)].

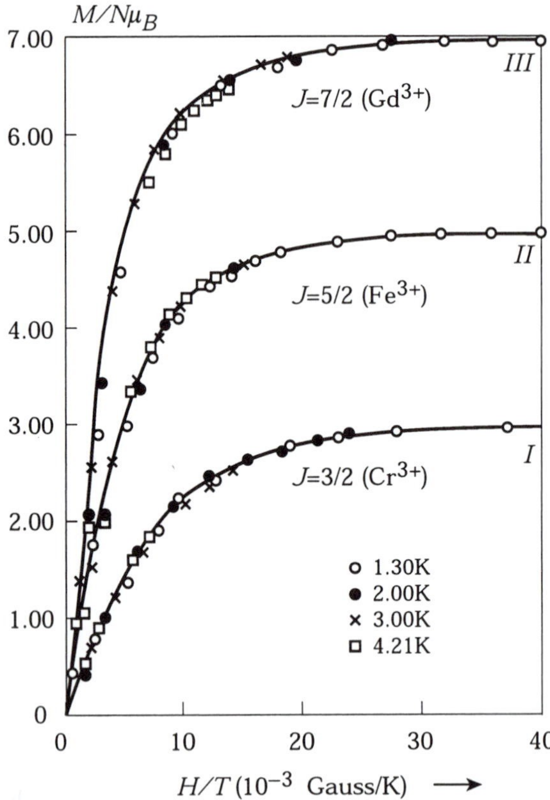

Fig. 2.7.1 Graphs of the average magnetization per ion in units of μ_B as a function of H/T. The measurements were made at different temperatures, as noted in the figure. In all the cases $g = 2$. Taken from W. E. Henry, *Phys. Rev.* **88**, 561 (1952).

The specific heat with the interesting structure (2.5.18) is very hard to measure experimentally, because it requires separating the magnetic effect from the numerous extra contributions originating from the lattice vibrations, the nonmagnetic electrons, and the like, which obscure the picture.

Summary

We have treated the special case of a paramagnet having N degrees of freedom (spins of ions), which can take two values (the projections along the direction of the magnetic field), or a larger, but finite, number of values.

We have used it to exemplify the microcanonical ensemble, representing an isolated system — a system with strictly constant energy. In such an ensemble, all states have equal probability: equal to the inverse of the total number of states with the same energy.

In order to define the temperature of a system, we must think of it as part of a larger isolated system. The isolated system is then composed of two systems, both of which are macroscopic; one of them is much larger than the other. The two subsystems — of which the larger one is a heat bath — exchange energy until they reach equilibrium.

Equilibrium is attained when the total energy is distributed between them in such a way that the number of states is maximized.

At the maximum, the energy derivative of the entropy in the system and in the heat bath are equal. This common value is found to be $\beta = 1/kT$, from the relation between the quantity of heat and the entropy.

The states of the small system are not restricted by conservation of energy. Instead, at a given temperature, the probabilities of its states are proportional to the Boltzmann factor $e^{-E/kT}$. This is the canonical ensemble.

All this has a nice experimental verification shown in Fig. 2.7.1.

Self-assessment exercises

Exercise 1

Solution on page 213

The number of states of the system composed of subsystems a and b in Eq. (2.4.3), Γ_T, has a maximum at $E_a = \bar{E}_a$. Calculate $\bar{E}_a$ and $\bar{E}_b$ for the special case in which $H_a = H_b$.

Exercise 2

Solution on page 213

(a) Prove that the specific heat at constant field (2.5.18) may also be calculated from

$$c_H = \frac{T}{n}\left(\frac{\partial S}{\partial T}\right)_H .$$

(b) Describe the similarities and differences between the paramagnet and the ideal gas.

Exercise 3

Solution on page 214

It is possible to describe the behavior of certain materials in nature by spins with magnetic moment m which can point in three possible directions, all in one plane, as depicted in the figure below.

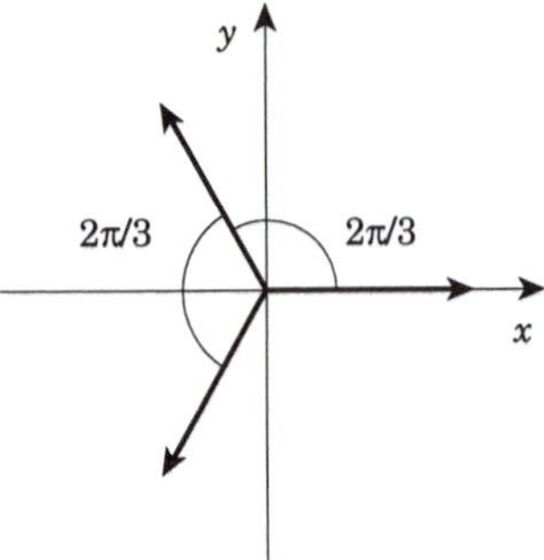

(a) Calculate the energy of a single spin, in each of the above states, in the presence of an external magnetic field $\mathbf{H} = H\hat{\mathbf{x}}$.

(b) A paramagnet whose constituents are spins of the type described above is at temperature T. There is no interaction between the spins. Find its partition function when an external magnetic field is applied along x.

(c) Find the average magnetization per spin in the paramagnet described in (b).

Calculate the value of the magnetization using two methods:

(1) As a canonical weighted average of the moment of a single spin.

(2) By direct calculation from the partition function.

(d) Calculate the magnetic susceptibility per spin of the paramagnet. Does the Curie law apply (namely, $\chi \propto 1/T$)?

(e) Calculate the average energy (per spin) of this paramagnet.

(f) Calculate its specific heat.

Exercise 4 *Solution on page 217*

Repeat the solution of Exercise 3 for a material for which the magnetic moment m has *four* possible directions as in the figure below. The field is still in the x direction.

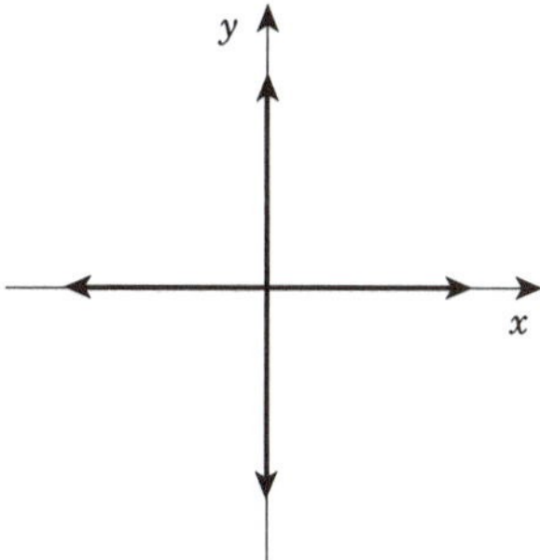

Exercise 5 *Solution on page 219*

Calculate the probability for a paramagnet of ions with $J = \frac{1}{2}$ at temperature T and magnetic field H to have energy E.

Exercise 6 *Solution on page 219*

The probability for the drunk in the random walk of Part I to be found at a distance R from his initial location, after N steps, has the same mathematical structure as the probability for a paramagnet of N spins to have magnetization M.

(a) Calculate the probability for a drunk who is walking in one dimension and having a step of length L to be found at a distance x from his initial location after N steps.

(b) Show that when $N \gg 1$ but $\frac{x}{L} \ll N$, the probability which you calculated in (a) goes over to a Gaussian distribution. Use Stirling's formula.

(c) Calculate the width (standard deviation) of the distribution, Δx, and its relative width, $\Delta x/NL$.

Solutions to exercises in the text

Solution 0.1 *Exercise on page 126*

(a) For an isothermal process we may use Eq. (2.0.5) to compute δQ. In a process with initial volume V_i, final volume V_f and (constant) temperature T we obtain

$$Q_T = \int_T \delta Q = \int_{V_i}^{V_f} Nk\frac{T}{V}dV = NkT \ln\left(\frac{V_f}{V_i}\right), \qquad \text{(i)}$$

where the subscript on Q denotes the variable kept fixed.

(b) In an isochoric (constant V) process we calculate analogously

$$Q_V = \int_V \delta Q = \int_{T_i}^{T_f} \frac{f}{2}NkdT = \frac{f}{2}Nk(T_f - T_i). \qquad \text{(ii)}$$

(c) In an isobaric (constant P) process none of the terms in Eq. (2.0.5) vanishes, but the equation of state (2.0.3) provides a relation between the differentials: $PdV = NkdT$. So we obtain

$$Q_P = \int_P \delta Q = \int_{T_i}^{T_f} \left(1+\frac{f}{2}\right) NkdT = \left(1+\frac{f}{2}\right) Nk(T_f - T_i). \qquad \text{(iii)}$$

If instead the volume is the integration variable, one obtains the equivalent result

$$Q_P = \int_P \delta Q = \int_{V_i}^{V_f} \left(1+\frac{f}{2}\right) PdV = \left(1+\frac{f}{2}\right) P(V_f - V_i). \qquad \text{(iv)}$$

Solution 0.2 *Exercise on page 126*

(a) Combining the results of the previous solution one finds that

$$Q = \oint \delta Q = Nk\left[T_2 - T_1 - T_1 \ln\left(\frac{V_2}{V_1}\right)\right], \qquad \text{(i)}$$

where V_2 is the volume at the end of the isobaric process. Since there is a linear relation between volume and temperature in an isobaric process, we may write the result (i) in terms of temperatures only:

$$Q = Nk\left[T_2 - T_1 - T_1 \ln\left(\frac{T_2}{T_1}\right)\right]. \qquad \text{(ii)}$$

We write Q in Eq. (ii) in the form

$$Q = NkT_1\left[\frac{T_2}{T_1} - 1 - \ln\left(\frac{T_2}{T_1}\right)\right]. \qquad \text{(iii)}$$

This is an increasing function of T_2/T_1 for $T_2/T_1 \geq 1$ (which one verifies by noting that its derivative is always positive in this interval) with a minimum $Q = 0$ at $T_2/T_1 = 1$. Hence Q is positive in this cycle.

(b) If the cycle is reversed Q changes sign and becomes negative.

(c) If δQ were an exact differential we would have found $Q = 0$, since the integration over a closed path would then have vanished.

Solution 0.3 *Exercise on page 126*

(a) In the isobaric process we have

$$W_P = \int_P \delta W = \int_{V_1}^{V_2} P dV = P(V_2 - V_1). \qquad \text{(i)}$$

In the isochoric process $W_V = 0$ and in the isothermal step

$$W_T = \int_T \delta W = \int_{V_2}^{V_1} \frac{NkT}{V} dV = -NkT_1 \ln\left(\frac{V_2}{V_1}\right). \qquad \text{(ii)}$$

Adding the three contributions one finds that

$$W = \oint \delta W = P(V_2 - V_1) - NkT_1 \ln\left(\frac{V_2}{V_1}\right). \qquad \text{(iii)}$$

(b) Using the equation of state, Eq. (2.0.3), one notes that this is the same result for Q of Eq. (i) of the previous solution. The reason is that the energy is a function of state and so its integral over any process depends only on the initial and final states:

$$\int dE = E_f - E_i\,.$$

Thus, its contribution over a full cycle vanishes and $W = Q$.

(c) In an adiabatic processes $\delta Q = 0$ and hence

$$\delta W = -dE = -\frac{f}{2} NkdT ,$$

which is integrated to obtain

$$W_{\text{adiabatic}} = -\frac{f}{2} Nk(T_f - T_i) . \tag{iv}$$

(d) Substituting $\delta Q = 0$ in Eq. (2.0.5) we find that

$$VT^{f/2} = \text{const} . \tag{v}$$

Solution 0.4 *Exercise on page 127*

We express δQ in terms of the variables T and P. Using the equation of state (2.0.3) we transform the infinitesimal work to the form

$$PdV = NkdT - VdP .$$

Then, using the expression for E, Eq. (2.0.4), we use Eq. (2.0.2) to write

$$\delta Q = dE + PdV = \frac{f}{2} NkdT + (NkdT - VdP) = Nk\left(1 + \frac{f}{2}\right) dT - VdP ,$$

leading to Eq. (2.0.9).

Solution 0.5 *Exercise on page 127*

For an adiabatic process we have from Eq. (v) of Solution 0.3

$$VT^{f/2} = \text{const} , \tag{i}$$

or

$$TV^{2/f} = \text{const} . \tag{ii}$$

Expressing T in terms of P and V by the equation of state gives, for a fixed number of particles,

$$PV^{\gamma} = \text{const} , \tag{iii}$$

with $\gamma = 1 + \frac{2}{f}$. See also Solution 1.8b in Part I.

Solution 0.6 *Exercise on page 128*

(a) We write Eq. (2.0.5) replacing $\delta Q = TdS$ to obtain

$$dS = Nk\left(\frac{f}{2}\frac{dT}{T} + \frac{dV}{V}\right). \tag{i}$$

The right hand side of Eq. (i) is now an exact differential [unlike the right hand side of Eq. (2.0.5)] and it can be integrated to give

$$S = Nk\ln(bVT^{f/2}), \tag{ii}$$

where b is a constant. This constant may be taken out of the logarithmic function to become an additive constant. As written above, S is not an extensive function because of the V inside the logarithm. It is made extensive by writing $b = c/N$, which (at constant N) adds but a constant to S.

(b) We can use Eq. (ii) at both end points, where the volumes are V_i and V_f (and the temperatures cancel), to obtain

$$\Delta S = Nk\ln\left(\frac{V_f}{V_i}\right). \tag{iii}$$

Note that one could obtain the same result by taking Q_T/T from Eq. (i) in Solution 0.1.

(c) Using Eq. (ii) at both end points, where the temperatures are T_i and T_f, one obtains

$$\Delta S = \frac{f}{2}Nk\ln\left(\frac{T_f}{T_i}\right). \tag{iv}$$

(d) If the entropy remains constant, $dS = 0$ and also $\delta Q = 0$. The process is therefore adiabatic. For an ideal gas Eq. (ii) holds and hence during the process the product $VT^{f/2}$ remains constant [see also Exercise 0.3(d)] or

$$TV^{\gamma-1} = \text{const}. \tag{v}$$

Solution 0.7 *Exercise on page 129*

Using Eq. (2.0.17) the enthalpy of an ideal gas is found to be

$$\mathsf{H} = \left(1 + \frac{f}{2}\right)NkT,$$

which gives, via (2.0.19), the correct value for C_P, Eq. (2.0.9).

Solution 0.8 *Exercise on page 130*

We use Eqs. (2.0.21). The first of these equations gives the identification of the temperature:

$$\frac{1}{T} = \left(\frac{\partial S}{\partial E}\right)_V = \frac{3}{2}\frac{Nk}{E}. \tag{i}$$

The second gives

$$\frac{P}{T} = \left(\frac{\partial S}{\partial V}\right)_E = \frac{Nk}{V}. \tag{ii}$$

These are just the relations between energy and temperature for a monoatomic ideal gas and its equation of state. Equation (2.0.22) is the same as (2.0.12) written in terms of the energy and with $f = 3$.

Solution 0.9 *Exercise on page 130*

It is possible to use Eq. (2.0.23) as the fundamental relation and calculate the chemical potential as

$$\mu = \left(\frac{\partial E}{\partial N}\right)_{S,V}. \tag{i}$$

However, this way requires the explicit form of the function $E(S, V, N)$, which can be obtained by inversion of (2.0.22). Instead we use Eq. (2.0.23) in the form

$$dS = \frac{1}{T}(dE + PdV - \mu dN) \tag{ii}$$

and calculate

$$\mu = -T\left(\frac{\partial S}{\partial N}\right)_{E,V} = -kT\left[\ln\left(\frac{aVE^{3/2}}{N^{5/2}}\right) - \frac{5}{2}\right]. \tag{iii}$$

Expressing the energy in terms of N and T and absorbing the 5/2 in the constant a, one obtains (2.0.24).

Solution 0.10 *Exercise on page 131*

In order to calculate $F(T, V, N)$ one needs E and S in terms of T, V and N. These are found in Eqs. (2.0.4) and (2.0.12). We substitute into the definition (2.0.25) to obtain the required expression for the free energy,

$$F = -NkT\left[\ln\left(\frac{cVT^{3/2}}{N}\right) - \frac{3}{2}\right]. \tag{i}$$

Now we calculate the chemical potential as

$$\mu = \left(\frac{\partial F}{\partial N}\right)_{T,V} = -kT\left[\ln\left(\frac{cVT^{3/2}}{N}\right) - \frac{5}{2}\right]. \tag{ii}$$

Comparison with Eq. (2.0.24) gives $b = ce^{-5/2}$, and thus

$$F = -NkT\left[\ln\left(\frac{bVT^{3/2}}{N}\right) + 1\right]. \tag{iii}$$

The entropy of the ideal gas, which was used to find the free energy explicitly, is rederived by applying (2.0.27):

$$S = -\left(\frac{\partial F}{\partial T}\right)_{V,N}.$$

Solution 0.11 *Exercise on page 132*

(a) Using the free energy, Eq. (2.0.27), one obtains

$$S = \frac{4a}{3}VT^3.$$

(b) Similarly,

$$P = \frac{a}{3}T^4.$$

This is the form of the equation of state of a photon gas. Note that the pressure is independent of the volume.

(c) To calculate the energy we use Eq. (2.0.25) to write

$$E = F + TS = aVT^4.$$

(d) Since the free energy is independent of N, the chemical potential vanishes. This means that photons may be freely emitted or absorbed by the walls of the container.

Solution 0.12 *Exercise on page 132*

The equation of state of a photon gas (1.1.7) is

$$PV = \frac{1}{3}E. \tag{i}$$

In an adiabatic compression, all of the work goes to changing the internal energy, i.e.

$$PdV = -dE\,. \tag{ii}$$

From Eq. (i) we obtain

$$dE = 3PdV + 3VdP\,. \tag{iii}$$

Adding up Eqs. (ii) and (iii),

$$4PdV + 3VdP = 0$$

$$\Downarrow$$

$$\frac{4}{3}\frac{dV}{V} + \frac{dP}{P} = 0\,. \tag{iv}$$

The solution of Eq. (iv) is

$$PV^{4/3} = \text{const}\,. \tag{v}$$

Equation (v) is the required adiabatic equation.

Solution 0.13 *Exercise on page 132*

(a) The first of the Maxwell relations is satisfied because

$$\left(\frac{\partial P}{\partial N}\right)_{T,V} = \frac{kT}{V}\,, \qquad \left(\frac{\partial \mu}{\partial V}\right)_{T,N} = -\frac{kT}{V}\,.$$

In order to verify the other two, we use Eq. (iii) in Solution 0.10 to write the entropy in the form

$$S = -\left(\frac{\partial F}{\partial T}\right)_{V,N} = Nk\left[\ln\left(\frac{bVT^{3/2}}{N}\right) + \frac{5}{2}\right], \tag{i}$$

as well as the expressions for $P(T,V,N)$ [Eq. (ii) of Solution 0.8] and $\mu(T,V,N)$ [Eq. (ii) of Solution 0.10].

(b) For the photon gas nothing depends on N. Therefore the first two Maxwell relations are empty, i.e. $0 = 0$. Using the results of (a) and (b) in Solution 0.11, the third relation is verified.

Solution 0.14 *Exercise on page 133*

(a) Deriving from Eq. (2.0.31) the mixed second derivatives, one finds that

$$\left(\frac{\partial P}{\partial \mu}\right)_{T,V} = \left(\frac{\partial N}{\partial V}\right)_{T,\mu},$$

$$\left(\frac{\partial S}{\partial \mu}\right)_{T,V} = \left(\frac{\partial N}{\partial T}\right)_{V,\mu},$$

$$\left(\frac{\partial S}{\partial V}\right)_{T,\mu} = \left(\frac{\partial P}{\partial T}\right)_{V,\mu}.$$

(b) To calculate $\Omega(T, V, \mu)$ one uses the definition (2.0.29) with the explicit relation for $N(\mu)$ obtained from Eq. (2.0.24). The result is

$$\Omega = -N(\mu)kT = -kbVT^{5/2}e^{\mu/kT}.$$

Solution 1.1 *Exercise on page 137*

We compare the entropy changes, as given in Eqs. (2.1.8) and (2.1.10):

$$dS = \frac{1}{T}dE + \frac{M}{T}dH, \tag{i}$$

$$dS = \frac{1}{T}dE^* - \frac{H}{T}dM, \tag{ii}$$

with the differentials of S, which is

$$dS = \left(\frac{\partial S}{\partial E}\right)_H dE + \left(\frac{\partial S}{\partial H}\right)_E dH, \tag{i)$'$}$$

$$dS = \left(\frac{\partial S}{\partial E^*}\right)_M dE^* + \left(\frac{\partial S}{\partial H}\right)_{E^*} dM. \tag{ii)$'$}$$

We obtain from Eqs. (i) and (i)′ (E and H are independent variables)

$$\left.\begin{aligned}\left(\frac{\partial S}{\partial E}\right)_H &= \frac{1}{T},\\ \left(\frac{\partial S}{\partial H}\right)_E &= \frac{M}{T}.\end{aligned}\right. \tag{iii}$$

And from Eqs. (ii) and (ii)′ (E^* and M are independent variables) we obtain

$$\left(\frac{\partial S}{\partial E^*}\right)_M = \frac{1}{T}, \qquad \left(\frac{\partial S}{\partial M}\right)_{E^*} = -\frac{H}{T}. \tag{iv}$$

The second order mixed derivatives of (iii) and (iv) give the Maxwell relations:

$$\frac{\partial^2 S}{\partial E \partial H} = \left[\frac{\partial(1/T)}{\partial H}\right]_E = \left[\frac{\partial(M/T)}{\partial E}\right]_H, \qquad \frac{\partial^2 S}{\partial E^* \partial M} = \left[\frac{\partial(1/T)}{\partial M}\right]_{E^*} = -\left[\frac{\partial(H/T)}{\partial E^*}\right]_M. \tag{v}$$

Solution 2.1 *Exercise on page 144*

Because the average of a sum is the sum of averages,

$$\langle M \rangle = \left\langle \sum_{i=1}^{N} \mu_B \sigma_i \right\rangle = \mu_B \sum_i \langle \sigma_i \rangle, \tag{i}$$

and $\langle \sigma_i \rangle$ does not depend on the spin's index i, we obtain

$$\langle M \rangle = \mu_B N \langle \sigma \rangle. \tag{ii}$$

The result (ii) is (2.2.15). This means that (2.2.15) is a reasonable result, since it is derivable from simple general considerations.

If there are N_a spins in the subsystem, the same argument will give

$$\langle M_a \rangle = \mu_B N_a \langle \sigma \rangle.$$

Solution 2.2 *Exercise on page 145*

In order to find the probability for spin i to have a given value for its projection along the z direction, σ_i, and for spin j to have the value σ_j, we have to evaluate the ratio of the number of configurations in which spins i and j assume these specified values, and the total number of configurations, which is of course 2^N.

The number of configurations in which spin i has the specified projection σ_i and spin j the projection σ_j, is the sum of the number of all configurations, in which these two spins have these specific values, whereas all others can have every possible value. We start from the fact that the number of configurations with all spin projections specified, $Q(\sigma_1,\ldots,\sigma_N) = 2^N P(\sigma_1,\ldots,\sigma_N)$. To obtain the total number of states with specified σ_i and σ_j, $Q(\sigma_i,\sigma_j)$, we perform the sum:

$$Q(\sigma_i,\sigma_j) = \sum_{\{\sigma\}}'' Q(\sigma_1,\ldots,\sigma_i,\ldots,\sigma_j,\ldots,\sigma_N)$$

$$= 2^N \sum_{\{\sigma\}}'' P(\sigma_1,\ldots,\sigma_N)\,, \tag{i}$$

in which the double prime indicates that while $N-2$ spins different from i and j assume, each, its two values ± 1, the projections of spins i and j are kept fixed.

The probability, $P(\sigma_i,\sigma_j)$, for spin i to have component σ_i and spin j component σ_j, is therefore

$$P(\sigma_i,\sigma_j) = \frac{1}{2^N} Q(\sigma_i,\sigma_j) \tag{ii}$$

and substituting Eq. (i) for Q we obtain Eq. (2.2.19).

Solution 3.1 *Exercise on page 148*

The number of states with energy E is given by Eq. (2.3.4):

$$\Gamma(E) = \frac{N!}{N_+!N_-!}\,, \tag{i}$$

where the relationship between the number of spins N_+ and N_- (in the direction of the field and in the opposite direction, respectively) and the energy E is given by Eqs. (2.3.2) and (2.3.3):

$$E = -\mu_{\rm B} H(N_+ - N_-)\,, \tag{ii}$$

and the total number of spins N is

$$N = N_+ + N_-\,. \tag{iii}$$

From Eqs. (ii) and (iii) we obtain

$$\begin{aligned} N_+ &= \frac{N}{2} - \frac{E}{2\mu_B H}\,, \\ N_- &= \frac{N}{2} + \frac{E}{2\mu_B H}\,. \end{aligned} \tag{iv}$$

Substituting (iv) in (i) we obtain

$$\Gamma(E) = \frac{N!}{\left(\frac{N}{2} - \frac{E}{2\mu_B H}\right)!\left(\frac{N}{2} + \frac{E}{2\mu_B H}\right)!}\,. \tag{v}$$

Solution 3.2 *Exercise on page 149*

The calculation of $P(\sigma_1 = -1)$ is similar to the calculation of $P(\sigma_1 = 1)$. Also here σ_1 is constant, but this time we have to calculate the number of states of $N-1$ spins, with an excess along $+z$, equal to $q+1$. In these states there are N_+ spins as in the calculation of $\Gamma(E)$ and $N_- - 1$ spins along $-z$ (since spin σ_1 is directed along $-z$).

The required number of states is therefore the number of possibilities of choosing $N_- - 1$ out of $N-1$:

$$\frac{(N-1)!}{(N_- - 1)!N_+!}\,,$$

so that

$$P(\sigma_1 = -1) = \frac{1}{\Gamma(E)}\frac{(N-1)!}{(N_- - 1)!N_+!} = \frac{N_-}{N}\,,$$

where we have used Eq. (2.3.4) for Γ.

Instead of N_- we substitute the expression we obtained for it in Solutions 3.1 (iv), to obtain

$$P(\sigma_1 = -1) = \frac{1}{2} + \frac{E/N}{2\mu_B H}\,.$$

Alternatively, one can calculate the probability as a ratio between the number of spins with projection -1 and the total number of spins.

Solution 3.3 *Exercise on page 150*

Since in our model the spin has only two possible values, we can write

$$P(1) + P(-1) = 1\,. \tag{i}$$

Equation (2.2.15) takes the form

$$\langle M\rangle = \mu_{\rm B}N[P(1) - P(-1)]$$

or

$$P(1) - P(-1) = \frac{\langle M\rangle}{\mu_{\rm B}N}\,. \tag{ii}$$

Equations (i) and (ii) are equations in two unknowns, $P(1)$ and $P(-1)$, whose solution is

$$\begin{aligned} P(1) \; &= \frac{1}{2} + \frac{\langle M\rangle}{2\mu_{\rm B}N}\,, \\ P(-1) &= \frac{1}{2} - \frac{\langle M\rangle}{2\mu_{\rm B}N}\,. \end{aligned} \tag{iii}$$

Since $E = -H\langle M\rangle$ [see Eq. (2.3.6)], we immediately obtain Eqs. (2.3.7) and (2.3.8).

Solution 3.4 *Exercise on page 151*

In order to obtain Eqs. (2.3.10) we calculate the number of states in which there are two constant spins, whereas all the other spins can have every possible value, under the constraint that there are a total of N_+ spins along the field and N_- spins in the opposite direction.

(a) Calculation of $P(+1,+1)$.
Among the $N-2$ spins $N_+ - 2$ are directed along the field and N_- in the opposite direction. The number of possible states is thus

$$\frac{(N-2)!}{(N_+-2)!N_-!}\,.$$

Hence

$$P(+1,+1) = \frac{1}{\Gamma(E)}\frac{(N-2)!}{(N_+-2)!N_-!} = \frac{N_+(N_+-1)}{N(N-1)}\,.$$

(b) Calculation of $P(+1,-1)$ or $P(-1,+1)$.
Among the $N-2$ spins N_+-1 are directed along the field and $N_- - 1$ in the opposite direction. The number of possible states is

$$\frac{(N-2)!}{(N_+-1)!(N_--1)!}\,.$$

Hence

$$P(+1,-1) = P(-1,+1) = \frac{1}{\Gamma(E)} \frac{(N-2)!}{(N_+-1)!(N_--1)!} = \frac{N_-N_+}{N(N-1)}.$$

(c) Calculation of $P(-1,-1)$.
Among the $N-2$ spins N_+ are directed along the field and $N_- - 2$ in the opposite direction. The number of possible states is

$$\frac{(N-2)!}{N_+!(N_--2)!},$$

so that

$$P(-1,-1) = \frac{N_-(N_--1)}{N(N-1)}.$$

Solution 3.5 *Exercise on page 152*

Since the magnetization is identical in all states with the same energy, there are no states with magnetization different from $\mu_B q$, so that the deviation of M from its average vanishes, and hence $(\Delta M)^2 = 0$. However, even though we already know the answer we will calculate $(\Delta M)^2$ in a direct manner, as this method of calculation is typical of many other cases and is worth knowing. You have already seen a simple example of such a calculation in Exercise 1.15 of Part I.

The average square deviation of M from its average is

$$\begin{aligned}(\Delta M)^2 &= \langle (M - \langle M \rangle)^2 \rangle = \langle (M^2 - 2M\langle M \rangle + \langle M \rangle^2) \rangle \\ &= \langle M^2 \rangle - \langle M \rangle^2 . \end{aligned} \tag{i}$$

$\langle M \rangle$ is given in Eq. (2.3.6) or (2.3.9):

$$\langle M \rangle = \mu_B q . \tag{ii}$$

Calculation of $\langle M^2 \rangle$:
From Eq. (2.2.10)

$$M = \mu_B \sum_{i=1}^{N} \sigma_i .$$

Hence

$$\begin{aligned}\langle M^2 \rangle &= \sum_{\{\sigma\}} \left(\mu_B \sum_{i=1}^{N} \sigma_i \right)^2 P(\sigma_1, \ldots, \sigma_N) \\ &= \sum_{\{\sigma\}} \mu_B^2 \sum_{i=1}^{N} \sum_{j=1}^{N} \sigma_i \sigma_j P(\sigma_1, \ldots, \sigma_N) . \end{aligned}$$

■ We have used the fact that

$$\left(\sum_k x_k\right)^2 = \left(\sum_i x_i\right)\left(\sum_j x_j\right) = \sum_i \sum_j x_i x_j .$$

Hence

$$\begin{aligned}
\langle M^2 \rangle &= \mu_B^2 \sum_{i=1}^{N} \sum_{j=1}^{N} \sum_{\{\sigma\}} \sigma_i \sigma_j P(\sigma_1, \dots, \sigma_N) \\
&= \mu_B^2 \sum_{i=1}^{N} \sum_{\substack{j=1 \\ j \neq i}}^{N} \sum_{\{\sigma\}} \sigma_i \sigma_j P(\sigma_1, \dots, \sigma_N) \\
&\quad + \mu_B^2 \sum_{i=1}^{N} \sum_{\{\sigma\}} \sigma_i^2 P(\sigma_1, \dots, \sigma_N) \equiv \Sigma_1 + \Sigma_2 ,
\end{aligned}$$

where we have separated between the cases with $i \neq j$ ($\sum_1$) and those with $i = j$ ($\sum_2$).

(a) To calculate $\sum_1$ we rewrite the inner sum in the form

$$\begin{aligned}
\sum_{\{\sigma\}} \sigma_i \sigma_j P(\sigma_1, \dots, \sigma_N) &= \sum_{\sigma_i, \sigma_j} \sigma_i \sigma_j \sum_{\{\sigma\}}{}'' P(\sigma_1, \dots, \sigma_N) \\
&= \sum_{\sigma_i, \sigma_j} \sigma_i \sigma_j P(\sigma_i, \sigma_j) ,
\end{aligned}$$

where we have used the notations of Chap. 2 and Eq. (2.2.19).

$P(\sigma_i, \sigma_j)$ has already been found [Eqs. (2.3.10)], and since it is independent of i and j we obtain

$$\begin{aligned}
\Sigma_1 &= \mu_B^2 \sum_{i=1}^{N} \sum_{\substack{j=1 \\ j \neq i}}^{N} [P(+1,+1) - P(+1,-1) - P(-1,+1) + P(-1,-1)] \\
&= \mu_B^2 N(N-1) \left[\frac{(N_+ - 1)N_+}{N(N-1)} - 2\frac{N_+ N_-}{N(N-1)} + \frac{(N_- - 1)N_-}{N(N-1)} \right] .
\end{aligned}$$

■ Note that

$$\sum_{i=1}^{N} \sum_{\substack{j=1 \\ j \neq i}}^{N} 1 = N(N-1) .$$

Hence

$$\Sigma_1 = \mu_B^2(N_+^2 - N_+ - 2N_+N_- + N_-^2 - N_-)$$
$$= \mu_B^2[(N_+ - N_-)^2 - (N_+ + N_-)] = \mu_B^2(q^2 - N)\,. \qquad \text{(iii)}$$

(b) Calculation of Σ_2:
Since $\sigma_i^2 = 1$ for all i and $\sum_{\{\sigma\}} P(\sigma_1, \ldots, \sigma_N) = 1$,

$$\Sigma_2 = \mu_B^2 \sum_{i=1}^{N} \sum_{\{\sigma\}} \sigma_i^2 P(\sigma_1, \ldots, \sigma_N) = \mu_B^2 N\,. \qquad \text{(iv)}$$

From (iii) and (iv) we obtain

$$\langle M^2 \rangle = \Sigma_1 + \Sigma_2 = \mu_B^2(q^2 - N) + \mu_B^2 N = \mu_B^2 q^2\,.$$

Substituting (v) and (ii) into (i), we get

$$(\Delta M)^2 = \mu_B^2 q^2 - \mu_B^2 q^2 = 0\,, \qquad \text{(v)}$$

as anticipated.

Solution 3.6 *Exercise on page 152*

In order to calculate $P(\sigma_1 = +1, \sigma_2 = +1, \sigma_3 = +1)$, we have to calculate the number of states with $N-3$ spins, having an excess in the $+z$ direction equal to $q - 3$. This number is

$$\frac{(N-3)!}{(N_+ - 3)!N_-!}\,,$$

so that

$$P(+1, +1, +1) = \frac{1}{\Gamma(E)} \frac{(N-3)!}{(N_+ - 3)!N_-!} = \frac{N_+(N_+ - 1)(N_+ - 2)}{N(N-1)(N-2)}\,.$$

In the thermodynamic limit both N_+ and $N \to \infty$, and

$$P(+1, +1, +1) \to \left(\frac{N_+}{N}\right)^3\,.$$

In the solution to Exercise 3.1 we found that

$$N_+ = \frac{N}{2} - \frac{E}{2\mu_B H} = \frac{N}{2}\left(1 - \frac{E/N}{\mu_B H}\right),$$

so that in the thermodynamic limit

$$P(+1,+1,+1) = \frac{1}{8}\left(1 - \frac{E/N}{\mu_B H}\right)^3 .$$

Solution 3.7 *Exercise on page 154*

(a) From Eqs. (2.3.4) and (2.3.5)

$$\Gamma(E) = \frac{N!}{\left(\frac{N}{2} - \frac{E}{2\mu_B H}\right)!\left(\frac{N}{2} + \frac{E}{2\mu_B H}\right)!} = \frac{N!}{N_+!N_-!}.$$

From the definition of the function S

$$\frac{1}{k}S \equiv \ln\Gamma(E) = \ln N! - \ln N_+! - \ln N_-! . \tag{i}$$

Applying Stirling's formula for large n, Eq. (2.3.12), we obtain

$$\ln n! \simeq \left(n + \frac{1}{2}\right)\ln n - n + \frac{1}{2}\ln(2\pi) . \tag{ii}$$

Assuming that N, N_+ and N_- are all large enough, we substitute their approximation (ii) in (i), to obtain

$$\begin{aligned}\frac{1}{k}S &\simeq \left(N + \frac{1}{2}\right)\ln N - N - \left(N_+ + \frac{1}{2}\right)\ln N_+ + N_+ \\ &\qquad - \left(N_- + \frac{1}{2}\right)\ln N_- + N_- - \frac{1}{2}\ln(2\pi) \\ &= \left(N + \frac{1}{2}\right)\ln N - \left(N_+ + \frac{1}{2}\right)\ln N_+ - \left(N_- + \frac{1}{2}\right)\ln N_- \\ &\quad - \frac{1}{2}\ln(2\pi) .\end{aligned} \tag{iii}$$

We now write

$$N_\pm = N \cdot \frac{N_\pm}{N} , \tag{iv}$$

and then

$$\frac{1}{k}S = \left(N+\frac{1}{2}\right)\ln N - \left(N_+ + \frac{1}{2}\right)\left(\ln N + \ln\frac{N_+}{N}\right)$$
$$- \left(N_- + \frac{1}{2}\right)\left(\ln N + \ln\frac{N_-}{N}\right) - \frac{1}{2}\ln(2\pi)$$
$$= -\left(N_+ + \frac{1}{2}\right)\ln\frac{N_+}{N} - \left(N_- + \frac{1}{2}\right)\ln\frac{N_-}{N} - \frac{1}{2}\ln(2\pi N)\,, \quad \text{(v)}$$

where the last transition was made after we noticed that from all the terms proportional to $\ln N$ the only one left was $-\frac{1}{2}\ln N$.

But (see Solution 3.1)

$$\begin{aligned} N_+ &= \frac{N}{2} - \frac{E}{2\mu_B H}\,, \\ N_- &= \frac{N}{2} + \frac{E}{2\mu_B H} \end{aligned} \quad \text{(vi)}$$
$$\Downarrow$$
$$\frac{N_\pm}{N} = \frac{1}{2} \mp \frac{E/N}{2\mu_B H}\,.$$

Substituting (vi) in (v) we obtain

$$\frac{1}{k}S \simeq -\left(\frac{N}{2} - \frac{E}{2\mu_B H}\right)\ln\left(\frac{1}{2} - \frac{E/N}{2\mu_B H}\right)$$
$$- \left(\frac{N}{2} + \frac{E}{2\mu_B H}\right)\ln\left(\frac{1}{2} + \frac{E/N}{2\mu_B H}\right) - \frac{1}{2}\ln(2\pi N)\,, \quad \text{(vii)}$$

where we have further neglected the term $\frac{1}{2}$ compared to $N_\pm$. Finally, we note that the first two terms on the right hand side are proportional to N whereas the third one is proportional to $\ln N$ only. For $N \to \infty$ the third term is negligible compared to the first two, so that we find (2.3.13).

(b) Equation (2.3.13) is not sufficiently accurate for the calculation of the number of states for large, but not macroscopic, values of N even though Stirling's approximation is already quite accurate. To this end we must return all the "branches we cut" in the process of obtaining (2.3.13), i.e. use Eq. (v) of (a) above:

$$\frac{1}{k}\tilde{S} = -\frac{1}{2}\ln(2\pi N) - \left(\frac{N+1}{2} - \frac{E}{2\mu_B H}\right)\ln\left(\frac{1}{2} - \frac{E/N}{2\mu_B H}\right)$$
$$- \left(\frac{N+1}{2} + \frac{E}{2\mu_B H}\right)\ln\left(\frac{1}{2} + \frac{E/N}{2\mu_B H}\right)\,. \quad \text{(viii)}$$

The following table summarizes the results of the calculation with $e^{S/k}$ [Eq. (2.3.13)] and with $e^{\tilde{S}/k}$ [Eq. (viii) here] for $N = 10$.

$E/\mu_B H$	−10	−8	−6	−4	−2	0	2	4	6	8	10
$e^{S/k}$	1	25.8	149.0	449.7	837.2	1024.0	837.2	449.7	149.0	25.8	1
$e^{\tilde{S}/k}$	–	10.9	47.0	123.8	215.6	258.4	215.6	123.8	47.0	10.9	–

Note the good agreement between the results in the second row of the table and the results of Table 3.1. Nevertheless, at the edges of the spectrum it is not possible to use $\tilde{S}$ as it diverges there. This does not mean that (2.3.13) is wrong, but that it is valid only in the macroscopic limit, $N \to \infty$.

Solution 3.8 *Exercise on page 155*

We use the notation $\epsilon = E/\mu_B H$ for short, and write (2.3.13) in the form

$$\begin{aligned}\frac{1}{k}S &= -\frac{N}{2}\left(1-\frac{\epsilon}{N}\right)\left[\ln\left(1-\frac{\epsilon}{N}\right)-\ln 2\right]\\ &\quad -\frac{N}{2}\left(1+\frac{\epsilon}{N}\right)\left[\ln\left(1+\frac{\epsilon}{N}\right)-\ln 2\right]\\ &= N\ln 2-\frac{N}{2}\left[\left(1-\frac{\epsilon}{N}\right)\ln\left(1-\frac{\epsilon}{N}\right)+\left(1+\frac{\epsilon}{N}\right)\ln\left(1+\frac{\epsilon}{N}\right)\right]. \qquad \text{(i)}\end{aligned}$$

Now, if $N \gg 1$ we may approximate the logarithmic function by the first two terms in its power expansion:

$$\frac{1}{k}S \simeq N\ln 2-\frac{N}{2}\left[\left(1-\frac{\epsilon}{N}\right)\left(-\frac{\epsilon}{N}-\frac{\epsilon^2}{2N^2}\right)+\left(1+\frac{\epsilon}{N}\right)\left(\frac{\epsilon}{N}-\frac{\epsilon^2}{2N^2}\right)\right]. \qquad \text{(ii)}$$

Keeping terms up to the second order in ϵ we obtain

$$\frac{1}{k}S \simeq N\ln 2-\frac{\epsilon^2}{2N} \qquad \text{(iii)}$$

or

$$\Gamma = e^{S/k} \simeq Ce^{-\epsilon^2/2N}, \qquad \text{(iv)}$$

which is Eq. (2.3.14). The constant C is not 2^N, as might be expected from Eq. (iii), since we neglected too many terms depending on N on our way to Eq. (iv). If we use instead of (2.3.13) the more exact result we obtained in the answer to the previous exercise [Eq. (viii)], then we should add $-\frac{1}{2}\ln(2\pi N)$ to the right hand side of Eq. (i).

Thus we obtain, instead of Eq. (iii),

$$\frac{1}{k}\tilde{S} \simeq \ln\left(\frac{2^N}{\sqrt{2\pi N}}\right)-\frac{\epsilon^2}{2N}, \qquad \text{(v)}$$

so that

$$C = \frac{2^N}{\sqrt{2\pi N}} \,. \tag{vi}$$

The same result is also obtained from the normalization condition:

$$C \int_{-\infty}^{\infty} \exp\left(-\frac{\epsilon^2}{2N}\right) d\epsilon = 2^N \,, \tag{vii}$$

since the value of the integral on the left hand side is $\sqrt{2\pi N}$.

■ Note that we integrated over energies from $-\infty$ to ∞, although clearly our approximation is valid only when ϵ/N is very small. But this means that most of the area of the graph is concentrated around $\epsilon = 0$ and the exponential function decays extremely fast, so that the "tail" does not affect the sum.

Solution 4.1 *Exercise on page 158*

We calculate $P(+1)$ and $P(-1)$ by solving two equations in two unknowns:

The sum of the probabilities satisfies

$$P(+1) + P(-1) = 1 \,. \tag{i}$$

Equation (2.4.8) reads

$$\frac{P(+1)}{P(-1)} = e^{2\mu_B H\beta} \tag{ii}$$

$$\Downarrow$$

$$\frac{P(+1)}{1 - P(+1)} = e^{2\mu_B H\beta}$$

$$\Downarrow$$

$$P(+1) = \frac{e^{2\mu_B H\beta}}{1 + e^{2\mu_B H\beta}} \,.$$

With the help of (i):

$$P(-1) = \frac{1}{1 + e^{2\mu_B H\beta}} \,.$$

Multiplying both by $e^{-\mu_B H\beta}$:

$$\Downarrow$$

$$P(+1) = \frac{e^{\mu_B H\beta}}{2\cosh(\mu_B H\beta)} \,,$$

$$P(-1) = \frac{e^{-\mu_B H\beta}}{2\cosh(\mu_B H\beta)} \,.$$

$P(\pm 1)$ as calculated here are the same probabilities given in Eqs. (2.3.7) and (2.3.8). The difference is that now they are expressed not as functions of E and H but as functions of β and H. By comparing the two forms it is possible to obtain the relationship (2.4.6) between E and β.

Solution 4.2 *Exercise on page 160*

The energy distribution (2.4.14) is a Gaussian distribution, which we write as

$$\Gamma_T(E_a) = C \exp\left[-\frac{(E_a - \bar{E}_a)^2}{2N\alpha^2}\right],$$

where α remains finite in the thermodynamic limit (check!). The simplest way to estimate the width of a Gaussian curve is to measure the distance to the energy value for which the distribution decreases to half of its maximal height.

Thus, we search for an energy E_a satisfying

$$\frac{C}{2} = C \exp\left[-\frac{(E_a - \bar{E}_a)^2}{2N\alpha^2}\right]$$

or

$$\ln 2 = \frac{(E_a - \bar{E}_a)^2}{2N\alpha^2}$$

$$\Downarrow$$

$$E_a = \bar{E}_a \pm (2N\alpha^2 \ln 2)^{1/2} .$$

This means that the width of the curve is

$$\Delta E_a \simeq 2(2N\alpha^2 \ln 2)^{1/2} \sim N^{1/2} .$$

As the average energy is proportional to the number of spins in the system, we have

$$\frac{\Delta E}{\bar{E}_a} \sim \frac{N^{1/2}}{N} \sim N^{-1/2} .$$

The calculation may, of course, be performed in a more exact manner, if we calculate

$$(\Delta E_a)^2 = \langle (E_a - \bar{E}_a)^2 \rangle ,$$

$$\langle (E_a - \bar{E}_a)^2 \rangle = \frac{\int_{-\infty}^{\infty} (E_a - \bar{E}_a)^2 \exp\left[-(E_a - \bar{E}_a)^2/2N\alpha^2\right] dE_a}{\int_{-\infty}^{\infty} \exp\left[-(E_a - \bar{E}_a)^2/2N\alpha^2\right] dE_a} .$$

Calculating the integral by the methods we developed in Part I, we obtain

$$(\Delta E_a)^2 = N\alpha^2 .$$

Hence

$$\Delta E_a \sim N^{1/2} .$$

We again use the fact that $\bar{E}_a \sim N$, so that

$$\frac{\Delta E_a}{\bar{E}_a} \sim \frac{1}{N^{1/2}}.$$

Solution 4.3 *Exercise on page 162*

(a) $\langle\sigma\rangle = (+1)P(+1) + (-1)P(-1) = P(+1) - P(-1)$.
Using the results of Solution 4.1 for $P(+1)$ and $P(-1)$ we obtain

$$\langle\sigma\rangle = \frac{e^{\mu_B H\beta}}{2\cosh(\mu_B H\beta)} - \frac{e^{-\mu_B H\beta}}{2\cosh(\mu_B H\beta)} = \tanh(\mu_B H\beta). \qquad \text{(i)}$$

Since the N spins are independent of one another, we have [Eq. (2.2.15)]

$$M = \mu_B N\langle\sigma\rangle,$$

and hence

$$M = \mu_B N \tanh(\mu_B H\beta). \qquad \text{(ii)}$$

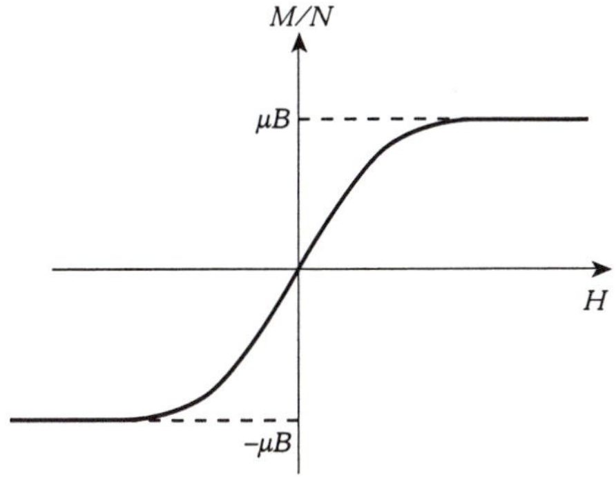

(b) Equation (2.4.6) provides us with the relationship between energy and temperature, and since $E = -MH$, also with a relationship between the magnetization and the temperature,

$$e^{2\mu_B H\beta} = \frac{N\mu_B + M}{N\mu_B - M},$$

and, if we solve for M, we get

$$M = \mu_B N \tanh(\mu_B H\beta).$$

(c) From (2.4.21) and (2.4.19) we obtain

$$\frac{H}{T} = -\frac{\partial S}{\partial M} = \frac{k}{2\mu_B}\ln\frac{1 + M/N\mu_B}{1 - M/N\mu_B}.$$

Now we can invert the relationship and express M in terms of all the other quantities exactly as in (b), and obtain (ii) again.

(d) Equation (2.4.24) can simply be obtained from $E = -MH$ using M from (a) or (b) above.

Solution 5.1 *Exercise on page 166*

The constant C in Eq. (2.5.1a) is a normalization constant, which ensures that

$$\sum_{\{\sigma\}} P(\sigma_1, \ldots, \sigma_n) = 1\,, \tag{i}$$

where $\{\sigma\}$ denotes the summation over all possible configurations, with $\sigma_i = \pm 1$ for each i.

■ Reminder: Eq. (i) is necessary if P is to be interpreted as the probability of a configuration.

That is, we require that

$$\sum_{\{\sigma\}} C \exp\left(\beta\mu_{\text{B}} H \sum_{i=1}^{n} \sigma_i\right) = 1\,,$$

or

$$C^{-1} = \sum_{\{\sigma\}} \exp\left(\beta\mu_{\text{B}} H \sum_{i=1}^{n} \sigma_i\right)\,. \tag{ii}$$

The energy of a given configuration is

$$E(\sigma_1, \ldots, \sigma_n) = -\mu_{\text{B}} H \sum_{i=1}^{n} \sigma_i\,.$$

Writing the summation over $\{\sigma\}$ in (ii) in detail, we obtain

$$C^{-1} = \sum_{\sigma_1=\pm 1} \cdots \sum_{\sigma_n=\pm 1} e^{-\beta E(\sigma_1, \ldots, \sigma_n)}\,.$$

Solution 5.2 *Exercise on page 168*

The average magnetization is given by

$$\begin{aligned}
\langle M \rangle &= \sum_{\{\sigma\}} \left[\mu_{\text{B}} \sum_{i=1}^{n} \sigma_i\right] P(\sigma_1, \ldots, \sigma_n) \\
&= \frac{1}{Z} \sum_{\{\sigma\}} \left[\mu_{\text{B}} \sum_{i=1}^{n} \sigma_i\right] \exp\left(\beta\mu_{\text{B}} H \sum_{j=1}^{n} \sigma_j\right),
\end{aligned} \tag{i}$$

where

$$Z = \sum_{\{\sigma\}} \exp\left(\beta \mu_{\rm B} H \sum_{i=1}^{n} \sigma_i\right). \tag{ii}$$

Since the sum in (i) is

$$\frac{1}{\beta}\frac{\partial Z}{\partial H},$$

we immediately obtain

$$\langle M \rangle = \frac{1}{\beta}\frac{1}{Z}\frac{\partial Z}{\partial H} = \frac{1}{\beta}\frac{\partial \ln Z}{\partial H}.$$

Solution 5.3 *Exercise on page 168*

If f is a function of $x = \beta H$ only, then the following relations hold:

$$\frac{\partial f}{\partial \beta} = \frac{\partial f}{\partial x}\frac{\partial x}{\partial \beta} = H\frac{\partial f}{\partial x}, \quad \frac{\partial f}{\partial H} = \frac{\partial f}{\partial x}\frac{\partial x}{\partial H} = \beta\frac{\partial f}{\partial x}.$$

If $f(x) = \ln Z(x)$, then

$$\frac{\partial \ln Z}{\partial \beta} = H\frac{\partial \ln Z}{\partial x}, \quad \frac{\partial \ln Z}{\partial H} = \beta\frac{\partial \ln Z}{\partial x}$$

$$\Downarrow$$

$$\frac{\partial \ln Z}{\partial \beta} = \frac{1}{\beta}H\frac{\partial \ln Z}{\partial H}.$$

And from Eqs. (2.5.7) and (2.5.8) we obtain

$$\langle E \rangle = -H\langle M \rangle.$$

Solution 5.4 *Exercise on page 171*

We substitute in the definition of the heat capacity per spin at constant field (2.5.17),

$$c_H = \frac{1}{n}\left(\frac{\partial E}{\partial T}\right)_H, \tag{i}$$

the average energy of the system, E, as given by Eq. (2.5.13):

$$E = -n\mu_{\rm B} H \tanh(\beta\mu_{\rm B} H). \tag{ii}$$

Differentiating with respect to T we obtain

$$\begin{aligned}\left(\frac{\partial E}{\partial T}\right)_H &= \frac{d\beta}{dT}\left(\frac{\partial E}{\partial \beta}\right)_H = -k\beta^2\left(\frac{\partial E}{\partial \beta}\right)_H \\ &= k\beta^2 n\mu_{\rm B} H\frac{\mu_{\rm B} H}{\cosh^2(\beta\mu_{\rm B} H)} = \frac{n\mu_{\rm B}^2 H^2}{kT^2\cosh^2(\beta\mu_{\rm B} H)}.\end{aligned} \tag{iii}$$

From (i) and (iii) we obtain

$$c_H = \frac{\mu_B^2 H^2}{kT^2 \cosh^2(\beta\mu_B H)} .$$

Solution 5.5 *Exercise on page 172*

The magnetization per spin at a constant temperature was plotted in Solution 4.3. See also Fig. 2.5.1. The plot at constant field is:

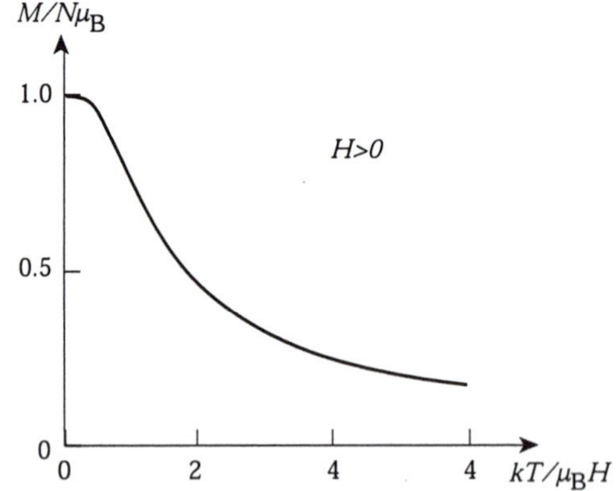

Note that this graph corresponds to the region $H > 0$ in Solution 4.3 or $x > 0$ in Fig. 2.5.1, and that the abscissa variable here is $1/x$.

Solution 5.6 *Exercise on page 172*

(a) The entropy of the paramagnet as a function of M is given by Eq. (2.4.18), which we rearrange and rewrite in a slightly different manner:

$$S = -\frac{Nk}{2}\left[\ln\left(1-\left(\frac{M}{N\mu_B}\right)^2\right) + \frac{M}{N\mu_B}\ln\left(\frac{N\mu_B + M}{N\mu_B - M}\right) - 2\ln 2\right] . \tag{i}$$

Substituting the explicit expression (2.5.14) for M as a function of T and H, we obtain

$$S(T,H) = -\frac{Nk}{2}\left[\ln(1-\tanh^2 x) + \tanh x \ln\left(\frac{1+\tanh x}{1-\tanh x}\right) - 2\ln 2\right] , \tag{ii}$$

where $x \equiv \beta\mu_B H$ and N is the number of spins.

Now, using the identities

$$1 - \tanh^2 x = \frac{1}{\cosh^2 x} ,$$

$$\frac{1+\tanh x}{1-\tanh x} = e^{2x} ,$$

we obtain

$$\begin{aligned} S(T,H) &= Nk[\ln(2\cosh x) - x\tanh x] \\ &= Nk\left\{\ln\left[2\cosh\left(\frac{\mu_B H}{kT}\right)\right] - \frac{\mu_B H}{kT}\tanh\left(\frac{\mu_B H}{kT}\right)\right\} . \end{aligned} \tag{iii}$$

(b) The entropy per spin as a function of temperature (at a constant field) is depicted in the following graph:

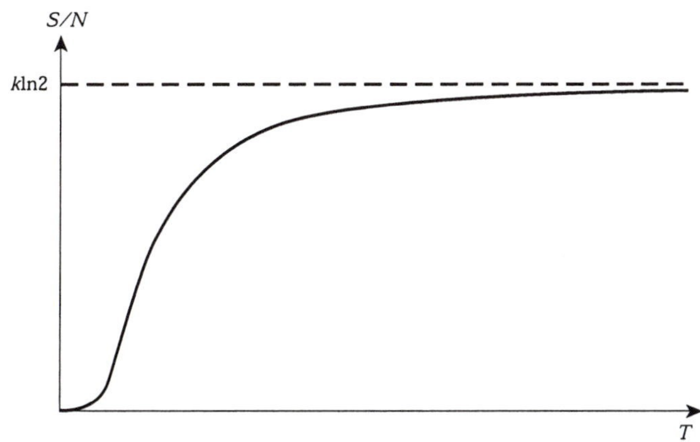

(c) The entropy per spin as a function of the field (at constant temperature):

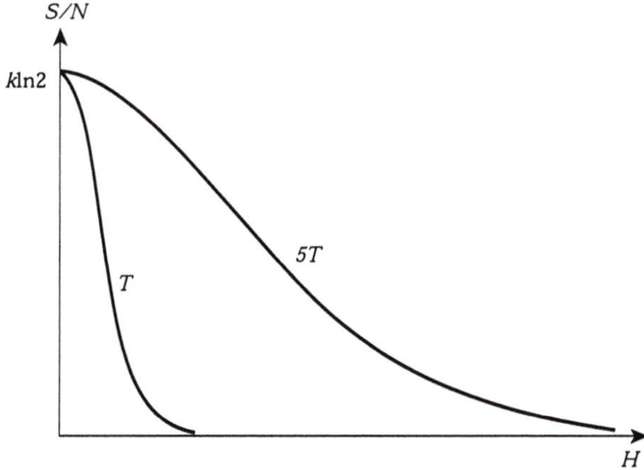

Solution 5.7 *Exercise on page 173*

(a) First we calculate, in analogy to Eq. (2.5.11), the partition function per ion:

$$z = \sum_{\sigma_i=-2J}^{2J} e^{\beta g \mu_B H \sigma_i/2} . \tag{i}$$

The summation in (i) can be computed, as a geometric series whose first term is $e^{\beta g\mu_B HJ}$. Denote $x = \beta g\mu_B H$; then

$$z = \sum_{\sigma_i=-2J}^{2J} e^{x\sigma_i/2} = \frac{e^{x(J+1)} - e^{-xJ}}{e^x - 1}. \tag{ii}$$

■ Reminder:

$$1 + q + q^2 + \cdots + q^{n-1} = \frac{q^n - 1}{q - 1}.$$

Note that σ_i changes in steps of 2.

Multiplying the numerator and the denominator in the last fraction by $e^{-x/2}$, we obtain

$$z = \frac{e^{x(J+1/2)} - e^{-x(J+1/2)}}{e^{x/2} - e^{-x/2}} = \frac{\sinh(J+1/2)x}{\sinh(x/2)}. \tag{iii}$$

If the number of spins is N, then the system's partition function is obtained by (2.5.12):

$$Z = \left[\frac{\sinh(J+1/2)x}{\sinh(x/2)}\right]^N. \tag{iv}$$

(b) The average magnetization per spin is calculated in a manner similar to that which brought us (in the case of spin $\frac{1}{2}$) to Eq. (2.5.14). Recall that Eq. (2.5.8) is completely general. We continue to express Z in terms of x, and in order to obtain the derivative with respect to H we use the chain rule.

$$\begin{aligned}\frac{\langle M\rangle}{N} &= \frac{1}{N}\frac{1}{\beta}\frac{\partial \ln Z}{\partial H} = \frac{1}{N}\frac{1}{\beta}\frac{d\ln Z}{dx}\frac{\partial x}{\partial H}\\ &= g\mu_B \frac{d}{dx}\ln\left[\frac{\sinh(J+1/2)x}{\sinh(x/2)}\right]\\ &= g\mu_B\left[\frac{(J+1/2)\cosh(J+1/2)x}{\sinh(J+1/2)x} - \frac{1/2\cosh(x/2)}{\sinh(x/2)}\right].\end{aligned} \tag{v}$$

We define Brillouin's function by

$$B_J(x) \equiv \frac{1}{J}\left[\left(J+\frac{1}{2}\right)\coth\left(J+\frac{1}{2}\right)x - \frac{1}{2}\coth\left(\frac{x}{2}\right)\right] \tag{vi}$$

and write (v) in the shortened form

$$\frac{\langle M\rangle}{N} = g\mu_B J B_J(x)$$

or, using the definition of x,

$$\frac{\langle M \rangle}{N} = g\mu_{\mathrm{B}} J B_J(\beta g \mu_{\mathrm{B}} H) \,. \tag{vii}$$

In order to plot a graph of $\langle M \rangle / N$ as a function of H, we have to analyze the behavior of the Brillouin function $B_J(x)$. Since it is an odd function, it is sufficient to study the region $x \geq 0$. For $x \to \infty$ it is easy to see that $B_J(x) \to 1$. This means that the saturation value of $\langle M \rangle / N$ is $g\mu_{\mathrm{B}} J$. At first sight it seems as if $B_J(x)$ diverges for $x \to 0$. However, we know that for $J = \frac{1}{2}$, $B_J(x) = \tanh x$, which behaves like x at small x, so we have to study its behavior near the origin more carefully. We expand $B_J(x)$ around $x = 0$, and write

$$\coth y = \frac{e^y + e^{-y}}{e^y - e^{-y}} = \frac{2 + y^2 + \ldots}{2y + y^3/3 + \ldots} \,.$$

For $y \ll 1$ it is possible to write

$$\coth y \simeq \frac{1 + y^2/2}{y(1 + y^2/6)} \,.$$

But, for $y \ll 1$,

$$\frac{1}{1 + y^2/6} \simeq 1 - \frac{y^2}{6} \,,$$

so that we obtain

$$\coth y \simeq \frac{1}{y} \left(1 + \frac{y^2}{2} \right) \left(1 - \frac{y^2}{6} \right) \,,$$

and by neglecting terms of order higher than y^2,

$$\coth y \simeq \frac{1}{y} \left(1 + \frac{y^2}{3} \right) = \frac{1}{y} + \frac{y}{3} \,. \tag{viii}$$

Hence, for small x the function $B_J(x)$ takes the form

$$\begin{aligned} B_J(x) &\simeq \frac{1}{J} \left\{ \left(J + \frac{1}{2} \right) \left[\frac{1}{(J + 1/2)x} + \frac{1}{3} \left(J + \frac{1}{2} \right) x \right] - \frac{1}{2} \left(\frac{2}{x} + \frac{x}{6} \right) \right\} \\ &= \frac{1}{J} \left[\frac{1}{3} \left(J + \frac{1}{2} \right)^2 x - \frac{1}{12} x \right] = \frac{x}{3J} (J^2 + J) = \left(\frac{J + 1}{3} \right) x \end{aligned} \tag{ix}$$

and this is the behavior we already know for the case $J = \frac{1}{2}$. Now we still have to check if indeed nothing special happens between $x = 0$ and $x \to \infty$. To this end we calculate the derivative:

$$\frac{dB_J}{dx} = -\frac{(J + 1/2)^2}{J \sinh^2 (J + 1/2)x} + \frac{1}{4J \sinh^2 x/2} \,.$$

A brief study reveals that dB_J/dx is positive for all x, so $B_J(x)$ is a monotonic increasing function in the region under discussion. Hence $\langle M\rangle/N$ as a function of H will look as in the following figure:

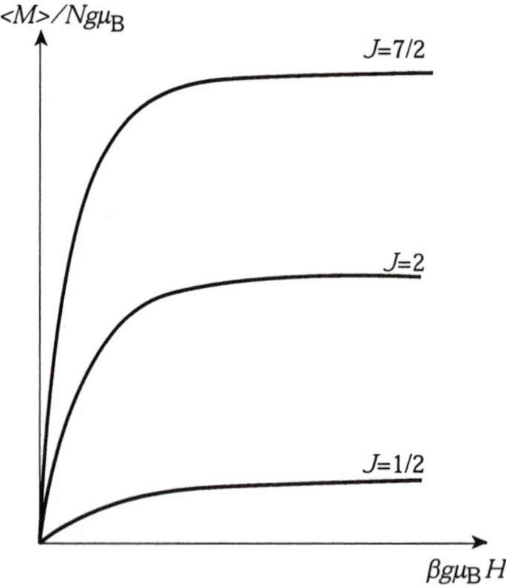

All we have used to arrive at the qualitative form of these curves is that the value at which they saturate increases in proportion to J and their slope at the origin in proportion to $J(J+1)$ [Eqs. (vii) and (ix)].

(c) We obtain the susceptibility from the behavior of $\langle M\rangle$ at small fields. From Eqs. (vii) and (ix) we immediately find

$$\frac{\langle M\rangle}{N} \approx g\mu_{\mathrm{B}} J \cdot \frac{J+1}{3} \cdot \beta g\mu_{\mathrm{B}} H = (g\mu_{\mathrm{B}})^2 \frac{J(J+1)}{3kT} H\,, \tag{x}$$

and the susceptibility is

$$\chi = (g\mu_{\mathrm{B}})^2 \frac{J(J+1)}{3kT}\,. \tag{xi}$$

Substituting $J = \frac{1}{2}$ and $g = 2$, we obtain the result (2.5.16).

(d) We calculate the specific heat at a constant field using Eqs. (2.5.17) and (2.5.7):

$$\begin{aligned} c_H &= \frac{1}{N}\left(\frac{\partial E}{\partial T}\right)_H = -\frac{1}{N} k\beta^2 \left(\frac{\partial E}{\partial \beta}\right)_H \\ &= \frac{1}{N} k\beta^2 \frac{\partial}{\partial \beta}\left(\frac{\partial \ln Z}{\partial \beta}\right) = k\beta^2 \frac{\partial^2 \ln z}{\partial \beta^2}\,, \end{aligned} \tag{xii}$$

where we have substituted $Z = z^N$. Next we note that

$$\frac{\partial^2 \ln z}{\partial \beta^2} = (g\mu_{\rm B} H)^2 \frac{d^2 \ln z}{dx^2} \,.$$

But we have already computed the derivative $d \ln z/dx$ in (v). Using (v) and (vii) we obtain

$$\frac{d \ln z}{dx} = J B_J(x) \,,$$

$$\frac{d^2 \ln z}{dx^2} = J \frac{dB_J(x)}{dx} = -\frac{(J+1/2)^2}{\sinh^2(J+\frac{1}{2})x} + \frac{1/4}{\sinh^2 x/2} \,,$$

so that finally

$$c_H = k(\beta g \mu_{\rm B} H)^2 \left[\frac{1}{4 \sinh^2\left(\frac{1}{2}\beta g \mu_{\rm B} H\right)} - \frac{(J+1/2)^2}{\sinh^2((J+1/2)\beta g \mu_{\rm B} H)} \right] .$$

Substituting $J = \frac{1}{2}, g = 2$ we obtain (2.5.18).

Solution 6.1 *Exercise on page 175*

If $E = 0$ we have equal probabilities for each state

$$P(+1) = P(-1) = \frac{1}{2} \,,$$

and then

$$\frac{S}{k} = -N \left(\frac{1}{2} \ln \frac{1}{2} + \frac{1}{2} \ln \frac{1}{2} \right) = N \ln 2 \,.$$

This means that the assumption that the probabilities for a spin to be along the direction of the field or in the opposite direction are equal, brings us from (2.6.6) to (2.6.4). In this case the thermal energy easily overcomes the interaction energy of the spins with the field, so that there is no preferred direction: the temperature is very high and disorder dominates.

Solution 6.2 **Solution on page 176**

Substituting the probabilities (2.5.5) into Eq. (2.6.7) we obtain

$$\begin{aligned} S &= -kZ^{-1} \sum_{\alpha=1}^{m} (-\beta E_\alpha - \ln Z) e^{-\beta E_\alpha} \\ &= k\beta Z^{-1} \sum_{\alpha=1}^{m} E_\alpha e^{-\beta E_\alpha} + kZ^{-1} \ln Z \sum_{\alpha=1}^{m} e^{-\beta E_\alpha} \,. \end{aligned}$$

The first term includes the average energy $\langle E \rangle$. In the second term there appears the sum of probabilities which is of course 1. Hence

$$S = k\beta\langle E\rangle + k \ln Z ,$$

which is the required result.

Solutions to self-assessment exercises

Solution 1 *Exercise on page 180*

From Eq. (2.4.5a) and from the explicit expression for β as a function of E, Eq. (2.4.6), we obtain the following equation for $\bar{E}_a$:

$$\frac{1}{2\mu_B H_a}\ln\frac{\mu_B H_a N_a - \bar{E}_a}{\mu_B H_a N_a + \bar{E}_a} = \frac{1}{2\mu_B H_b}\ln\frac{\mu_B H_b N_b - E + \bar{E}_a}{\mu_B H_b N_b + E - \bar{E}_a}\,.$$

If $H_a = H_b = H$, the coefficients of the logarithms cancel and their arguments must be equal. With a little algebra we obtain

$$\bar{E}_a = \frac{N_a}{N_a + N_b}E\,.$$

$\bar{E}_b$ has to make up the difference between $\bar{E}_a$ and E, so that

$$\bar{E}_b = \frac{N_b}{N_a + N_b}E\,.$$

You may have been able to guess in advance that the energies at equilibrium are distributed in direct proportion to the sizes of the two systems, so that each of them has the same energy per degree of freedom. However, it is nice to find yet another verification that we are treading on solid ground.

Solution 2 *Exercise on page 180*

(a) Since $\delta Q = TdS$, the specific heat at constant field is $T(\partial S/\partial T)_H$, and the specific heat per degree of freedom is

$$c_H = \frac{T}{n}\left(\frac{\partial S}{\partial T}\right)_H\,.$$

We now use the expression that we found for $S(T, H)$ in Exercise 5.6(a):

$$S(T,H) = nk[\ln(2\cosh x) - x\tanh x]\,, \qquad x \equiv \frac{\mu_B H}{kT}\,.$$

Hence

$$\frac{\partial S}{\partial T} = -\frac{x}{T}\frac{dS}{dx} = \frac{nkx^2}{T}\frac{1}{\cosh^2 x},$$

so that

$$c_H = k\frac{x^2}{\cosh^2 x},$$

and this result is identical to Eq. (2.5.18).

(b)

	Ideal gas	Paramagnet
Extensive variable	V	M
Intensive variable	P	$-H$
	$TdS = dE + PdV$	$TdS = dE^* - HdM$
	$TdS = d\mathsf{H} - VdP$	$TdS = dE + MdH$
Enthalpy	$\mathsf{H} = E + PV$	$E^* = E + MH$
	$E = \frac{3}{2}PV$	$E^* \equiv 0$
Equation of state	$PV = NkT$	$M = N\mu_B \tanh\left(\frac{\mu_B H}{kT}\right)$
		(High T) $M \simeq N\frac{\mu_B H}{kT}$
Adiabatic system	$dS = 0$	$dS = 0$
	$dE = -PdV$	$dM = 0$ or $H = 0$
	$PV^\gamma = \text{const}$	$M = \text{const}$ or $H = 0$

Solution 3 *Exercise on page 180*

(a) The energy of a magnetic moment, $\mathbf{m}$, in a field $\mathbf{H}$ is given by

$$\epsilon = -\mathbf{m}\cdot\mathbf{H} = -mH\cos\alpha\,. \qquad \text{(i)}$$

Substituting the value of α into each of the possible states, we find

$$\epsilon(\alpha) = \begin{cases} -mH\ , & \alpha = 0\,, \\ \dfrac{mH}{2}\ , & \alpha = \dfrac{2\pi}{3}\,, \\ \dfrac{mH}{2}\ , & \alpha = \dfrac{4\pi}{3}\,. \end{cases} \qquad \text{(ii)}$$

(b) Since the temperature is given, it is natural to use the canonical ensemble. The partition function is

$$Z(\beta, H) = \sum_{\substack{\{\text{all}\\ \text{states}\}}} e^{-\beta E(\text{state})}\,. \qquad \text{(iii)}$$

The same argument that led to (2.5.12), which is based solely on the fact that the spins are independent, will yield for the paramagnetic system under discussion as well,

$$Z(\beta, H) = [z(\beta, H)]^N \,, \tag{iv}$$

where N is the number of spins in the system, and z is the single spin partition function. In this case

$$z(\beta, H) = \sum_{\alpha} e^{-\beta\epsilon(\alpha)} \,, \tag{v}$$

with $\alpha = 0, 2\pi/3, 4\pi/3$.

The energy per spin has already been calculated in (a), Eq. (ii), and thus

$$z(\beta, H) = e^{\beta mH} + 2e^{-\beta mH/2} \tag{vi}$$

$$\Downarrow$$

$$Z(\beta, H) = (e^{\beta mH} + 2e^{-\beta mH/2})^N \,. \tag{vii}$$

(c) In the absence of an external field there will be an equal population in each of the three possible states, and hence a vanishing magnetization.

Since the field is acting in the x direction, the spins aligned in the x direction have a lower energy, so that they have a higher probability. Thus it is expected that the net magnetization will also be in the x direction. (Note: there is a complete symmetry between the directions y and $-y$. Hence the magnetization of this system cannot have a component along y.) We shall verify this conclusion quantitatively using both methods mentioned.

(1) The contribution to the magnetization in the x direction, of each of the states, is

$$m_x(\alpha) = m \cos \alpha$$

$$\Downarrow$$

$$\begin{cases} m_x(0) = m \,, \\ m_x\left(\dfrac{2\pi}{3}\right) = m_x\left(\dfrac{4\pi}{3}\right) = m\cos\left(\dfrac{2\pi}{3}\right) = -\dfrac{m}{2} \,. \end{cases} \tag{viii}$$

The contribution along y is

$$m_y(\alpha) = m\cos\left(\alpha - \frac{\pi}{2}\right)$$

$$\Downarrow$$

$$\begin{cases} m_y(0) = 0\,, \\ m_y\left(\frac{2\pi}{3}\right) = m\sqrt{\frac{3}{2}}\,, \\ m_y\left(\frac{4\pi}{3}\right) = -m\sqrt{\frac{3}{2}}\,. \end{cases} \tag{ix}$$

The probability of a given one-particle state, α, is given by

$$P_\alpha(\beta, H) = \frac{e^{-\beta\epsilon(\alpha)}}{z(\beta, H)}\,, \tag{x}$$

where the denominator, given in Eq. (v) or (vi), guarantees that the sum of probabilities is 1. In order to obtain the magnetization, we multiply the contribution of each state, Eqs. (viii) and (ix), by the probability of that state (x) and sum over the products:

$$\langle m_x(\beta, H)\rangle = \sum_\alpha m_x(\alpha)P_\alpha(\beta, H) = m\frac{e^{\beta mH} - e^{-\beta mH/2}}{e^{\beta mH} + 2e^{-\beta mH/2}}\,, \tag{xi}$$

$$\langle m_y(\beta, H)\rangle = \sum_\alpha m_y(\alpha)P_\alpha(\beta, H) = 0\,. \tag{xii}$$

This is of course the average magnetization per spin.

(2) Since the field is in the x direction we can use (2.5.8) and write the average magnetization per spin in the form

$$\langle m_x\rangle = \frac{1}{\beta N}\frac{\partial \ln Z}{\partial H}\,, \tag{xiii}$$

where Z is given in Eq. (vii); from which

$$\ln Z = N\ln(e^{\beta mH} + 2e^{-\beta mH/2}) \tag{xiv}$$

and hence

$$\langle m_x\rangle = \frac{1}{\beta}\frac{e^{\beta mH} - e^{-\beta mH/2}}{e^{\beta mH} + 2e^{-\beta mH/2}}\beta m\,, \tag{xv}$$

which is the result we obtained in (xi).

(d) Since the field is weak, we use the approximation $e^x \simeq 1 + x$ for the numerator and the denominator in Eq. (xi), and hence

$$\langle m_x \rangle \simeq m \frac{3\beta m H/2}{3} = \frac{\beta m^2 H}{2}, \qquad \text{(xvi)}$$

and from here

$$\chi = \frac{m^2}{2kT}, \qquad \text{(xvii)}$$

i.e. $\chi \propto 1/T$, and Curie's law is recovered.

(e) The average energy per spin may be calculated either from Eq. (2.5.7) or simply from $\langle \epsilon \rangle = -\langle m_x \rangle H$. Both methods yield

$$\langle \epsilon \rangle = -mH \frac{e^{\beta m H} - e^{-\beta m H/2}}{e^{\beta m H} + 2e^{-\beta m H/2}} = -mH \left(1 - \frac{3}{2 + e^{3\beta m H/2}}\right). \qquad \text{(xviii)}$$

(f) The specific heat per spin at a constant field is

$$c_H = \left(\frac{\partial \langle \epsilon \rangle}{\partial T}\right)_H = 2k \frac{(3\beta m H/2)^2}{(2e^{-3\beta m H/4} + e^{3\beta m H/4})^2} \qquad \text{(xix)}$$

Solution 4 *Exercise on page 181*

(a) The energy of a magnetic moment in an external field H in the x direction

$$\epsilon(\alpha) = -\mathbf{m} \cdot \mathbf{H} = -mH \cos\alpha,$$

where in this case

$$\alpha = 0, \frac{\pi}{2}, \pi, \frac{3\pi}{2}$$
$$\Downarrow$$
$$\begin{aligned} \epsilon(0) &= -mH, \\ \epsilon\left(\frac{\pi}{2}\right) &= 0, \\ \epsilon(\pi) &= mH, \\ \epsilon\left(\frac{3\pi}{2}\right) &= 0. \end{aligned} \qquad \text{(i)}$$

(b) For this paramagnet as well the partition function is a product of single particle partition functions:

$$Z(\beta, H) = [z(\beta, H)]^N,$$

where N is the number of spins in the system.

This time the single spin states are characterized by four angles. Hence

$$z(\beta, H) = e^{\beta m H} + e^{-\beta m H} + 2, \qquad \text{(ii)}$$

so that

$$Z(\beta, H) = (e^{\beta m H} + e^{-\beta m H} + 2)^N. \qquad \text{(iii)}$$

(c) Calculation of the average magnetization per spin:

(1) In this case only the states $\alpha = 0$ and $\alpha = \pi$ contribute to the magnetization along x, since the other two states have zero projections along x. With a field in the x direction the magnetization does not have a component along y here as well.

The average magnetization per spin is therefore

$$\begin{aligned}\langle m_x \rangle &= m_x(\alpha = 0)P_{\alpha=0}(\beta, H) + m_x(\alpha = \pi)P_{\alpha=\pi}(\beta, H) \\ &= m\frac{e^{\beta mH} - e^{-\beta mH}}{e^{\beta mH} + e^{-\beta mH} + 2}\,. \end{aligned} \tag{iv}$$

(2) Differentiating the partition function we obtain

$$\langle m_x \rangle = \frac{1}{\beta N}\frac{\partial \ln Z}{\partial H} = \frac{1}{\beta}\frac{\partial \ln z}{\partial H} = m\frac{e^{\beta mH} - e^{-\beta mH}}{e^{\beta mH} + e^{-\beta mH} + 2}\,.$$

(d) A power expansion of the numerator and the denominator in Eq. (iv), along with the approximation $e^x \simeq 1 + x$ for $x \ll 1$, gives the magnetization for weak fields:

$$\langle m_x \rangle \simeq \frac{\beta m^2}{2}H\,,$$

so that

$$\chi = \frac{m^2}{2kT}\,. \tag{v}$$

In this case as well Curie's law is valid and we obtain the same result as in Exercise 3.

(e) Using Eq. (2.5.7) and the result (ii), we obtain

$$\langle \epsilon \rangle = -\frac{\partial \ln z}{\partial \beta} = -mH\frac{e^{\beta mH} - e^{-\beta mH}}{e^{\beta mH} + e^{-\beta mH} + 2} = -mH\tanh\left(\frac{\beta mH}{2}\right)\,, \tag{vi}$$

which means that, as expected for a paramagnet,

$$\langle \epsilon \rangle = -\langle m_x \rangle H\,.$$

(f) The heat capacity per spin at a constant external field is

$$c_H = \left(\frac{\partial \langle \epsilon \rangle}{\partial T}\right)_H\,, \tag{vii}$$

where $\langle\epsilon\rangle$ is given in Eq. (vi). Differentiating we obtain

$$c_H = 2k\frac{(\beta mH/2)^2}{\cosh^2(\beta mH/2)}\,.$$

Solution 5 *Exercise on page 181*

The probability for a given state, α, with energy $E_\alpha = E$ is $Z^{-1}e^{-\beta E}$ [Eq. (2.5.5)]. However, there are many states with the same energy E. Their number $\Gamma(E)$ is given by (2.3.5). Thus the probability of finding the system in a state with energy E is

$$P(E) = Z^{-1}e^{-\beta E}\Gamma(E)\,. \tag{i}$$

Now $\Gamma(E)$ is related to the entropy by $\Gamma = e^{S/k}$, so that

$$P(E) = Z^{-1}e^{(TS-E)/kT}\,. \tag{ii}$$

In order to obtain the explicit dependence on E we have to substitute for Z the expression (2.5.12) with (2.5.11) and (2.3.13) for S. Note that it is not correct to use the expression for $S(T,H)$ which we obtained in Exercise 5.6, as this expression is obtained under the assumption that $E = \langle E\rangle$, whereas here we are interested precisely in arbitrary values of E which may be different from the average.

■ Recall the warning at the end of Sec. 5.2.

As the explicit form of $P(E)$ is complicated and not very helpful, we leave the result in the form of Eq. (ii).

It is worth noting that the probability is maximal not when S is maximal but when the combination $TS-E$ is. More on this in the coming part.

Solution 6 *Exercise on page 181*

(a) Suppose that out of the N steps made by the drunk N_+ were to the right and N_- to the left. If the length of each step is L, then he will be located at a distance

$$x_N = (N_+ - N_-)L \equiv qL\,. \tag{i}$$

We also have

$$N_+ + N_- = N\,. \tag{ii}$$

Note that if N is even, then q must also be even, and if N is odd, so is q. Hence q changes in steps of 2 between $-N$ and N. From Eqs. (i) and (ii) we have

$$N_+ = \frac{N+q}{2}, \quad N_- = \frac{N-q}{2}\,. \tag{iii}$$

We are thus interested in the probability of q having a certain value. This probability is equal to the probability that out of N steps N_+ will be to the right $[(1/2)^{N_+}]$ times the probability for there to be N_- steps to the left $[(1/2)^{N_-}]$ times the number of possibilities of distributing N steps into two groups $N!/(N_+!N_-!)$. Making use of Eq. (iii), we find

$$P(q) = \left(\frac{1}{2}\right)^N \frac{N!}{N_+!N_-!} = \frac{N!}{2^N \left(\frac{N+q}{2}\right)! \left(\frac{N-q}{2}\right)!}, \tag{iv}$$

and this is the probability for the drunk to be located at a distance $x = qL$ from the initial point. Note that $P(q)$ is different from zero only if q is an integer of the same parity as N.

(b) We use Stirling's formula, in the form

$$\ln n! \simeq n \ln n - n + \frac{1}{2}\ln(2\pi n). \tag{v}$$

We write $\ln P$ using (v), without neglecting the last term:

$$\begin{aligned}
\ln P &= \ln N! - \ln N_+! - \ln N_-! - N \ln 2 \\
&\simeq \left[N \ln N - N + \frac{1}{2}\ln(2\pi N)\right] - \left[N_+ \ln N_+ - N_+ + \frac{1}{2}\ln(2\pi N_+)\right] \\
&\quad - \left[N_- \ln N_- - N_- + \frac{1}{2}\ln(2\pi N_-)\right] - N \ln 2 \\
&= \left(N + \frac{1}{2}\right)\ln N - \left(N_+ + \frac{1}{2}\right)\ln N_+ - \left(N_- + \frac{1}{2}\right)\ln N_- \\
&\quad - N \ln 2 - \frac{1}{2}\ln 2\pi,
\end{aligned} \tag{vi}$$

where the fact that $N_+ + N_- = N$ led to the cancellation of the terms linear in N, N_+ and N_-. One term $\frac{1}{2}\ln 2\pi$ was canceled as well. Now, we use the fact that we are interested in the behavior of $P(q)$ near $q = 0$, which is the average of q. In this case Eq. (iii) implies that N_+ and N_- are very close to $\frac{1}{2}N$. We thus write

$$\ln N_\pm = \ln N - \ln 2 + \ln\left(1 \pm \frac{q}{N}\right). \tag{vii}$$

Taking only the first two terms in the expansion of the logarithm, we obtain

$$\ln N_\pm \simeq \ln N - \ln 2 \pm \frac{q}{N} - \frac{1}{2}\left(\frac{q}{N}\right)^2. \tag{viii}$$

- $\ln(1+x) = x - \frac{x^2}{2} + \frac{x^3}{3} - \dots$

We now substitute everything into (vi) and obtain

$$\ln P \simeq \left(N+\frac{1}{2}\right)\ln N - \left(N_+ + \frac{1}{2}\right)\left(\ln N - \ln 2 + \frac{q}{N} - \frac{q^2}{2N^2}\right)$$
$$- \left(N_- + \frac{1}{2}\right)\left(\ln N - \ln 2 - \frac{q}{N} - \frac{q^2}{2N^2}\right)$$
$$- N\ln 2 - \frac{1}{2}\ln 2\pi\,. \qquad \text{(ix)}$$

Note that of all the terms proportional to $\ln N$ only $-\frac{1}{2}\ln N$ is left, and, after neglecting $1/2N^2$ with respect to $1/N$, of all the terms that include q the only ones left are

$$\frac{(N_- - N_+)q}{N} + \frac{(N_+ + N_- + 1)q^2}{2N^2} = \left(-\frac{1}{2N} + \frac{1}{2N^2}\right)q^2 \simeq -\frac{q^2}{2N}\,.$$

And of all the terms that contain $\ln 2$, only $\ln 2$ is left. Thus we obtain

$$\ln P \simeq -\frac{q^2}{2N} - \frac{1}{2}\ln N + \ln 2 - \frac{1}{2}\ln 2\pi = \ln\frac{2}{\sqrt{2\pi N}} - \frac{q^2}{2N}\,. \qquad \text{(x)}$$

We have therefore obtained a Gaussian distribution:

$$P(q) = \frac{2}{\sqrt{2\pi N}}e^{-q^2/2N}\,. \qquad \text{(xi)}$$

The distribution function of x is obtained by substituting $q = x/L$ and by further dividing (xi) by 2 due to the fact that q changes by steps of 2, so that $\Delta x = 2L\Delta q$. From here

$$f(x) = \frac{1}{\sqrt{2\pi NL^2}}\exp\left(-\frac{x^2}{2NL^2}\right)\,. \qquad \text{(xii)}$$

(c) Since $\langle x\rangle = 0$,

$$(\Delta x)^2 = \langle x^2\rangle\,.$$

We have already calculated such integrals many times along this course, and the result is always one over twice the coefficient of the square of the random variable. Hence

$$\Delta x = \sqrt{NL^2} = L\sqrt{N}$$

and the relative width is

$$\frac{\Delta x}{NL} = \frac{1}{\sqrt{N}}\,.$$

Part III

Statistical Physics and Thermodynamics

Introduction

In Part II we saw that the canonical ensemble emerges naturally from the ensemble that characterizes an isolated system (the microcanonical ensemble). There it was demonstrated in the special case of a simple system — the paramagnet, but the argument can be generalized in a natural way to any system. The canonical description is the most useful one, as in most practical cases it is possible to control the temperature and not the energy of the system. Indeed we will use the canonical ensemble, as a starting point in the analysis of most systems we shall consider.

In this part we will employ the canonical ensemble to investigate the properties of several simple systems. Instead of deducing from the properties of the microcanonical ensemble the validity of the description based on the canonical ensemble, we shall accept it as a commandment. As a compensation we will dedicate the first chapter of this part to considerations that will make the transition smoother. First, we will show that the assumption of a canonical ensemble fits well the laws of thermodynamics. We will then deal with a collection of quantum oscillators, in which each degree of freedom has an infinite number of states that are discrete. This is a generalization of the paramagnet in which each degree of freedom has a finite number of states.

Next, we go on to discuss gases. As opposed to the paramagnet and the quantum oscillator, the degrees of freedom here, the coordinates and momenta of the molecules, are continuous variables. Several generalizations of the methods described in the preceding part are needed, in order to obtain the familiar results of dilute gases. Here as well we will check the correspondence with known results. We will show that the kinetic theory derives from statistical mechanics. We will deduce the ideal gas law as well as Dalton's law, and verify the consistency of our results, which will not be an easy matter, as we will find out. Finally, we will deal with fluctuations of thermodynamic quantities and the conditions under which they can be neglected, i.e. the question of the width of the distribution of thermodynamic quantities in the canonical ensemble.

Chapter 1

The Canonical Ensemble and Thermodynamics

1.1 The partition function and the internal energy

No harm will be done if we repeat once more the rule that an uncompromising test which every description of statistical mechanics must pass is its conformity with the laws of thermodynamics. The status of the latter is much stronger than that of the dynamical models or of the methods we formulate to calculate averages.

In the canonical description, the system is described by its microscopic states, which we denote by the index i. The innocent index i may be a state of a paramagnet (one of the 2^N values taken by the spins, denoted as $\{\sigma\}$ in Part II), or the collection of coordinates and momenta of all gas molecules. Each state i is assigned an energy E_i. The energy can be given by an expression like (2.2.9) for a paramagnet, or by the sum of kinetic and potential energies of all of the molecules in the gas, having coordinates and momenta that characterize the state i.

■ The classical physicist needs all the coordinates and momenta in order to characterize a state (see Chaps. 3 and 4 ahead). The quantum physicist will remark that if coordinates as well as momenta are needed, then a compromise will have to be made in the precision with which they are specified.

"Canonical" means that:

(a) In the ensemble all states with the same number of particles and the same volume are allowed.
(b) The relative probability for a state i to appear in the ensemble is

$$e^{-\beta E_i}\,,$$

where $\beta = 1/kT$.

In other words, the ratio of the probabilities for the appearance of two states i and j is

$$\frac{P(i)}{P(j)} = e^{-\beta(E_i - E_j)} . \tag{3.1.1}$$

■ We are treating i as if it were a discrete index, in order not to complicate the discussion. In a gas, for instance, this is not the case. However, the required generalizations are very simple, and will appear in the following chapters.

In Chap. 5 of Part II we mentioned that the partition function has an especially important role. We define it anew for a gas of particles:

partition function

$$Z(T,V,N) = \sum_i e^{-\beta E_i} . \tag{3.1.2}$$

On the left hand side we have emphasized the fact that the sum over states is a function of the temperature, the volume and the number of particles, since these are held fixed. The appearance of these variables and not others hints at a possible relation between Z and the Helmholtz free energy F. See e.g. Sec. 0.3, Part II. But why rush?

■ The volume is a natural variable when one is considering a gas or moving particles (liquid or solid). In the discussion of the paramagnet in Part II, the spins are fixed in their positions and are insensitive to the volume. The partition function there depends on T, H and N, which are held constant when summing over all the states of the system.

Exercise 1.1

Explain where is the dependence on V and N on the right hand side of (3.1.2).

Solution on page 292

The definition (3.1.2) holds for every system. Thus, we can express the average energy of the system in terms of its partition function using an expression identical to (2.5.7). That expression was obtained indeed in the special case of the paramagnet. But the same argument can be used again, since the transition from (2.5.6) to (2.5.7) did not depend on any particular feature of the system.

Exercise 1.2

Prove that (3.1.2) implies that

$$\langle E \rangle = -\frac{\partial \ln Z}{\partial \beta} = kT^2 \frac{\partial \ln Z}{\partial T} . \tag{3.1.3}$$

Solution on page 292

We identify $\langle E\rangle$ with the internal energy which in the thermodynamic context is denoted by the letter E or sometimes, in other texts, U (see also the remark at the end of Sec. 5.2 of Part II). $\langle E\rangle$ is of course a state function, in the thermodynamic sense, since if the volume, the number of particles and the temperature are given then $\langle E\rangle$ is determined.

Now, we continue to obtain the rest of the thermodynamic quantities using equations like (3.1.3). Once we have obtained these quantities, we will have the connection between the microscopic laws of the system and its thermodynamic properties; we will have reached an understanding of the statistical origin of the laws of thermodynamics.

1.2 Thermodynamic work

Since we have already found out how to express the internal energy in terms of the partition function, the next step is to obtain an expression for the thermodynamic work in terms of the partition function, and to find the correspondence with the first law of thermodynamics. Then we go on to identify the entropy and the Helmholtz free energy.

In order to identify the thermodynamic work, we first note that the work performed by a system is related to the variation of external parameters. In a gas, for instance, the work is related to the volume. The work can be related to a change in the position of the system, if it is located in an external force field. In the example of the paramagnet in Part II: if the external field H varies from one spatial position to another, then a change in the system's position is accompanied by a change of magnitude dH in the field, and will therefore involve performing work according to Eq. (2.1.6). The problem before us, therefore, is to identify and to formulate the relationship between the variation of the external parameters and the work performed by the system.

Suppose that we have identified such an external parameter that will, in most cases, depend on an external body, which affects the energies of the system's states (for example: the walls of a container, the source of a magnetic or a gravitational field, etc.). We denote it as X. The work that the system performs will therefore be performed on this body.

■ In the general case the partition function will depend on the variables T, X, N and the right hand side of Eq. (3.1.2) will be $Z(T, X, N)$.

The energy of each microscopic state depends, therefore, on X; namely,

$$E_i = E_i(X)\,. \tag{3.1.4}$$

This work is related to the existence of a "force":

$$F_i = -\frac{\partial E_i}{\partial X} \,. \tag{3.1.5}$$

■ The "force" in Eq. (3.1.5) is a force in a generalized thermodynamic sense and may also be pressure or magnetization.

It is important to make sure that the meaning of Eq. (3.1.5) is indeed clear: the system + the external body "tends," as usual, to reduce the energy. Suppose that the system is in a state i. If the increase in X involves a decrease in its energy, the system will tend to realize this change, i.e. it will apply a "force" F_i on the external body directed along X. F_i will then be positive, and $\partial E_i/\partial X$ negative. A similar argument can be performed in the case where E_i is an increasing function of X. The body will "want" to move in the direction of $-X$ in order to decrease the energy. That is, a negative "force" will act upon it.

Now, if the force is given by (3.1.5), then the work done by the system in a state i on the external body, when the external coordinate changes from X to $X + dX$, is obtained by the product of the force and dX or, from a different point of view, from the change in its energy:

$$\delta W_i = F_i dX = -\frac{\partial E_i(X)}{\partial X} dX \,. \tag{3.1.6}$$

The thermodynamic work performed by the macroscopic system will be, like every thermodynamic quantity, an average of δW_i over the canonical ensemble

$$\delta W = Z^{-1} \sum_i (\delta W_i) e^{-\beta E_i} \,. \tag{3.1.7}$$

Substituting (3.1.6) in (3.1.7), using the chain rule for the derivatives and carrying out the summation, we obtain

$$\delta W = \frac{1}{\beta} \frac{\partial \ln Z}{\partial X} dX \,. \tag{3.1.8}$$

Equation (3.1.8) tells us, therefore, that once we have obtained the partition function it is possible to calculate from it the thermodynamic work accompanying every change of the external parameter, by multiplying the change dX by the thermodynamic "force" $\beta^{-1}\partial \ln Z/\partial X$. It is important to note that (3.1.8) expresses the work performed by the system in a *quasi-static* process: we required that the infinitesimal process of varying X be performed, while the system has a definite temperature and the change dX induces a continuous change in the microscopic states. Only then are we allowed to use the probabilities of the canonical ensemble in order to calculate δW as in Eq. (3.1.7).

Exercise 1.3

Prove (3.1.8).

Solution on page 293

■ Note that Eq. (3.1.8) does *not* imply that

$$\delta W = -\frac{\partial E}{\partial X} dX \,.$$

Before we proceed, we exemplify the arguments made in the two cases we have already met: the paramagnet and the ideal gas.

Paramagnet

In this case the energy of a microscopic state i of N spins is given by

$$E_i = -\mu_B H \cdot (\sigma_1 + \sigma_2 + \ldots + \sigma_N) \,. \tag{3.1.9}$$

■ Here i denotes the microscopic states that were denoted in Part II by $(\sigma_1, \sigma_2, \ldots, \sigma_N)$.

The "force" is

$$\frac{\partial E_i}{\partial H} = -\mu_B \cdot (\sigma_1 + \sigma_2 + \ldots + \sigma_N) \tag{3.1.10}$$

and this is actually the magnetization in state i, M_i.

Thus, the work performed by the system in *state* i, when the magnetic field changes from H to $H + dH$, will be

$$\delta W_i = -\frac{\partial E_i}{\partial H} dH \tag{3.1.11}$$

and the thermodynamic work is obtained by averaging:

$$\delta W = -Z^{-1} \sum_i \frac{\partial E_i}{\partial H} e^{-\beta E_i} dH \,. \tag{3.1.12}$$

This sum is nothing but $-\frac{1}{\beta}\frac{\partial Z}{\partial H}$, so that

$$\delta W = \frac{1}{\beta}\frac{1}{Z}\frac{\partial Z}{\partial H} dH = \frac{1}{\beta}\frac{\partial \ln Z}{\partial H} dH \,, \tag{3.1.13}$$

and of course this is the result we wanted. In order to identify the meaning of $\beta^{-1} \partial \ln Z / \partial H$ we turn to Part II, to find that this is precisely the magnetization [see Eq. (2.5.8)]. Hence we can write

$$\delta W = M dH \,, \tag{3.1.14}$$

which is the result (2.1.6), obtained using similar arguments in the previous part.

Actually, we could have obtained this result by averaging (3.1.10) using the canonical probabilities P_i:

$$\left\langle \frac{\partial E}{\partial H} \right\rangle = \sum_i \frac{\partial E_i}{\partial H} P_i = -\mu_{\rm B} \sum_{\{\sigma\}} (\sigma_1 + \ldots + \sigma_N) P(\sigma_1, \ldots, \sigma_N)\,.$$

An ideal gas in a container

In spite of its apparent simplicity, this case requires special attention. The reason for this is that the role of the external parameter X is played by the volume whereas the energy of a molecule in the container apparently depends only upon its speed, since $\epsilon = \frac{1}{2}mv^2$. Now, if E_i is independent of V, then δW_i vanishes [see (3.1.6)], so clearly the thermodynamic average δW also vanishes. This leads to the paradoxical conclusion that the volume change of an ideal gas does not require any work.

In order to avoid this trap, we note that the energy of a molecule in the container has an additional term responsible for the confinement of the molecule to the container. This addition may be thought of as a potential that vanishes inside the container and increases sharply to infinity on the sides of the container (see e.g. Fig. 3.4.1 in Sec. 4.4 below). But although it is possible to continue from here in the usual manner, it is more convenient to calculate the thermodynamic work directly with the help of momentum conservation arguments. We, therefore, inquire into the relationship between the volume change of a gas in a container and the resulting work, and convince ourselves that $\beta^{-1}\partial \ln Z/\partial V$ is in fact the pressure. We do this by considering a piston that can move along x, as depicted in Fig. 3.1.1, and assume, for the sake of simplicity, that the system is one dimensional, so that all the velocities are directed along x.

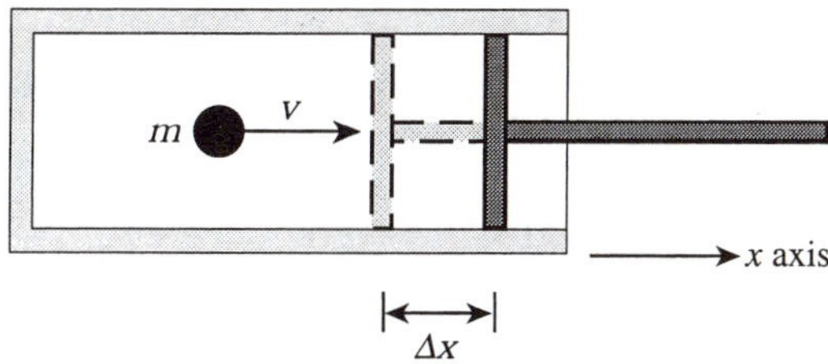

Fig. 3.1.1 A gas molecule transferring momentum to a piston.

We start by considering a single particle, of velocity v. The microscopic state of the particle will be characterized by the position of the particle and by its velocity. Clearly, if the particle hits the piston it will recoil and change its velocity, and momentum will be transferred to the piston as well as energy. This momentum and energy transfer will be calculated in the following way (compare with the calculation of the pressure in Part I and note the differences): the piston's position changes by $\Delta x = u\Delta t$ (see Fig. 3.1.1), where u is its speed during that interval. If the particle's recoil is elastic, the momentum it transfers to the piston is

$$\Delta p = 2m(v-u)\,. \tag{3.1.15}$$

Exercise 1.4

Prove that a particle that collides with a moving piston transfers momentum to it according to (3.1.15), and recoils from it with velocity $-v+2u$.

Solution on page 293

Since the velocity of the recoiled particle is reduced owing to the collision, it loses energy of order

$$-\Delta\epsilon = \frac{m}{2}v^2 - \frac{m}{2}(2u-v)^2 = 2mu(v-u)\,, \tag{3.1.16}$$

and assuming that the piston's velocity is not too high (namely $u \ll v$),

$$\Delta\epsilon \simeq -2muv\,. \tag{3.1.17}$$

Observe that when the piston is motionless ϵ does not change.

We now want to calculate the total energy loss of the gas as the piston advances by Δx during time Δt. The energy loss contributed by the molecules of velocity v is obtained, of course, by multiplying $\Delta\epsilon$ by the number of molecules whose distance from the piston is at most $v\Delta t$ (here as well we assume that $u \ll v$). If $n(v)$ is the number of molecules of velocity v per unit volume (see Chap. 1 of Part I), then the energy lost by them is

$$(\Delta E)_v = -(2muv)[n(v)Av\Delta t] = -2mv^2 n(v)A\Delta x\,. \tag{3.1.18}$$

The contribution of all the molecules (and not only of molecules with velocity v) to the energy loss is obtained by summing over (3.1.18):

$$\Delta E = -2m\sum v^2 n(v)A\Delta x\,, \tag{3.1.19}$$

where the sum is carried out over all velocities whose direction is towards the piston, namely over all $v > 0$. We should stress here the principal difference between the discussion here and the discussion of the pressure in Part I. Here $n(v)$ is *not* the Maxwell–Boltzmann distribution but a distribution that characterizes a *specific microscopic state*, i. It is therefore proper to add an index i to $n(v)$ and ΔE.

If in addition we note that $A\Delta x = \Delta V$, we obtain

$$\frac{\partial E_i}{\partial V} = -2m \sum v^2 n_i(v)\,. \tag{3.1.20}$$

The expression on the right is nothing but the pressure of the gas [see Eqs. (1.1.2) and (1.1.3)] *in the microscopic state* i, so that

$$\frac{\partial E_i}{\partial V} = -P_i\,. \tag{3.1.21}$$

In this way we have overcome the main difficulty and found that the volume independence of the energy in the microscopic state i is only apparent, and that actually the fact that the molecules of the gas are confined to move inside a container of volume V gives rise to a volume dependence, as expressed by Eq. (3.1.21). Since the energy of each microscopic state depends on the volume, clearly the partition function must also depend on it.

From here the continuation is clear and is carried on precisely as in the previous case:

$$\delta W = -Z^{-1} \sum \frac{\partial E_i}{\partial V} e^{-\beta E_i} dV = \frac{1}{\beta} \cdot \frac{\partial \ln Z}{\partial V} dV\,. \tag{3.1.22}$$

To identify the meaning of $\beta^{-1}\partial \ln Z/\partial V$ we note that the middle expression in (3.1.22) may be written, using (3.1.21), in the form

$$Z^{-1} \sum P_i e^{-\beta E_i}\,.$$

Since $Z^{-1}e^{-\beta E_i}$ are the canonical probabilities, this expression is the average pressure, P. Hence we obtain, as for the relation between the magnetization and the partition function, a relation between the the pressure and the partition function:

$$P = \frac{1}{\beta}\frac{\partial \ln Z}{\partial V}\,. \tag{3.1.23}$$

1.3 Entropy, free energy, the first and second laws

In the previous section we found that the thermodynamic work performed by a system is related to the change of the internal energy in individual microscopic states. The average change of internal energy resulting from a change in an external macroscopic parameter is the thermodynamic work.

This is, of course, the thermodynamic generalization of the energy conservation law from mechanics. However, in addition to the average change of the internal energy, which we may denote by $\langle dE \rangle$, we may consider another quantity, the change of the average internal energy, which we denote by $d\langle E \rangle$ or dE. Even though $\langle dE \rangle + \delta W = 0$ by definition, in general $dE + \delta W \neq 0$, as in thermodynamics. A mathematical justification for the inequality sign is the fact that the canonical probabilities depend, via the energies, on the macroscopic external parameter X, whose variation characterizes the process (see Exercise 1.3 and the remark that follows), and thus $\langle dE \rangle \neq d\langle E \rangle$. We identify the sum $dE + \delta W$ with the heat transferred to the system in a quasistatic process, δQ, and obtain the first law of thermodynamics [Eq. (2.0.1)]:

$$dE + \delta W = \delta Q \,. \tag{3.1.24a}$$

The next step is to obtain the entropy in terms of the partition function. To this end we have to verify that it is indeed possible to write δQ of Eq. (3.1.24a) in the form that is required by the second law of thermodynamics:

$$\delta Q = T dS \,. \tag{3.1.24b}$$

First, we shall find that as a result of a change of the external parameter by dX and a change in temperature expressed by $d\beta$, in a quasistatic process,

$$dE + \delta W = \left(\frac{1}{\beta} \frac{\partial \ln Z}{\partial X} - \frac{\partial^2 \ln Z}{\partial \beta \partial X} \right) dX - \frac{\partial^2 \ln Z}{\partial \beta^2} d\beta \,. \tag{3.1.25}$$

Exercise 1.5

(a) Prove Eq. (3.1.25).
(b) Show that the right hand side of Eq. (3.1.25) cannot be an exact differential of a state function.

Solution on page 294

Now, given (3.1.24), we ask: Is it possible to convert the right hand side of (3.1.25) into an exact differential by multiplying by β? And indeed

$$\beta(dE+\delta W) = \left(\frac{\partial \ln Z}{\partial X} - \beta \frac{\partial^2 \ln Z}{\partial \beta \partial X} \right) dX - \beta \frac{\partial^2 \ln Z}{\partial \beta^2} d\beta = k^{-1} dS \,, \tag{3.1.26}$$

where

$$k^{-1}S = \ln Z - \beta \frac{\partial \ln Z}{\partial \beta} \tag{3.1.27}$$

(k is the Boltzmann constant).

Exercise 1.6

Show that the right hand side of (3.1.26) is an exact differential, and prove (3.1.27).

Solution on page 295

We have thus identified the entropy, and obtained its relation with the canonical partition function, $Z(T, X, N)$. Lastly, we obtain the Helmholtz free energy $F(T, X, N)$, defined in thermodynamics (see Part II, Sec. 0.3) by

Helmholtz free energy

$$F = E - TS\,. \tag{3.1.28}$$

To this end we note that the second term on the right hand side of Eq. (3.1.27) is none other than βE [see Eq. (3.1.3)]. Hence,

$$S = k \ln Z + \frac{E}{T}\,. \tag{3.1.29}$$

Comparing (3.1.29) with (3.1.28), we obtain

$$F = -kT \ln Z = -\beta^{-1} \ln Z\,. \tag{3.1.30}$$

Exercise 1.7

As is known from thermodynamics, the entropy is obtained from the free energy by [see Eq. (2.0.27)]

$$S = -\left(\frac{\partial F}{\partial T}\right)_{X,N}\,.$$

Start from the expression (3.1.30) for F, and verify Eq. (3.1.27).

Solution on page 296

We have seen here one of the convincing successes of statistical mechanics (according to Gibbs): statistical mechanics created a link between the conservation of energy in the microscopic theory and the first law of thermodynamics. It was then found that the average values of the internal energy, the work and the heat, defined naturally within statistical mechanics and averaged, satisfy the second law of thermodynamics. In this manner we have obtained the internal energy, the work, the heat and the entropy, and finally the free energy in terms of the canonical partition function. We know from thermodynamics that given F, a complete description of the macroscopic properties of the system is possible. Since it is possible to calculate F from Z [Eq. (3.1.30)], the calculation of Z in the canonical ensemble provides us with all of the thermodynamic information on the system.

1.4 The paramagnet — revisited

To end this chapter we implement the results we obtained in the simple case of the paramagnet.

In Chap. 5 of Part II, we saw that the canonical partition function of the paramagnet of spin 1/2, can be written as (2.5.11) and (2.5.12), namely

$$Z(T,H,N) = \left[2\cosh\left(\frac{\mu_B H}{kT}\right)\right]^N , \tag{3.1.31}$$

where N is the number of spins in the system.

The Helmholtz free energy is, therefore,

$$F(T,H,N) = -kT\ln Z = -NkT\ln\left(2\cosh\frac{\mu_B H}{kT}\right) \tag{3.1.32}$$

and the entropy

$$S = -\left(\frac{\partial F}{\partial T}\right)_{H,N} = Nk\left[\ln\left(2\cosh\frac{\mu_B H}{kT}\right) - \frac{\mu_B H}{kT}\tanh\frac{\mu_B H}{kT}\right] . \tag{3.1.33}$$

Exercise 1.8

Show that from F and S it is possible to obtain E in Eq. (2.5.13), with the help of thermodynamic relationships.

Solution on page 296

Exercise 1.9

Show that from S it is possible to obtain the expression for the specific heat, at a constant field H, given in (2.5.18).

Solution on page 296

Exercise 1.10

Is it possible to obtain from F the entropy as a function of the energy, the field and the number of spins? That is, is it possible to pass from $F(T,H,N)$ to $S(E,H,N)$ as given by Eq. (2.3.13)?

Solution on page 297

Exercise 1.11

If the energy of the microscopic states changes by the addition of a constant which is independent of the state

$$E_i \to E_i + C\,,$$

how do the partition function, the average energy, the free energy and the magnetization change?

Solution on page 298

1.5 On the statistical meaning of the free energy

In an isolated system in which the energy is constant, a state of thermodynamic equilibrium is attained when the entropy is maximal. You have seen an example of this in Part II, Chap. 4, where we discussed two systems a and b (of spins, for example) which are isolated from the rest of the universe but which interact thermally with each other. In this case the entropy of each of the systems is a function of the magnetic field, the number of spins and the energy of each system where the fact that the systems are isolated is expressed by the constancy of $E_a + E_b$. Nevertheless, the energy of each system is free to change. Equilibrium between the two systems is attained at the most probable macroscopic state of the composite system, where the total entropy attains its maximum. If we choose the energy E_a as a variable that characterizes the partition of the total energy between the two systems, the condition for equilibrium is

$$\frac{\partial}{\partial E_a}[S(E_a, H_a, N_a) + S(E - E_a, H_b, N_b)] = 0\,. \qquad (3.1.34)$$

Actually this line of argument is completely general, and may be applied to any system, provided the entropy is expressed in terms of the appropriate variables (volume instead of magnetic field, pressure instead of magnetization, etc.).

Now we ask, what happens in a nonisolated system which instead has a given temperature, namely that is free to exchange energy with a heat bath. What is the quantity that attains its maximum at the most probable state?

Clearly the entropy (or the number of states Γ) is related to the answer, since the larger the entropy, the larger the probability of the macroscopic state. But here there is an additional factor acting in the opposite direction, which is the Boltzmann factor, $e^{-\beta E}$. This factor expresses the fact that lower energy states have higher probability.

Hence the probability for the system to be in a macroscopic state of energy E is

$$P(E) = Z^{-1}e^{-\beta E}\Gamma(E)\,. \tag{3.1.35}$$

Now, we can write $\Gamma = e^{S/k}$, and from here it is immediately clear that the most probable state is the one in which the sum $S - E/T$ is maximal. We have seen therefore that a system that is held at a constant temperature finds its way to equilibrium subject to two opposing factors: the entropy's tendency to increase and the energy's tendency to decrease. The compromise is that equilibrium is determined by the combination $S - E/T$.

But this is nothing more than the free energy in disguise:

$$S - \frac{E}{T} = -\frac{F}{T}\,, \tag{3.1.36}$$

so that the probability for a system coupled to a heat bath to have energy E is

$$P(E) = Z^{-1}e^{-\beta F(E)}\,. \tag{3.1.37}$$

Note that we have suppressed the dependence of F and Z on all other variables, such as V, N, T.

The most probable state of a system in which all the variables are fixed except for the energy is therefore the state in which $-F(E)$ is maximal or $F(E)$ is minimal. In this case, one has

$$\frac{1}{T} = \frac{\partial S}{\partial E}\,, \tag{3.1.38}$$

which is the relation we use to pass from the microcanonical ensemble to the canonical one.

Exercise 1.12

Prove (3.1.38).

Solution on page 298

The free energy is therefore the fundamental function of the canonical ensemble and satisfies

$$dF = dE - TdS - SdT = -SdT - \delta W\,. \tag{3.1.39}$$

This is the origin of the name "free energy." In a process that occurs without the transfer of heat, a system can perform work at the expense of its internal energy. But if the process takes place at a constant temperature, it must be accompanied by the transfer of heat, so that the work is performed at the expense of the *free energy*: $\delta W = -(dF)_T$.

At the temperature of absolute zero the free energy becomes identical to the internal energy, so that any changes in the free energy are equivalent to internal energy changes, as expected from a state without any thermal fluctuations. At higher temperatures, the free energy decreases with T, as is easy to see from the relation [obtained from (3.1.39)]

$$\frac{\partial F}{\partial T} = -S\,, \tag{3.1.40}$$

since the entropy is always nonnegative.

In order to illustrate the fact that the free energy attains its minimum at equilibrium, we return to the paramagnet which is coupled to a heat bath at temperature T. The free energy of a *configuration* of the paramagnet having energy E is

$$\begin{aligned} F(E) = E + kT\Bigg\{&\left(\frac{N}{2} - \frac{E}{2\mu_{\mathrm{B}}H}\right)\ln\left(\frac{1}{2} - \frac{E/N}{2\mu_{\mathrm{B}}H}\right) \\ &+ \left(\frac{N}{2} + \frac{E}{2\mu_{\mathrm{B}}H}\right)\ln\left(\frac{1}{2} + \frac{E/N}{2\mu_{\mathrm{B}}H}\right)\Bigg\}\,, \end{aligned} \tag{3.1.41}$$

where we have used Eq. (2.3.13) for S.

$F(E)$ as a function of E and the probability $P(E)$ of a configuration of energy E are depicted in Fig. 3.1.2, for a system with $N = 200$. Note that the minimum of the free energy and the maximum of the probability are attained at the same energy.

Exercise 1.13

(a) Verify that the energy at which $F(E)$ attains its minimum in Fig. 3.1.2 is in fact consistent with the known energy of a paramagnet obtained, for example, in Exercise 1.8.
(b) Compare the minimum of $F(E)$ as well against Eq. (3.1.32).
(c) Show that from the requirement that $F(E)$ be minimal at thermodynamic equilibrium the known expressions for E and F are obtained.

Solution on page 299

Another quantity that should be mentioned here is the chemical potential. We reached the concept of temperature in a natural manner through the requirement that the equilibrium state between two isolated systems, that are in thermal contact and can exchange energy, be the state in which the entropy of the composite system is maximal. In this state the quantity $\partial S/\partial E$ must be equal in the two systems. We identified this quantity as $1/T$.

chemical potential

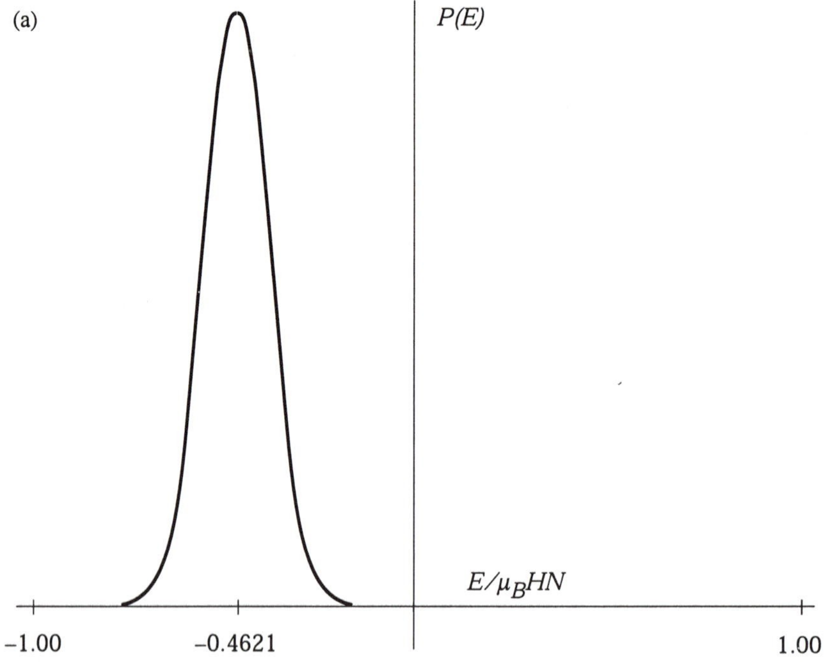

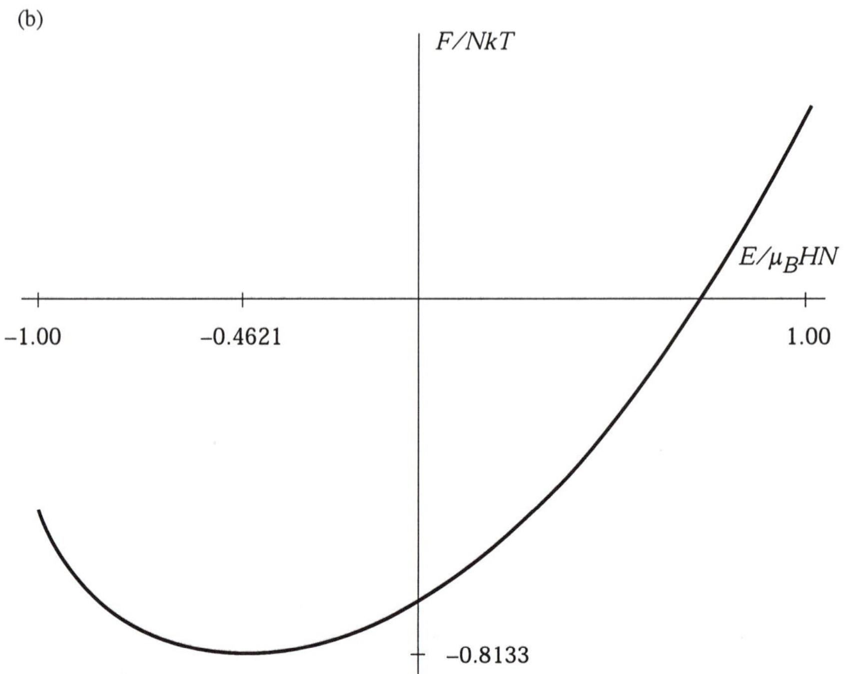

Fig. 3.1.2 (a) The probability of a configuration of energy E, for a paramagnet with $N = 200$ where $\mu_B H/kT = 0.5$. The units of the vertical axis are arbitrary. (b) The free energy of a configuration of energy E. Note that the minimum free energy is at the most probable energy.

We now ask the analogous question concerning the number of particles. Suppose that both systems contain particles that are free to move (for instance, gas molecules) and that a partition separates the two systems but allows the transfer of particles between them. We shall also assume, for simplicity, that the two systems contain the same type of particles.

The number of states of such a system depends of course on the energy, the volume and the number of particles, and hence so does the entropy $S(E, V, N)$. The total entropy of the composite system depends on all six variables $E_a, V_a, N_a, E_b, V_b, N_b$, but only three will be independent, as the total energy, the total number of particles N and the total volume V remain constant.

From the requirement of maximum entropy we obtain not only the condition of equal temperatures (and the condition of equal pressures by differentiating with respect to V_a, for example, if V_a varies) but also the condition that the chemical potentials of the two systems should be equal, namely $\mu_a = \mu_b$ where μ is defined by [see Eq. (2.0.23)]

$$\mu = -T\left(\frac{\partial S}{\partial N}\right)_{E,V} . \tag{3.1.42}$$

Exercise 1.14

Prove that the chemical potentials are equal at equilibrium between the systems.

Solution on page 300

We now complicate matters a bit, and assume that both systems are not isolated from the rest of the universe but are at equilibrium with a heat bath at temperature T. In this case equilibrium is attained when the free energy is minimal, and in the same way as for the isolated system we obtain $\mu_a = \mu_b$, where [see e.g. Eq. (2.0.27)]

$$\mu = \left(\frac{\partial F}{\partial N}\right)_{T,V} . \tag{3.1.43}$$

In order to better understand the meaning of the chemical potential, let us also check what happens in a state that is not an equilibrium state. In such a state the free energy of the system is not minimal, and hence the system tends to decrease it by redistributing N between a and b. Suppose that the number of particles in system a changes (increases or decreases) by dN_a. The free energy will change by

$$dF = \left(\frac{\partial F_a}{\partial N_a} + \frac{\partial F_b}{\partial N_a}\right) dN_a = (\mu_a - \mu_b) dN_a . \tag{3.1.44}$$

Recall that since the free energy must decrease, $dF < 0$.

This means that N_a increases ($dN_a > 0$) as long as $\mu_a < \mu_b$. In other words, particles will flow from b to a if the chemical potential of b is larger than that of a. We have therefore found that the chemical potential determines the direction of particle flow between systems just as the temperature determines the direction of energy flow: particles will flow from the higher chemical potential to the lower one.

We demonstrate this property with the help of the ideal gas. In Chap. 5, Eq. (3.5.9) below, we find that the chemical potential of an ideal gas is given by

$$\mu = kT \ln \left[\left(\frac{h^2}{2\pi m k T} \right)^{3/2} n \right] = kT \ln \left[\frac{h^3}{(2\pi m)^{3/2}} \cdot \frac{P}{(kT)^{5/2}} \right] , \qquad (3.1.45)$$

where P is the pressure of the gas substituted from the equation of state. It is clear from this that the chemical potential increases with the density of the gas or with its pressure, and thus that molecules of the gas will flow from regions of high density to regions of lower density or from regions of high pressures to those of low pressures.

▪ Compare the discussion here to the discussion of self-diffusion — Secs. 3.3–3.5 of Part I.

Finally, we note that the magnitude of the chemical potential is of no physical significance; only the *differences* in the chemical potential are meaningful, as is the case for the energy.

Chapter 2

Harmonic Oscillator and Einstein Solid

2.1 Microscopic states

In Part II we have treated the paramagnet in a magnetic field in which each degree of freedom, each spin, had a finite number of energy levels: two states for spin $1/2$, or $2J+1$ states for spin J. This kind of quantization is absent in the classical case, where the degrees of freedom have a continuum of values. Before treating such a problem, we treat an intermediate case, in which the degrees of freedom take discrete values, but the number of different values is infinite. The case we shall treat here has also some practical applications for the calculation of the specific heat of solids.

For us, the sole importance of the quantum aspect is that the energy levels of the microscopic states take discrete values. It may be worth mentioning once more that the fact that in quantum theory the energy levels are discrete, makes the statistical interpretation of such systems more natural than their classical counterparts.

The degree of freedom, replacing the spin from Part II, is a harmonic oscillator. The system we shall discuss is a lattice which has at each of its sites a harmonic oscillator with angular frequency (or angular velocity) ω in place of the magnetic moment in Part II. You may think of this system as a lattice of masses m, each connected at the lattice site to a spring with spring constant K, so that $\omega = \sqrt{K/m}$. The interest in such a system stems from the fact that in many respects this is a good approximation for a crystal whose atoms are oscillating around their equilibrium positions. When the oscillations are small their motion is approximately harmonic.

Before turning to calculate the partition function of the canonical ensemble, in accordance with the discussion in Chap. 1 of this part, we must discuss the states of a single oscillator. Because the system we are thinking of is a lattice of atoms, the oscillators discussed here are quantum oscillators. The quantum discussion of the harmonic oscillator may be summarized for our purposes by the fact that the energy levels of the one-dimensional harmonic oscillator are characterized by a nonnegative integer n:

$$\epsilon_n = \left(n + \frac{1}{2}\right)\hbar\omega, \quad n = 0, 1, 2, 3 \ldots \tag{3.2.1}$$

Note that all the values of ϵ_n are positive, just as for a classical oscillator, but this is where the similarity ends. The amplitude A of a classical harmonic oscillator can take a continuum of values, and hence its energy, $\epsilon = \frac{1}{2}m\omega^2 A^2$, takes a continuum of values, starting from the minimal value $\epsilon = 0$. The energies of the quantum oscillator are quantized in intervals of $\hbar\omega$, starting from the minimal value, which is *not zero* but $\epsilon = \frac{1}{2}\hbar\omega$.

As n increases, the energy of the oscillator increases. It is thus natural to refer to n as the degree of excitation of the oscillator. n characterizes the state of the oscillator just as the number σ used for the magnetic moment in Part II.

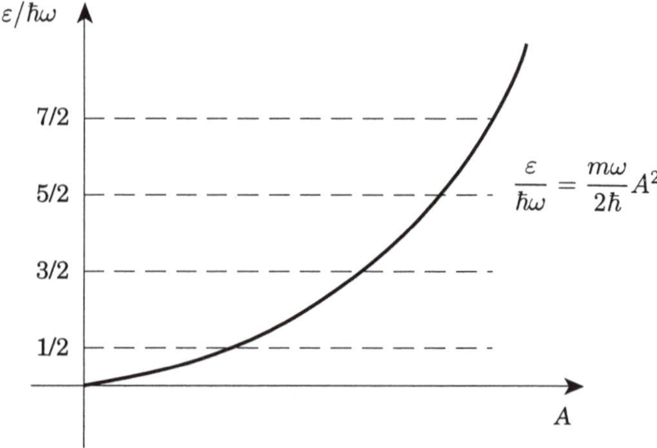

Fig. 3.2.1 The energy levels of a harmonic oscillator: Dashed lines — the quantized levels; full line — energy of classical oscillator vs amplitude A.

The next step is to investigate the microscopic states of a lattice of N harmonic oscillators as a whole. Since each oscillator is characterized by a degree of excitation n, the microscopic state of N oscillators will be characterized by N degrees of excitation $(n_1, n_2, n_3, \ldots, n_N)$, similar to the specification of all the spin values in terms of the N numbers $(\sigma_1, \sigma_2, \ldots, \sigma_N)$ in the paramagnet in Part II. The energy of a microscopic state in which oscillator 1 has a degree of excitation n_1, the second a degree of excitation n_2 and so on, will be the sum of all the oscillator energies:

$$E(n_1, n_2, \ldots, n_N) = \epsilon(n_1) + \epsilon(n_2) + \cdots + \epsilon(n_N)\,. \tag{3.2.2}$$

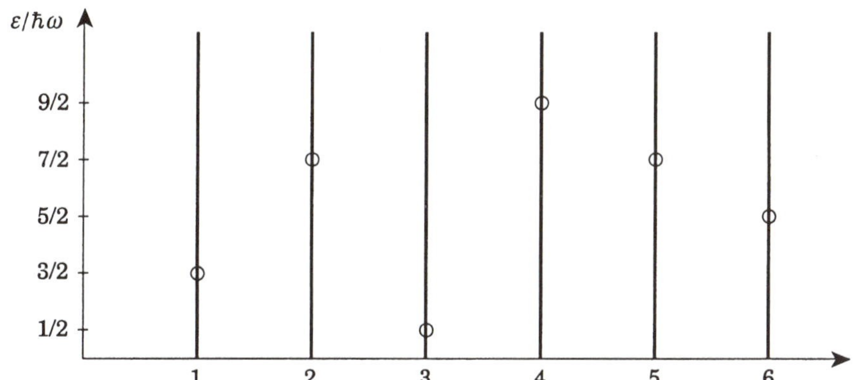

Fig. 3.2.2 A state of a system of six oscillators: $n_1 = 1, n_2 = 3, n_3 = 0, n_4 = 4, n_5 = 3, n_6 = 2$.

All this is, of course, based on the assumption that the N oscillators are unaffected by each other, and that all are vibrating at the same frequency in an independent manner. Figure 3.2.2 schematically describes a state of a system with $N = 6$.

2.2 Partition function for oscillators

We now write the partition function and obtain

$$Z = \sum_{n_1,n_2,\ldots,n_N} e^{-\beta E(n_1,n_2,\ldots,n_N)} = \sum_{n_1,n_2,\ldots,n_N} e^{-\beta\epsilon(n_1)} e^{-\beta\epsilon(n_2)} \ldots e^{-\beta\epsilon(n_N)}, \tag{3.2.3}$$

and each summation variable varies between 0 and ∞. We can write the sum of products as a product of sums, in precisely the same way as was done for Eq. (2.5.9):

$$Z = \left(\sum_{n_1=0}^{\infty} e^{-\beta\hbar\omega(n_1+1/2)}\right)\left(\sum_{n_2=0}^{\infty} e^{-\beta\hbar\omega(n_2+1/2)}\right) \times \cdots \times \left(\sum_{n_N=0}^{\infty} e^{-\beta\hbar\omega(n_N+1/2)}\right). \tag{3.2.4}$$

Since all the factors in the product on the right hand side of (3.2.4) are identical, we write the partition function in the form

$$Z = z^N, \tag{3.2.5}$$

where z may be thought of as a partition function of a single oscillator:

$$z = \sum_{n=0}^{\infty} e^{-\beta\epsilon_n} = \sum_{n=0}^{\infty} e^{-\beta\hbar\omega(n+1/2)}. \tag{3.2.6}$$

Exercise 2.1

In order to demonstrate the equality of (3.2.3) and (3.2.4), let us study the system of three identical oscillators, each of them having three energy levels. Verify for this case that (3.2.3) and (3.2.4) are indeed identical.

Solution on page 300

Our problem is therefore reduced to calculating the partition function of a single oscillator which is nothing more than the summation of a geometric series:

$$z = e^{-\beta\hbar\omega/2} \sum_{n=0}^{\infty} e^{-\beta\hbar\omega n} = \frac{e^{-\beta\hbar\omega/2}}{1 - e^{-\beta\hbar\omega}} . \tag{3.2.7}$$

■ Reminder: the sum of an infinite geometric series

$$\sum_{n=0}^{\infty} x^n = \frac{1}{1-x}, \quad x < 1 .$$

Now we can use the general formula (3.1.3) to calculate the average energy of an oscillator. The result is

$$\langle \epsilon \rangle = \frac{\hbar\omega}{2} + \frac{\hbar\omega}{e^{\hbar\omega/kT} - 1} . \tag{3.2.8}$$

Exercise 2.2

Prove (3.2.8).

Solution on page 302

Another interesting question is the average value of n, the degree of excitation of the oscillators. The answer is

$$\langle n \rangle = \frac{1}{e^{\hbar\omega/kT} - 1} , \tag{3.2.9}$$

Bose–Einstein distribution

which is the celebrated Bose–Einstein distribution. More on this in the parts that follow.

Exercise 2.3

Prove (3.2.9) in two ways:

(a) By a direct calculation of the average.
(b) From (3.2.8).

Solution on page 302

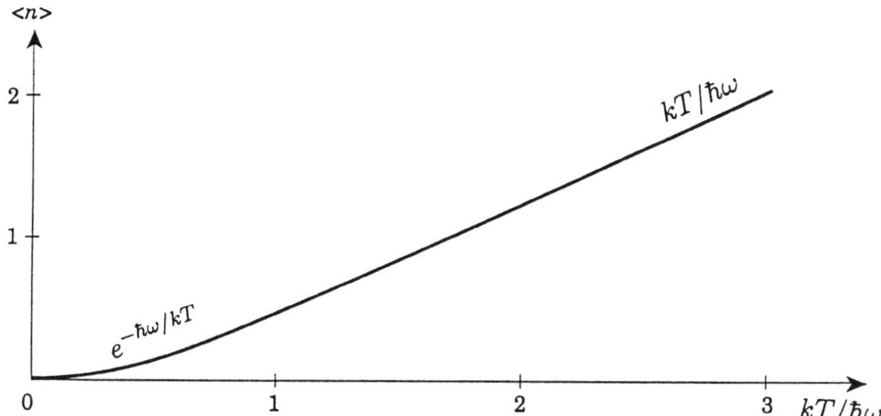

Fig. 3.2.3 The temperature dependence of the average degree of excitation of a harmonic oscillator.

Analyzing the temperature dependence of $\langle n \rangle$ (Fig. 3.2.3), we find that

$$\langle n \rangle \sim \begin{cases} e^{-\hbar\omega/kT}, & kT \ll \hbar\omega \text{ (low temperatures)}, \\ \dfrac{kT}{\hbar\omega}, & kT \gg \hbar\omega \text{ (high temperatures)}. \end{cases} \tag{3.2.10}$$

Exercise 2.4

Verify that (3.2.9) has the asymptotic behaviors (3.2.10).

Solution on page 303

Namely, at low temperatures $\langle n \rangle$ vanishes faster than any power of T. This means that the oscillator cannot be excited from its ground state $n = 0$ since the energy kT supplied by the heat bath is insufficient to overcome the difference $\hbar\omega$ to the state $n = 1$. In contrast, at high temperatures states may be excited up to a certain value of n, given approximately by the thermal energy divided by $\hbar\omega$. Thus, we may expect that

$$\langle n \rangle \sim \frac{kT}{\hbar\omega}, \qquad e^{-\hbar\omega/kT} \qquad kT/\hbar\omega \qquad kT/\hbar\omega$$

which is the relation (3.2.10) for high T.

phonons

■ In the professional jargon, instead of saying that the oscillator is excited to degree of excitation n, we say that there are n "bosons" or "phonons," each with energy $\hbar\omega$. The calculation of $\langle n \rangle$ is described in this language to be the average number of phonons as a function of T. Equation (3.2.10) states that the phonons disappear as $T \to 0$ and rapidly grow in number with increasing temperature.

Exercise 2.5

(a) Calculate the free energy of a harmonic oscillator as well as its entropy, and sketch their temperature dependence.

(b) If the frequency of the oscillator is 10^{13} s^{-1}, what is the transition temperature from low temperature behavior to high temperature behavior?

Solution on page 304

The discerning reader will probably have noticed that our discussion in this section began with a system of N oscillators and drifted after Eq. (3.2.5) to a discussion of a single oscillator and to the calculation of its thermal averages, which are, at least at first sight, of doubtful significance. To make things clear we note that the single oscillator upon which our discussion concentrated may be thought of as an oscillator in thermodynamic equilibrium with a heat bath with which it may exchange energy. You may, for example, picture for yourself as a concrete example a situation in which a single harmonic (quantum) oscillator resides in a container full of gas at a given temperature. The concepts of single particle partition functions and of thermal averages with respect to such ensembles are to be understood from this point of view. Such a description is made possible by the absence of interaction between the oscillators.

2.3 Einstein's solid

We now describe a model for the vibrations of a solid. If the crystal is made up of N atoms, the motion of each of them has three independent components. The atoms cannot move freely; instead they vibrate about equilibrium positions, which determine the geometric structure of the crystal. The possible vibrations of each atom are described by a model of three harmonic oscillators, so that N atoms are equivalent to $3N$ harmonic oscillators.

To make things simple, we choose, as Einstein did, the same frequency for all the oscillators. This simplifying assumption can and must be improved, as was done by Debye. However, we shall continue with the simplified model for the time being, namely the model of $3N$ oscillators, all of which have the same frequency, ω.

If the $3N$ oscillators do not affect one another, we may describe a microscopic state of the system by $3N$ numbers n_α $(\alpha = 1, \ldots, 3N)$, which describe the state of excitation of oscillator number α (or the number of bosons of type α; see Fig. 3.2.2).

From now on we can use the partition function, obtained in the preceding sections, for calculating the relevant thermodynamic quantities, such as the average energy, free energy and specific heat. The only change we will have to make is to change the meaning of N: N, which was the number of oscillators, will here denote the number of atoms in the crystal, so that we replace N by $3N$.

The partition function of our crystal is

$$Z = z^{3N}\,, \tag{3.2.11}$$

where z is as given by (3.2.7).

Using (3.1.30) we find for the free energy

$$F = -kT\ln Z = -3NkT\ln z = 3N\left[\frac{\hbar\omega}{2} + kT\ln(1 - e^{-\beta\hbar\omega})\right]\,. \tag{3.2.12}$$

Of course, we could have obtained this result by multiplying f as calculated in Exercise 2.5, by $3N$. In this way we also obtain the average energy:

$$\langle E\rangle = -\frac{\partial \ln Z}{\partial\beta} = -3N\frac{\partial \ln z}{\partial\beta} = 3N\left(\frac{\hbar\omega}{2} + \frac{\hbar\omega}{e^{\beta\hbar\omega} - 1}\right)\,. \tag{3.2.13}$$

We now investigate the behavior of the average energy at high temperatures and at low temperatures:

(a) At high temperatures $kT \gg \hbar\omega$, or $\beta\hbar\omega \ll 1$, so that

$$e^{\beta\hbar\omega} \simeq 1 + \beta\hbar\omega + \frac{1}{2}(\beta\hbar\omega)^2\cdots\,,$$

and expanding the denominator on the right hand side of Eq. (3.2.13) in powers of $\beta\hbar\omega$ we find, as in (3.2.10),

$$\langle E\rangle \simeq 3NkT\,, \tag{3.2.14}$$

which is the equipartition law of classical kinetic theory (see Secs. 1.3 and 1.6 of Part I): Each degree of freedom of a harmonic oscillator has a kinetic energy term $\frac{1}{2}mv^2$, and a potential energy term $\frac{1}{2}Kx^2$. As we will find in Sec. 4.3, each of these terms contributes $\frac{1}{2}kT$. Hence $3N$ oscillators have an average energy of $3NkT$.

It is important to note also that the result (3.2.14) does not involve Planck's constant, $\hbar$. Thus, its origin cannot be quantum-mechanical.

A different way of understanding this result is to interpret it as the limit $\hbar \to 0$ of the quantum energy Eq. (3.2.13), at constant temperature. If $\hbar \to 0$, the separation between successive energy levels vanishes and the classical continuum returns. Since only the ratio $\hbar\omega/kT$ determines the behavior (3.2.14), the limit of large T is equivalent to the limit of small $\hbar$.

(b) At low temperatures $kT \ll \hbar\omega$, or $\beta\hbar\omega \gg 1$. In this case

$$e^{\beta\hbar\omega} \gg 1\,,$$

so that

$$\frac{1}{e^{\beta\hbar\omega}-1} \simeq e^{-\beta\hbar\omega}\,.$$

and the average energy is

$$\langle E\rangle \simeq 3N\hbar\omega\left(e^{-\hbar\omega/kT}+\frac{1}{2}\right)\,. \tag{3.2.15}$$

This means that at low temperatures the average energy tends very rapidly to the minimal value, $3N\frac{\hbar\omega}{2}$, allowed by quantum theory. The excited states are "frozen."

From (3.2.13), with N taken as Avogadro's number, it is possible to calculate the molar specific heat:

$$C = \frac{\partial\langle E\rangle}{\partial T} = \frac{3R(\hbar\omega)^2}{(kT)^2}\frac{e^{\hbar\omega/kT}}{(e^{\hbar\omega/kT}-1)^2}\,, \tag{3.2.16}$$

where R is the gas constant.

Exercise 2.6

(a) Deduce (3.2.16) from (3.2.13).
(b) Deduce (3.2.16) from the entropy.

Solution on page 305

We can obtain the behavior of the specific heat at high temperatures and at low temperatures by checking the two limits of (3.2.16) or (simpler) from Eqs. (3.2.14) and (3.2.15), respectively. Thus, at high temperatures

$$C \simeq 3R\,, \tag{3.2.17}$$

and at low temperatures

$$C \simeq 3R\left(\frac{\hbar\omega}{kT}\right)^2 e^{-\hbar\omega/kT}\,. \tag{3.2.18}$$

Dulong–Petit law

The result (3.2.17) is the Dulong–Petit law. Dulong and Petit found (in 1819) that many solids have a constant molar specific heat of $3R$. The fact, found later, that as the temperature decreases so does the specific heat, caused grave difficulties for classical physics. This phenomenon was also found at intermediate temperatures, but only in hard crystals. These are different manifestations of the "heat capacity problem" which we have already met in Part I (Sec. 1.3) on the molecular level. Einstein's model, as presented here, explains the decrease of the specific heat with decreasing temperature — Eq. (3.2.16). The decrease with stiffness is also explained, since the frequency ω of a stiffer crystal will be higher. But high and low temperatures are determined, as we have seen, by the ratio of kT and $\hbar\omega$, so that if $\hbar\omega$ is large, the crystal reaches the behavior that is characteristic of low temperatures in a temperature range in which a softer crystal behaves classically.

Exercise 2.7

(a) If the oscillators have angular frequency $\omega = 10^7$ s^{-1}, how much heat must be supplied to a mole of crystal in order to raise its temperature by 10^{-2} degrees in the vicinity of 1 K, and in the vicinity of 100 K?

▪ A crystal with such a frequency is extremely soft and unrealistic. The crystal appearing in (b) has a realistic frequency.

(b) Repeat (a), with $\omega = 10^{12}$ s^{-1}.

Solution on page 306

The characteristic temperature for each material is the *Einstein temperature*, $\Theta_{\rm E} = \frac{\hbar\omega}{k}$, which we can use to rewrite Eq. (3.2.16) in the form

Einstein temperature

$$C = 3R\left(\frac{\Theta_{\rm E}}{T}\right)^2 \frac{\exp(\Theta_{\rm E}/T)}{[\exp(\Theta_{\rm E}/T) - 1]^2}\,. \tag{3.2.19}$$

For lead $\Theta_{\rm E} \simeq 90$ K, whereas for diamond $\Theta_{\rm E} \simeq 2000$ K. Hence at room temperature the specific heat of lead will behave classically, according to the equipartition law, or the Dulong–Petit law. The diamond will behave in a very different manner; its specific heat will be much smaller than the classical value.

Both cases can be identified in Fig. 3.2.4, which depicts the specific heat as a function of $T/\Theta_{\rm E}$. For lead at room temperature $T/\Theta_{\rm E} \simeq 3$ and $C/R \simeq 3$. In contrast, for diamond at the same temperature $T/\Theta_{\rm E} \simeq 0.15$, and the specific heat is small.

A study of Fig. 3.2.4 reveals that Einstein's theory indeed gives a qualitative description of the temperature variation of the specific heat and constitutes an interpolation between the classical, high temperature, value ($3R$ per mole) and zero, as the temperature tends to zero. However, the quantitative correspondence with the experimental results is not very good. The reason for this is the unrealistic assumption that all the oscillators have the same frequency. Debye constructed a theory that accounts correctly for the vibrations of the crystal, taking into account the fact that the motion of one atom will, necessarily, affect its neighbors. Thus the vibrations which must be taken into account are the collective vibrations of many atoms, vibrational modes, having many different frequencies. When the frequency distribution is used to modify Einstein's theory, the result is in very good quantitative agreement with experiment, as may be seen in Fig. 3.2.4. More on the Debye model in the next part.

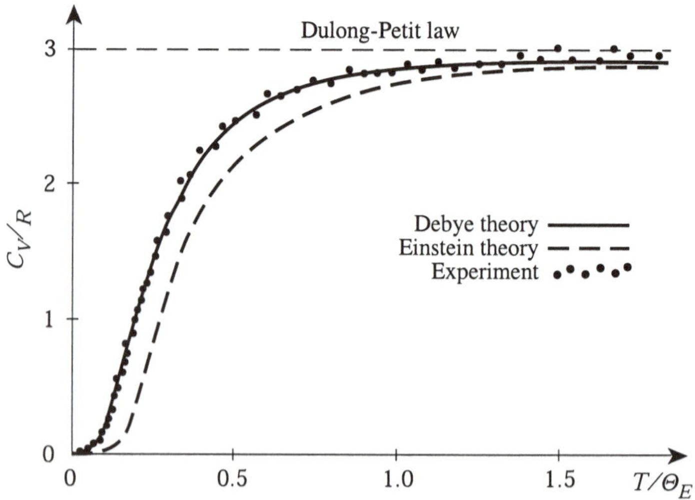

Fig. 3.2.4 The specific heat vs temperature — theory and experimental results. Although Einstein's theory describes the qualitative behavior well, quantitatively it is the Debye theory that is accurate.

Finally, we note here that what interested Einstein was not a full quantitative theory of crystals, but the sensational fact that the extension of Planck's assumption concerning the quantization of the modes of electromagnetic radiation to the vibrations of matter suffices to solve, in principle, the nagging problem of the decrease of the specific heat of solids with decreasing temperature.

Chapter 3

Statistical Mechanics of Classical Systems

3.1 Statistical mechanics of a single particle

So far we have treated quantum systems whose states may be enumerated, so that it was easy to assign them probabilities. When dealing with classical systems (surprisingly!) an extension of these concepts is required, but the extension does not require more than the probability densities introduced in the kinetic theory (Part I, Chap. 1).

We open with the following question: What is required in order to characterize the state of a particle?

The particle itself is defined by a mass (possibly also an electric charge, nuclear charge, etc.), and might be confined to a box. But it can be anywhere inside the box. We therefore need, at least, to know the particle's position in order to characterize its state.

At every point inside the box there may be forces acting (forces of gravity, electric forces, etc.). The particle's motion will be determined by Newton's laws, Maxwell's laws, etc. But even given the particle's position and the forces acting upon it, its velocity remains undetermined. We know from mechanics that the particle's initial position and initial velocity must be given in order to determine its motion under the influence of the given forces. Thus, as part of the characterization of the state of the particle, its velocity must be given. It turns out that the discussion is more symmetrical if instead of the velocity the momentum is used. In the cases we shall treat, this change amounts merely to multiplication by the mass. However, in more complicated cases involving, for example, magnetic forces that depend on the velocity, the relation is less simple. Anyway, the position and momentum at a given instant characterize the particle's state. The acceleration, for instance, is determined by Newton's laws: It does not add independent information.

Given the particle's position and velocity, it is possible to calculate its energy, which is

$$E = \frac{1}{2}m\mathbf{v}^2 + U(\mathbf{r}) = \frac{\mathbf{p}^2}{2m} + U(\mathbf{r}), \tag{3.3.1}$$

i.e. the sum of the kinetic energy $mv^2/2$, or $p^2/2m$, and the potential energy $U(\mathbf{r})$ which depends on the position of the particle.

Thus, apparently, we can speak of the canonical ensemble of a single particle in a region with a potential U, i.e. of a particle at equilibrium with a heat bath and under the influence of a potential U. The probability of finding the particle in a state $(\mathbf{r}, \mathbf{p})$ (namely at point $\mathbf{r}$ with momentum $\mathbf{p}$) would be proportional to the Boltzmann factor:

$$\exp\left\{-\beta\left[\frac{\mathbf{p}^2}{2m} + U(\mathbf{r})\right]\right\}. \tag{3.3.2}$$

But how do we sum over states?

The answer is that since $\mathbf{r}$ and $\mathbf{p}$ are continuous variables, we are to discuss the *probability density* and not the probability. That is, we must say that the probability for the particle to be found in a small volume dV around $\mathbf{r}$, and for its momentum to be found in a small volume $d\tau$ around $\mathbf{p}$, is proportional to

$$\exp\left\{-\beta\left[\frac{\mathbf{p}^2}{2m} + U(\mathbf{r})\right]\right\} dV\, d\tau\,. \tag{3.3.3}$$

■ In Part I $d\tau$ denoted the volume element in *velocity space*. Here it denotes the volume element in *momentum space*.

Hence, instead of summing over states we shall have to *integrate* over all the values of position and momentum that are allowed by the conditions of the problem.

phase space

The picture is, therefore, that we have a six-dimensional space (this space is called phase space): three dimensions for position and three for momentum. This space is divided into small cells of size $dV\,d\tau$, which is a six-dimensional volume element. The probability for the particle to be in one of the cells is proportional to the volume of the cell and to the Boltzmann factor, corresponding to the total energy of the particle in that cell.

It is simpler, of course, to illustrate the phase space when its dimensionality is less than 6. If the particle moves in one dimension, it is described by one coordinate x, and one velocity v, and therefore one momentum p. The phase space of such a particle will be two-dimensional. This space is depicted in Fig. 3.3.1.

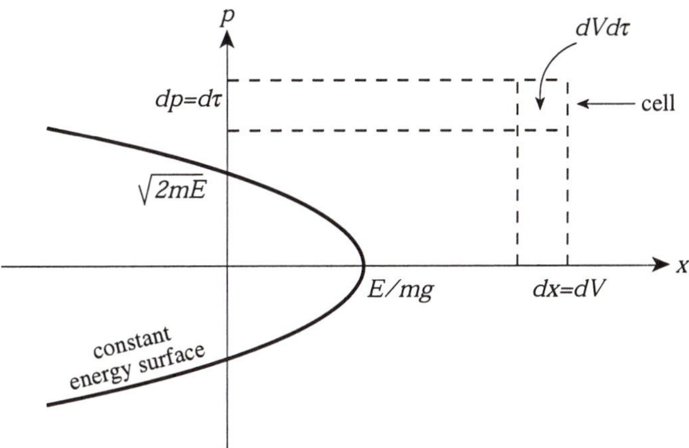

Fig. 3.3.1 The phase space of a particle whose motion in a gravitational field is confined to one dimension.

If the particle has mass m, and is attracted by a gravitational field along $-x$ with acceleration g, then its energy is

$$E = \frac{p^2}{2m} + mgx\,. \tag{3.3.4}$$

The parabola in the figure describes a surface of constant energy.

Exercise 3.1

The energy of a classical harmonic oscillator is given by

$$E = \frac{p^2}{2m} + \frac{1}{2}Kx^2\,, \tag{3.3.5}$$

where the time dependence of x is

$$x = x_0 \cos\omega t\,, \tag{3.3.6}$$

with $\omega = \sqrt{\frac{K}{m}}$.

Sketch the phase space, and describe in it the trajectory of a particle that moves according to (3.3.6).

Solution on page 307

The calculation of the averages will be performed as in Part I. That is, if $A(\mathbf{r},\mathbf{p})$ is some function of the particle's position and momentum (an observable of its state), then

$$\langle A\rangle = \frac{1}{z_c}\int A(\mathbf{r},\mathbf{p})\exp\left\{-\beta\left[\frac{\mathbf{p}^2}{2m} + U(\mathbf{r})\right]\right\} dV\,d\tau\,, \tag{3.3.7}$$

where the normalization constant may be thought of as a partition function of a single particle:

$$z_c = \int \exp\left\{-\beta\left[\frac{\mathbf{p}^2}{2m} + U(\mathbf{r})\right]\right\} dV\, d\tau\,. \tag{3.3.8}$$

Examples

(a) The average kinetic energy of the particle: equipartition. In this case $A(\mathbf{r},\mathbf{p}) = \frac{\mathbf{p}^2}{2m}$,

$$\left\langle \frac{\mathbf{p}^2}{2m} \right\rangle = \frac{\int \frac{\mathbf{p}^2}{2m} e^{-\beta\mathbf{p}^2/2m} d^3p \int e^{-\beta U(\mathbf{r})} d^3r}{\int e^{-\beta\mathbf{p}^2/2m} d^3p \int e^{-\beta U(\mathbf{r})} d^3r}$$

$$= -\frac{\partial}{\partial\beta} \ln\left[\int e^{-\beta\mathbf{p}^2/2m} d^3p\right] = \frac{3}{2}kT\,. \tag{3.3.9}$$

What has happened here? First, we have noticed that for our specific choice of A, both integrals, in the numerator and in the denominator, can be decomposed as a product — one integral over the position and another over the momentum. We wrote $dV = d^3r$, $d\tau = d^3p$, and noted that the integral over the position cancels out. Using the methods we developed in Sec. 1.6, we immediately obtain the result (3.3.9) — the equipartition law for the kinetic energy, namely $\frac{1}{2}kT$ per momentum component.

Exercise 3.2

Calculate $\langle p_x^2/2m \rangle$. How is this result connected with equipartition?

Solution on page 308

(b) The average potential energy $A(\mathbf{r},\mathbf{p}) = U(\mathbf{r})$. We find that

$$\langle U(\mathbf{r}) \rangle = \frac{\int U(\mathbf{r}) e^{-\beta U(\mathbf{r})} d^3r}{\int e^{-\beta U(\mathbf{r})} d^3r}\,. \tag{3.3.10}$$

Note that this time the integral over the momentum cancels out and we obtain

$$\langle U(\mathbf{r}) \rangle = -\frac{\partial}{\partial\beta} \ln\left[\int e^{-\beta U(\mathbf{r})} d^3r\right]\,. \tag{3.3.11}$$

We cannot proceed any further without making some assumptions concerning U. However, we can deduce that the right hand side will depend on the temperature, the volume in which the particle is confined (since the integral is only over the coordinates inside the volume in which the particle is contained) and the parameters which characterize U. Actually, the volume to which the particle is confined is also one of the parameters of the potential, since it is possible to describe the confinement of the particle to a certain region of space with the help of a "step potential," as already mentioned in Sec. 1.2. Usually, however, it is customary to write explicitly only the potential that is unrelated to the walls of the container and to represent the container by the limits of the integration over the position.

If the potential U limits the particle to a spatial region that is very small compared to the size of the container, then it is possible to perform the calculation as if the container extended to infinity. Thus, for example, if

$$U(\mathbf{r}) = \frac{1}{2}Kr^2\,, \tag{3.3.12}$$

namely the potential of a harmonic oscillator, then the energy of the particle increases rapidly with its distance from $\mathbf{r} = 0$ (see Fig. 3.3.2). Hence its probability to be far from $\mathbf{r} = 0$ becomes very small, unless the temperature is very high.

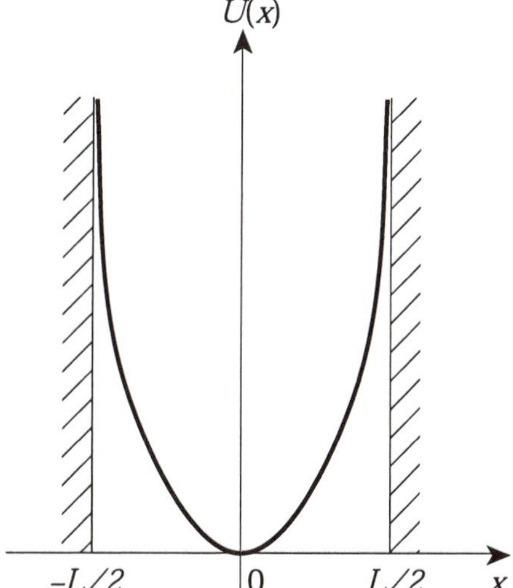

Fig. 3.3.2 Harmonic potential of a particle in a box.

Exercise 3.3

A three-dimensional box of length L contains a particle in a harmonic "potential well" (3.3.12), at the center of the box.

(a) At what temperature will the particle begin to feel the existence of the walls?
(b) What will $\langle U(\mathbf{r})\rangle$ be if we extend L to infinity?

Solution on page 309

In a similar manner, given any function of the particle's state $A(\mathbf{r},\mathbf{p})$, it is possible to calculate the average $\langle A\rangle$. The integrals may get more complicated, but the principle remains the same.

3.2 Statistical mechanics of a classical gas

Usually there is not much interest in the statistical mechanics of a single particle, but rather in a system of many particles — a macroscopic system. Suppose, therefore, that in a box of volume V there are N particles. The particles may be under the influence of some external field (other than the container) and they may interact with one another. What is required in order to characterize the macroscopic state of the system?

The answer is, of course, that we require the coordinates and momenta of each and every particle of the gas. If we label the particles by numbers from 1 to N, we require all N pairs of vectors $(\mathbf{r}_1,\mathbf{p}_1),(\mathbf{r}_2,\mathbf{p}_2),\ldots,$ $(\mathbf{r}_N,\mathbf{p}_N)$ at a given instant in order to specify a state of a system.

■ Each component of $\mathbf{r}_i$ can take any value in the confining volume. Each component of $\mathbf{p}_i$ can vary between $-\infty$ and ∞.

We have found, therefore, that it is possible to think of the state of a system of N particles as a point in a $6N$-dimensional space — the system's phase space. The temporal evolution of the system, namely the variation of $(\mathbf{r}_1,\mathbf{p}_1),\ldots,(\mathbf{r}_N,\mathbf{p}_N)$ with time, corresponds from this point of view to a continuous transition from point to point in phase space, i.e. to a motion along a curve in phase space. But here we are not interested in the details of the trajectories but, as already mentioned several times, in various averages along them. To this end we must calculate the energy of a given microscopic state. This will be made up of a kinetic energy,

$$E_k = \frac{1}{2}m_1v_1^2 + \frac{1}{2}m_2v_2^2 + \ldots + \frac{1}{2}m_Nv_N^2 = \sum_{i=1}^{N}\frac{1}{2m_i}p_i^2\,, \tag{3.3.13}$$

where m_i is the mass of particle number i, and a potential energy that depends, in most cases, on the positions of the particles:

$$U = \sum_{i=1}^{N} U_1(\mathbf{r}_i) + \sum_{i,j=1}^{N} U_2(\mathbf{r}_i, \mathbf{r}_j) + \ldots \tag{3.3.14}$$

The expression for the potential energy is to be read in the following manner: the potential energy is the sum of potential energies experienced by each particle separately (as a result of external forces), of potential energies originating from the mutual forces between pairs of particles, potential energies originating from the mutual interactions of three particles, etc. In order to clarify (3.3.14) we discuss two simple cases.

Examples

(a) Suppose that N particles, which do not interact with one another, are contained in a box in a gravitational field. In this case only the first contribution to the potential energy will appear — the single particle potential. What is it composed of? Well, each particle will feel the gravitational force, independently of the position of all other particles and of its own velocity. Hence U will have the form

$$U = \sum_{i=1}^{N} U_1(\mathbf{r}_i) = \sum_{i=1}^{N} m_i g z_i\,, \tag{3.3.15}$$

where we have chosen the z axis to be "up."

(b) Suppose that the particles are charged, and that the charge of the ith particle is q_i. In this case, the electrostatic potential energy of particle number j, due to particle number i, is

$$U_2(\mathbf{r}_i, \mathbf{r}_j) = \frac{kq_iq_j}{|\mathbf{r}_i - \mathbf{r}_j|}\,, \tag{3.3.16}$$

i.e. the product of the charges divided by the distance between then. Here, k is the electrostatic constant.

Thus, the total potential energy is

$$U = \sum_{\substack{i,j \\ (i<j)}} \frac{kq_iq_j}{|\mathbf{r}_i - \mathbf{r}_j|}\,, \tag{3.3.17}$$

where the sum is organized in such a way that it excludes self-interactions and that the potential of a pair of particles is not counted twice.

Given the energy of a microscopic state of the system, we can assign to it a probability in phase space, where again the space is partitioned into small cells. Since this time the space has $6N$ dimensions, the cells are small, $6N$-dimensional volumes. The probability for the system to be in a microscopic state, inside the infinitesimal volume around the point $(\mathbf{r}_1, \mathbf{p}_1), (\mathbf{r}_2, \mathbf{p}_2), \ldots, (\mathbf{r}_N, \mathbf{p}_N)$, is proportional to

$$\exp\left[-\beta E(\mathbf{r}_1, \mathbf{p}_1, \mathbf{r}_2, \mathbf{p}_2, \ldots, \mathbf{r}_N, \mathbf{p}_N)\right] dV_1 d\tau_1 dV_2 d\tau_2 \ldots dV_N d\tau_N \,. \quad (3.3.18)$$

This is the probability for particle 1 to be in the small volume dV_1 around the point $\mathbf{r}_1$, and for its momentum to be in the volume $d\tau_1$ around the momentum vector $\mathbf{p}_1$; for particle 2 to be in the volume dV_2 around the point $\mathbf{r}_2$, and in the volume $d\tau_2$ around $\mathbf{p}_2$; and so on. Compare with Eq. (1.1.50).

The normalization factor of the distribution is of course the partition function of the classical system:

$$Z_c = \int (dV d\tau)^N e^{-\beta E(\mathbf{r}_i, \mathbf{p}_i)} \,, \quad (3.3.19)$$

where, for brevity, we have written the energy of the microscopic state as $E(\mathbf{r}_i, \mathbf{p}_i)$, and the infinitesimal volume as $(dV d\tau)^N$.

▪ Note: the dimensions of the "classical" single-particle partition function z_c are (length × momentum)^3, and those of Z_c are (length × momentum)^{3N}, in contrast to the quantum cases of discrete degrees of freedom, for which the partition function is dimensionless.

Now that we have defined the probability density of every microscopic state it is possible to calculate the average of any observable, i.e. of every function of $\mathbf{r}_i$ and $\mathbf{p}_i$.

Given such a function,

$$A(\mathbf{r}_i, \mathbf{p}_i) = A(\mathbf{r}_1, \mathbf{p}_1; \mathbf{r}_2, \mathbf{p}_2; \ldots; \mathbf{r}_N, \mathbf{p}_N) \,, \quad (3.3.20)$$

the average of A is

$$\langle A \rangle = \frac{1}{Z_c} \int (dV d\tau)^N A(\mathbf{r}_i, \mathbf{p}_i) e^{-\beta E(\mathbf{r}_i, \mathbf{p}_i)} \,. \quad (3.3.21)$$

These are all simple generalizations of Eqs. (3.3.7) and (3.3.8), etc., of the previous section.

Chapter 4

Statistical Mechanics of an Ideal Gas

4.1 The ideal gas

The classical system that is simplest to treat, using the methods formulated in Chap. 3, is the ideal gas. This system is defined by the requirement that the collection of particles be confined to a given volume, and the particle energies be purely kinetic. As a second step, it is possible to add external forces, but interactions between particles are still neglected. Such a description is reasonable for a dilute gas of neutral particles. The forces between neutral particles are short-ranged, and when the gas is dilute the probability for one particle to be found close to another, to feel the force exerted by it, is very small.

ideal gas

■ In a gas at standard conditions the mean free path is much larger than the intermolecular distance, and thus also much larger than the characteristic size of a molecule. See Exercise 3.5 of Part I.

The first important result: If the energy of the system can be written as a sum of energies, each depending only on the state of one particle, then the probability for a given state of the system is given as a product of the probabilities of the separate particles, i.e. if

$$E(\mathbf{r}_1, \mathbf{p}_1, \mathbf{r}_2, \mathbf{p}_2, \ldots, \mathbf{r}_N, \mathbf{p}_N) = \sum_{i=1}^{N} \epsilon_i(\mathbf{r}_i, \mathbf{p}_i), \tag{3.4.1}$$

then

$$\begin{aligned} &\exp[-\beta E(\mathbf{r}_1, \mathbf{p}_1, \mathbf{r}_2, \mathbf{p}_2, \ldots, \mathbf{r}_N, \mathbf{p}_N)] dV_1 d\tau_1 \ldots dV_N d\tau_N \\ &= \exp[-\beta\epsilon_1(\mathbf{r}_1, \mathbf{p}_1)] dV_1 d\tau_1 \cdot \exp[-\beta\epsilon_2(\mathbf{r}_2, \mathbf{p}_2)] dV_2 d\tau_2 \cdot \\ &\quad \ldots \exp[-\beta\epsilon_N(\mathbf{r}_N, \mathbf{p}_N)] dV_N d\tau_N . \end{aligned} \tag{3.4.2}$$

In other words, in the canonical distribution single particle states are independent of each other.

■ As we saw in Part II [Eq. (2.3.11)], in a *microcanonical* ensemble there can exist dependence between the single particle probabilities, even though there are no interaction forces between them. It is, however, negligibly small.

The second result: If the particles are identical, the partition function is a power of a single particle partition function. Namely,

$$Z_c = z_c^N \,, \tag{3.4.3}$$

where z_c is as given in Eq. (3.3.8).

■ Equation (3.4.3) will be corrected in Chap. 5 below — Eq. (3.5.3).

Exercise 4.1

Prove (3.4.3).

Solution on page 309

Given the partition function, the free energy is determined by Eq. (3.1.30). From the free energy it is possible to deduce all the thermodynamic properties of the system. That was the moral of Chap. 1.

Thus, we now calculate z_c from Eq. (3.3.8), for the case of a particle moving freely in the entire volume of the box. In the calculation the integral separates into a product of an integral over position and an integral over momentum:

$$z_c = \int dV e^{-\beta U(\mathbf{r})} \cdot \int d\tau e^{-\beta \mathbf{p}^2/2m} \,. \tag{3.4.4}$$

Since $\mathbf{r}$ is confined to the volume of the box and the particle is free, $U = 0$, the first integral is equal to V — the volume of the box. The second integral is a Gaussian integral, the like of which we calculated in Part I (Exercise 1.13, for instance). Hence

$$z_c = V \left(\frac{2\pi m}{\beta} \right)^{3/2} = V(2\pi mkT)^{3/2} \,. \tag{3.4.5}$$

Exercise 4.2

Prove (3.4.5).

Solution on page 310

Substituting (3.4.5) into the expression for the free energy (3.1.30), we obtain

$$F = -\frac{1}{\beta} \ln Z_c = -\frac{N}{\beta} \ln z_c = -NkT \left[\ln V + \frac{3}{2} \ln(2\pi mkT) \right] \,. \tag{3.4.6}$$

From F we can immediately obtain the pressure and the entropy, as

$$P = -\left(\frac{\partial F}{\partial V}\right)_{T,N} = \frac{NkT}{V} \tag{3.4.7}$$

and

$$S = -\left(\frac{\partial F}{\partial T}\right)_{V,N} = Nk\left[\ln V + \frac{3}{2}\ln(2\pi mkT) + \frac{3}{2}\right]. \tag{3.4.8}$$

Note that (3.4.7) is the ideal gas equation of state!

It is of interest to recall the way in which the volume enters the partition function. As already mentioned in Sec. 1.2, the volume does not appear explicitly in the kinetic energy of the particles; nevertheless the partition function depends upon it. Here it enters into the partition function via the limits of the integration over the particles' positions.

From the expression for S it is possible to obtain the specific heat at constant volume:

$$C_V = T\left(\frac{\partial S}{\partial T}\right)_{V,N} = \frac{3}{2}Nk\,, \tag{3.4.9}$$

and at constant pressure:

$$C_P = T\left(\frac{\partial S}{\partial T}\right)_{P,N} = \frac{5}{2}Nk\,. \tag{3.4.10}$$

Exercise 4.3

Prove (3.4.9) and (3.4.10).

Solution on page 310

Exercise 4.4

Use the partition function to obtain the average internal energy. How is it related to the specific heat?

Solution on page 311

4.2 Mixtures of ideal gases — Dalton's law

Consider a system which is a mixture of M gases, i.e. there are M groups of particles, with different masses, for instance. In group number j there are N_j particles, of mass m_j. The whole mixture is confined to a box of volume V.

The energy of the system is given by Eq. (3.4.1), written in a slightly different form. In group number j particle number i will have energy ϵ_{ij}.

The energy of the jth group will be given by the sum over i of ϵ_{ij}, from 1 to N_j. Therefore, the total energy will be

$$E = \sum_{j=1}^{M} \sum_{i=1}^{N_j} \epsilon_{ij} \,. \tag{3.4.11}$$

All particles of all gases share the same constraint, namely they are confined to the box. The difference is in the kinetic energies: Every particle in group number j, which has momentum p, has kinetic energy $p^2/2m_j$.

The first result that we can deduce is that instead of (3.4.3),

$$Z_c = (z_1)^{N_1} (z_2)^{N_2} \cdot \ldots \cdot (z_M)^{N_M} \,. \tag{3.4.12}$$

■ Equation (3.4.12) has a correction in Chap. 5 — Eq. (3.5.4).

Exercise 4.5

Prove (3.4.12).

Solution on page 311

Hence the free energy, (3.4.6), is replaced by

$$F = -kT(N_1 \ln z_1 + N_2 \ln z_2 + \ldots + N_M \ln z_M) \,. \tag{3.4.13}$$

As we observed, the difference between the different z_j is only in the particles' mass. That is,

$$z_j = V(2\pi m_j kT)^{3/2} \,. \tag{3.4.14}$$

Substituting in (3.4.13), we obtain

$$\begin{aligned} F &= -kT \left\{ N_1 \left[\ln V + \frac{3}{2} \ln(2\pi m_1 kT) \right] + N_2 \left[\ln V + \frac{3}{2} \ln(2\pi m_2 kT) \right] \right. \\ &\qquad \left. + \cdots + N_M \left[\ln V + \frac{3}{2} \ln(2\pi m_M kT) \right] \right\} \\ &= -kT \left[N \ln V + \frac{3}{2} \sum_{j=1}^{M} N_j \ln(2\pi m_j kT) \right] , \end{aligned} \tag{3.4.15}$$

where we have used the fact that

$$\sum_{j=1}^{M} N_j = N \,,$$

which is the total number of particles. From the expression for F it is possible to obtain the pressure, as in (3.4.7):

$$P = -\left(\frac{\partial F}{\partial V}\right)_{T,N} = \frac{(N_1 + N_2 + \dots N_M)kT}{V} = \frac{NkT}{V}. \qquad (3.4.16)$$

That is, the pressure depends only on the total number of particles in the container and not on the number of particles of each type. However, note the form of the intermediate expression in (3.4.16). This expression is the sum of the pressures of each of the constituents had it filled the container alone:

$$P = \sum_{j=1}^{M} P_j, \quad P_j = \frac{N_j kT}{V}. \qquad (3.4.17)$$

This is Dalton's law: The total pressure of the mixture is equal to the sum of the partial pressures.

■ See also Exercise 1.5 of Part I.

Exercise 4.6

Show that the specific heat of the mixture is the sum of partial specific heats. Is this true of the entropy?

Solution on page 312

4.3 Maxwell–Boltzmann distribution and equipartition

In Chap. 1 we have shown that the central relations of thermodynamics can be fully identified in statistical mechanics. Were we to ask a similar question with respect to the possibility of identifying the kinetic theory of gases in the formulation of statistical mechanics, we would not be able to answer in the same generality, since the validity of the kinetic theory is not general but is restricted to dilute gases. It is to be expected that the central results of the kinetic theory be derivable from the statistical mechanics of ideal gases. We shall treat, therefore, an ideal gas that can reside in an external potential field $U(\mathbf{r})$.

First, we discuss the momentum and position distributions of a single particle. Without limiting the generality of our discussion, we choose particle number 1. The general distribution of the system's states is given by (3.4.2), after normalization. In other words, the (joint) probability for particle number 1 to be in $dV_1 d\tau_1$ around $(\mathbf{r}_1, \mathbf{p}_1)$, and for particle number 2 to be in $dV_2 d\tau_2$ around $(\mathbf{r}_2, \mathbf{p}_2)$, and so forth, is

$$P(\mathbf{r}_1, \mathbf{p}_1; \ldots; \mathbf{r}_N, \mathbf{p}_N) dV_1 d\tau_1 dV_2 d\tau_2 \ldots dV_N d\tau_N$$
$$= Z_c^{-1} e^{-\beta\epsilon(\mathbf{r}_1,\mathbf{p}_1)} e^{-\beta\epsilon(\mathbf{r}_2,\mathbf{p}_2)} \ldots e^{-\beta\epsilon(\mathbf{r}_N,\mathbf{p}_N)}$$
$$\times dV_1 d\tau_1 dV_2 d\tau_2 \ldots dV_N d\tau_N \,. \tag{3.4.18}$$

Clearly, if we sum the probabilities over all the particle states, namely integrate P over all $6N$ variable, we will obtain 1.

In order to obtain the probability for a given particle (say, number 1) to be in $dV_1 d\tau_1$ around $(\mathbf{r}_1, \mathbf{p}_1)$, we must allow all the other particles to be in the whole volume and to possess every possible momentum. Namely,

$$P(\mathbf{r}_1, \mathbf{p}_1) = \int P(\mathbf{r}_1, \mathbf{p}_1; \ldots; \mathbf{r}_N, \mathbf{p}_N) dV_2 d\tau_2 \ldots dV_N d\tau_N \,. \tag{3.4.19}$$

Exercise 4.7

Prove that $P(\mathbf{r}_1, \mathbf{p}_1)$ is a normalized probability distribution.

Solution on page 313

In order to calculate $P(\mathbf{r}_1, \mathbf{p}_1)$ and to obtain its explicit form, we shall use the fact that the partition function satisfies $Z_c = z_c^N$, so that the right hand side factors into a product of

$$P(\mathbf{r}_1, \mathbf{p}_1; \ldots; \mathbf{r}_N, \mathbf{p}_N) = \frac{e^{-\beta\epsilon(\mathbf{r}_1,\mathbf{p}_1)}}{z_c} \cdots \frac{e^{-\beta\epsilon(\mathbf{r}_N,\mathbf{p}_N)}}{z_c} \,. \tag{3.4.20}$$

Thus the integration in (3.4.19) over the $N-1$ vectors of position and momentum, of particles $2, \ldots, N$, will factor into a product of $N-1$ identical integrals of the form

$$z_c^{-1} \int \exp[-\beta\epsilon(\mathbf{r}, \mathbf{p})] dV d\tau = 1 \,. \tag{3.4.21}$$

Hence we obtain

$$P(\mathbf{r}_1, \mathbf{p}_1) = z_c^{-1} \exp\left\{-\beta\left[\frac{\mathbf{p}_1^2}{2m} + U(\mathbf{r}_1)\right]\right\} , \tag{3.4.22}$$

which is the Maxwell–Boltzmann distribution formulated in terms of momenta. [Cf. Eq. (1.1.50).]

Another result of the kinetic theory is easily obtained here. This is the *equipartition theorem*: every variable of phase space on which the energy depends quadratically, contributes $\frac{1}{2}kT$ to the average energy (the

equipartition theorem

variable may be any component of the coordinate or of the momentum of any particle).

We have used this theorem without proof in Sec. 1.3 of Part I, referring to the proof to be given here, and to the conclusion to be derived from it.

Proof: Let us calculate the average energy contributed by one of the momentum components, p_{1x}, for instance. p_{1x} represents here a variable that appears in the expression for the energy in the form ap_{1x}^2 and does not appear anywhere else. The total energy of the state is written therefore as the sum

$$E = ap_{1x}^2 + E' , \tag{3.4.23}$$

where E' is the rest of the energy and it depends on all the variables of phase space except p_{1x}. We have chosen the momentum as an example because it usually appears in the energy as a separate quadratic term. The constant a is then of course $1/2m$. A quadratic dependence of the energy upon the coordinates is less common, and is characteristic only of a harmonic potential.

Now, since p_{1x} appears in the energy only once, its probability distribution function is proportional to $\exp(-\beta ap_{1x}^2)$. It is of course possible to obtain this by integrating the total distribution function over all variables except p_{1x}.

The average energy corresponding to p_{1x} is thus

$$\langle ap_{1x}^2 \rangle = \frac{\int dp_{1x} \exp(-\beta ap_{1x}^2) ap_{1x}^2}{\int dp_{1x} \exp(-\beta ap_{1x}^2)} . \tag{3.4.24}$$

Exercise 4.8

Derive (3.4.24) directly from the distribution function in phase space.

Solution on page 313

We have already calculated expressions such as the one on the right hand side of (3.4.24) many times (see, for example, Exercise 3.2), and the result is indeed $\frac{1}{2}kT$, as the theorem asserts.

Finally we note that, as we have already seen in Solution 4.8, the equipartition theorem is valid for every system provided that we can identify a variable that satisfies Eq. (3.4.23). A good example of such a variable is the momentum of an atom of a gas molecule. If the forces acting between this atom and the other atoms are independent of its momentum, then the conditions of the theorem are satisfied, and thus the average kinetic energy of each atom in the molecule is given by the equipartition theorem, namely $\frac{1}{2}kT$ per component and a total of $\frac{3}{2}kT$. This was the answer we gave to question (a) in Sec. 1.3 of Part I, and here at last it is justified.

4.4 Ideal gas of quantum particles

In our discussion of the ensemble of harmonic oscillators in Chap. 2, we treated them as a quantum system but we also dealt with the question of the behavior of an ensemble of classical oscillators. The result is that the classical behavior is obtained from the quantum results in the limit of high temperatures. On the other hand, our discussion of ideal gases has so far completely ignored the possible quantum aspects related to the particle's motion in the container. We would therefore end this chapter by considering the effects that the quantization of energy has on the ideal gas laws. We shall see that apart from a new point of view, the addition of quantization does little, and we obtain again the ideal gas equation (3.4.7).

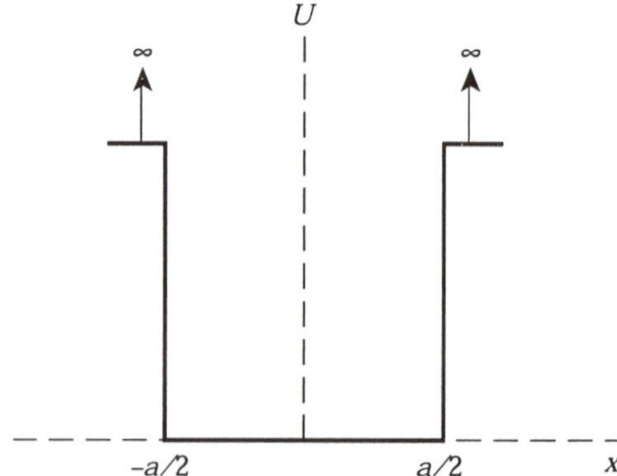

Fig. 3.4.1 A potential well with height tending to infinity.

In light of all our experience so far, it is clear that our problem comes down to calculating the partition function of a single particle of mass m that is confined to a container of volume V. For the sake of simplicity we assume that the container is a box of sides a, b and c. The particle's confinement to the box can be described by a potential that vanishes inside the box and jumps to infinity at the walls. Because the motion of the particle occurs in three independent directions, we start by considering the simple case of a particle executing a one-dimensional motion in an "infinite potential well," as in Fig. 3.4.1. Then we pass on to the three-dimensional case, just as we did in Chap. 2 in our discussion of the Einstein solid. The quantum states of a particle moving in such a potential are described by standing de Broglie waves that vanish on the sides of the well, $x = \pm a/2$. The allowed wavelengths are quantized according to

$$\lambda_n = \frac{2a}{n}, \quad n = 1, 2, 3, \dots, \tag{3.4.25}$$

and the corresponding energy values are thus

$$\epsilon_n = \frac{h^2 n^2}{8ma^2} , \quad n = 1, 2, 3 \dots , \tag{3.4.26}$$

where m is the particle's mass and h is the Planck constant.

The particle's states in the well are thus characterized by a natural number n similarly to the states of the harmonic oscillator. The difference is of course in the dependence of ϵ on n. The particle's motion in the three-dimensional box can be considered as a sum of three independent one-dimensional motions each taking place inside a well as the one depicted in Fig. 3.4.1. Thus, the condition (3.4.25) must be satisfied independently in each of the directions, and to each direction of motion there will correspond a natural number of its own. The energy values will thus be

$$\epsilon_{npq} = \frac{h^2}{8m} \left(\frac{n^2}{a^2} + \frac{p^2}{b^2} + \frac{q^2}{c^2} \right) , \quad n, p, q = 1, 2, 3, \dots , \tag{3.4.27}$$

which means that a single particle state in a well is determined by three natural numbers (n, p, q).

The next step is of course the calculation of the single particle partition function:

$$\begin{aligned} z &= \sum_{npq} \exp(-\beta\epsilon_{npq}) \\ &= \sum_{n=1}^{\infty} \exp\left(-\frac{\beta h^2 n^2}{8ma^2}\right) \sum_{p=1}^{\infty} \exp\left(-\frac{\beta h^2 p^2}{8mb^2}\right) \sum_{q=1}^{\infty} \exp\left(-\frac{\beta h^2 q^2}{8mc^2}\right) . \end{aligned} \tag{3.4.28}$$

Note again that owing to the independence of the motions in each of the three dimensions, the partition function has factored to a product of three one-dimensional partition functions.

We are left with the calculation of a one-dimensional partition function, and we choose the motion along the x direction:

$$z_x = \sum_{n=1}^{\infty} \exp\left(-\frac{\beta h^2 n^2}{8ma^2} \right) . \tag{3.4.29}$$

Calculating the sum on the right hand side of (3.4.29) is no simple matter. Fortunately we are mostly interested in the case in which the box is much larger than the thermal de Broglie wavelength, λ_T, of the atoms of the gas.

This quantity is the wavelength, λ_T, of a gas particle with momentum determined by the average thermal kinetic energy per degree of freedom,

$kT/2$. If we use the de Broglie relation

$$p_x = \frac{h}{\lambda}, \tag{3.4.30}$$

then from $p_x^2/2m = kT/2$ we have

$$\lambda_T = \sqrt{\frac{h^2}{mkT}}. \tag{3.4.31}$$

Exercise 4.9

(a) Helium atoms are confined to a box with $a = 1$ cm. Calculate λ_T/a at 10 K and at 300 K.

(b) At what temperature does λ_T become comparable to the size of the box?

Solution on page 314

The exponent in the terms of the sum in Eq. (3.4.29) can be written as $\lambda_T^2 n^2/8a^2$. Since λ_T/a is very small, the difference between successive terms in this sum is also very small and the sum can be approximated by an integral, as we now explain. The terms in the sum can be regrouped as follows: define a variable $s = \frac{\lambda_T}{2a}n$, where n is the running index in the sum. Even in a small interval Δs there is a very large number of terms of the sum (different values of n), since the corresponding $\Delta n = \frac{2a}{\lambda_T}\Delta s$ is made large by the enormous value of a/λ_T. See Exercise 4.9. In each interval Δs, the terms of the sum are equal to a very good approximation. The sum can therefore be written as

$$z_x = \sum_s e^{-\frac{1}{2}s^2}\left(\frac{2a}{\lambda_T}\right)\Delta s \xrightarrow[\Delta s \to 0]{} \frac{2a}{\lambda_T}\int_0^\infty e^{-\frac{1}{2}s^2} ds\,. \tag{3.4.32}$$

The limit in the above equation is the definition the Riemann integral.

Converting s into a momentum variable, via Eq. (3.4.25) and the de Broglie relation (1), we write

$$s = \frac{\lambda_T}{\lambda_n} = \frac{\lambda_T p_x}{h}$$

and z_x becomes

$$z_x = \frac{2a}{h}\int_0^\infty \exp\left(-\frac{\lambda_T^2 p_x^2}{2h^2}\right) dp_x = \frac{a}{h}\int_{-\infty}^\infty \exp\left(-\frac{\beta p_x^2}{2m}\right) dp_x\,, \tag{3.4.33}$$

where the last transition has been made with the help of the fact that the integrand is an even function. Although p_x was defined as a positive quantity (since n is positive), we have found that the sum over the positive values of n is equivalent to an integral over all values of p_x, positive as well as negative. That means that a state with a given n is equivalent to two momentum states: one positive and one negative.

Hence the partition function of a single quantum-mechanical particle in the three-dimensional box will be [within the approximation which led to (3.4.32) and (3.4.33)] a product of one-dimensional partition functions:

$$\begin{aligned} z &= z_x z_y z_z \\ &= \frac{abc}{h^3} \int_{-\infty}^{\infty} \exp\left(-\frac{\beta p_x^2}{2m}\right) dp_x \int_{-\infty}^{\infty} \exp\left(-\frac{\beta p_y^2}{2m}\right) dp_y \int_{-\infty}^{\infty} \exp\left(-\frac{\beta p_z^2}{2m}\right) dp_z \\ &= \frac{V}{h^3} \int \exp\left(-\frac{\beta \mathbf{p}^2}{2m}\right) d^3p \,. \end{aligned} \tag{3.4.34}$$

In the last transition we have used the fact that the product abc is the volume of the box V, as well as the fact that the product of the three integrals can be written as a three-dimensional integral over all momentum space. We have also used the notation $d^3p = dp_x dp_y dp_z$ instead of $d\tau$. The integral on the right hand side of (3.4.34) is the same integral that appeared in the calculation of the classical partition function, e.g. Eq. (3.4.4) and Exercise 4.2.

The quantum-mechanical point of view does not change, therefore, the classical result except for making the partition function dimensionless by introducing the Planck constant, which possesses precisely the dimensions required for this purpose. See the remark after Eq. (3.3.19) as well. However, we must stress that what we have done here is restricted to the quantum effects due exclusively to the quantization of the motion of a single particle. As we shall see in Part V, quantum mechanics requires additional changes that are expressed by the fact that the canonical partition function of an ideal gas cannot be written as the product of independent single particle partition functions. These changes lead to fundamental changes in the laws of ideal gases at low temperatures and/or high densities. At regular conditions the results we shall obtain in Part V reduce to the results of the classical treatment, so that all our discussions on ideal gases in this part remain valid in a very wide region of densities and temperatures.

Finally, we make several remarks.

(a) The fact that we have again obtained the classical partition function should not come as a great surprise, since we calculated the sum in (3.4.28) only in the approximation that the thermal de Broglie wavelength is very small or, in other words, in the limit in which quantum effects are negligible. This is also the limit of high temperatures.

(b) Indeed, we obtained (3.4.34) for a box-shaped container, but in fact it is clear that the result cannot depend on the container's shape as long as the volume is large enough. Hence this result is valid for any shape of the container, precisely as in the classical case.

(c) Concerning the partition function's dependence on the volume: In Sec. 4.1 we have seen that the volume dependence of the partition function of a classical gas enters through the integration limits. Conversely, the energies of the quantum states depend explicitly on the volume through the dimensions of the box [Eq. (3.4.27)]. Thus the relationship between the thermodynamic work and the volume change, found in Sec. 1.2, is now simpler to analyze and the long discussion made there for the calculation of $\partial E/\partial V$ in the classical case becomes redundant.

Exercise 4.10

Obtain δW arising from a volume change dV of a cube-shaped vessel that contains an ideal gas of quantum particles.

Solution on page 314

(d) From the result (3.4.34) it is inferred that also in the quantum case it is possible to formulate the partition function in terms of the classical variables of phase space — position and momentum. Moreover, the canonical probabilities are of the same form as in the classical case. The difference is that the quantum partition function is in principle a discrete sum and not a continuous integral. Instead of (3.4.28) we may therefore write a sum over cells in phase space:

$$z = \sum_k \exp\left(-\frac{\beta p_k^2}{2m}\right)\frac{\Delta V_k \Delta \tau_k}{h^3}\,, \tag{3.4.35a}$$

where the factor $\Delta V_k \Delta \tau_k / h^3$ describes the *number* of states (and not their density) in cell number k. This implies that a quantum state of a single particle takes up a volume of h^3 in phase space. Hence the quantum state of a system of N particles will take up a volume of h^{3N}. In the classical calculation of Z the size of the cell could be arbitrarily small. Here it must contain at least one quantum state. However, the volume of the cell must still be small enough for the continuum approximation of the sum as an integral to be valid. In this case we can, therefore, turn (3.4.35a) into an integral, and if the particles are affected by an external potential $U(\mathbf{r})$, we obtain for the single particle partition function

$$z = \frac{1}{h^3} \int dV d\tau \exp\left\{-\beta \left[\frac{\mathbf{p}^2}{2m} + U(\mathbf{r})\right]\right\} . \qquad (3.4.35b)$$

This equation is identical to (3.3.8) except for the factor h^{-3}. This is the classical limit of the quantum partition function, and the sole witness to the quantum origin of (3.4.35b) is the factor h^{-3}. Instead of (3.3.19) and (3.3.21) we obtain, therefore,

$$Z = \int \left(\frac{dV d\tau}{h^3}\right)^N e^{-\beta E(\mathbf{r}_i,\mathbf{p}_i)} , \qquad (3.4.36a)$$

$$\langle A \rangle = \frac{1}{Z} \int \left(\frac{dV d\tau}{h^3}\right)^N A(\mathbf{r}_i, \mathbf{p}_i) e^{-\beta E(\mathbf{r}_i,\mathbf{p}_i)} . \qquad (3.4.36b)$$

(e) The astute reader has probably noticed that in the equations we obtained for the free energy and the entropy [(3.4.6) and (3.4.8)] there is a mismatch of units and as a consequence the logarithmic functions appearing in them are not functions of dimensionless arguments. This mismatch stems from the fact that the classical partition function (3.4.4) or (3.4.5) is not dimensionless. The quantum partition function, in contrast, is naturally dimensionless even in the classical limit, as is clear from Eqs. (3.4.35) and (3.4.36), owing to the appearance of the Planck constant in the expression for the volume element in phase space. Usually, this difference is not significant as most thermodynamic quantities are obtained from derivatives of the free energy and entropy, so that the additional terms depending on Planck's constant cancel out and disappear. However, this is not always the case, and, moreover, the equations are more "aesthetic" when the logarithmic functions have dimensionless arguments. Thus, from now on we shall write

$$z = \frac{V(2\pi mkT)^{3/2}}{h^3} \qquad (3.4.37)$$

so that

$$F = -NkT \ln z = -NkT \left[\ln V + \frac{3}{2} \ln\left(\frac{2\pi mkT}{h^2}\right)\right] , \qquad (3.4.38)$$

$$S = Nk \left[\ln V + \frac{3}{2} \ln\left(\frac{2\pi mkT}{h^2}\right) + \frac{3}{2}\right] . \qquad (3.4.39)$$

■ We have written $\ln z$ as a sum of logarithmic functions, the same as before, but it is always possible to turn these sums into logarithmic functions of a product of dimensionless variables.

This is the first correction to the form of the free energy and entropy of an ideal gas. In the next chapter we shall see that another correction is required.

Chapter 5

The Gibbs Paradox and the Third Law

5.1 Two difficulties

(a) *Extensivity*

In thermodynamics, the free energy and entropy are extensive quantities. This means that doubling the system leads to a doubling of the free energy and the entropy. If we now check the expressions obtained for F and S, (3.4.38) and (3.4.39), we find that they are not extensive since when $V \to 2V$ and $N \to 2N$ we obtain

$$F \to 2F - 2NkT \ln 2\,,$$

$$S \to 2S + 2Nk \ln 2\,.$$

This problem, whose solution we discuss in the following section, is known as the Gibbs paradox.

The same problem appears in a different guise in the chemical potential. As recalled [Eq. (3.1.43)], the chemical potential is the derivative of the state function with respect to the number of particles, and must be an intensive variable, i.e. independent of the system's magnitude. But, calculating it from (3.4.38) we obtain

$$\mu = -kT\left[\ln V + \frac{3}{2}\ln\left(\frac{2\pi mkT}{h^2}\right)\right]\,, \tag{3.5.1}$$

which is not an intensive quantity.

(b) *The third law of thermodynamics*

One way of stating the third law of thermodynamics is that the entropy tends to zero as $T \to 0$. (See e.g. Part II, Sec. 0.4) Not only does the entropy Eq. (3.4.39) not tend to zero, it diverges when $T \to 0$. We return to this problem in Sec. 5.3. Here we only note that this problem finds its solution in the framework of the quantum theory.

Exercise 5.1

Both problems mentioned here do not appear in calculating the properties of the paramagnet or of the Einstein solid. Verify this.

Solution on page 314

5.2 The Gibbs paradox and its resolution

The first difficulty mentioned in the preceding section was presented in a dry, formal manner: The entropy of an ideal gas, as obtained from classical statistical mechanics, is not an extensive quantity. But the same problem can be presented in a more intuitive manner.

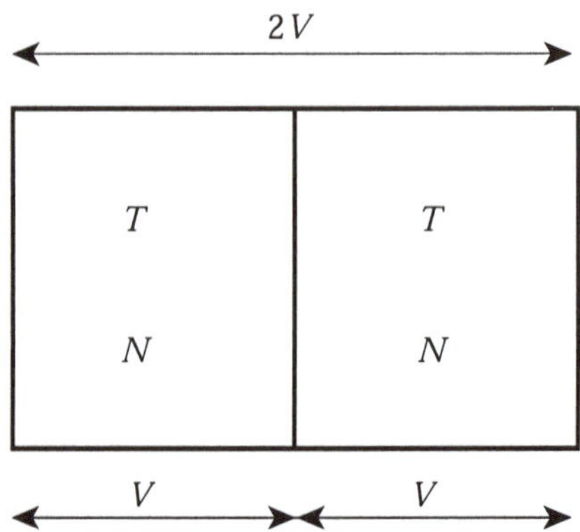

Fig. 3.5.1 The double container and the Gibbs paradox.

Imagine a container divided into two identical parts by a partition (Fig. 3.5.1). On both sides of the partition there is the same number of identical particles and the same temperature. We perform an experiment in which the partition is removed. Such a process is definitely reversible. The removal of the partition will not change the temperature, and there is no reason for the number of particles to change on either side. Hence, according to thermodynamics, the entropy should not change upon the removal of the partition.

But the entropy in the presence of the partition is the sum of the entropies of the two sides, which is given by twice (3.4.39),

$$S = 2Nk\left[\ln V + \frac{3}{2}\ln\left(\frac{2\pi mkT}{h^2}\right) + \frac{3}{2}\right], \qquad (3.5.2a)$$

but in the absence of the partition, the entropy is obtained from (3.4.39) by replacing V and N with $2V$ and $2N$:

$$S' = 2Nk \left[\ln 2V + \frac{3}{2} \ln \left(\frac{2\pi mkT}{h^2} \right) + \frac{3}{2} \right] . \qquad (3.5.2b)$$

Hence there is a change in entropy, given by

$$\Delta S = 2Nk \ln 2 \neq 0 ,$$

which is the paradox.

It is not hard to find the origin of the paradox. To this end we shall change the conditions a bit, and assume that there is a different type of gas on each side of the partition, but the volumes, the temperatures and the numbers are identical. Removing the partition this time is not a reversible process, since clearly particles of one type will cross over to the other side and vice versa. The gases will get mixed, and returning the partition to its place will not return the system to its original state. It is, of course, not surprising that removing the partition leads to an increase in the entropy. The increase in the entropy is related to the mixing, which increases the disorder.

Exercise 5.2

Describe a process by which it is possible to separate two types of molecules, each to its original side of the partition. Does this process return the system to its original state?

Solution on page 315

It turns out, therefore, that there exists a contradiction between (a) our feeling that returning the partition to its place, in the case of identical gases on both sides of the container, returns the original state, and (b) the fact that we treat the particles as if it were possible to distinguish between them. Indeed, it is clear that upon the removal of the partition, particles cross from one side to the other, and that upon its return the particles on both sides will not be the same ones as before.

Since we stand by (a), based on experience, we have no option but to give up (b).

If indeed it is not possible to distinguish between molecules (or atoms) of the same gas, so that doubling the volume is simply doubling the entropy, then we have not calculated the partition function, Z, correctly. When we integrate over the position and momentum of each of the molecules, along with each state there also appears the state in which two molecules have interchanged their position and momentum. In the integration, each molecule covers the entire volume, and its momentum assumes all allowed values. Hence, in counting different states in Z there appears the state in which particle number i is in state $(\mathbf{r}_i, \mathbf{p}_i)$ and particle number j in $(\mathbf{r}_j, \mathbf{p}_j)$, as well as the state in which particle number j has $(\mathbf{r}_i, \mathbf{p}_i)$ and particle number i has $(\mathbf{r}_j, \mathbf{p}_j)$. In Fig. 3.5.2 two states are depicted, differing from each other only in a permutation of two particles.

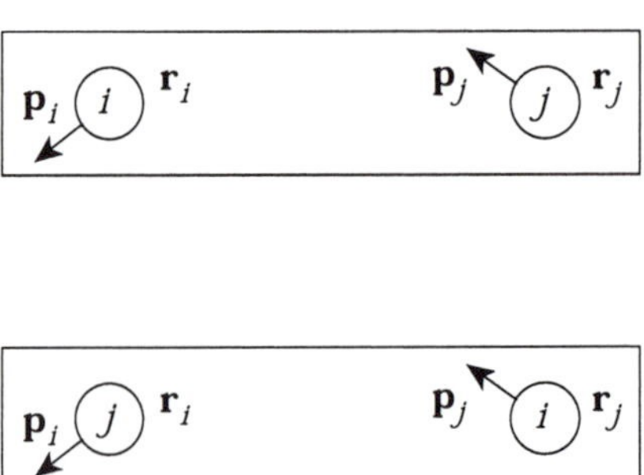

Fig. 3.5.2 Two states differing only by the permutation of two particles.

But if there is no way of distinguishing between these particles, then two such states must be counted as a single state. This means that we have counted too many states in Z. Along with every state $(\mathbf{r}_1, \mathbf{p}_1; \ldots; \mathbf{r}_N, \mathbf{p}_N)$ there appear all the states in which all the particles have the same list of coordinates and momenta, but the order of particles is different. Namely, each microscopic state $(\mathbf{r}_1, \mathbf{p}_1; \ldots; \mathbf{r}_N, \mathbf{p}_N)$ must be identified with the $N!$ states obtained by changing the order of the particles.

Fortunately it is very easy to correct the mistake: two states differing from each other only by changing the position of the particles have the same energy, and hence the same probability. In each microscopic state it is possible to change the positions of the particles in $N!$ ways, where the velocities remain unchanged. Each of these $N!$ states, which are to be counted as a single state, has contributed to the partition function, as calculated so far, whereas there should have been only one contribution due to all of them. But since all $N!$ terms are equal, and all terms in the canonical partition function have an equal number of particles, it suffices to divide by $N!$ in order to obtain the corrected partition function. Thus, instead of (3.4.3) we write

$$Z = \frac{1}{N!} z^N \tag{3.5.3}$$

and instead of (3.4.12) we write

$$Z = \frac{(z_1)^{N_1}}{N_1!} \cdot \frac{(z_2)^{N_2}}{N_2!} \cdot \ldots \cdot \frac{(z_M)^{N_M}}{N_M!} \tag{3.5.4}$$

in the case where there are M groups of molecules, and it is not possible to distinguish between molecules belonging to the same group, but it is possible to distinguish between molecules of different groups.

The concept of a microscopic state has thus undergone a significant change following this discussion. We can continue to denote a state by the $2N$ vectors $(\mathbf{r}_1, \mathbf{p}_1; \ldots; \mathbf{r}_N, \mathbf{p}_N)$ but this notation does not signify that particle number 1 is located at $\mathbf{r}_1$ with momentum $\mathbf{p}_1$, etc., but the fact that there is one particle in position $\mathbf{r}_1$ with momentum $\mathbf{p}_1$, etc. In order to calculate the partition function employing the methods that we devised, we imagine that it is possible to distinguish between particles and then correct, dividing by $N!$.

Exercise 5.3

Show that the above change does not affect the calculation of thermal averages.

Solution on page 316

Exercise 5.4

Deduce Eq. (3.5.4).

Solution on page 317

Exercise 5.5

Explain why there is no need to correct the partition function of the paramagnet and the system of oscillators.

Solution on page 317

What will the form of the free energy and entropy be? Well, after substituting (3.5.3) into (3.1.30) and using the explicit form of the single particle partition function (3.4.37), we obtain

$$F = -NkT\left[\ln V + \frac{3}{2}\ln\left(\frac{2\pi mkT}{h^2}\right) - \frac{1}{N}\ln N!\right]. \tag{3.5.5}$$

In the thermodynamic limit, when N is very large, we can approximate the last term on the right hand side of (3.5.5) using Stirling's formula (2.3.12) to write

$$\frac{1}{N}\ln N! \simeq \ln N - 1\,, \tag{3.5.6}$$

and hence Eq. (3.5.5) reduces to

$$F = -NkT\left[\ln\left(\frac{V}{N}\right) + \frac{3}{2}\ln\left(\frac{2\pi mkT}{h^2}\right) + 1\right]. \tag{3.5.7}$$

The entropy is derived from the newly found F and instead of (3.4.39) we obtain

$$S = Nk\left[\ln\left(\frac{V}{N}\right) + \frac{3}{2}\ln\left(\frac{2\pi mkT}{h^2}\right) + \frac{5}{2}\right], \tag{3.5.8}$$

both of which are of course extensive, as expected.

It is important to note that even though we have shown in Exercise 5.3 that the change in the partition function, (3.5.3), does not affect the thermal averages, the free energy F, the entropy S and the chemical potential μ (see below) all change. This implies that these quantities cannot be expressed as averages.

Exercise 5.6

Prove, by using the free energy and the entropy, that the pressure and the specific heat are not affected by Gibbs' correction.

Solution on page 317

We can also write the chemical potential that replaces (3.5.1). The result is

$$\mu = kT\left[\ln n - \frac{3}{2}\ln\left(\frac{2\pi mkT}{h^2}\right)\right], \tag{3.5.9}$$

where $n\,(= N/V)$ denotes the number of particles per unit volume. This is, of course, an intensive expression.

Exercise 5.7

Deduce Eq. (3.5.9).

Solution on page 318

And, finally, in a mixture of M different types of molecules,

$$F = -kT\sum_{j=1}^{M} N_j\left[\ln\left(\frac{V}{N_j}\right) + \frac{3}{2}\ln\left(\frac{2\pi m_j kT}{h^2}\right) + 1\right]. \tag{3.5.10}$$

Exercise 5.8

Prove that Dalton's law remains valid regardless of the changes in F.

Solution on page 318

Exercise 5.9

Calculate the expression for the entropy of a mixture of gases.

Solution on page 319

Exercise 5.10

What is the change in entropy due to the removal of the partition if in Fig. 3.5.1 there are different types of gases on both sides of the partition? Consider the case in which the number of molecules and the volume on each side are different.

Solution on page 319

5.3 Remarks on the third law of thermodynamics

The macroscopic formulation of the third law of thermodynamics — Nerst's law — is sometimes presented in the following form: It is impossible to cool a system down to absolute zero. And sometimes in the following equivalent form: The entropy of a system at zero temperature does not depend on the parameters that characterize the system's equilibrium state.

The assertion that the two forms are equivalent relies on the fact that if the entropy depends on external parameters, it is possible to continue adiabatic cooling and to lower its temperature in finite steps, even near absolute zero.

We have already seen that the entropy we obtained for an ideal gas, (3.4.39), does not obey this law. The corrected entropy according to Gibbs, (3.5.8), is no better in this sense. A hint of the direction in which we have to search for the answer is the fact that in a system at low temperatures quantum effects must be taken into account. This hint is supported by the fact that the entropy of the Einstein solid, derived in Exercises 2.5 and 2.6, goes to zero quite fast as $T \to 0$. Consider, therefore, the entropy of a quantum system at temperatures tending to zero. The fact that the system is quantum-mechanical is expressed by the fact that the microscopic energies of the system are discrete. We denote by E_i the energy of the microscopic state i. Of course, there can be many microscopic states with the same energy. Such a state of affairs is called a degeneracy, and the number of microscopic states with energy E is denoted by $g(E)$, and called the "degree of degeneracy," or simply the number of states at energy E. **degeneracy**

In Chap. 6 of Part II we saw [Eq. (2.6.7)] that the entropy can be written in the form

$$S = -k \sum_i P_i \ln P_i \,, \tag{3.5.11a}$$

where

$$P_i = \frac{e^{-\beta E_i}}{\sum_n e^{-\beta E_n}} \tag{3.5.11b}$$

is the probability of the microscopic state i.

If E_0 is the system's lowest energy, we can also write P_i in the form

$$P_i = \frac{e^{-\beta(E_i - E_0)}}{\sum_n e^{-\beta(E_n - E_0)}} \,. \tag{3.5.12}$$

Thus for each state not corresponding to the lowest energy, $E_i > E_0$, we find in the limit $T \to 0$ ($\beta \to \infty$) that $P_i \to 0$, because $E_i - E_0$ is positive. There is, of course, one exception: the states for which $E_i = E_0$. For these states we find in the limit $T \to 0$ that $P_i \to 1/g_0$, where g_0 is the degeneracy of E_0, i.e. the number of different microscopic states with energy E_0.

This is so since for these states the numerator in (3.5.12) is unity whereas the denominator contains, in addition to the sum of zeros, g_0 terms each contributing unity, and a total of g_0 in all. We have, therefore, obtained

$$P_i = \begin{cases} 0\,, & E_i \neq E_0\,, \quad T \to 0\,, \\ 1/g_0\,, & E_i = E_0\,, \quad T \to 0\,. \end{cases} \tag{3.5.13}$$

Now, we can calculate the entropy in the limit $T \to 0$ by direct substitution in (3.5.11a):

$$S \to k \ln g_0 \quad , \; T \to 0\,. \tag{3.5.14}$$

Exercise 5.11

Prove Eq. (3.5.14).

Solution on page 319

If the degeneracy of the ground state is not too large, and this is the common case, S/k tends to a value of order unity when $T \to 0$. Such a value is negligible compared to the typical values of S/k at temperatures above absolute zero which are of order N (since S is extensive and proportional to N), and hence we can say that S tends to 0, in such common systems, when $T \to 0$. In doing this we have identified the microscopic, statistical source of the third law.

We immediately identify two exceptions to the third law:

(1) Classical systems, in which $E_j - E_0$ can be arbitrarily small. In such systems when $T \to 0$, or $\beta \to \infty$, there always exists a j such that $\beta(E_j - E_0)$ will be finite, and then (3.5.13) will no longer be valid, nor will (3.5.14). This actually happens in an ideal gas; see Eq. (3.5.8).

(2) Quantum systems, in which the degeneracy of the ground state is very high, so that the right hand side of (3.5.14) is of the order of the number of the system's degrees of freedom.

Exercise 5.12

Calculate the entropy of a system of N classical three-dimensional harmonic oscillators ("classical Einstein solid"), and show that it diverges at $T \to 0$ and that quantum considerations "save the situation."

A similar phenomenon occurs in a classical paramagnet (see Self-Assessment Exercise 8).

Solution on page 320

5.4 Summary

As a result of Gibbs correction we obtain for an ideal gas

$$z = \frac{V(2\pi mkT)^{3/2}}{h^3} ,$$

$$Z = \frac{1}{N!} z^N ,$$

$$F = -NkT \left[\ln \left(\frac{(2\pi mkT)^{3/2}}{nh^3} \right) + 1 \right] ,$$

$$S = Nk \left[\ln \left(\frac{(2\pi mkT)^{3/2}}{nh^3} \right) + \frac{5}{2} \right] ,$$

$$\mu = -kT \ln \left[\frac{(2\pi mkT)^{3/2}}{nh^3} \right] .$$

In spite of the appearance of Planck's constant, these results are valid only as long as quantum effects are negligible, i.e. far enough from absolute zero.

Chapter 6

Fluctuations and Thermodynamic Quantities

So far we have assumed in all the calculations that the distributions of the physical quantities are very narrow around the average so that, at thermodynamic equilibrium, these variables always obtain their values at the maximum of the distribution. To justify this assumption, beyond the pictorial justification given in Secs. 3.3 and 4.1–4.3 of Part II, we shall discuss deviations from the average of two typical quantities, the magnetization and the energy, and demonstrate that indeed the relative width of their distributions tends to zero in the thermodynamic limit.

6.1 Paramagnet: fluctuations in the magnetization

The average magnetic moment of a paramagnet at equilibrium with a heat bath at temperature T is

$$\langle M\rangle = \frac{1}{Z}\sum_i M_i e^{-\beta E_i} = \frac{1}{\beta}\frac{\partial \ln Z}{\partial H}, \tag{3.6.1}$$

where M_i is the magnetization in the microscopic state i and E_i is the energy of that state [see (2.5.8) and Exercise 5.2 of Part II].

The magnetization varies with the magnetic field H. Namely, $\langle M\rangle$ is a function of H. The derivative of $\langle M\rangle$ with respect to H is the susceptibility,

$$\chi = \frac{1}{N}\frac{\partial\langle M\rangle}{\partial H}. \tag{3.6.2}$$

response coefficient

This quantity measures the ease with which the magnetization (per spin) can be affected by varying the field. It is an example of a whole family of quantities that are known as *response coefficients*.

■ Equation (3.6.2) is a generalization of the definition (2.5.15), which was given for small H, where $\langle M\rangle$ is a linear function of H. Note: N was denoted there by n.

Clearly, when χ is large, small changes in H will lead to large changes in $\langle M\rangle$. It is then to be expected that the distribution of M around the average is not narrow. In other words, states with different M's do not differ significantly in their probabilities. Thus one expects that the deviations around the average will be significant. Namely, there must be a relationship between the susceptibility and the width of the distribution of M, around the average $\langle M\rangle$.

This is indeed the case. It turns out that χ is proportional to the average square deviation of M from its average.

In order to establish this, we recall that the energy of the microscopic state E_i in (3.6.1) is

$$E_i = -HM_i\,. \tag{3.6.3}$$

Hence, differentiating (3.6.1) with respect to H we obtain for χ in (3.6.2)

$$\begin{aligned} N\chi = \frac{\partial\langle M\rangle}{\partial H} &= \frac{\beta}{Z}\sum_i M_i^2 e^{-\beta E_i} - \frac{1}{Z^2}\frac{\partial Z}{\partial H}\sum_i M_i e^{-\beta E_i} \\ &= \beta(\langle M^2\rangle - \langle M\rangle^2) \end{aligned}$$

and hence

$$N\chi = \beta\langle (M - \langle M\rangle)^2\rangle = \frac{(\Delta M)^2}{kT}\,. \tag{3.6.4}$$

Except for the temperature factor, the right hand side measures the average square deviation of the magnetization from its average value. The larger the square deviation, the larger the susceptibility, and vice versa. The quantity $\Delta M \equiv \sqrt{(\Delta M)^2}$ is also called the *fluctuation* of M. In the language of probability theory, ΔM is the standard deviation of the distribution of M and $(\Delta M)^2$ is its variance. **fluctuation**

■ See also Exercise 3.5 of Part II.

Equation (3.6.4) leads to several important conclusions:

(a) The susceptibility of a paramagnet must be positive: $(\Delta M)^2$ is the average of a positive quantity, and hence χ is positive.
(b) The relative width of the distribution of the magnetization, namely $\Delta M/\langle M\rangle$, tends to zero in the thermodynamic limit ($N \to \infty$), which justifies the assertion that M is always at its average value, $\langle M\rangle$.

In order to show that conclusion (b) is indeed valid, we write

$$\frac{\Delta M}{\langle M\rangle}=\frac{\sqrt{NkT\chi}}{\langle M\rangle}=\frac{\sqrt{kT\chi}}{\langle\mu_{\rm B}\rangle}\cdot\frac{1}{\sqrt{N}}\,, \tag{3.6.5}$$

where $\langle\mu_{\rm B}\rangle$ denotes the average magnetization per spin (Chap. 4, Part II) and we use Eq. (2.2.15), $\langle M\rangle=\mu_{\rm B}N\langle\sigma\rangle$. This means that if χ is finite, then when $N\to\infty$, the width of the distribution tends to 0.

A word of caution: For a paramagnet at $H=0$, $\langle\sigma\rangle=0$ and the right hand side of Eq. (3.6.5) diverges. The difficulty is not very grave. In the presence of the smallest field, $\langle M\rangle$ becomes large, proportional to N. [See Eq. (2.5.14), Part II.] But even at $H=0$ strictly, what we find in Eq. (3.6.4) is that the standard deviation of M is of order $\sqrt{N}$, while any thermodynamically significant quantity is proportional to N. Hence, this width is irrelevant in the thermodynamic limit.

We will meet a similar phenomenon in the next section, for the fluctuations in energy, and see that they are proportional to the specific heat. In a paramagnet, the fluctuations in energy are the same as the fluctuations in the magnetization, because the two are proportional to each other.

6.2 Energy fluctuations and the specific heat

The relationship we found in the previous section between the fluctuation in the magnetization and the susceptibility is one example of the relationship between fluctuations and response coefficients. We will now see another example.

The average energy in a canonical system is

$$\langle E\rangle=\frac{1}{Z}\sum_i E_i e^{-\beta E_i}=-\frac{\partial\ln Z}{\partial\beta}\,, \tag{3.6.6}$$

which is, of course, extensive. The specific heat (per particle) measures the energy's responsiveness to change in the temperature. It is, therefore, a kind of susceptibility of the energy.

$$c=\frac{1}{N}\frac{\partial\langle E\rangle}{\partial T}\,. \tag{3.6.7}$$

That is, the easier it is to change $\langle E\rangle$ by changing T, the larger is the specific heat and it is to be expected that the distribution around the average energy will be wider. We show this explicitly:

$$Nc=\frac{\partial\langle E\rangle}{\partial T}=\frac{1}{kT^2}(\langle E^2\rangle-\langle E\rangle^2)\,,$$

and from here

$$Nc = \frac{\overline{(\Delta E)^2}}{kT^2} \, . \tag{3.6.8}$$

We have found, therefore, that the fluctuations in energy are proportional to the response of the energy to changes in the temperature, namely to the specific heat, just as the fluctuations in the magnetization are proportional to the response of the magnetization to changes in the magnetic field, namely the susceptibility.

Exercise 6.1

Complete the deduction of (3.6.8).

Solution on page 321

We obtain again a similar dependence as a function of N:

$$\frac{\Delta E}{\langle E \rangle} = \frac{\sqrt{NckT^2}}{N\langle \epsilon \rangle} = \frac{\sqrt{ck}T}{\langle \epsilon \rangle} \cdot \frac{1}{\sqrt{N}} \, , \tag{3.6.9}$$

where $\Delta E = \sqrt{\overline{(\Delta E)^2}}$ and $\langle \epsilon \rangle$ is the average energy per degree of freedom. That is, the specific heat is positive, like the susceptibility, and its fluctuations increase with the specific heat, as expected. In the thermodynamic limit ($N \to \infty$) the width of the distribution becomes negligible compared to the average. Almost all of the systems in the canonical ensemble have the same energy, $\langle E \rangle$. From here also it is inferred that the canonical description is equivalent to the microcanonical description.

Exercise 6.2

Calculate the relative energy fluctuation of an ideal gas $\Delta E/E$.

Solution on page 321

6.3 Summary

(a) The fluctuation of an extensive thermodynamic quantity is proportional to the response of this quantity to changes in its conjugate variable.

(b) As a result of this relationship it is possible to directly measure the fluctuation.

(c) All the response coefficients are positive.

(d) In the thermodynamic limit, the widths of the distributions of extensive quantities become negligible with respect to the average values (except for extreme cases — a divergence of the response, or the vanishing of the average).

Self-assessment exercises

Exercise 1 *Solution on page 322*

Gibbs free energy

(a) Show that the Gibbs free energy,

$$G = E - TS + PV\,,$$

can be derived from the partition function by

$$G = kTV^2 \frac{\partial}{\partial V}\left(\frac{\ln Z}{V}\right).$$

(b) Show that the enthalpy,

$$\mathsf{H} = E + PV\,,$$

is given by

$$\mathsf{H} = \frac{1}{\beta}\left(V\frac{\partial \ln Z}{\partial V} - \beta\frac{\partial \ln Z}{\partial \beta}\right).$$

Exercise 2 *Solution on page 322*

What is the form of the partition function Z of a crystal, described by a model in which the harmonic oscillators have different energies from one another? What is the expression for the free energy?

Exercise 3 *Solution on page 323*

We define a bounded harmonic oscillator by energy levels like the ones given in Eq. (3.2.1),

$$E_n = \epsilon_0 + n\epsilon\,,$$

with the additional condition

$$n < n_0\,.$$

(a) Calculate the average energy and the average degree of excitation of the bounded oscillator at temperature T.
(b) Calculate the entropy and the specific heat of a bounded oscillator.
(c) Compare with the harmonic oscillator. When can we expect, from qualitative considerations, a difference between the two, and when should they behave similarly.
(d) What is the asymptotic behavior of the specific heat at low temperatures and at high temperatures?

Exercise 4 *Solution on page 325*

The table below contains values of the specific heat of crystalline sodium.

T (K)	1	2	4	6	8	10	12	16	20	25
C_V (cal/mole · K)	0.0001	0.0009	0.0072	0.024	0.058	0.114	0.18	0.46	0.91	1.78

(a) The asymptotic behavior of C_V of a solid in the Einstein model, at low temperatures, is given by (3.2.18):

$$C_V \simeq 3R\left(\frac{\hbar\omega}{kT}\right)^2 e^{-\hbar\omega/kT}\,.$$

Draw a graph of C_V, as a function of $kT/\hbar\omega$, and find Θ_E and ω using the value of C_V at 10 K.
(b) Use the graph from (a) to describe C_V as a function of T for the $\Theta_E = \hbar\omega/k$ you obtained there. Add to the graph the experimental values given in the table.

Is it possible to assert that the values in the table can be described by the graph?
(c) Do the temperatures in the table justify a low temperature approximation (namely, is $T \ll \Theta_E$?)? Sketch a graph of the ratio C_V(Einstein)/ C_V(table) as a function of T.

Exercise 5 *Solution on page 326*

In the Debye model (Part IV, Sec. 3.3) it is found that the specific heat per mole, at low temperatures, is given by

$$C_V = \frac{12\pi^4}{5}R\left(\frac{T}{\Theta_D}\right)^3\,.$$

Using the table from the previous exercise, determine Θ_D from the value of C_V at $T = 10$ K.

Draw in a single figure C_V of the Debye approximation and C_V of the Einstein approximation as a function of T, for the Θ_D determined above, and the experimental values in the table.

Exercise 6 *Solution on page 327*

Suppose that the following explicit formulas for the partition function were found (say, in an excavation):

(a)

$$Z = V^N (2\pi mkT)^{5N/2} ,$$

(b)

$$Z = (V - Nb)^N (2\pi mkT)^{3N/2} e^{aN^2/VkT} .$$

For both cases calculate the equation of state and the specific heat at constant volume.

From the results identify the systems described by these partition functions.

Exercise 7 *Solution on page 330*

Calculate the average energy, the equation of state, and the specific heat of a gas of extremely relativistic particles satisfying $\epsilon = cp$, where c is the speed of light.

Exercise 8 *Solution on page 331*

Classical paramagnet: We return to the system of magnetic ions from Part II. This time we treat the magnetic moment of each ion as a classical vector, with constant magnitude, μ, which can point in any direction. Namely, the tip of the magnetic moment can point in any direction or, in other words, can be found at any point on a sphere.

The summation over the different states of the system goes over, naturally, to an integration over all the directions along which the magnetic moment can point (see figure). Such a direction is described by the two angles of the spherical system, θ and ϕ. The integration over all the directions is carried out using the fact that an infinitesimal surface element on a sphere of radius R is given by

$$da = R^2 \sin\theta d\theta d\phi .$$

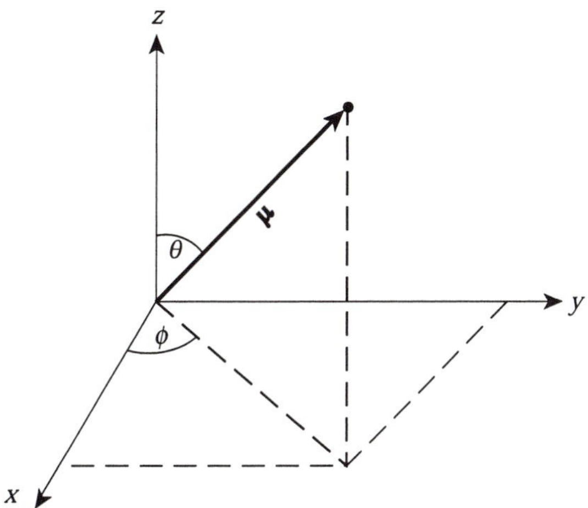

(a) Write the energy of a system of N spins in a magnetic field H directed along the z axis, as a function of the angles.
(b) Calculate the canonical partition function.
(c) Calculate the average magnetization, the average energy and the entropy.
(d) Compare with the discrete paramagnet and show that the magnetization and the energy obtained are identical to the results of the discrete paramagnet in the limit where $J \to \infty$.
(e) Does the third law of thermodynamics apply here?

Exercise 9 *Solution on page 334*

(a) Show that the work performed by an ideal gas at temperature T as it expands from volume V_1 to volume V_2 is equal to the decrease ΔF in its free energy.
(b) Is ΔF larger or smaller than the decrease in the internal energy of the gas? What is the origin of the difference?

Exercise 10 *Solution on page 334*

(a) Calculate the specific heat of a paramagnet and the fluctuations of the energy and show that the relation (3.6.8) indeed holds.
(b) If $N = 100$ and $\mu H = 0.01\text{eV}$, at what temperature will the average energy lose its thermodynamic meaning?

Solutions to exercises in the text

Solution 1.1

Exercise on page 227

The dependence on N is hidden inside E_i, which is the sum of energies of N particles, and also implicitly in the sum over states $\sum_i$. The dependence on V is in the summation over all possible positions as well as in the energies of the particles. To the kinetic energy of a classical particle we must add the condition which confines its motion to the volume V, and this can be done by adding a "step potential" which is zero inside the container and infinite outside. The energy of a quantum particle of mass m inside a three-dimensional box of side d is quantized in quanta of $\hbar^2/8md^2$, which, of course, leads to a dependence on the volume d^3.

Solution 1.2

Exercise on page 227

$\langle E \rangle$ is given by definition as

$$\langle E \rangle = \sum_i E_i P_i \,, \tag{i}$$

where P_i is the probability of microscopic state i. This probability, as we have seen, is proportional to $e^{-\beta E_i}$. Namely (after normalization),

$$P_i = \frac{e^{-\beta E_i}}{\sum_i e^{-\beta E_i}} \equiv \frac{e^{-\beta E_i}}{Z} \,, \tag{ii}$$

where we have used the definition (3.1.2) of Z.

The average of E is thus written as

$$\begin{aligned}\langle E \rangle &= \sum_i E_i \frac{e^{-\beta E_i}}{Z} = \frac{1}{Z} \sum_i E_i e^{-\beta E_i} = \frac{1}{Z} \left(-\frac{\partial}{\partial \beta} \sum_i e^{-\beta E_i} \right) \\ &= -\frac{1}{Z} \frac{\partial Z}{\partial \beta} = -\frac{\partial \ln Z}{\partial \beta} \,.\end{aligned} \tag{iii}$$

But $\beta = 1/kT$ or $T = 1/\beta k$. We replace the differentiation with respect to β by a differentiation with respect to T in the following manner:

$$\frac{\partial}{\partial\beta} = \frac{dT}{d\beta}\frac{\partial}{\partial T} = -\frac{1}{k\beta^2}\frac{\partial}{\partial T} = -kT^2\frac{\partial}{\partial T},$$

and obtain from (iii), for the average of E

$$\langle E\rangle = kT^2\frac{\partial \ln Z}{\partial T}. \qquad \text{(iv)}$$

Solution 1.3 *Exercise on page 230*

The work performed by the macroscopic system upon the external body is an average of (3.1.6) over the canonical ensemble. Namely,

$$\delta W = \frac{1}{Z}\sum_i \left(-\frac{\partial E_i}{\partial X}dX\right) e^{-\beta E_i}.$$

But

$$\frac{\partial}{\partial X}(e^{-\beta E_i}) = -\beta e^{-\beta E_i}\frac{\partial E_i}{\partial X},$$

so that we may write

$$\delta W = \frac{1}{Z}\sum_i \frac{1}{\beta}\frac{\partial}{\partial X}(e^{-\beta E_i})dX.$$

Interchanging the differentiation and the summation we obtain

$$\delta W = \frac{1}{Z}\frac{1}{\beta}\frac{\partial}{\partial X}\left(\sum_i e^{-\beta E_i}\right)dX = \frac{1}{\beta}\frac{1}{Z}\frac{\partial Z}{\partial X}dX = \frac{1}{\beta}\frac{\partial \ln Z}{\partial X}dX.$$

Solution 1.4 *Exercise on page 232*

We denote by m and M the masses of the molecule and of the piston, respectively, by v and u their respective velocities before the collision, and by v' and u' their velocities after. p and P are the momenta of the molecule and of the piston before the collision, and p' and P' after.

For an elastic collision the kinetic energy is conserved:

$$\frac{p^2}{2m} + \frac{P^2}{2M} = \frac{p'^2}{2m} + \frac{P'^2}{2M} \qquad \text{(i)}$$

or

$$\frac{p^2 - p'^2}{2m} = -\frac{P^2 - P'^2}{2M}. \qquad \text{(ii)}$$

Making use of the conservation of momentum,

$$p + P = p' + P' \Rightarrow p - p' = -(P - P'), \qquad \text{(iii)}$$

dividing (ii) by (iii) we obtain another first order equation:

$$\frac{p+p'}{m} = -\frac{P+P'}{M}\,. \tag{iv}$$

Equations (iii) and (iv) are two equations for the two unknowns p' and P'. The quantity of interest is p', for which we find

$$p' = \frac{2mP-(M-m)p}{M+m}\,. \tag{v}$$

Now, the momentum Δp gained by the piston is equal to the momentum lost by the molecule, and thus

$$\Delta p = p - p' = \frac{2m(v-u)}{1+m/M}\,, \tag{vi}$$

and since the mass of the molecule is negligible compared to the mass of the piston $m/M \ll 1$, we obtain (3.1.15). The molecule's velocity after the collision is obtained using (v):

$$v' = \frac{p'}{m}\,, \tag{vii}$$

and for $m \ll M$,

$$v' \simeq 2u - v\,.$$

Solution 1.5 *Exercise on page 234*

(a) The work performed by the system due to a change dX in the parameter is given by Eq. (3.1.8):

$$\delta W = \frac{1}{\beta}\frac{\partial \ln Z}{\partial X}dX\,. \tag{i}$$

The internal energy E is given by (3.1.3):

$$E = -\frac{\partial \ln Z}{\partial \beta}\,. \tag{ii}$$

The energy change dE is made of a part that is due to the change in temperature, and a part that is due to the volume change:

$$dE = \frac{\partial E}{\partial X}dX + \frac{\partial E}{\partial \beta}d\beta = -\frac{\partial^2 \ln Z}{\partial \beta \partial X}dX - \frac{\partial^2 \ln Z}{\partial \beta^2}d\beta\,. \tag{iii}$$

From (i) and (iii) we obtain

$$dE + \delta W = \left(\frac{1}{\beta}\frac{\partial \ln Z}{\partial X} - \frac{\partial^2 \ln Z}{\partial \beta \partial X}\right)dX - \frac{\partial^2 \ln Z}{\partial \beta^2}d\beta\,, \tag{iv}$$

which is Eq. (3.1.25).

(b) For the right hand side of (3.1.25) to be the change of a state function it should be an exact differential.

The condition for the expression $M(x,y)dx + N(x,y)dy$ to be an exact differential is

$$\frac{\partial M}{\partial y} = \frac{\partial N}{\partial x} \Rightarrow \frac{\partial N}{\partial x} - \frac{\partial M}{\partial y} = 0\,. \qquad \text{(v)}$$

(See e.g. Part II, Chap. 0.) Writing the right hand side of (3.1.25) in the form

$$M dX + N d\beta\,, \qquad \text{(vi)}$$

we have to calculate

$$\begin{aligned}\frac{\partial N}{\partial X} - \frac{\partial M}{\partial \beta} &= -\frac{\partial^3 \ln Z}{\partial \beta^2 \partial X} - \frac{1}{\beta}\frac{\partial^2 \ln Z}{\partial \beta \partial X} + \frac{1}{\beta^2}\frac{\partial \ln Z}{\partial X} + \frac{\partial^3 \ln Z}{\partial \beta^2 \partial X} \\ &= -\frac{\partial}{\partial \beta}\left(\frac{1}{\beta}\frac{\partial \ln Z}{\partial X}\right)\,.\end{aligned}$$

This expression does not vanish in general, hence (vi), and consequently the right hand side of Eq. (3.1.25), is not an exact differential.

Solution 1.6 *Exercise on page 235*

First we write (3.1.26) as

$$\beta(dE + \delta W) = M dX + N d\beta \qquad \text{(i)}$$

and calculate

$$\frac{\partial N}{\partial X} - \frac{\partial M}{\partial \beta} = -\beta \frac{\partial^3 \ln Z}{\partial X \partial \beta^2} - \frac{\partial^2 \ln Z}{\partial \beta \partial X} + \frac{\partial}{\partial \beta}\left(\beta \frac{\partial^2 \ln Z}{\partial X \partial \beta}\right) = 0\,.$$

Thus (i) is an exact differential, and we must seek the function from which it is derived. We shall denote this function by S/k. We note that

$$N = -\beta \frac{\partial^2 \ln Z}{\partial \beta^2} = \frac{\partial}{\partial \beta}\left(\ln Z - \beta \frac{\partial \ln Z}{\partial \beta}\right)\,. \qquad \text{(ii)}$$

The coefficient M of dX in (3.1.26) can be derived from the same function:

$$M = \frac{\partial}{\partial X}\left(\ln Z - \beta \frac{\partial \ln Z}{\partial \beta}\right)\,, \qquad \text{(iii)}$$

so $k^{-1}S$ is given by Eq. (3.1.27).

Solution 1.7 *Exercise on page 235*

$$S = -\left(\frac{\partial F}{\partial T}\right)_{X,N} = k\frac{\partial}{\partial T}(T\ln Z) = k\ln Z + kT\frac{\partial \ln Z}{\partial T}\,.$$

Here we change variables from T to β, and as

$$T\frac{\partial \ln Z}{\partial T} = -\beta\frac{\partial \ln Z}{\partial \beta}$$

we obtain

$$S = k\ln Z - k\beta\frac{\partial \ln Z}{\partial \beta}\,,$$

which is the required equation.

Solution 1.8 *Exercise on page 236*

The Helmholtz free energy is $F = E - TS$. Substituting for F and S the expressions (3.1.32) and (3.1.33), respectively, we obtain

$$\begin{aligned} E &= F + TS = -NkT\ln\left(2\cosh\frac{\mu_B H}{kT}\right) \\ &+NkT\left[\ln\left(2\cosh\frac{\mu_B H}{kT}\right) - \frac{\mu_B H}{kT}\tanh\frac{\mu_B H}{kT}\right] \\ &= -N\mu_B H\tanh\frac{\mu_B H}{kT}\,, \end{aligned}$$

as we obtained in (2.5.13).

Remark: In (2.5.13) n denotes the number of spins in the system.

Solution 1.9 *Exercise on page 236*

The specific heat at constant field can be found from the energy as in (2.5.17) or, alternatively, with the help of the entropy. Since $\delta Q = TdS$, the specific heat at constant field will be [cf. Eq. (2.0.13)]

$$c_H = \frac{1}{N}T\left(\frac{\partial S}{\partial T}\right)_{H,N}\,.$$

Differentiating S in Eq. (3.1.33) and substituting it here, yields

$$\begin{aligned} c_H &= kT\left[\frac{(-\mu_B H/kT^2)\sinh(\mu_B H/kT)}{\cosh(\mu_B H/kT)} + \frac{\mu_B H}{kT^2}\tanh\frac{\mu_B H}{kT}\right. \\ &\left. - \frac{\mu_B H}{kT}\cdot\frac{-\mu_B H/kT^2}{\cosh^2(\mu_B H/kT)}\right] = \frac{\mu_B^2 H^2}{kT^2\cosh^2(\mu_B H/kT)}\,, \end{aligned}$$

which is Eq. (2.5.18).

Solution 1.10 *Exercise on page 236*

We first obtain $S(T, H, N)$ from F, as in Eq. (3.1.33), and then convert the dependence on T into a dependence on E. To this end we use the relation between E and T as given by (2.5.13); see Solution 1.8 as well.

We also use the identity

$$\frac{1+\tanh x}{1-\tanh x} = e^{2x}, \tag{i}$$

which is obtained by writing

$$\tanh x = \frac{\sinh x}{\cosh x} = \frac{e^x - e^{-x}}{e^x + e^{-x}} = \frac{e^{2x}-1}{e^{2x}+1},$$

and solving for e^{2x}. Equation (i) implies that

$$\begin{aligned} x &= \frac{1}{2}\ln(1+\tanh x) - \frac{1}{2}\ln(1-\tanh x) \\ &= \frac{1}{2}\ln\left(\frac{1}{2}+\frac{1}{2}\tanh x\right) - \frac{1}{2}\ln\left(\frac{1}{2}-\frac{1}{2}\tanh x\right). \end{aligned} \tag{ii}$$

In addition, since $\cosh^2 x - \sinh^2 x = 1$ we can write

$$\cosh x = (1-\tanh^2 x)^{-1/2} = (1+\tanh x)^{-1/2}(1-\tanh x)^{-1/2},$$

which implies that

$$\begin{aligned} \ln\cosh x &= -\frac{1}{2}\ln(1+\tanh x) - \frac{1}{2}\ln(1-\tanh x) \\ &= -\frac{1}{2}\ln\left(\frac{1}{2}+\frac{1}{2}\tanh x\right) - \frac{1}{2}\ln\left(\frac{1}{2}-\frac{1}{2}\tanh x\right) - \ln 2. \end{aligned} \tag{iii}$$

Now, for convenience, we denote $x = \mu_B H/kT$ in the equation for the entropy (3.1.33), obtained from F, and find

$$\begin{aligned} S &= Nk[\ln(2\cosh x) - x\tanh x] \\ &= -Nk\left[\frac{1}{2}\ln\left(\frac{1}{2}+\frac{1}{2}\tanh x\right) + \frac{1}{2}\ln\left(\frac{1}{2}-\frac{1}{2}\tanh x\right)\right. \\ &\quad \left. +\frac{1}{2}\ln\left(\frac{1}{2}+\frac{1}{2}\tanh x\right)\tanh x - \frac{1}{2}\ln\left(\frac{1}{2}-\frac{1}{2}\tanh x\right)\tanh x\right] \\ &= -Nk\left[\left(\frac{1}{2}+\frac{1}{2}\tanh x\right)\ln\left(\frac{1}{2}+\frac{1}{2}\tanh x\right)\right. \\ &\quad \left. +\left(\frac{1}{2}-\frac{1}{2}\tanh x\right)\ln\left(\frac{1}{2}-\frac{1}{2}\tanh x\right)\right], \end{aligned} \tag{iv}$$

where we have used the identities (ii) and (iii).

From the expression for the energy

$$E = -N\mu_B H \tanh\frac{\mu_B H}{kT}$$

we obtain

$$\tanh x = -\frac{E/N}{\mu_B H}, \tag{v}$$

and then S in (iv) takes the form

$$S = -k\left[\left(\frac{N}{2} - \frac{E}{2\mu_B H}\right)\ln\left(\frac{1}{2} - \frac{E/N}{2\mu_B H}\right) + \left(\frac{N}{2} + \frac{E}{2\mu_B H}\right)\ln\left(\frac{1}{2} + \frac{E/N}{2\mu_B H}\right)\right],$$

which is Eq. (2.3.13).

Solution 1.11 *Exercise on page 237*

First, we calculate Z_C with the shifted energies:

$$Z_C = \sum_i e^{-\beta(E_i + C)} = e^{-\beta C}\sum_i e^{-\beta E_i} = Ze^{-\beta C},$$

where Z is the partition function of the system with the energies E_i. The average energy, (3.1.3), is

$$\langle E\rangle_C = -\frac{\partial \ln Z_C}{\partial \beta} = -\frac{\partial}{\partial\beta}(\ln Z - \beta C) = -\frac{\partial \ln Z}{\partial \beta} + C = \langle E\rangle + C.$$

Namely, the average energy has changed by C, i.e. by the same constant that we added to the microscopic energies.

The new free energy F_C [Eq. (3.1.30)],

$$F_C = -\frac{1}{\beta}\ln Z_C = -\frac{1}{\beta}(\ln Z - \beta C) = F + C,$$

and the magnetization (2.5.8) is unchanged:

$$\langle M\rangle_C = \frac{1}{\beta}\frac{\partial \ln Z_C}{\partial H} = -\frac{\partial F_C}{\partial H} = -\frac{\partial F}{\partial H} = \langle M\rangle.$$

Solution 1.12 *Exercise on page 238*

$F(E) = E - TS$ is minimal when

$$\frac{\partial F}{\partial E} = 0.$$

The dependence on E appears in the first linear term and is also hidden in the entropy and thus, since T is constant,

$$\frac{\partial F}{\partial E} = 1 - T\frac{\partial S}{\partial E} = 0\,,$$

and from here (3.1.38) is immediately obtained.

Solution 1.13 *Exercise on page 239*

(a) Substituting $\mu_{\mathrm{B}}H/kT = 0.5$ into (2.5.13) we obtain

$$\frac{\langle E\rangle}{\mu_{\mathrm{B}}HN} = -\tanh(0.5) = -0.4621\,,$$

which is exactly the same result as obtained from Fig. 3.1.2.

(b) Substituting into Eq. (3.1.32) we obtain

$$\frac{F}{NkT} = -\ln[2\cosh(0.5)] = -0.8133\,,$$

and this is exactly the value obtained from the figure or from direct substitution of $\langle E\rangle$ of (a) into Eq. (3.1.41).

(c) The requirement that F be minimal yields an equation for the energy at the minimum,

$$0 = \frac{\partial F}{\partial E} = 1 - \frac{kT}{2\mu_{\mathrm{B}}H}\ln\left(\frac{\mu_{\mathrm{B}}HN - E}{\mu_{\mathrm{B}}HN + E}\right), \tag{i}$$

so that

$$\frac{\mu_{\mathrm{B}}HN - E}{\mu_{\mathrm{B}}HN + E} = \exp\left(\frac{2\mu_{\mathrm{B}}H}{kT}\right) \tag{ii}$$

or

$$E = -\mu_{\mathrm{B}}HN\tanh\left(\frac{\mu_{\mathrm{B}}H}{kT}\right), \tag{iii}$$

and this is the familiar expression for the average energy of the paramagnet at temperature T. In order to obtain the value of F at equilibrium, we have to substitute into (3.1.41) the expression for E which mimimizes $F(E)$, namely (iii).

Denoting

$$\frac{\mu_{\mathrm{B}}H}{kT} = x \tag{iv}$$

we can write F in the form

$$\begin{aligned} F = NkT\Big\{&-x\tanh x + \frac{1}{2}(1-\tanh x)\ln\left[\frac{1}{2}(1-\tanh x)\right] \\ &+\frac{1}{2}(1+\tanh x)\ln\left[\frac{1}{2}(1+\tanh x)\right]\Big\} = NkT\Big\{-x\tanh x \\ &+\frac{1}{2}\ln\left[\frac{1}{4}(1-\tanh^2 x)\right] + \frac{1}{2}\tanh x\ln\left(\frac{1+\tanh x}{1-\tanh x}\right)\Big\}\,. \end{aligned} \tag{v}$$

We now use the identities

$$1 - \tanh^2 x = \frac{1}{\cosh^2 x}, \qquad \text{(vi)}$$

$$\frac{1 + \tanh x}{1 - \tanh x} = e^{2x}, \qquad \text{(vii)}$$

and obtain

$$F = NkT\left[-x\tanh x - \frac{1}{2}\ln(4\cosh^2 x) + x\tanh x\right]$$

$$= -NkT\ln(2\cosh x), \qquad \text{(viii)}$$

and this is exactly Eq. (3.1.32).

Solution 1.14 *Exercise on page 241*

The requirement of maximum entropy implies that

$$\frac{\partial}{\partial N_a}[S(E_a, V_a, N_a) + S(E_b, V_b, N_b)] = 0.$$

Performing the differentiation and using the fact that $\partial N_b / \partial N_a = -1$, we obtain

$$\frac{\partial S_a}{\partial N_a} - \frac{\partial S_b}{\partial N_b} = 0$$

or

$$\frac{\partial S_a}{\partial N_a} = \frac{\partial S_b}{\partial N_b};$$

namely, the equality of the quantity $\partial S / \partial N$ determines the equilibrium between the systems.

Now

$$\frac{\partial S}{\partial N} = -\frac{\mu}{T}$$

and since the temperatures must be equal so do the chemical potentials.

Solution 2.1 *Exercise on page 246*

The partition function in our example is obtained from Eq. (3.2.3):

$$Z = \sum_{n_1, n_2, n_3} e^{-\beta E(n_1, n_2, n_3)},$$

where each of the summation variables takes the values 0, 1 and 2.

Now, the energy of a given state described by the triplet (n_1, n_2, n_3) is

$$E(n_1, n_2, n_3) = \hbar\omega\left(\frac{3}{2} + n_1 + n_2 + n_3\right) = \frac{3}{2}\hbar\omega + \hbar\omega(n_1 + n_2 + n_3).$$

Note that the first term, $\frac{3}{2}\hbar\omega$, is common to all states and is independent of (n_1, n_2, n_3). Hence

$$e^{-\beta E(n_1,n_2,n_3)} = e^{-3\beta\epsilon/2} e^{-\beta(n_1+n_2+n_3)\epsilon},$$

where we have denoted $\hbar\omega = \epsilon$.

In order to find the partition function, we have to sum over all states of the system: Each oscillator has three possible states. Thus a system of three oscillators has $3^3 = 27$ microscopic states which are described by all the triplets $(n_1; n_2, n_3)$ in the table below.

(0,0,0)						
(0,0,1)	(0,1,0)	(1,0,0)				
(0,0,2)	(0,2,0)	(2,0,0)	(0,1,1)	(1,0,1)	(1,1,0)	
(0,1,2)	(0,2,1)	(1,0,2)	(1,2,0)	(2,0,1)	(2,1,0)	(1,1,1)
(0,2,2)	(2,0,2)	(2,2,0)	(1,1,2)	(1,2,1)	(2,1,1)	
(1,2,2)	(2,1,2)	(2,2,1)				
(2,2,2)						

Note that all the states that appear in the same row have the same energy.

To conclude,

$$Z = e^{-3\beta\epsilon/2}(1 + 3e^{-\beta\epsilon} + 6e^{-2\beta\epsilon} + 7e^{-3\beta\epsilon} + 6e^{-4\beta\epsilon} + 3e^{-5\beta\epsilon} + e^{-6\beta}).$$

Let us check if indeed this result is obtained from (3.2.4), i.e. from the cube of the single particle partition function. For a single oscillator

$$z = \sum_{n=0}^{2} e^{-\beta\epsilon_n} = e^{-\beta\epsilon/2}(1 + e^{-\beta\epsilon} + e^{-2\beta\epsilon}),$$

$$\begin{aligned} z^3 &= e^{-3\beta\epsilon/2}\Big(1 + e^{-3\beta\epsilon} + e^{-6\beta\epsilon} + 3e^{-\beta\epsilon} + 3e^{-2\beta\epsilon} + 3e^{-2\beta\epsilon} + 3e^{-4\beta\epsilon} \\ &\quad + 3e^{-4\beta\epsilon} + 3e^{-5\beta\epsilon} + 6e^{-3\beta\epsilon}\Big) \\ &= e^{-3\beta\epsilon/2}\Big(1 + 3e^{-\beta\epsilon} + 6e^{-2\beta\epsilon} + 7e^{-3\beta\epsilon} + 6e^{-4\beta\epsilon} + 3e^{-5\beta\epsilon} \\ &\quad + e^{-6\beta\epsilon}\Big) = Z, \end{aligned}$$

where we have used the identity

$$(1 + x + y)^3 = 1 + x^3 + y^3 + 3x + 3y + 3x^2 + 3y^2 + 3x^2y + 3xy^2 + 6xy.$$

Equations (3.2.3) and (3.2.4) are indeed equal.

Solution 2.2 *Exercise on page 246*

The average energy is given by Eq. (3.1.3):

$$\langle E\rangle = -\frac{\partial \ln Z}{\partial\beta} = -N\frac{\partial \ln z}{\partial\beta},$$

where the partition function of a single oscillator, z, is given by Eq. (3.2.7):

$$z = e^{-\beta\hbar\omega/2}(1-e^{-\beta\hbar\omega})^{-1}.$$

Hence the average energy per oscillator $\langle E\rangle/N$ can be derived directly from the single particle partition function:

$$\begin{aligned}\langle\epsilon\rangle &= -\frac{\partial}{\partial\beta}\ln\frac{e^{-\beta\hbar\omega/2}}{1-e^{-\beta\hbar\omega}} = -\frac{\partial}{\partial\beta}\left[-\beta\hbar\omega/2 - \ln(1-e^{-\beta\hbar\omega})\right]\\ &= \frac{\hbar\omega}{2} + \frac{\hbar\omega}{e^{\hbar\omega/kT}-1},\end{aligned}$$

where we have substituted $\beta = 1/kT$.

Solution 2.3 *Exercise on page 246*

(a) Direct calculation of $\langle n\rangle$:
First we derive the probability for a given oscillator to be in the nth energy level, in a way similar to the calculation of the probability for a given magnetic moment to be found in state $\sigma = +1$ or $\sigma = -1$ in Part II. To do this we recall that the probability of a microscopic state i, given by the set of excitation numbers $(n_1, n_2, \ldots, n_N)$, is

$$P(n_1, n_2, \ldots, n_N) = Z^{-1}e^{-\beta E(n_1,n_2,\ldots,n_N)}.$$

Now, since E is the sum of single oscillator energies ϵ_n and Z is the product of N identical factors $Z = z^N$, we can write the probability for a microscopic state as the product of N probabilities:

$$P(n_1, n_2, \ldots, n_N) = P(n_1)P(n_2)\cdot\ldots\cdot P(n_N),$$

where

$$P(n) = z^{-1}e^{-\beta\epsilon_n}.$$

Using the excitation probabilities we calculate the average according to

$$\langle n\rangle = z^{-1}\sum_{n=0} ne^{-\beta\epsilon_n}.$$

Substituting ϵ_n from (3.2.1) and z from (3.2.6) with the notation $\epsilon = \hbar\omega$, we obtain

$$\langle n\rangle = \frac{e^{-\beta\epsilon/2}\sum_{n=0}^{\infty} ne^{-\beta n\epsilon}}{e^{-\beta\epsilon/2}\sum_{n=0}^{\infty} e^{-\beta n\epsilon}} .$$

To calculate the right hand side we use a trick similar to the one used to prove Eq. (3.1.3) (Solution 1.2). Note that the numerator is the derivative of the denominator with respect to the variable $x = \beta\epsilon$. We thus write

$$\langle n\rangle = \frac{\sum_{n=0}^{\infty} ne^{-xn}}{\sum_{n=0}^{\infty} e^{-xn}} = -\frac{d}{dx}\ln\left(\sum_{n=0}^{\infty} e^{-xn}\right) = -\frac{d}{dx}\ln\left(\frac{1}{1-e^{-x}}\right)$$

$$= \frac{e^{-x}}{1-e^{-x}} = \frac{1}{e^{x}-1} .$$

Substituting $x = \beta\epsilon$, $\beta = 1/kT, \epsilon = \hbar\omega$, we obtain Eq. (3.2.9):

$$\langle n\rangle = \frac{1}{e^{\hbar\omega/kT}-1} .$$

(b) Calculation of $\langle n\rangle$ from (3.2.8).
Averaging Eq. (3.2.1) for a given oscillator yields

$$\langle\epsilon\rangle = \left\langle\frac{\hbar\omega}{2}\right\rangle + \langle n\hbar\omega\rangle = \frac{\hbar\omega}{2} + \hbar\omega\langle n\rangle .$$

This result must be equal to (3.2.8), and hence

$$\langle n\rangle = \frac{1}{e^{\hbar\omega/kT}-1} ,$$

as obtained above.

Solution 2.4 *Exercise on page 247*

(a) At low temperatures $\hbar\omega/kT \gg 1$ and $e^{\hbar\omega/kT} \gg 1$. Hence

$$\langle n\rangle \sim \frac{1}{e^{\hbar\omega/kT}} = e^{-\hbar\omega/kT} .$$

(b) At high temperatures $\hbar\omega/kT \ll 1$, so it is possible to use the power expansion of the exponential function,

$$e^{x} \simeq 1 + x + \frac{x^2}{2!} + \dots ,$$

and to take only the first two terms,

$$e^{\hbar\omega/kT} \simeq 1 + \frac{\hbar\omega}{kT},$$

so that

$$\langle n \rangle \sim \frac{1}{1 + \hbar\omega/kT - 1} = \frac{kT}{\hbar\omega}.$$

Solution 2.5 *Exercise on page 248*

(a) The free energy of a single harmonic oscillator is F/N:

$$f = \frac{F}{N} = -kT \ln z,$$

where z is the partition function (3.2.7):

$$z = \frac{e^{-\beta\hbar\omega/2}}{1 - e^{-\beta\hbar\omega}}.$$

Hence

$$f = \frac{\hbar\omega}{2} + kT \ln(1 - e^{-\hbar\omega/kT}).$$

The entropy of a single harmonic oscillator

$$s = \frac{S}{N} = -\frac{\partial f}{\partial T} = -k \ln(1 - e^{-\hbar\omega/kT}) - kT \frac{(-\hbar\omega/kT^2)e^{-\hbar\omega/kT}}{1 - e^{-\hbar\omega/kT}}$$

$$= -k \left[\ln(1 - e^{-\hbar\omega/kT}) - \frac{\hbar\omega}{kT} \frac{1}{e^{\hbar\omega/kT} - 1} \right].$$

It is worth noting that we could have obtained the same result by using the relation $S = (E - F)/T$ [(3.1.28) or (3.1.29)].

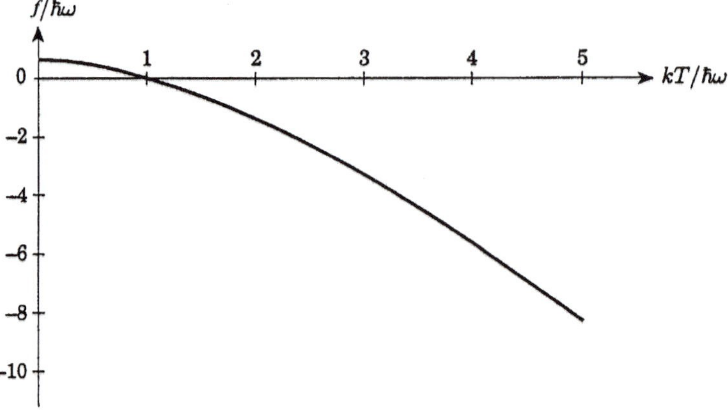

The free energy (in units of $\hbar\omega$) of a single harmonic oscillator as a function of the temperature (in units of $\hbar\omega/k$).

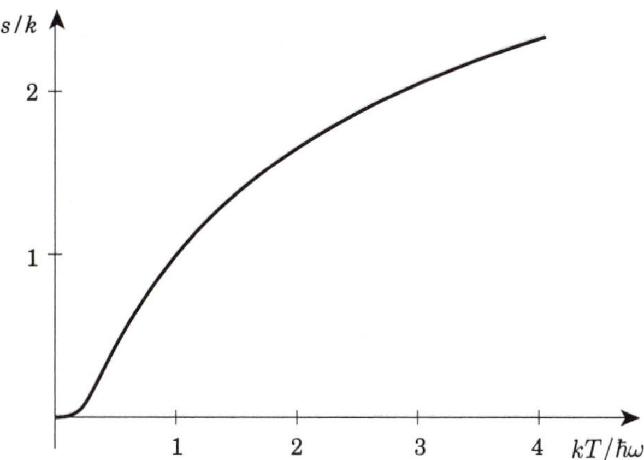

The entropy (in units of k) of a single harmonic oscillator as a function of the temperature (in units of $\hbar\omega/k$).

(b) The transition temperature between the two types of behavior is in the region in which the quantity $kT/\hbar\omega$ is not very large and not very small, namely $kT/\hbar\omega \simeq 1$. This can also be seen from both graphs above. The transition temperature is thus

$$T \simeq \frac{\hbar\omega}{k} \simeq \frac{10^{-34} \times 10^{13} \times 2\pi}{1.4 \times 10^{-23}} \simeq 450 \text{ K}.$$

Solution 2.6 *Exercise on page 250*

(a) Differentiating the average energy with respect to the temperature in Eq. (3.2.13) gives the specific heat:

$$C = \frac{\partial\langle E\rangle}{\partial T} = 3N\hbar\omega \frac{\hbar\omega/kT^2 e^{\hbar\omega/kT}}{(e^{\hbar\omega/kT}-1)^2}.$$

The molar specific heat is the specific heat calculated for $N = N_0$ atoms, where N_0 is the Avogadro number. Recalling that $N_0 k = R$, we obtain

$$C = 3R\left(\frac{\hbar\omega}{kT}\right)^2 \frac{e^{\hbar\omega/kT}}{(e^{\hbar\omega/kT}-1)^2}.$$

(b) The entropy S can be calculated by differentiating the free energy given in (3.2.12):

$$S = -\left(\frac{\partial F}{\partial T}\right),$$

or with the help of the relation $S = (E - F)/T$, but the simplest way is to use the result from Exercise 2.5.

$$S = -3Nk\left[\ln(1 - e^{\hbar\omega/kT}) - \frac{\hbar\omega/kT}{e^{\hbar\omega/kT}-1}\right].$$

Now, since $\delta Q = TdS$, the specific heat can also be derived from the entropy [Eq. (2.0.14)].

$$C = T\left(\frac{\partial S}{\partial T}\right)_N = -\ 3NkT\left[\frac{(-\hbar\omega/kT^2)e^{-\hbar\omega/kT}}{1-e^{-\hbar\omega/kT}} + \frac{\hbar\omega/kT^2}{e^{\hbar\omega/kT}-1}\right.$$

$$\left. - \frac{(\hbar\omega/kT)^2(1/T)e^{\hbar\omega/kT}}{(e^{\hbar\omega/kT}-1)^2}\right],$$

in which the first two terms cancel. We denote $N_0k = R$, and obtain the molar specific heat

$$C = 3R\left(\frac{\hbar\omega}{kT}\right)^2 \frac{e^{\hbar\omega/kT}}{(e^{\hbar\omega/kT}-1)^2}\,.$$

Solution 2.7 *Exercise on page 251*

(a) $\omega = 10^7\ \mathrm{s}^{-1}$:
The amount of heat (per mole) required in order to raise the temperature of the crystal by ΔT is

$$\Delta Q = C\Delta T\,,$$

where C is the specific heat (per mole) of the crystal. Around $T = 1$ K

$$\frac{\hbar\omega}{kT} \simeq \frac{10^{-34}\times 10^7}{1.4\times 10^{-23}\times 1} \simeq 7\times 10^{-5} \ll 1\,.$$

This is, therefore, a very high temperature for this crystal, and it is possible to approximate (3.2.16) by (3.2.17):

$$C \simeq 3R \simeq 25\ \mathrm{J/K}\,.$$

For $\Delta T = 10^{-2}$ K

$$\Delta Q = 3R\Delta T \simeq 0.25\ \mathrm{J}\,.$$

Around $T = 100$ K we certainly have

$$\frac{\hbar\omega}{kT} \ll 1\,,$$

so that there as well

$$\Delta Q \simeq 0.25\ \mathrm{J}\,.$$

In both cases we have used the high temperature approximation.

(b) $\omega = 10^{12}\ \mathrm{s}^{-1}$:
In this case, for $T = 100$ K,

$$\frac{\hbar\omega}{kT} \simeq 7 \times 10^{-2} \ll 1\,.$$

Still 100 K is a high temperature. Hence we have again

$$\Delta Q = C\Delta T \simeq 3R\Delta T \simeq 0.25\ \mathrm{J}\,.$$

But, for $T = 1$ K,

$$\frac{\hbar\omega}{kT} \simeq 7\,.$$

Hence we must use the low temperature approximation to obtain

$$C \simeq 3R\left(\frac{\hbar\omega}{kT}\right)^2 e^{-\hbar\omega/kT} = 3R \cdot 49 \cdot e^{-7} \simeq 1.25\ \mathrm{J/K}$$

and

$$\Delta Q \simeq 1.25 \times 10^{-2}\ \mathrm{J}\,.$$

Solution 3.1 *Exercise on page 255*

The energy of a harmonic oscillator has the form

$$E = \frac{p^2}{2m} + \frac{1}{2}m\omega^2x^2\,. \tag{i}$$

A particle that moves in that potential will conserve its energy, and hence its trajectory will be a curve of constant E. Curves of constant energy in phase space (x, p) are ellipses. We can identify the elliptic structure by rewriting (i) in the form

$$\frac{p^2}{2mE} + \frac{x^2}{2E/m\omega^2} = 1\,.$$

The axes of the ellipse are thus

$$a = \sqrt{2mE}\,,$$

$$b = \sqrt{2E/m\omega^2}\,.$$

This description of the ellipse does not expose the time dependence. This is provided by the parametric description of the ellipse, namely writing x and p as functions of a quantity that varies from one point to another along the particle's trajectory in phase space. This variable is of course the time, and if we choose the time such that at $t = 0$ x is maximal, we obtain

$$x = x_0 \cos \omega t\,,$$

$$p = m\dot{x} = -m\omega x_0 \sin \omega t\,.$$

x_0, of course, is not another free variable. It is the axis of the ellipse, determined by the energy. Substituting x and p in (i) gives

$$x_0 = b = \sqrt{\frac{2E}{m\omega^2}}\,.$$

The next figure illustrates several trajectories of the (classical) harmonic oscillator in phase space.

Each of five radial lines intersects all the trajectories at points corresponding to one of five times: $(0, \frac{1}{8}T, \frac{1}{6}T, \frac{1}{4}T, \frac{1}{2}T)$.

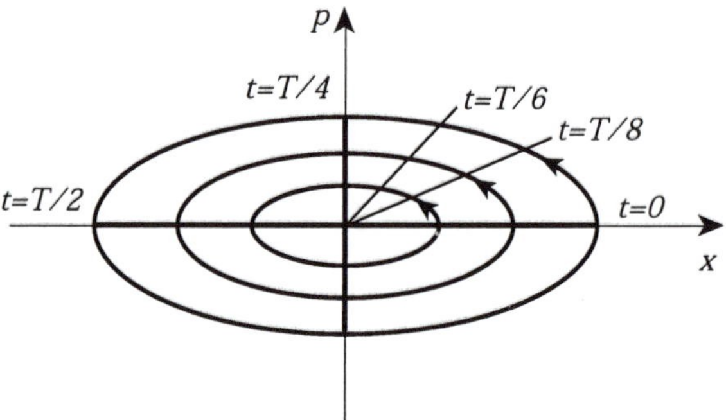

Solution 3.2 *Exercise on page 256*

The average kinetic energy corresponding to the motion along x is

$$\left\langle \frac{p_x^2}{2m} \right\rangle = \frac{\int (p_x^2/2m) e^{-\beta p^2/2m} d^3p \int e^{-\beta U(\mathbf{r})} d^3r}{\int e^{-\beta p^2/2m} d^3p \int e^{-\beta U(\mathbf{r})} d^3r}$$

$$= \frac{\int (p_x^2/2m) e^{-\beta p^2/2m} d^3p}{\int e^{-\beta p^2/2m} d^3p}$$

$$= \frac{\int (p_x^2/2m) e^{-\beta p_x^2/2m} dp_x \int e^{-\beta p_y^2/2m} dp_y \int e^{-\beta p_z^2/2m} dp_z}{\int e^{-\beta p_x^2/2m} dp_x \int e^{-\beta p_y^2/2m} dp_y \int e^{-\beta p_z^2/2m} dp_z}$$

$$= \frac{\int (p_x^2/2m) e^{-\beta p_x^2/2m} dp_x}{\int e^{-\beta p_x^2/2m} dp_x}\,.$$

We have already calculated such expressions in Chap. 1 of Part I. The result is

$$\left\langle \frac{p_x^2}{2m} \right\rangle = \frac{1}{2}kT\,,$$

which is in accord with the equipartition law (an energy of $kT/2$ per degree of freedom).

Solution 3.3 *Exercise on page 258*

(a) The particle's potential is

$$U(r) = \frac{1}{2}Kr^2\,.$$

As long as the vibrational amplitude of the particle is less than $L/2$ or, alternatively, its total energy is less than $\frac{1}{2}K(L/2)^2$, it will not "feel" the existence of the walls of the box.

In order for the particle to reach the wall of the box, it needs at least this amount of energy. If the origin of this energy is thermal, then

$$kT = \frac{1}{2}K\left(\frac{L}{2}\right)^2$$

or

$$T = \frac{KL^2}{8k}\,.$$

(b) If the box becomes infinite, the average of $U(\mathbf{r})$ is given by (3.3.10) or (3.3.11), where the integration over the particle's position extends from $-\infty$ to $+\infty$, in each of the three spatial directions.

$$\langle U(\mathbf{r})\rangle = \frac{\int (Kr^2/2)e^{-\beta Kr^2/2}dV}{\int e^{-\beta Kr^2/2}dV} = -\frac{\partial}{\partial\beta}\ln\left(\int dV e^{-\beta Kr^2/2}\right)\,.$$

Since the integration over $\mathbf{r}$ is over the entire space, it is possible to calculate the last expression as in Chap. 1 of Part I. See also Eq. (1.1.60). The integral has the dimensions of volume or $(\text{length})^3$, and since the only quantity with the dimensions of length in the integral is $1/\sqrt{\beta K}$, it is clear that the integral is proportional to $(\beta K)^{-3/2}$. Differentiating with respect to β yields

$$\langle U(\mathbf{r})\rangle = \frac{3}{2}kT\,.$$

Solution 4.1 *Exercise on page 262*

Since the particles are identical, the functions $\epsilon_1(\mathbf{r}_1, \mathbf{p}_1)$, $\epsilon_2(\mathbf{r}_2, \mathbf{p}_2)$, etc., have the same dependence upon the position of the particles and upon their momentum, and hence the partition function will take the form

$$Z_c = \int dV_1 d\tau_1 \dots dV_N d\tau_N e^{-\beta[\epsilon(\mathbf{r}_1,\mathbf{p}_1)+\dots+\epsilon(\mathbf{r}_N,\mathbf{p}_N)]} .$$

The exponent is a product of exponents of the type $e^{-\beta\epsilon(\mathbf{r},\mathbf{p})}$, one factor for each particle. As a result the integral is a product of integrals and we can write

$$Z_c = \int dV_1 d\tau_1 e^{-\beta\epsilon(\mathbf{r}_1,\mathbf{p}_1)} \int dV_2 d\tau_2 e^{-\beta\epsilon(\mathbf{r}_2,\mathbf{p}_2)} \cdot \ldots \cdot \int dV_N d\tau_N e^{-\beta\epsilon(\mathbf{r}_N,\mathbf{p}_N)} .$$

All of the integrals in the product are identical, as $\mathbf{r}_i$, $\mathbf{p}_i$ are variables of integration, and thus

$$Z_c = \left[\int dV d\tau e^{-\beta\epsilon(\mathbf{r},\mathbf{p})}\right]^N = z^N .$$

Solution 4.2 *Exercise on page 262*

The single particle partition function (3.4.4) is

$$z_c = \int dV e^{-\beta U(\mathbf{r})} \int d\tau e^{-\beta p^2/2m} .$$

The first integral is over the volume of the box. Since the particle moves freely within the box, namely $U(\mathbf{r}) = 0$ in the region of integration, we have

$$\int dV e^{-\beta U(\mathbf{r})} = \int_{\text{box}} dV = V .$$

For the calculation of the second integral we write $A = \beta/2m$, and then we have to calculate an integral of the form

$$I = \int dp_x dp_y dp_z e^{-A(p_x^2+p_y^2+p_z^2)} ,$$

whose value, as found for instance in Exercise 1.13 of Part I, is $(\pi/A)^{3/2}$. Hence

$$I = \left(\frac{2\pi m}{\beta}\right)^{3/2} ,$$

so that

$$z_c = V\left(\frac{2\pi m}{\beta}\right)^{3/2} = V(2\pi mkT)^{3/2} .$$

Solution 4.3 *Exercise on page 263*

The specific heat at constant volume is obtained directly from the temperature derivative of $S(T,V,N)$ in Eq. (3.4.8):

$$\left(\frac{\partial S}{\partial T}\right)_{V,N} = \frac{3Nk}{2T} \Rightarrow C_V = \frac{3}{2}Nk .$$

In contrast, to obtain the specific heat at constant pressure from the entropy, we first have to express the entropy as a function of the pressure instead of the volume. This is done using the ideal gas equation (3.4.7). We thus obtain

$$S(T,P,N) = Nk\left\{\ln\left[\frac{(2\pi m)^{3/2}(kT)^{5/2}N}{P}\right] + \frac{3}{2}\right\}$$

and from here

$$\left(\frac{\partial S}{\partial T}\right)_{P,N} = \frac{5Nk}{2T} \;\Rightarrow\; C_P = \frac{5}{2}Nk\,.$$

Solution 4.4 *Exercise on page 263*

The internal energy is obtained from the derivative of the partition function [Eq. (3.1.3)]:

$$\langle E\rangle = -\frac{\partial}{\partial\beta}\ln Z = -N\frac{\partial}{\partial\beta}\ln z = NkT^2\frac{\partial}{\partial T}\ln z = \frac{3}{2}NkT\,.$$

Since in an isochoric process $\delta E = \delta Q$,

$$C_V = \left(\frac{\partial E}{\partial T}\right)_{N,V}\,.$$

And indeed differentiation yields the previous result. In contrast, C_P cannot be simply related to the energy since $\delta E \neq \delta Q$ in an isobaric process. For the terminology see Part II, Chap. 0.

Solution 4.5 *Exercise on page 264*

The partition function of a mixture of gases is written in the form

$$Z_c = \int dV_1 d\tau_1 \ldots dV_N d\tau_N e^{-\beta E}\,,$$

where E is the energy in Eq. (3.4.11).

Since the energy is a sum of the energies of each particle separately, and there are no terms connecting the position variables and the momentum variables, the integral factors into a product as in Solution 4.1, this time of different factors:

$$Z_c = \left[\int dV_1 d\tau_1 \exp\left(-\frac{\beta p_1^2}{2m_1}\right)\right]^{N_1}\left[\int dV_2 d\tau_2 \exp\left(-\frac{\beta p_2^2}{2m_2}\right)\right]^{N_2}$$
$$\ldots\left[\int dV_M d\tau_M \exp\left(-\frac{\beta p_M^2}{2m_M}\right)\right]^{N_M}\,.$$

Each of the factors in the product can be identified as the partition function of a single particle of type j raised to the power of the number of particles of that type N_j.

Thus we have obtained

$$Z_c = (z_1)^{N_1}(z_2)^{N_2}\dots(z_M)^{N_M} \,,$$

where

$$z_j = \int dV d\tau \exp\left(-\frac{\beta p^2}{2m_j}\right) .$$

Solution 4.6 *Exercise on page 265*

The free energy of the mixture is

$$F = -kT\left[N\ln V + \frac{3}{2}\sum_{j=1}^{M} N_j \ln(2\pi m_j kT)\right] .$$

and the entropy

$$S = -\left(\frac{\partial F}{\partial T}\right)_{V,N_i} = Nk\left(\frac{3}{2} + \ln V\right) + \frac{3}{2}k\sum_{j=1}^{M} N_j \ln(2\pi m_j kT) .$$

The specific heat at constant volume is

$$C_V = T\left(\frac{\partial S}{\partial T}\right)_{V,N_i} = \frac{3}{2}kT\sum_{j=1}^{M}\frac{N_j}{T} = \sum_{j=1}^{M} C_{V_j} \,,$$

where $C_{V_j} = \frac{3}{2}N_j k$, as we obtained for a monocomponent gas in Eq. (3.4.9); namely, the specific heat of a mixture of gases is equal to the sum of the partial specific heats.

The specific heat at constant pressure is obtained in a similar way (see also Solution 4.3), and again

$$C_P = T\left(\frac{\partial S}{\partial T}\right)_{P,N_i} = \sum_{j=1}^{M} C_{P_j} \,,$$

where

$$C_{P_j} = \frac{5}{2}N_j kT .$$

The expression for the entropy can be written in the form

$$S = \sum_{j=1}^{M} N_j k\left[\frac{3}{2} + \ln V + \frac{3}{2}\ln(2\pi m_j kT)\right] = \sum_{j=1}^{M} S_j \,,$$

where S_j is the partial entropy of the jth constituent of the gas — the expression we obtained in Eq. (3.4.8). Namely, the entropy of the mixture is equal to the sum of partial entropies. Actually, we could have started from this result to derive the additivity of both specific heats.

Solution 4.7 *Exercise on page 266*

We have to show that

$$\int P(\mathbf{r}_1, \mathbf{p}_1) dV_1 d\tau_1 = 1 .$$

Using (3.4.19) the left hand side can be written as

$$\int \left[\int P(\mathbf{r}_1, \mathbf{p}_1, \ldots, \mathbf{r}_N, \mathbf{p}_N) dV_2 d\tau_2 \ldots dV_N d\tau_N \right] dV_1 d\tau_1$$

$$= \int P(\mathbf{r}_1, \mathbf{p}_1, \ldots, \mathbf{r}_N, \mathbf{p}_N) dV_1 d\tau_1 \ldots dV_N d\tau_N = 1 ,$$

since $P(\mathbf{r}_1, \mathbf{p}_1, \ldots, \mathbf{r}_N, \mathbf{p}_N)$ has been defined in (3.4.18) as normalized over the full N-particle space.

Solution 4.8 *Exercise on page 267*

The probability distribution function of *any* system, not necessarily an ideal gas, is

$$P(\mathbf{r}_1, \mathbf{p}_1, \ldots, \mathbf{r}_N, \mathbf{p}_N) = Z_c^{-1} \exp\left[-\beta E(\mathbf{r}_1, \mathbf{p}_1, \ldots, \mathbf{r}_N, \mathbf{p}_N)\right] .$$

Now, the assumption (3.4.23) implies that

$$P = Z_c^{-1} \exp(-\beta a p_{1x}^2 - \beta E') .$$

Hence the average of $a p_{1x}^2$ will be

$$\langle a p_{1x}^2 \rangle = \frac{\int dV_1 d\tau_1 \ldots dV_N d\tau_N a p_{1x}^2 \exp(-\beta p_{1x}^2) \exp(-\beta E')}{\int dV_1 d\tau_1 \ldots dV_N d\tau_N \exp(-\beta p_{1x}^2) \exp(-\beta E')} .$$

In the denominator and the numerator there appears a product of $6N$ integrals (over $3N$ position variables and $3N$ momentum variables).

The integrals in the denominator and in the numerator are equal, except for the integral over the x component of the momentum of particle number 1 which factors out. We therefore obtain

$$\langle a p_{1x}^2 \rangle = \frac{\int dp_{1x} a p_{1x}^2 \exp(-\beta a p_{1x}^2)}{\int dp_{1x} \exp(-\beta a p_{1x}^2)} ,$$

which is Eq. (3.4.24).

Solution 4.9 *Exercise on page 270*

(a) We have $\lambda_T/a = \sqrt{\frac{h^2}{mka^2}} \cdot T^{-\frac{1}{2}}$. In SI units: $h = 6.6 \times 10^{-34}$; $k = 1.38 \times 10^{-23}$; $m_{\text{He}} = 6.7 \times 10^{-27}$ and $a = 10^{-2}$, we have

$$\lambda_T/a = 2.17 \times 10^{-8} \times T^{-\frac{1}{2}} .$$

Hence: at $T = 300$ K, $\lambda_T/a = 1.3 \cdot 10^{-9}$;
at $T = 10$ K, $\lambda_T/a = 7.1 \cdot 10^{-9}$.

(b) $\lambda_T/a = 1$ implies that $T = 4.7 \cdot 10^{-16}$ K.

Solution 4.10 *Exercise on page 272*

The energy levels of a single particle of mass m inside a cubic container of volume V are

$$\epsilon_{npq} = \frac{h^2}{8mV^{2/3}}(n^2 + p^2 + q^2)$$

and the energy of a microscopic state of N such particles is

$$E(n_1, p_1, q_1, \ldots, n_N, p_N, q_N) = \frac{h^2}{8mV^{2/3}}(n_1^2+p_1^2+q_1^2+\ldots+n_N^2+p_N^2+q_N^2) .$$

Since E explicitly depends on the volume, we can repeat precisely the same arguments that brought us from (3.1.6) to (3.1.8) (with $X = V$) and write

$$\delta W = -Z^{-1} \sum_{(n_1,p_1,q_1,\ldots)} \frac{\partial E(n_1, p_1, q_1, \ldots)}{\partial V} e^{-\beta E(n_1,p_1,q_1,\ldots)} dV = \frac{1}{\beta}\frac{\partial \ln Z}{\partial V} dV .$$

Solution 5.1 *Exercise on page 276*

(a) *Paramagnet*

The free energy of the paramagnet has the form (3.1.32):

$$F = -NkT \ln\left[2\cosh\left(\frac{\mu_{\text{B}}H}{kT}\right)\right] .$$

F in this case is proportional to N and is independent of the volume. Namely, it is definitely extensive:

$$F(2N) = 2F(N) .$$

The same is true of its derivatives with respect to intensive parameters. Among others, the entropy $S = -(\partial F/\partial T)_{H,N}$ must be extensive. This may be seen explicitly from the expression we obtained for the entropy, (3.1.33):

$$S = Nk\left\{\ln\left[2\cosh\left(\frac{\mu_B H}{kT}\right)\right] - \left(\frac{\mu_B H}{kT}\right)\tanh\left(\frac{\mu_B H}{kT}\right)\right\},$$

which is also proportional to N and independent of the volume.

The behavior of the entropy at low temperatures is obtained by noting that $2\cosh x \sim e^x, \tanh x \to 1$ for $x \to \infty$, and thus we obtain in this limit

$$S \simeq Nk\left[\ln\exp\left(\frac{\mu_B H}{kT}\right) - \frac{\mu_B H}{kT}\right] = 0\,.$$

(b) *System of oscillators*

The free energy (3.2.12) is

$$F = 3N\left[\frac{\hbar\omega}{2} + kT\ln(1 - e^{-\beta\hbar\omega})\right]\,.$$

As in the case of the paramagnet, the extensivity is due to the explicit proportionality to N and the independence of the volume. The same applies to S, which was calculated in Solution 2.6(b):

$$S = -3Nk\left[\ln(1 - e^{-\hbar\omega/kT}) - \frac{\hbar\omega/kT}{e^{\hbar\omega/kT} - 1}\right]\,.$$

Now, at low temperatures the first term tends to $\ln 1 = 0$. The second term is of the form xe^{-x} ($x = \hbar\omega/kT$) and in the limit $x \to \infty$ vanishes as well. Thus, here as well, when $T \to 0$, $S \to 0$. See also the figure in Solution 2.5.

Solution 5.2 *Exercise on page 277*

We start from a state in which two gases, of type A and type B, are in volumes, V_A and V_B, respectively at equal pressure, P, and temperature, T.

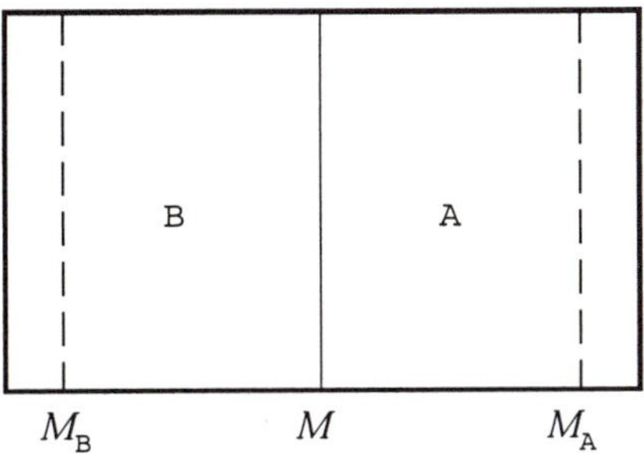

When the partition M is removed, the gases mix and a uniform state forms in the joint volume. This is Joule's expansion: no work is performed, the energy does not change, and hence the temperature does not change. The pressure of each of the two gases in the combined volume is

$$P_{\mathrm{A}} = P\frac{V_{\mathrm{A}}}{V_{\mathrm{A}} + V_{\mathrm{B}}}\,,$$

$$P_{\mathrm{B}} = P\frac{V_{\mathrm{B}}}{V_{\mathrm{A}} + V_{\mathrm{B}}}\,,$$

respectively, and hence the total pressure (according to Dalton's law) remains

$$P_{\mathrm{A}} + P_{\mathrm{B}} = P\,.$$

We now insert, as described in the figure, two partitions, M_{A} and M_{B}. M_{A} is only permeable to molecules of type A, and M_{B} is only permeable to molecules of type B. We separate the two gases by bringing the partitions M_{A} and M_{B} to the position of the partition M. Obviously, after the partitions reach this location, each gas will be in its original volume. But work has been performed upon both of them — the work required to decrease the volume of each of the gases and to increase its pressure. This work raises the temperature of both gases, and they do not return to their original state.

Solution 5.3 *Exercise on page 279*

In a canonical ensemble, the probability of the microscopic state i is

$$P_i = \frac{1}{Z}e^{-\beta E_i}\,.$$

The partition function is the normalization coefficient of the probability.

The average of a physical quantity A is given by (3.4.36b):

$$\langle A\rangle = Z^{-1}\int\left(\frac{dV\,d\tau}{h^3}\right)^N A(\mathbf{r}_i,\mathbf{p}_i)e^{-\beta E(\mathbf{r}_i,\mathbf{p}_i)}\,,$$

where the correction that we have introduced necessitates changing the integration region in phase space to the region of states that are not connected by the exchange of identical particles. Since the exchange of identical particles does not change the value of A (if this were not so, they would not be identical), the reduction of the region of integration is equivalent to a division by $N!$ of the integral performed over the entire phase space. Thus in calculating $\langle A\rangle$ we can leave the region of integration unchanged and correct for this by dividing the integral by $N!$.

The average of A will be given by the ratio between the corrected integral and the corrected Z. But the two are multiplied by an identical factor, and thus the new average is equal to the original one.

Solution 5.4 *Exercise on page 279*

It is possible to distinguish between molecules of different groups; however, it is not possible to distinguish between molecules of the same group. That is, in each group j there are N_j particles that are indistinguishable from one another, and thus each microscopic state has $N_j!$ internal permutations which are to be considered as a single state.

In the calculation of (3.4.12) we counted a total number of states that is $N_1!N_2!N_3!\ldots N_M!$ times larger than the number of microscopic states. Hence, the corrected partition function is obtained by dividing (3.4.12) by this number:

$$Z = \frac{(z_1)^{N_1}}{N_1!}\frac{(z_2)^{N_2}}{N_2!}\cdots\frac{(z_M)^{N_M}}{N_M!}\,.$$

Solution 5.5 *Exercise on page 279*

The paramagnet is a system of N spins in fixed positions. Hence the position of each spin does not count as a degree of freedom, as witnessed by the fact that the energy of each spin does not include a kinetic term. Such spins cannot exchange positions. Therefore, the summation over the states of the spins does not include a summation over their positions, so that no allowance has to be made for states in which the positions of two spins have been exchanged.

The situation in the system of oscillators is similar. Here each oscillator does have a kinetic term; however, each oscillator (atom) vibrates around its own center of force, and again it is impossible to exchange two oscillators.

Solution 5.6 *Exercise on page 280*

The pressure is given by the volume derivative of the free energy:

$$P = -\left(\frac{\partial F}{\partial V}\right)_{T,N}\,.$$

Substituting Eq. (3.5.7), we obtain again the equation of state of an ideal gas.

The reason why there is no change in P is that the additive change in F, due to the Gibbs correction, (3.5.5), is independent of the volume.

Similarly,

$$C_V = T\left(\frac{\partial S}{\partial T}\right)_{V,N}\,,\quad C_P = T\left(\frac{\partial S}{\partial T}\right)_{P,N}\,.$$

Substituting (3.5.8) we obtain

$$C_V = \frac{3}{2}Nk\,,$$

which is the usual result.

In order to calculate C_P we have to express S in terms of T, P and N instead of T, V and N, and this we do as in Solution 4.3 using the equation of state.

$$S(T,P,N) = Nk\left\{\ln\left[\frac{(2\pi m)^{3/2}(kT)^{5/2}}{Ph^3}\right] + \frac{3}{2}\right\}$$

and thus we obtain

$$C_P = \frac{5}{2}Nk\,.$$

The difference between S in (3.5.8) and S in (3.4.39) is independent of T, and hence C_V and C_P do not change.

Solution 5.7 *Exercise on page 280*

$$F = -NkT\left[\ln\left(\frac{V}{N}\right) + \frac{3}{2}\ln\left(\frac{2\pi mkT}{h^2}\right) + 1\right],$$

$$\mu = \left(\frac{\partial F}{\partial N}\right)_{T,V} = -kT\left[\ln\left(\frac{V}{N}\right) + \frac{3}{2}\ln\left(\frac{2\pi mkT}{h^2}\right) + 1 - 1\right]$$
$$= -kT\left[\ln\left(\frac{V}{N}\right) + \frac{3}{2}\ln\left(\frac{2\pi mkT}{h^2}\right)\right].$$

Denoting $n = N/V$ (the density of molecules) we obtain (3.5.9).

Solution 5.8 *Exercise on page 280*

The free energy of a mixture of M different types of molecules (3.5.10) is

$$F = -kT\sum_{j=1}^{M} N_j\left[\ln\left(\frac{V}{N_j}\right) + \frac{3}{2}\ln\left(\frac{2\pi m_j kT}{h^2}\right) + 1\right]$$

and the pressure is

$$P = -\left(\frac{\partial F}{\partial V}\right)_{T,N} = kT\sum_{j=1}^{M}\left(\frac{N_j}{V}\right) = \sum_{j=1}^{M} N_j\frac{kT}{V} = \sum_{j=1}^{M} P_j\,,$$

where P_j is the partial pressure of the gas of type j.

Solution 5.9 *Exercise on page 280*

The entropy is obtained from the free energy:

$$\begin{aligned} S &= -\left(\frac{\partial F}{\partial T}\right)_{V,N} \\ &= k\sum_{j=1}^{M} N_j \left[\ln\left(\frac{V}{N_j}\right) + \frac{3}{2}\ln\left(\frac{2\pi m_j kT}{h^2}\right) + 1\right] + kT\sum_{j=1}^{M}\frac{3}{2}N_j\frac{1}{T} \\ &= k\sum_{j=1}^{M} N_j \left[\ln\left(\frac{V}{N_j}\right) + \frac{3}{2}\ln\left(\frac{2\pi m_j kT}{h^2}\right) + \frac{5}{2}\right]. \end{aligned}$$

This is of course the sum of the partial entropies of each of the gases.

Solution 5.10 *Exercise on page 281*

The entropy before the removal of the partition is

$$S = k\sum_{j=1}^{2} N_j \left[\ln\left(\frac{V_j}{N_j}\right) + \frac{3}{2}\ln\left(\frac{2\pi m_j kT}{h^2}\right) + \frac{5}{2}\right],$$

where V_j and N_j represent the volume and the number of molecules on each side of the partition.

After the removal of the partition, the volume of each of the gases is $V = V_1 + V_2$, and the entropy is

$$S' = k\sum_{j=1}^{2} N_j \left[\ln\left(\frac{V}{N_j}\right) + \frac{3}{2}\ln\left(\frac{2\pi m_j kT}{h^2}\right) + \frac{5}{2}\right].$$

The change in entropy is

$$\Delta S = S' - S = k\sum_j N_j \ln\left(\frac{V}{V_j}\right) = k\left(N_1 \ln\frac{V}{V_1} + N_2 \ln\frac{V}{V_2}\right),$$

and of course $\Delta S > 0$.

Solution 5.11 *Exercise on page 282*

The only contribution to the right hand side of (3.5.11a) is from the g_0 states with $E_i = E_0$, all of which have the same probability, $1/g_0$. Hence the sum consists of g_0 identical terms $(1/g_0)\ln g_0$ so that in the limit $T \to 0$

$$S \to -kg_0\left[\frac{1}{g_0}\ln g_0\right] = -k\ln\left(\frac{1}{g_0}\right) = k\ln g_0\,.$$

Solution 5.12 *Exercise on page 283*

First we have to calculate the partition function of the classical (three-dimensional) harmonic oscillator z_c. The partition function of N classical oscillators will, of course, be $(z_c)^N$. z_c can be calculated directly from (3.4.35b):

$$z_c = \frac{1}{h^3} \int dV\, d\tau \exp\left[-\beta \left(\frac{p^2}{2m} + \frac{m\omega^2}{2} r^2 \right) \right] . \qquad \text{(i)}$$

This integral is a product of two three-dimensional Gaussian integrals (see Solution 4.2). From the integration over the momentum we obtain $(2\pi mkT)^{3/2}$ and from the integration over space, $(2\pi kT/m\omega^2)^{3/2}$, so that the overall result is

$$z_c = \left(\frac{2\pi kT}{h\omega} \right)^3 = \left(\frac{kT}{\hbar\omega} \right)^3 . \qquad \text{(ii)}$$

Now, the free energy is

$$F_c = -NkT \ln z_c = -3NkT \ln \left(\frac{kT}{\hbar\omega} \right) \qquad \text{(iii)}$$

and the entropy

$$S_c = -\left(\frac{\partial F_c}{\partial T} \right)_N = 3Nk \left[\ln \left(\frac{kT}{\hbar\omega} \right) + 1 \right] . \qquad \text{(iv)}$$

This expression, of course, diverges when $T \to 0$.

In order to see the behavior of the quantum Einstein solid we use the result of Exercise 2.5:

$$S = -3Nk \left[\ln(1 - e^{-\hbar\omega/kT}) - \frac{\hbar\omega}{kT} \frac{1}{e^{\hbar\omega/kT} - 1} \right] . \qquad \text{(v)}$$

At high temperatures

$$e^{\hbar\omega/kT} - 1 \simeq \frac{\hbar\omega}{kT} , \qquad 1 - e^{-\hbar\omega/kT} \simeq \frac{\hbar\omega}{kT} , \qquad \text{(vi)}$$

and we recover the classical result:

$$S \simeq -3Nk \left[\ln \left(\frac{\hbar\omega}{kT} \right) - 1 \right] = 3Nk \left[\ln \left(\frac{kT}{\hbar\omega} \right) + 1 \right] . \qquad \text{(vii)}$$

At low temperatures we find this time that the entropy is bounded and tends to zero, as can also be seen in the figure in Solution 2.5. When $T \to 0$, we have

$$e^{-\hbar\omega/kT} \ll 1 , \quad e^{\hbar\omega/kT} \gg 1 . \qquad \text{(viii)}$$

Thus, the first term in the brackets of (v) vanishes as $e^{-\hbar\omega/kT}$. The second term becomes

$$S \simeq 3Nk \cdot \left(\frac{\hbar\omega}{kT}\right) \exp\left(-\frac{\hbar\omega}{kT}\right), \tag{ix}$$

since the exponential dominates the denominator and when $T \to 0$, $S \to 0$.

Solution 6.1 *Exercise on page 287*

The average energy is

$$\langle E\rangle = \frac{1}{Z}\sum_i E_i e^{-\beta E_i}$$

and the specific heat is

$$\begin{aligned} Nc &= \frac{\partial\langle E\rangle}{\partial T} = -\frac{1}{kT^2}\frac{\partial\langle E\rangle}{\partial\beta} \\ &= -\frac{1}{kT^2}\left(-\frac{1}{Z^2}\frac{\partial Z}{\partial\beta}\sum_i E_i e^{-\beta E_i} - \frac{1}{Z}\sum_i E_i^2 e^{-\beta E_i}\right). \end{aligned}$$

We now note that the average energy appears in the first term once as an explicit sum and a second time in the form $-\partial \ln Z/\partial\beta$, hence it is proportional to $\langle E\rangle^2$. The second term is $\langle E^2\rangle$. Hence

$$Nc = \frac{1}{kT^2}(\langle E^2\rangle - \langle E\rangle^2) = \frac{1}{kT^2}\langle(E - \langle E\rangle)^2\rangle = \frac{(\Delta E)^2}{kT^2}.$$

Solution 6.2 *Exercise on page 287*

The average energy per molecule of an ideal gas is

$$\langle\epsilon\rangle = \frac{3}{2}kT$$

and the specific heat per molecule (at constant volume) is

$$c = \frac{3}{2}k.$$

Hence the substitution into (3.6.9) yields

$$\frac{\Delta E}{E} = \sqrt{\frac{2}{3N}}.$$

Solutions to self-assessment exercises

Solution 1 *Exercise on page 288*

(a) The Gibbs free energy is

$$G = E - TS + PV = F + PV\,.$$

Substituting the expressions for F and P, (3.1.30) and (3.1.23), we obtain

$$G = -kT\ln Z + kTV\frac{\partial \ln Z}{\partial V}\,.$$

Since $\ln Z$ appears once with a derivative and once without it, we can write G in the form

$$G = kT\left(V\frac{\partial \ln Z}{\partial V} - \ln Z\frac{\partial V}{\partial V}\right) = kTV^2\frac{\partial}{\partial V}\left(\frac{\ln Z}{V}\right)\,.$$

(b) The enthalpy:

$$\mathsf{H} = E + PV\,.$$

Substituting (3.1.3) and (3.1.23) we obtain

$$\mathsf{H} = -\frac{\partial \ln Z}{\partial \beta} + \frac{1}{\beta}V\frac{\partial \ln Z}{\partial V} = \frac{1}{\beta}\left(V\frac{\partial \ln Z}{\partial V} - \beta\frac{\partial \ln Z}{\partial \beta}\right)\,.$$

Solution 2 *Exercise on page 288*

We denote the frequency of the oscillator number α by ω_α. Its energy ϵ_α takes the following values:

$$\epsilon_\alpha = \left(n_\alpha + \frac{1}{2}\right)\hbar\omega_\alpha\,, \qquad \text{(i)}$$

A microscopic state of a system of $3N$ oscillators is described by $3N$ excitation numbers $(n_1, n_2, \ldots, n_{3N})$. Its energy is

$$E(n_1, n_2, \ldots, n_{3N}) = \sum_{\alpha=1}^{3N}\epsilon_\alpha\,. \qquad \text{(ii)}$$

The partition function is the usual sum over all values of all n_α. It factors into a product of sums:

$$\begin{aligned}
Z &= \sum_{n_1,n_2,\ldots,n_{3N}} \exp[-\beta E(n_1,n_2,\ldots,n_{3N})] \\
&= \sum_{n_1,n_2,\ldots,n_{3N}} \exp\left[-\beta\hbar\omega_1\left(n_1+\frac{1}{2}\right)\right] \exp\left[-\beta\hbar\omega_2\left(n_2+\frac{1}{2}\right)\right]\cdot\ldots\cdot \\
&\qquad \times \exp\left[-\beta\hbar\omega_{3N}\left(n_{3N}+\frac{1}{2}\right)\right] \\
&= z_1 z_2 \ldots z_{3N}\,. \qquad \text{(iii)}
\end{aligned}$$

Each of the factors z_α on the right is a partition function of a single oscillator with frequency ω_α, given by (3.2.7):

$$z_\alpha = \frac{\exp(-\beta\hbar\omega_\alpha/2)}{1-\exp(-\beta\hbar\omega_\alpha)}\,. \qquad \text{(iv)}$$

The free energy will be

$$F = -kT\ln Z = -kT\sum_{\alpha=1}^{3N}\ln z_\alpha = \frac{\hbar}{2}\sum_{\alpha=1}^{3N}\omega_\alpha + kT\sum_{\alpha=1}^{3N}\ln(1-e^{-\beta\hbar\omega_\alpha})\,. \qquad \text{(v)}$$

Note that the first term is the sum of all the zero point (ground state) energies of all the oscillators and is a quantity which does not affect the thermodynamic quantities that are derived from the partition function such as the entropy, specific heat etc.

Solution 3 *Exercise on page 288*

The energy levels of the bounded oscillator are

$$E_n = \epsilon_0 + n\epsilon\,, \quad 0 \le n < n_0\,. \qquad \text{(i)}$$

The partition function of a single oscillator will be

$$z = e^{-\beta\epsilon_0}\sum_{n=0}^{n_0-1} e^{-\beta n\epsilon} = e^{-\beta\epsilon_0}\frac{1-e^{-\beta\epsilon n_0}}{1-e^{-\beta\epsilon}}\,, \qquad \text{(ii)}$$

where this time we have used the expression for the sum of a finite geometric series:

$$\sum_{n=0}^{n_0-1} x^n = \frac{1-x^{n_0}}{1-x}\,.$$

Note that the right hand side of (ii) tends to (3.2.7) when $n_0 \to \infty$.

(a) The average energy of the oscillator

$$\langle E\rangle = -\frac{\partial \ln z}{\partial\beta} = \epsilon_0 - \frac{\epsilon n_0 e^{-\beta\epsilon n_0}}{1-e^{-\beta\epsilon n_0}} + \frac{\epsilon e^{-\beta\epsilon}}{1-e^{-\beta\epsilon}}$$

$$= \epsilon_0 + \frac{\epsilon}{e^{\beta\epsilon}-1} - \frac{\epsilon n_0}{e^{\beta\epsilon n_0}-1}\,. \qquad \text{(iii)}$$

The average degree of excitation is easily calculated from the average energy, since

$$\langle E\rangle = \epsilon_0 + \langle n\rangle\epsilon\,.$$

Hence

$$\langle n\rangle = \frac{1}{e^{\beta\epsilon}-1} - \frac{n_0}{e^{\beta\epsilon n_0}-1}\,. \qquad \text{(iv)}$$

(b) The entropy

$$S = \frac{E-F}{T} = E/T + k\ln z$$

$$= k[\ln(1-e^{-\beta\epsilon n_0}) - \ln(1-e^{-\beta\epsilon})] - \frac{\epsilon}{T}\left(\frac{n_0}{e^{\beta\epsilon n_0}-1} - \frac{1}{e^{\beta\epsilon}-1}\right) \qquad \text{(v)}$$

and the specific heat

$$c = T\frac{\partial S}{\partial T} = -\beta\frac{\partial S}{\partial\beta} = k\beta^2\epsilon^2\left[\frac{e^{\beta\epsilon}}{(e^{\beta\epsilon}-1)^2} - n_0^2\frac{e^{\beta\epsilon n_0}}{(e^{\beta\epsilon n_0}-1)^2}\right]. \qquad \text{(vi)}$$

Note that all the quantities E, S, c refer here to a single oscillator.

(c) All of the expressions (ii)–(vi) give the corresponding quantities for the harmonic oscillator in the limit $n_0 \to \infty$. When the temperature is very low, the thermodynamic behavior of the oscillator is controlled by the lowest levels. If the number of thermally relevant levels is much smaller than n_0, it is expected that there will be no difference between the bounded oscillator and the regular oscillator. Namely, there is no qualitative difference when $kT \ll n_0\epsilon$. On the other hand, if the temperature is high, $kT \gg \epsilon$, all the states are equally important, and then the difference is significant.

(d) At low temperatures we obtain from (vi)

$$c \simeq k\left[\left(\frac{\epsilon}{kT}\right)^2 e^{-\epsilon/kT} - \left(\frac{n_0\epsilon}{kT}\right)^2 e^{-n_0\epsilon/kT}\right].$$

It is clear that when $n_0\epsilon/kT \gg 1$, the second term is negligible compared to the first one, and we obtain (3.2.18). At high temperatures, $\epsilon n_0/kT \ll 1$, the two terms in (vi) tend to the same limit, which is k, so that the specific heat tends to 0 instead of the classical value k of an unbounded oscillator.

Remark. If these calculations seem familiar it is because the bounded oscillator has the same structure of energy levels as the spin in a magnetic field. The sole difference is in the location of the zero level. Hence all the results obtained here could have been obtained from our discussion in Sec. 5.4 of Part II.

Solution 4 *Exercise on page 289*

At low temperatures, the asymptotic behavior of C_V in the Einstein model of a solid is given by

$$C_V \simeq 3R\left(\frac{\hbar\omega}{kT}\right)^2 e^{-\hbar\omega/kT}\,. \tag{i}$$

(a) We denote $x = kT/\hbar\omega$. A sketch of $C_V/3R$ as a function of x is given in the figure. $R \simeq 2$ cal/K. The temperature scaling of the horizontal axis is discussed in (b) and (c), below.

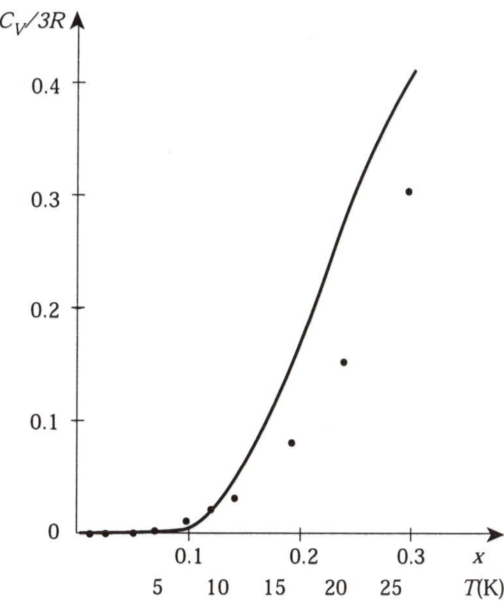

For $T = 10$ K, $C_V = 0.114$ cal/K, and thus we obtain an equation for x:

$$\frac{1}{x^2}e^{-1/x} = 0.0191\,.$$

This equation can be solved numerically, to give $x = 0.1227$. Hence

$$\Theta_E = \frac{10}{x} \simeq 81\text{ K}$$

and thus

$$\omega \simeq 10^{13}\ \mathrm{s}^{-1}\,.$$

This frequency is of the order of typical lattice frequencies.

(b) Scaling the horizontal axis according to $x = T/81$ K leads to the graph for C_V, now as a function of T, plotted in the figure. The points in the figure depict the experimental results given in the table. It is clear that not all the values in the table can be described by the graph.

(c) According to our determination, $\Theta_E = 81$ K, and thus the temperatures in the table justify a low temperature approximation. However, the Einstein approximation does not describe the experimental results well.

The ratio of C_V in the Einstein approximation to C_V in the table, as a function of T, is depicted in the next figure.

We see that when T decreases below 10 K the deviations of the theory from the experiment increase, instead of decreasing, as we would expect from the low temperature approximation. The Einstein approximation yields values that are much smaller than the experimental values.

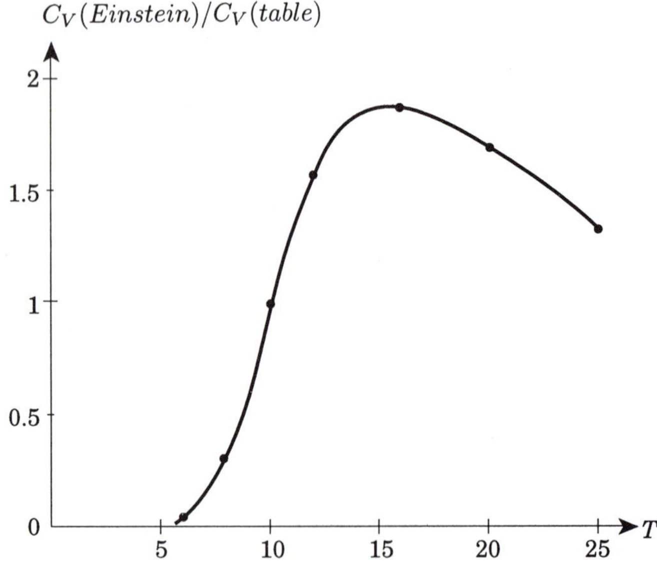

Solution 5 *Exercise on page 289*

In the Debye model (see Part IV, Sec. 3.3), the specific heat behaves at low temperatures according to

$$C_V = \frac{12\pi^4}{5} R \left(\frac{T}{\Theta_D}\right)^3 .$$

We calculate Θ_D from C_V in the table at $T = 10$ K:

$$\Theta_D = 10 \left(\frac{12\pi^4}{5} \cdot 2 \cdot \frac{1}{0.114} \right)^{1/3} \simeq 160 \text{ K}.$$

In the next figure both approximations are depicted along with the experimental results. A study of the figure reveals that the Debye approximation provides a good description of the experimental results. The Einstein approximation does not provide a good quantitative description even if we use the Debye value of 160 K for Θ_E.

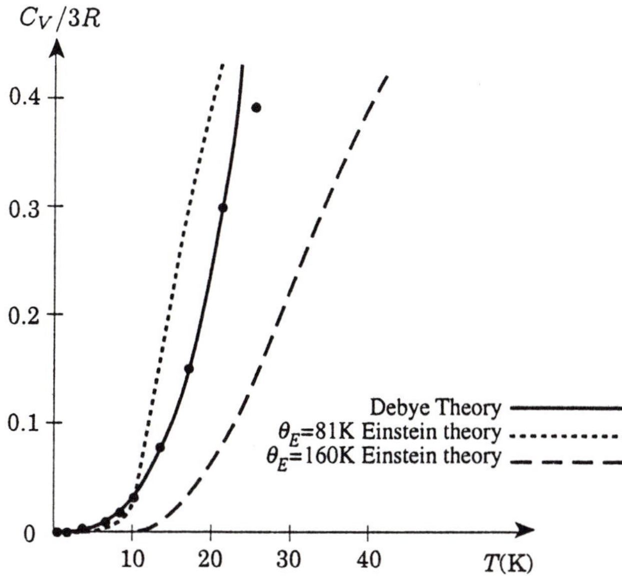

Solution 6 *Exercise on page 290*

(a) The partition function is

$$Z = V^N (2\pi mkT)^{5N/2}.$$

In order to find the equation of state we must calculate the pressure as a function of the volume and temperature. The pressure is given by

$$P = -\frac{\partial F}{\partial V} = kT \frac{\partial \ln Z}{\partial V} = \frac{NkT}{V}.$$

Namely, the equation of state

$$PV = NkT,$$

of an ideal gas.

In order to calculate the specific heat at constant volume, we first calculate the internal energy:

$$E = -\frac{\partial \ln Z}{\partial \beta} = kT^2 \frac{\partial \ln Z}{\partial T} = \frac{5}{2} NkT \,.$$

Comparing this expression to the internal energy obtained in Part I, we can interpret the result as the energy of N particles, as can also be deduced from the equation of state, each having average energy $\frac{5}{2}kT$, and hence five degrees of freedom. The gas is not monoatomic. A simple argument, that will be extended in the next part, leads to the conclusion that this is a diatomic gas, in which the atoms in the molecule are bound to each other in a rigid manner. A pair of atoms has, as we have seen in Part I, six degrees of freedom. But because the distance between the two atoms in the molecule is constant, one degree of freedom is frozen, leaving five degrees of freedom.

The specific heat at constant volume is given by

$$C_V = \left(\frac{\partial E}{\partial T}\right)_{V,N} = \frac{5}{2} Nk \,.$$

(b) The partition function is

$$Z = (V - Nb)^N (2\pi mkT)^{3N/2} e^{aN^2/VkT} \,. \tag{i}$$

The pressure is given by

$$P = kT \frac{\partial \ln Z}{\partial V} = \frac{NkT}{V - Nb} - \frac{N^2}{V^2} a \tag{ii}$$

hence the equation of state,

$$\left(P + \frac{N^2}{V^2} a\right)(V - Nb) = NkT \,, \tag{iii}$$

which is the Van der Waals equation.

This equation can be interpreted in the following manner: It is a gas of N particles whose "active" volume is less than the total volume V. The decrease in volume can be attributed to the fact that each molecule has a finite rigid spherical volume of magnitude b. Hence the free volume for each molecule is $V - Nb$.

The pressure factor is also modified with respect to the ideal gas. The fact that the measured pressure P is smaller than the pressure of free molecules moving in the active volume $V - Nb$ can be explained by the fact that there exist attraction forces between the molecules, beyond their rigid volume. These forces are derived from the potential energy of the molecules, which is depicted qualitatively in the next figure:

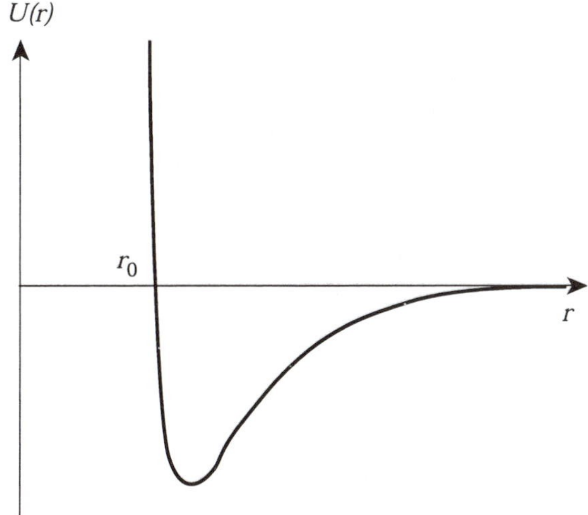

r_0 is the radius of the rigid sphere of a molecule, which determines b. Beyond r_0, the potential is attractive, which decreases the pressure. Next we study how this interpretation is expressed in the internal energy and in the specific heat.

The internal energy is

$$E = -\frac{\partial \ln Z}{\partial \beta} = kT^2 \frac{\partial \ln Z}{\partial T} = \frac{3}{2}NkT - \frac{N^2}{V}a \tag{iv}$$

and the specific heat is

$$C_V = \left(\frac{\partial E}{\partial T}\right)_V = \frac{3}{2}Nk\,. \tag{v}$$

Namely, this is a monoatomic gas — each atom contributes $\frac{3}{2}k$ to the specific heat.

The form of the internal energy indicates that there exist attraction forces that decrease the energy below its value for an ideal gas. Is the magnitude of the decrease in energy reasonable? In order to check this we approximate the potential in the figure as follows: Suppose that each molecule produces an equal attraction potential, of magnitude ϵ_0, at the position of every other molecule, which is not further away than some constant distance r_1. Thus, each molecule feels an average potential of attraction, which is proportional to the depth of the well ϵ_0 and to the density N/V. The higher the density, the more molecules there are within range r_1 of a certain molecule. Hence the contribution of a given molecule, due to the attraction of its neighbors, to the total energy will be

$$-\epsilon_0\left(\frac{N}{V}\right)\cdot r_1^3\,.$$

There are a total of N molecules. Hence the total contribution of the attraction to the energy will be

$$-\frac{1}{2}N\epsilon_0\left(\frac{N}{V}\right)r_1^3 = -\frac{N^2}{V}\frac{\epsilon_0 r_1^3}{2}\,. \tag{vi}$$

The 1/2 appears because we are counting the attraction energy between each pair of molecules twice. Comparing (vi) to the last term in (iv) we find that a has the meaning of the depth of the potential times the volume of its region of attraction.

Solution 7 *Exercise on page 290*

The single particle partition function is

$$z = h^{-3}\int dV d\tau e^{-\beta cp} = Vh^{-3}\int d^3pe^{-\beta cp} = 4\pi Vh^{-3}\int_0^\infty p^2e^{-\beta cp}dp\,. \tag{i}$$

The last transition was made after integrating over the angles in momentum space, which is equivalent to viewing $d\tau$ or d^3p as the volume of a spherical shell of thickness dp and radius p, that is $4\pi p^2dp$. Performing a change of variables to $x = \beta cp$, we obtain

$$z = \frac{4\pi V}{(\beta ch)^3}\int_0^\infty x^2e^{-x}dx\,. \tag{ii}$$

We still have to find the value of the integral on the right hand side of (ii). Looking it up in tables, or calculating it with the help of integration by parts, we find 2. In fact there is no need to calculate it, as the physical variables are not sensitive to the multiplication of the partition function by a constant. In any event we have obtained

$$z = \frac{8\pi V}{(\beta ch)^3}\,. \tag{iii}$$

The average energy is

$$E = -\frac{\partial \ln Z}{\partial \beta} = -N\frac{\partial \ln z}{\partial \beta} = \frac{3N}{\beta} = 3NkT\,. \tag{iv}$$

We have found, therefore, that the average energy of an extremely relativistic particle moving at a speed close to the speed of light is $3kT$, compared to $\frac{3}{2}kT$ for a nonrelativistic particle. The reason is that the energy of such a particle is not proportional to the square of its momentum but is linear in the momentum. The equation of state is

$$P = -\frac{\partial F}{\partial V} = kT\frac{\partial \ln Z}{\partial V} = NkT\frac{\partial \ln z}{\partial V} = \frac{N}{V}kT\,. \tag{v}$$

Thus, the equation of state for the relativistic gas is unchanged. (See also Part I, Exercise 1.2.)

The specific heat at constant volume is $(\partial E/\partial T)_V$, and we again obtain, as in (iv), a magnitude which is twice the nonrelativistic magnitude:

$$C_V = 3Nk\,. \qquad \text{(vi)}$$

For the calculation of the specific heat at constant pressure we shall use the enthalpy $\mathsf{H} = E + PV$ and the relation $C_P = (\partial H/\partial T)_P$. We thus obtain

$$\mathsf{H} = 3NkT + NkT = 4NkT \Rightarrow C_P = 4Nk\,. \qquad \text{(vii)}$$

Solution 8 *Exercise on page 290*

(a) We denote the magnetic moment by $\boldsymbol{\mu}$ and its magnitude by μ. The energy of a single magnetic moment in a field $\mathbf{H}$ is

$$\epsilon = -\boldsymbol{\mu}\cdot\mathbf{H}\,.$$

We assumed that all ions have magnetic moments of constant magnitude, and that the difference between them is in the orientation. Namely, the state of ion number i is characterized by ϕ_i and θ_i, which are angles that describe its orientation in space. The field is directed along the z axis, and the energy of ion number i is

$$\epsilon_i = -\mu H\cos\theta_i$$

and is independent of the azimuthal angle ϕ_i. The energy of the system of N ions, for a given microscopic state (specifying the orientation of every spin), is given by

$$E = -\mu H\sum_{i=1}^{N}\cos\theta_i\,.$$

(b) The partition function is written as an integral over all possible values of the $2N$ angles θ_i and ϕ_i, which factors into a product of integrals because there are no interactions between the different spins.

$$\begin{aligned} &Z(\beta,H,N)\\ &= \int_0^{2\pi}d\phi_1\int_0^{\pi}d\theta_1\sin\theta_1\ldots\int_0^{2\pi}d\phi_N\int_0^{\pi}d\theta_N\sin\theta_N e^{-\beta E}\\ &= \left(\int_0^{2\pi}d\phi\int_0^{\pi}d\theta\sin\theta e^{\beta\mu H\cos\theta}\right)^N = [z(\beta,H)]^N\,. \end{aligned}$$

We shall now calculate z.

The integration over ϕ simply gives 2π, as there is no dependence upon this angle. Hence

$$z = 2\pi \int_0^{\pi} d\theta \sin\theta e^{\beta\mu H \cos\theta} .$$

In calculating this integral we note that the integrand is a derivative of $e^{\beta\mu H \cos\theta}$. Hence

$$z = 2\pi \frac{e^{\beta\mu H} - e^{-\beta\mu H}}{\beta\mu H} = \frac{4\pi}{\beta\mu H} \sinh(\beta\mu H) .$$

And hence

$$Z(\beta, H, N) = \left[\frac{4\pi}{\beta\mu H} \sinh(\beta\mu H)\right]^N . \tag{i}$$

(c) The average magnetization is, according to Eq. (2.5.8),

$$M = \frac{1}{\beta}\frac{\partial \ln Z}{\partial H} = N\mu \left[\coth(\beta\mu H) - \frac{1}{\beta\mu H}\right] . \tag{ii}$$

The function in the square brackets is called Langevin's function:

$$L(x) \equiv \coth(x) - \frac{1}{x} ,$$

so that we may write

$$M = N\mu L(\beta\mu H) .$$

The average energy of the paramagnet is

$$E = -HM = -N\mu H L(\beta\mu H) .$$

The entropy is derived from the partition function, or from the free energy, using

$$S = -\left(\frac{\partial F}{\partial T}\right)_{H,N} = \frac{\partial}{\partial T}(kT \ln Z)_{H,N} . \tag{iii}$$

Substituting Z we obtain

$$\begin{aligned} S &= \frac{\partial}{\partial T}\left\{kTN \ln\left[\frac{4\pi}{\beta\mu H} \sinh(\beta\mu H)\right]\right\} \\ &= Nk\left\{\ln\left[\frac{4\pi}{\beta\mu H} \sinh(\beta\mu H)\right] + 1 - \beta\mu H \coth(\beta\mu H)\right\} . \end{aligned}$$

(d) In order to compare with the discrete paramagnet we check the behavior of the magnetization as a function of H or T. At low temperatures, or high fields, the coth in Eq. (ii) tends to 1 and the second term tends to 0, so the magnetization saturates: $M \to N\mu$, as in the quantum case. At high temperatures, or low fields, we must study the Langevin function $L(x)$ for $x \ll 1$, just as we studied the Brillouin function in Solution 5.7 of Part II. To this end we use the result obtained in Eq. (viii) in Solution 5.7 of Part II:

$$\coth x \simeq \frac{1}{x} + \frac{x}{3}\,, \qquad x \ll 1\,,$$

so

$$L(x) \simeq \frac{x}{3}\,, \qquad \text{for } x \ll 1\,,$$

which is again similar to the behavior of the Brillouin function. We thus find that for small $\mu H/kT$

$$M = \frac{N\mu^2}{3kT} H\,, \qquad \chi = \frac{\mu^2}{3kT}\,,$$

and this is Curie's law again.

We have found that the dependence of the magnetization upon the variable $\beta\mu H$ is rather similar to that obtained in the quantum case, in which each magnetic moment has a finite number of states.

Moreover, it is possible to show that the magnetization found here is the classical limit of the one found in Part II. The classical limit is obtained not when the temperatures is high or when the field weak, but when the quantum spin becomes a classical spin, namely when $J \gg 1$.

The quantum result we obtained was

$$M = Ng\mu_{\mathrm{B}} \left\{ \left(J + \frac{1}{2}\right) \coth\left[\left(J + \frac{1}{2}\right) \beta g\mu_{\mathrm{B}} H\right] - \frac{1}{2} \coth\left(\frac{\beta g\mu_{\mathrm{B}} H}{2}\right) \right\}.$$

If we compare the expressions for the susceptibility, it is clear that when J is very large it is possible to identify $g\mu_{\mathrm{B}}J$ with μ. Hence the limit $J \to \infty$ must be taken in such a manner that $g\mu_{\mathrm{B}}J$ remains constant (and equal to μ). Therefore, writing M with the help of μ and J we obtain

$$M = N \left\{ \left(\mu + \frac{\mu}{2J}\right) \coth\left[\left(\mu + \frac{\mu}{2J}\right) \beta H\right] - \frac{\mu}{2J} \coth\left(\frac{\beta\mu H}{2J}\right) \right\}.$$

In the limit $J \to \infty$ we can write

$$M = N\left[\mu\coth(\beta\mu H) - \frac{\mu}{2J}\cdot\frac{2J}{\beta\mu H}\right] = N\left[\mu\coth(\beta\mu H) - \frac{1}{\beta H}\right],$$

and this is the magnetization we found in (c) above.

(e) We shall check how S behaves at low temperatures.

If

$$x \equiv \beta\mu H \gg 1$$

then

$$S \simeq -Nk\ln x\,.$$

Namely, $S \to \infty$ for $x \to \infty$, and this is another example in which the third law does not apply to a classical system. (See discussion in Sec. 5.3.)

Solution 9 *Exercise on page 291*

(a) The work performed by the gas in an isothermal expansion is

$$W = \int_{V_1}^{V_2} PdV = NkT\int_{V_1}^{V_2}\frac{dV}{V} = NkT(\ln V_2 - \ln V_1)\,.$$

Since the process is isothermal the work is performed at the expense of the free energy (see Sec. 1.5), and the result must equal the decrease in the free energy. Indeed, using (3.5.7) we obtain

$$\begin{aligned}-\Delta F &= F_1 - F_2 = NkT\ln\left(\frac{V_2}{N}\right) - NkT\ln\left(\frac{V_1}{N}\right)\\ &= NkT(\ln V_2 - \ln V_1)\,.\end{aligned}$$

(b) The internal energy of the ideal gas depends only on the temperature, and hence it has not changed during the process: $\Delta E = 0$. The origin of the work is therefore the energy extracted from the heat bath, and converted into mechanical energy by the gas.

Solution 10 *Exercise on page 291*

(a) The specific heat of a paramagnet, calculated in Part II, Eq. (2.5.18), is

$$c_H = \frac{(\mu_B H)^2}{kT^2\cosh^2(\beta\mu_B H)}\,. \qquad \text{(i)}$$

To calculate the fluctuations we need the difference:

$$\langle E^2\rangle - \langle E\rangle^2 = Z^{-1} \sum_{\{\sigma_i=\pm 1\}} E^2(\sigma_1,\ldots,\sigma_N)e^{-\beta E(\sigma_1,\ldots,\sigma_N)} - \left[Z^{-1} \sum_{\{\sigma_i=\pm 1\}} E(\sigma_1,\ldots,\sigma_N)e^{-\beta E(\sigma_1,\ldots,\sigma_N)}\right]^2 .$$

This expression takes on a simpler form if we note that

$$\begin{cases} \langle E^2\rangle &= \dfrac{1}{Z}\dfrac{\partial^2 Z}{\partial\beta^2}, \\ \langle E\rangle^2 &= \dfrac{1}{Z^2}\left(\dfrac{\partial Z}{\partial\beta}\right)^2 \end{cases}$$

$$\Downarrow$$

$$\langle E^2\rangle - \langle E\rangle^2 = \frac{1}{Z}\frac{\partial^2 Z}{\partial\beta^2} - \frac{1}{Z^2}\left(\frac{\partial Z}{\partial\beta}\right)^2 = \frac{\partial}{\partial\beta}\left[\frac{1}{Z}\left(\frac{\partial Z}{\partial\beta}\right)\right] = \frac{\partial^2}{\partial\beta^2}(\ln Z).$$

Z, calculated in Part II, Eqs. (2.5.11) and (2.5.12), is

$$Z = [2\cosh(\beta\mu_{\mathrm{B}}H)]^N ,$$

and we obtain

$$\begin{aligned} \langle E^2\rangle - \langle E\rangle^2 &= N\frac{\partial^2}{\partial\beta^2}\ln[2\cosh(\beta\mu_{\mathrm{B}}H)] \\ &= N\mu_{\mathrm{B}}H\frac{\partial}{\partial\beta}\tanh(\beta\mu_{\mathrm{B}}H) = N\frac{(\mu_{\mathrm{B}}H)^2}{\cosh^2(\beta\mu_{\mathrm{B}}H)} . \qquad \text{(ii)} \end{aligned}$$

From (i) and (ii) we obtain (3.6.8), namely

$$\langle E^2\rangle - \langle E\rangle^2 = NkT^2 c_H .$$

(b) The average of the energy will lose its thermodynamic meaning when

$$\frac{\langle E^2\rangle - \langle E\rangle^2}{\langle E\rangle^2} \simeq 1 .$$

Namely, when

$$N\frac{(\mu_{\mathrm{B}}H)^2}{\cosh^2(\beta\mu_{\mathrm{B}}H)}\,\frac{1}{N^2(\mu_{\mathrm{B}}H)^2\tanh^2(\beta\mu_{\mathrm{B}}H)} = \frac{1}{N\sinh^2(\beta\mu_{\mathrm{B}}H)} \simeq 1$$

or

$$\sqrt{N}\sinh(\beta\mu_{\mathrm{B}}H) \simeq 1 .$$

With $N = 100$, $\mu H = 0.01\text{eV}$, we are searching for a solution for

$$\sinh x = 0.1\,,$$

where $x = \mu H/kT$. One finds (numerically) that

$$x \simeq 0.1\,.$$

The required temperature is thus

$$T = \frac{\mu_{\rm B} H}{kx} \simeq 1160 \text{ K}.$$

Part IV

From Ideal Gas to Photon Gas

Introduction

This part is a continuation and extension of Part III and is based on the implementation of the methods presented in the previous parts. In Chap. 1 the subject of ideal gases of molecules devoid of internal structure, analyzed in Part III, is extended to deal with diatomic molecules. We learn how to take into account the internal structure of the molecules, and its effects on the properties of the gases. We shall dwell upon the heat capacity problem which we encountered in Part I and resolve it here at last.

In Chap. 2 we go one step further and consider the case in which the molecules of the gas disintegrate into their constituents or participate in chemical reactions; also here the theory developed has much to say.

In Chap. 3 we return to the problem of the specific heat of solids, and learn how it is possible, with relative ease, to improve the Einstein model and obtain a good correspondence between the theoretical explanation and the experimental results. This chapter will prepare us for the fourth chapter, which deals with the thermodynamics of electromagnetic radiation. Since electromagnetic radiation can be treated as a collection of harmonic oscillators with different frequencies, it is possible to apply to it all the methods developed in Chap. 2 of Part III for systems of oscillators. The principal difficulty overcome in Chap. 4 is the demonstration that the electromagnetic radiation actually behaves as a system of free oscillators. From here on the road ahead is clear.

Chapter 1

An Ideal Gas of Molecules with Internal Degrees of Freedom

1.1 Center of mass and internal motions

In this chapter we continue the discussion of the ideal gas of molecules, i.e. a gas in which it is possible to neglect the forces between molecules. But in order to proceed a step further toward a more realistic description, we now take into account the internal structure of the molecules. We have already seen in Part III that if there are no forces between the molecules, the total energy of the gas is the sum of single molecule energies, and hence the partition function is a product of single molecule partition functions [see, for instance, Eqs. (3.5.3) and (3.5.4)]. Thus we shall concentrate on the energy of a single molecule.

The molecule is made up of q constituents — the nuclei of different atoms and electrons. We denote their positions by $\mathbf{r}_\alpha$ and their masses by m_α, where $\alpha = 1, \ldots, q$. The energy of a molecule has the typical form

$$E_{\text{mol}} = \sum_{\alpha=1}^{q} \frac{1}{2} m_\alpha \mathbf{v}_\alpha^2 + \frac{1}{2} \sum_{\substack{\alpha,\beta \\ \alpha \neq \beta}} U(\mathbf{r}_\alpha - \mathbf{r}_\beta) = \sum_{\alpha=1}^{q} \frac{\mathbf{p}_\alpha^2}{2m_\alpha} + \frac{1}{2} \sum_{\substack{\alpha,\beta \\ \alpha \neq \beta}} U(\mathbf{r}_\alpha - \mathbf{r}_\beta)\,, \tag{4.1.1}$$

where $\mathbf{v}_\alpha$ and $\mathbf{p}_\alpha$ are the velocity and momentum, respectively, of the αth particle in the molecule. $U(\mathbf{r}_\alpha - \mathbf{r}_\beta)$ is the potential energy due to the interaction of the pair of particles α and β inside the molecule — for example, an attractive force between an electron (negatively charged) and a nucleus (positively charged), or a repulsive electric force between two electrons, etc. The potential energies also depend on the types of the interacting particles, but at this stage we shall refrain from complicating the notations. The factor $\frac{1}{2}$ in the potential energy term compensates for double counting, since the sum over all values of α and β counts each pair twice. Note that the summation excludes the terms with $\alpha = \beta$, i.e. a particle does not interact with itself. See also Example (b), Sec. 3.2, Part III.

When in Part III we have treated each molecule as a point, the coordinate and momentum of this point corresponded to the center of mass of the molecule it represents. Hence, in turning to treat the effects of the internal structure of the molecules upcn the statistical mechanics of the system, it is natural that we distinguish between the center of mass variables which we have already treated and the internal variables (see Sec. 1.3 of Part I). We therefore decompose the total energy of a molecule into two parts: energy related to the motion of the center of mass and internal energy.

The center of mass is located at

$$\mathbf{R} = \frac{1}{M}\sum_{\alpha=1}^{q} m_\alpha \mathbf{r}_\alpha , \tag{4.1.2}$$

where $M = \sum_{\alpha=1}^{q} m_\alpha$ and the momentum associated with the center of mass is given by

$$\mathbf{P} = M\dot{\mathbf{R}} = \sum_{\alpha}^{q} \mathbf{p}_\alpha . \tag{4.1.3}$$

This is the momentum of the entire molecule.

The internal coordinates of a particle, namely the ones measured with respect to the center of mass, will be denoted by $\boldsymbol{\rho}_\alpha$, so that

$$\mathbf{r}_\alpha = \mathbf{R} + \boldsymbol{\rho}_\alpha . \tag{4.1.4}$$

The momentum of a particle relative to the center of mass is denoted by $\boldsymbol{\pi}_\alpha$ and is given by

$$\boldsymbol{\pi}_\alpha = m_\alpha \dot{\boldsymbol{\rho}}_\alpha . \tag{4.1.5}$$

Thus

$$\mathbf{p}_\alpha = \frac{m_\alpha}{M}\mathbf{P} + \boldsymbol{\pi}_\alpha . \tag{4.1.6}$$

The definition of the center of mass implies the two identities

$$\sum_{\alpha=1}^{q} m_\alpha \boldsymbol{\rho}_\alpha = 0 , \tag{4.1.7a}$$

$$\sum_{\alpha=1}^{q} \boldsymbol{\pi}_\alpha = 0 . \tag{4.1.7b}$$

Inserting Eq. (4.1.6) for $\mathbf{p}_\alpha$ we obtain for the kinetic energy of the molecule

$$\sum_{\alpha=1}^{q} \frac{\mathbf{p}_\alpha^2}{2m_\alpha} = \frac{\mathbf{P}^2}{2M} + \sum_{\alpha=1}^{q} \frac{\boldsymbol{\pi}_\alpha^2}{2m_\alpha} . \tag{4.1.8}$$

The first term on the right hand side is the kinetic energy related to the motion of the center of mass of the molecule, whereas the second term is the kinetic energy related to the internal motions of the different constituents.

Exercise 1.1

Prove Eqs. (4.1.6)–(4.1.8).

Solution on page 406

Next we deal with the potential energy.

The potential energies depend only on mutual distances $(\mathbf{r}_\alpha - \mathbf{r}_\beta)$, and thus do not depend on the coordinate of the center of mass $\mathbf{R}$, but only on the difference between the internal coordinates, $\boldsymbol{\rho}_\alpha - \boldsymbol{\rho}_\beta$. Thus, since there is no potential energy term dependent on $\mathbf{R}$, the molecule's center of mass moves freely.

The energy (4.1.1) therefore takes the form

$$E_{\text{mol}} = \frac{\mathbf{P}^2}{2M} + \sum_{\alpha=1}^{q} \frac{\boldsymbol{\pi}_\alpha^2}{2m_\alpha} + \frac{1}{2}\sum_{\alpha,\beta} U(\boldsymbol{\rho}_\alpha - \boldsymbol{\rho}_\beta) = \frac{\mathbf{P}^2}{2M} + E_I\,, \tag{4.1.9}$$

where E_I is the internal energy, i.e. the energy in the frame of reference of the molecule's center of mass. In all of our previous discussions of ideal gases we assumed that $E_I = 0$.

■ Note that as a result of (4.1.7) only $q-1$ of the momentum variables $\boldsymbol{\pi}_\alpha$ are independent. Similarly, only $q-1$ spatial variables $\boldsymbol{\rho}_\alpha$ are independent.

1.2 Kinematics of a diatomic molecule

At this stage we increase our resolving power and distinguish between the atoms inside the molecule, but we still do not take the electrons into account. For simplicity we treat a diatomic molecule. The positions of the two atoms will be denoted by $\mathbf{r}_1$ and $\mathbf{r}_2$, the momenta by $\mathbf{p}_1$ and $\mathbf{p}_2$, and the potential between them $U(\mathbf{r}_1 - \mathbf{r}_2)$ will have the typical form depicted in Fig. 4.1.1.

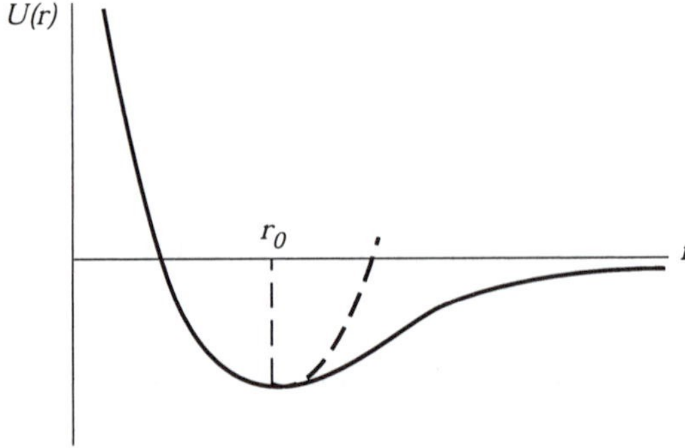

Fig. 4.1.1 Illustration of a typical potential in a diatomic molecule. r_0 is the equilibrium distance. The dashed line is the harmonic approximation.

The energy of the molecule (or of the two atoms) is

$$E_{\text{mol}} = \frac{\mathbf{p}_1^2}{2m_1} + \frac{\mathbf{p}_2^2}{2m_2} + U(\mathbf{r}_1 - \mathbf{r}_2). \tag{4.1.10}$$

The coordinates of the center of mass

$$\mathbf{R} = \frac{1}{M}(m_1\mathbf{r}_1 + m_2\mathbf{r}_2). \tag{4.1.11a}$$

The relative coordinates (in the center-of-mass reference frame) are

$$\boldsymbol{\rho}_1 = \mathbf{r}_1 - \mathbf{R}, \tag{4.1.11b}$$

$$\boldsymbol{\rho}_2 = \mathbf{r}_2 - \mathbf{R}, \tag{4.1.11c}$$

and the relative momenta $\boldsymbol{\pi}_\alpha$ are [see Eqs. (4.1.5) and (4.1.6)]

$$\boldsymbol{\pi}_1 = m_1\dot{\boldsymbol{\rho}}_1 = \mathbf{p}_1 - m_1\dot{\mathbf{R}} = \mathbf{p}_1 - \frac{m_1}{M}\mathbf{P}, \tag{4.1.12a}$$

$$\boldsymbol{\pi}_2 = \mathbf{p}_2 - \frac{m_2}{M}\mathbf{P}, \tag{4.1.12b}$$

where

$$\mathbf{P} = \mathbf{p}_1 + \mathbf{p}_2 = M\dot{\mathbf{R}}. \tag{4.1.13}$$

Using the explicit forms of $\boldsymbol{\rho}_1, \boldsymbol{\rho}_2, \boldsymbol{\pi}_1$ and $\boldsymbol{\pi}_2$ it is easy to verify that Eqs. (4.1.7) are actually satisfied, i.e.

$$m_1\boldsymbol{\rho}_1 + m_2\boldsymbol{\rho}_2 = 0, \tag{4.1.14a}$$

$$\boldsymbol{\pi}_1 + \boldsymbol{\pi}_2 = 0. \tag{4.1.14b}$$

In order to separate the center-of-mass variables from the internal variables, we express the energy with the help of $\mathbf{P}, \boldsymbol{\pi}_\alpha$ and $\boldsymbol{\rho}_\alpha$:

$$E_{\text{mol}} = \frac{\mathbf{P}^2}{2M} + \frac{\boldsymbol{\pi}_1^2}{2m_1} + \frac{\boldsymbol{\pi}_2^2}{2m_2} + U(\boldsymbol{\rho}_1 - \boldsymbol{\rho}_2), \tag{4.1.15}$$

and this is a particular case of Eq. (4.1.9).

We now perform several operations that are possible due to the two-body nature of the problem. We write $\boldsymbol{\pi}_1$ and $\boldsymbol{\pi}_2$ in the form

$$-\boldsymbol{\pi}_2 = \boldsymbol{\pi}_1 = \mu(\dot{\boldsymbol{\rho}}_1 - \dot{\boldsymbol{\rho}}_2), \tag{4.1.16}$$

where

$$\mu = \frac{m_1 m_2}{M}$$

is the reduced mass of the system.

Exercise 1.2

Prove Eq. (4.1.16).

Solution on page 407

The right hand side of Eq. (4.1.16) is the relative momentum and the argument of U in Eq. (4.1.15) is the relative position. We denote them by $\boldsymbol{\pi}_{12}$ and $\boldsymbol{\rho}_{12}$, respectively, so that

$$\boldsymbol{\pi}_{12} = \mu\dot{\boldsymbol{\rho}}_{12}\,. \tag{4.1.17}$$

We can now write the energy of the molecule as the sum of the energy of the center of mass and the energy of the relative motion:

$$E_{\text{mol}} = \frac{\mathbf{P}^2}{2M} + \frac{\boldsymbol{\pi}_{12}^2}{2\mu} + U(\boldsymbol{\rho}_{12})\,. \tag{4.1.18}$$

We now make three remarks that will be of use in what follows:

(a) The kinetic term in the internal energy depends on $\dot{\boldsymbol{\rho}}_{12}$. Namely, this is the rate of change of the vector that connects the two atoms. This vector can change in magnitude — molecular vibrations. Or, it can remain of constant magnitude and change its direction, namely a rotation of the molecule. See Fig. 4.1.2.

(b) The potential $U(\boldsymbol{\rho}_{12})$ is usually a *central potential*, which means that it depends only on the magnitude of the distance between the two nuclei, which we denote by ρ.

(c) The vibrations of the interatomic distance will be centered around a certain equilibrium distance, ρ_0. When the amplitude of the vibrations is small, it is possible to expand $U(\rho)$. Namely,

$$U(\rho) = U(\rho_0) + U'(\rho_0)(\rho-\rho_0) + \frac{1}{2}U''(\rho_0)(\rho-\rho_0)^2 + \cdots \tag{4.1.19}$$

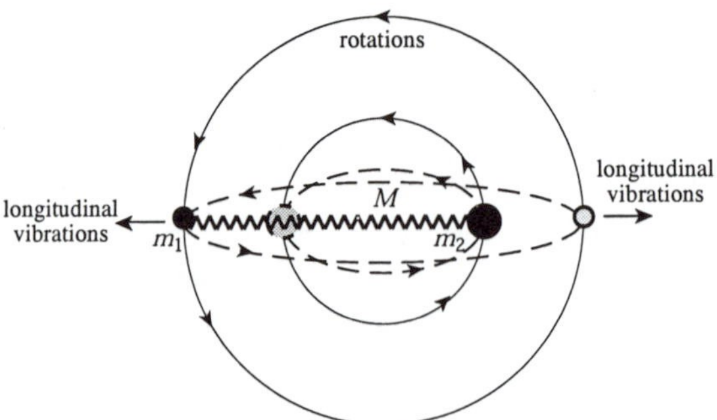

Fig. 4.1.2 Changes in the magnitude of the vector connecting the two atoms represent longitudinal vibrations of the molecule (like those of a spring). Changes in the direction of this vector represent rotations of the molecule relative to its center of mass.

The first term is independent of ρ and is simply a constant, which can be ignored. The second term is zero, since we assumed that ρ_0 is the equilibrium distance, and the potential energy must be minimal there, and hence $U'(\rho_0) = 0$. It is possible, therefore, to approximate the potential between the two atoms by a harmonic potential, centered on the equilibrium distance. This is the dashed line in Fig. 4.1.1.

Exercise 1.3

Consider a hypothetical molecule in which the potential between the two atoms is of the form

$$U(\rho) = \epsilon \left[\left(\frac{a}{\rho} \right)^2 - 2\frac{a}{\rho} \right] .$$

(a) What is the equilibrium distance in the molecule?
(b) What is the *binding energy* (the energy required to separate to infinite distance the two atoms from their equilibrium position)?
(c) Calculate the magnitude of the "spring constant," if a is 2 Å and ϵ is 2.5 eV.
(d) How much energy is required to change the interatomic distance by 5% of their equilibrium distance? What temperature is required to produce this change?
(e) What is the relative change in the molecular length at room temperature?

Solution on page 407

Exercise 1.4

Prove that in a diatomic molecule the following relationships exist between the relative position and momentum of the particles in the molecule and between their coordinates and momenta with respect to the laboratory frame of reference:

$$\mathbf{r}_1 = \mathbf{R} + \frac{m_2}{M}\boldsymbol{\rho}_{12}, \quad \mathbf{r}_2 = \mathbf{R} - \frac{m_1}{M}\boldsymbol{\rho}_{12},$$
$$\mathbf{p}_1 = \frac{m_1}{M}\mathbf{P} + \boldsymbol{\pi}_{12}, \quad \mathbf{p}_2 = \frac{m_2}{M}\mathbf{P} - \boldsymbol{\pi}_{12}.$$

Solution on page 408

Exercise 1.5

Write the total energy of a triatomic molecule as the sum of a free part related to the center-of-mass motion and an internal part. How many variables are required to characterize the molecule's state? (Disregard the motion of the electrons.)

Solution on page 409

1.3 Gas of general composite molecules

We discuss a gas of N molecules, each having an energy of the form (4.1.9). The energy of the whole system (recall that we assume that there are no forces between the molecules) in a given microscopic state is

$$E = \sum_{\nu=1}^{N} \frac{\mathbf{P}_\nu^2}{2M} + \sum_{\nu=1}^{N} \epsilon_\nu \,, \tag{4.1.20}$$

where $\mathbf{P}_\nu$ is the momentum of the center of mass of molecule number ν and ϵ_ν is the internal energy of molecule ν, which was denoted before by E_I.

According to Eq. (3.4.36a) with the Gibbs' correction, in a gas of classical molecules, the partition function is

$$Z = \frac{1}{h^{3N} N!} \int e^{-\beta E} d^{3N}R \; d^{3N}P d^{3N(q-1)}\rho \; d^{3N(q-1)}\pi \,, \tag{4.1.21}$$

where E is given by Eq. (4.1.20). This is a sum over all the system's states, which are described by three coordinates and three momenta of the center of mass of each molecule, along with the $3(q-1)$ internal coordinates and $3(q-1)$ internal momenta of each molecule.

A purely classical description leads to the equipartition law and to the difficulties in the calculation of the specific heat which we have encountered in Sec. 1.3 of Part I. (See also next section.) But we know that to deal with phenomena that are taking place inside a molecule one must apply quantum theory. The first term of Eq. (4.1.20) can still be treated classically, since the molecules are free to move inside a container of macroscopic dimensions. Recall Sec. 4.4 of Part III. The second term must be treated as a sum of discrete internal molecular energies, which are characterized according to quantum mechanics by the level number n. The internal energy of a molecule in level n will be denoted by $\epsilon(n)$. The values of these energies may in principle be obtained from the quantum-mechanical calculation or, alternatively, from experiment. The state of molecule number ν is therefore characterized by specifying the position and momentum of its center of mass, $\mathbf{R}_\nu, \mathbf{P}_\nu$, and its quantum state, n_ν. A microscopic state of the gas as a whole will be characterized, therefore, by

$$(\mathbf{R}_1, \mathbf{P}_1, n_1, \mathbf{R}_2, \mathbf{P}_2, n_2, \ldots, \mathbf{R}_N, \mathbf{P}_N, n_N)$$

and the energy in this state will be

$$E(\mathbf{R}_1, \mathbf{P}_1, n_1, \ldots, \mathbf{R}_N, \mathbf{P}_N, n_N) = \sum_{\nu=1}^{N} \frac{\mathbf{P}_\nu^2}{2M} + \epsilon_1(n_1) + \epsilon_2(n_2) + \cdots + \epsilon_N(n_N) \,. \tag{4.1.22}$$

The partition function of the system is obtained by summing over the discrete states and integrating over the continuous degrees of freedom:

$$Z = \frac{1}{h^{3N}N!}\int d^3R_1 d^3P_1 \ldots d^3R_N d^3P_N \sum_{n_1,\ldots,n_N} \times \exp[-\beta E(\mathbf{R}_1, \mathbf{P}_1, n_1, \ldots, \mathbf{R}_N, \mathbf{P}_N, n_N)] \,. \qquad (4.1.23)$$

The right hand side of (4.1.23) may be decomposed into a product of N factors of the form of a single molecule partition function:

$$z_\nu = \frac{1}{h^3}\int d^3R_\nu d^3P_\nu e^{-\beta P_\nu^2/2M} \sum_{n_\nu} e^{-\beta\epsilon_\nu(n_\nu)} \,. \qquad (4.1.24)$$

Since the molecules are identical, they all have the same sequence of energy levels. Namely, the energy of a molecule does not depend on the molecule's number ν but only on the level it occupies (Fig. 4.1.3), so that the right hand side of (4.1.23) decomposes into a product of N *identical* factors. We can therefore write the partition function as

$$Z = \frac{1}{N!}z^N \,. \qquad (4.1.25)$$

The single molecule partition function, z, can be further factored into the center-of-mass variables and the internal variables:

$$z(T,V,N) = z_{\text{c.m.}}(T,V,N)\zeta(T) \,, \qquad (4.1.26)$$

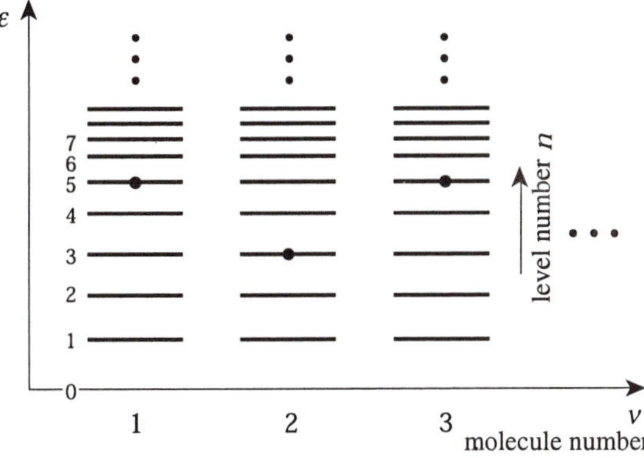

Fig. 4.1.3 A collection of identical molecules having the same energy levels. Molecules 1 and 3 in the figure, for instance, have the same energy, $\epsilon_1(5) = \epsilon_3(5) = \epsilon(5)$, whereas molecule 2 has $\epsilon_2(3) = \epsilon(3)$.

where

$$z_{\text{c.m.}} = \frac{1}{h^3}\int d^3R d^3P \exp\left(-\frac{\beta P^2}{2M}\right) = \frac{V(2\pi MkT)^{3/2}}{h^3}, \quad (4.1.27a)$$

$$\zeta = \sum_n e^{-\beta\epsilon(n)} = \sum_n e^{-\epsilon(n)/kT}. \quad (4.1.27b)$$

The internal structure of the molecules is expressed by the additional multiplicative factor, ζ, in the single molecule partition function. Since ζ depends only on the molecule's internal structure, we may refer to it as the internal partition function. In the following sections we shall calculate it explicitly for some simple cases.

internal partition function

We now calculate the free energy, and obtain Eq. (3.5.7) with an additional term depending on ζ:

$$F = -kT\ln Z = -NkT\left[\ln\frac{V}{N} + \frac{3}{2}\ln\left(\frac{2\pi MkT}{h^2}\right) + 1 + \ln\zeta\right]. \quad (4.1.28)$$

The equation of state

The pressure P is calculated from (4.1.28). The result is

$$P = -\frac{\partial F}{\partial V} = \frac{NkT}{V}, \quad (4.1.29a)$$

from which we obtain again the equation of state:

$$PV = NkT. \quad (4.1.29b)$$

It appears as if the internal degrees of freedom have no effect at all. But we are again faced with the questions that arose in Sec. 1.3 of Part I. Namely, how is it possible to decide which is the molecule and which are the constituents? After all we could have chosen to treat each nucleus and each electron separately!

This time we have an answer at hand: The reason why the equation does not change is that we assumed that ζ depends only on the temperature and *does not depend on the volume of the container.* Namely, the forces acting inside the molecule determine its size, and hence the distances between its different parts. When a part of the molecule hits the wall of the container, the other parts are found at a microscopic distance from the wall and do not wander freely in the container. This assumption will lose its validity as the temperature rises, since then the probability for very high energy states in the molecules to be occupied increases. Among the high energy states there will also be states of the disintegrated molecule, and in these states the energy depends on the volume. It goes without saying that in this case the constituents of the molecule come into play.

Entropy and internal energy per molecule

We use (4.1.28), and write

$$s = \frac{S}{N} = -\frac{1}{N}\left(\frac{\partial F}{\partial T}\right)_{V,N}$$
$$= k\left[\ln\frac{V}{N} + \frac{3}{2}\ln\left(\frac{2\pi MkT}{h^2}\right) + \frac{5}{2}\right] + \frac{d}{dT}(kT\ln\zeta)\,. \quad (4.1.30)$$

Comparing to (3.5.8), we find that there is an additional entropy per molecule of magnitude:

$$\Delta s = \frac{d}{dT}(kT\ln\zeta)\,. \quad (4.1.31)$$

The internal energy per molecule

$$\langle\epsilon\rangle = \frac{E}{N} = \frac{3}{2}kT + kT^2\frac{d}{dT}\ln\zeta\,. \quad (4.1.32)$$

Exercise 1.6

Deduce Eq. (4.1.32).

Solution on page 410

An important consequence is that although the form of the energy changes, owing to the internal structure of the molecules, the gas will still satisfy Joule's law of expansion. That is, since E depends only on T, expansion without the performance of work will occur without a change of temperature.

Additional results are the expression for the specific heat at constant volume,

$$C_V = \frac{3}{2}Nk + Nk\frac{d}{dT}\left(T^2\frac{d}{dT}\ln\zeta\right)\,, \quad (4.1.33)$$

and the relationship between the specific heats at constant pressure and at constant volume,

$$C_P - C_V = Nk\,, \quad (4.1.34)$$

as in a gas of pointlike molecules.

Exercise 1.7

Deduce Eqs. (4.1.33) and (4.1.34).

Solution on page 410

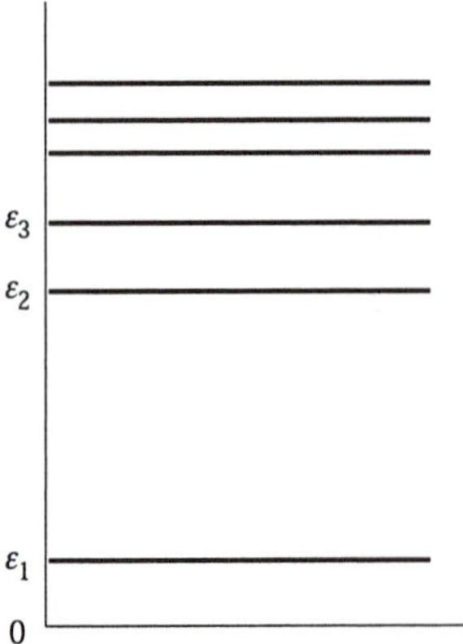

Fig. 4.1.4 The energy levels of the internal states in a molecular model.

Exercise 1.8

A model molecule has states whose energies are depicted in Fig. 4.1.4. The energy of the ground is zero and of the first excited state it is ϵ_1. Suppose $\epsilon_1 \ll \epsilon_2, \epsilon_3, \ldots$. Calculate the corrections to the entropy, the internal energy and the specific heat, resulting from the internal structure of the molecules. How do these corrections behave when $kT \ll \epsilon_1$?

Solution on page 411

The chemical potential

$$\mu = \left(\frac{\partial F}{\partial N}\right)_{T,V} = kT\left[\ln\frac{N}{V} - \frac{3}{2}\ln\left(\frac{2\pi MkT}{h^2}\right)\right] - kT\ln\zeta\,, \qquad (4.1.35)$$

differing from (3.5.9) by the term $-kT\ln\zeta$.

Monoatomic gas

So far we have treated the atoms as the fundamental building blocks of the molecules and ignored the internal structure of the atoms themselves. That internal structure, in a monoatomic gas, consists of internal degrees of freedom (electronic or nuclear). To take those into account one repeats the entire discussion from the beginning of this section, with the internal partition function (4.1.27b) now calculated as a sum over the electronic energy levels of the (monoatomic) molecule. Nevertheless, gases such as helium or argon, namely noble gases, are very well described as a gas of particles devoid of internal structure. The reason is that the first electronic excitation energy, ϵ_1, is about 10 eV above the ground state; namely, it corresponds to a temperature of 10^5 K. Thus, in a wide range of temperature, which may be considered very low with respect to this temperature, we may take only the first terms of Eq. (4.1.27b), which we write in the form

$$\zeta(T) = g_0 e^{-\beta\epsilon_0} + g_1 e^{-\beta\epsilon_1} + \cdots = e^{-\beta\epsilon_0}[g_0 + g_1 e^{-\beta(\epsilon_1-\epsilon_0)} + \cdots] . \quad (4.1.36)$$

Note that room temperature is safely within this range. Here g_0 and g_1 are the number of different atomic states having energies ϵ_0 and ϵ_1, respectively. These numbers are called the degrees of degeneracy or, simply, the degeneracies of the levels. **degeneracy**

■ The degeneracy of the ground state has already been mentioned in Sec. 5.3 of Part III.

In the helium atom, $g_0 = 1$. Since $kT \ll \epsilon_1 - \epsilon_0$, the second term of (4.1.36) is negligible, and the internal partition function is a multiplicative factor of $e^{-\beta\epsilon_0}$. The sole effect of such a factor is a constant shift of the energy levels, which does not change the thermodynamic quantities, as we have seen in Exercise 1.11 of Part III. If $g_0 = 1$, S and C_V will, therefore, be as in an ideal gas, and the chemical potential will change by a constant — ϵ_0.

Exercise 1.9

Prove that S and C_V of a monoatomic gas with $g_0 = 1$ are identical to those of a gas of particles with no internal structure, if the temperature is low with respect to the first electronic excitation energy. What about μ?

Solution on page 413

Exercise 1.10

Consider a monoatomic gas, under the same conditions as in Exercise 1.9, but with a doubly degenerate ground state. Calculate the change in S, C_V and in μ relative to their corresponding values in a gas of particles with no internal structure.

Solution on page 413

We find, therefore, that for the monoatomic gas there appears a characteristic temperature, of order 10^4–10^5 K, such that

$$k\Theta_e = \epsilon_1 . \quad (4.1.37)$$

Below this temperature the electronic degrees of freedom are *frozen* — there is not enough thermal energy to excite them. This, of course, is the answer to the heat capacity problem which we raised in Part I. Namely, the question why the electronic degrees of freedom do not contribute to the specific heat of gases. More on this in the following.

1.4 Diatomic gas: classical treatment

The next example is the diatomic molecule, which is a system composed of two nuclei, two large masses positively charged, and a group of electrons. The stability of the molecule implies that both nuclei "found out" that if they are separated by a distance ρ_0, then their energy is at a minimum — any increase or decrease of this distance increases the energy. The calculation leading to the determination of ρ_0 is extremely complicated, since it requires the solution of the quantum-mechanical dynamics of both nuclei and of all the electrons.

But, since the nuclei are very heavy compared to the electrons — their mass is several thousand times larger ($m_p/m_e \simeq 1840$) — it is possible to simplify the problem using the Born–Oppenheimer approximation (for more details see standard text on quantum mechanics), which treats the diatomic system (as well as the many-atomic) in two stages:

(a) First the nuclei are fixed in their positions, and one calculates the behavior of the system of electrons. A crucial part at this stage is the calculation of the dependence of the energy of the electrons upon the internuclear distance.

(b) Then the motion of the nuclei is treated as a slow motion with respect to the rapid motion of the electrons. The electrons always manage to accommodate themselves to the instantaneous state of the slow nuclei. The motion of the nuclei is determined by the dependence of the electronic energy upon the internuclear distance.

When performing stage (a) one finds that only the ground state of the system of electrons is required. The rest of the electronic states have excitation energies of the order of 1 eV and thus are frozen, up to temperatures of several thousand degrees. The ground state energy of the electronic system ϵ_0 depends on the distance ρ between the nuclei. This leads to the appearance of a potential energy $\epsilon_0(\rho)$, affecting the motion of the nuclei, which is the U that appears, for example, in Eq. (4.1.18).

If ϵ_0 has a sharp minimum at a distance ρ_0, it is possible to approximate $\epsilon_0(\rho)$ by the harmonic approximation (4.1.19). The energy of the molecule will then be given by

$$E_{\text{mol}} = \frac{\mathbf{P}^2}{2M} + \frac{\pi^2}{2\mu} + \frac{1}{2}K(\rho - \rho_0)^2 \,, \tag{4.1.38}$$

where we have used the notations $\boldsymbol{\pi}_{12} = \boldsymbol{\pi}$, $\boldsymbol{\rho}_{12} = \boldsymbol{\rho}$, $U''(\rho_0) = K$, and dropped $\epsilon_0(\rho_0)$, a constant which determines the ground state energy of the molecule.

■ Note: if ϵ_0 depends only on the distance between the nuclei and not on the direction, the potential depends only on $|\boldsymbol{\rho}| = \rho$.

The free energy will thus be, according to Eq. (4.1.28),

$$F = F_{\text{c.m.}} - NkT \ln \zeta \,, \tag{4.1.39}$$

where $F_{\text{c.m.}}$ has the same form as the free energy of a monoatomic gas [Eq. (3.5.7)]. The function ζ is given by (4.1.27b):

$$\zeta(T) = \sum_n e^{-\beta \epsilon_n} \,. \tag{4.1.40}$$

n specifies the quantum states of the system, whose energy is given by the last two terms of (4.1.38). That is the energy of internal motions E_I:

$$E_I = \frac{\boldsymbol{\pi}^2}{2\mu} + \frac{1}{2} K(\rho - \rho_0)^2 \,. \tag{4.1.41}$$

In other words, after the separation of the center-of-mass energy, we are left with an internal energy of a particle of mass μ, moving in a three-dimensional harmonic potential. The potential energy term causes radial vibrations around the equilibrium point $\rho = \rho_0$. If the vibrations are small their frequency is

$$\omega = \sqrt{\frac{K}{\mu}} \,. \tag{4.1.42}$$

In addition rotations are, of course, also allowed and we shall discuss them in the next section. Both types of motion are illustrated in Fig. 4.1.2.

The calculation of $\zeta(T)$ passes through a computation of the quantum energy levels of the molecule. We perform the quantum calculation in the next section. First, we note that if the temperature T is high with respect to the energies of the first excited states, it is possible to treat the variables $\boldsymbol{\pi}$ and $\boldsymbol{\rho}$ as classical variables. The condition imposed on the temperature is

$$kT \gg \hbar\omega \,. \tag{4.1.43}$$

A typical value of ω is $10^{13}\,\text{s}^{-1}$; $\hbar \simeq 10^{-34}\,\text{J}\cdot\text{s}$. Hence the classical approximation is justified for $kT \gg 10^{-21}$ J or $T \gg 100$ K. Yet we still assume that the temperature is too low to excite the electronic states.

To calculate the internal partition function we first write it as an integral over position and momentum:

$$\begin{aligned}\zeta(T) &= \frac{1}{h^3} \int d^3\pi d^3\rho \exp\left\{-\beta\left[\frac{\boldsymbol{\pi}^2}{2\mu} + \frac{1}{2}K(\rho - \rho_0)^2\right]\right\} \\ &= \left(\frac{2\pi\mu kT}{h^2}\right)^{3/2} \int d^3\rho \exp\left[-\frac{\beta K}{2}(\rho - \rho_0)^2\right] \,. \end{aligned} \tag{4.1.44}$$

In the second integral the integrand depends only on ρ and not on $\boldsymbol{\rho}$. Hence

$$\int d^3\rho \exp\left[-\frac{\beta K}{2}(\rho - \rho_0)^2\right] = 4\pi \int_0^\infty \rho^2 \exp\left[-\frac{\beta K}{2}(\rho - \rho_0)^2\right] d\rho \tag{4.1.45}$$

after performing the angular integrations. The last integral is calculated approximately. Since we know how to calculate the integral when the lower limit is $-\infty$ (instead of 0), we assume that the center of the bell-like curve is very far from the origin (compared to its width; see Fig. 4.1.5). This is the case if

$$\rho_0^2 \gg \frac{1}{\beta K}, \tag{4.1.46a}$$

or

$$kT \ll K\rho_0^2 . \tag{4.1.46b}$$

Having assumed this, the integral is well approximated by

$$\int_{-\infty}^{\infty} \rho^2 \exp\left[-\frac{\beta K}{2}(\rho - \rho_0)^2\right] d\rho \simeq \rho_0^2 \sqrt{\frac{2\pi}{\beta K}} . \tag{4.1.47}$$

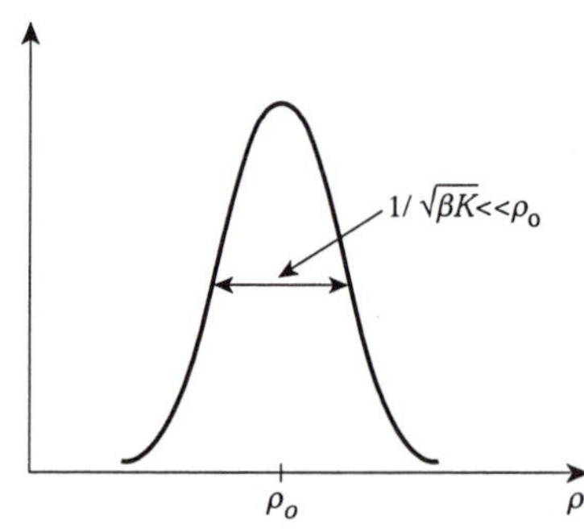

Fig. 4.1.5 When conditions (4.1.46) hold it is possible to treat the bell as if it extended from $-\infty$ to $+\infty$.

Exercise 1.11

Prove Eq. (4.1.47).

Solution on page 414

Before proceeding to calculate the partition function, we check the range of temperature to which condition (4.1.46) constrains us. Recalling that $\rho_0 \simeq 10^{-9}$ m, $\omega \simeq 10^{13}\,\mathrm{s}^{-1}$ and $\mu \simeq 10^{-26}$ kg (about 10 proton masses), we obtain

$$K\rho_0^2 = \mu\omega^2\rho_0^2 \simeq 10^{-18}\ \mathrm{J}, \tag{4.1.48}$$

namely $kT \ll 10^{-18}$ J or $T \ll 10^5$ K.

We have found that the temperature range for which both of the conditions (4.1.43) and (4.1.46) are satisfied is 10^3–10^4 K. The internal partition function in this temperature range is obtained by substituting (4.1.47) and (4.1.45) into (4.1.44):

$$\zeta(T) = \left(\frac{2\pi\mu kT}{h^2}\right)^{3/2} \cdot 4\pi\rho_0^2\sqrt{\frac{2\pi kT}{\mu\omega^2}} = \frac{2\mu\rho_0^2(kT)^2}{\hbar^3\omega} . \tag{4.1.49}$$

Substituting the last result into Eq. (4.1.39), we obtain the free energy:

$$F = F_{\text{c.m.}} - 2NkT\ln\left(\frac{kT}{C}\right) , \tag{4.1.50}$$

where C is a constant depending on the properties of the molecules (ω, ρ_0, μ) but not on T. Using (4.1.49) we also find that the specific heat at constant volume is given by

$$C_V = \frac{7}{2}Nk ; \tag{4.1.51}$$

namely, the specific heat of the gas of N diatomic molecules is larger than the specific heat of the gas were the atoms free ($3Nk$). The bond between the atoms increases the specific heat, as we hinted in Sec. 1.3 of Part I.

Exercise 1.12

(a) Calculate the internal energy of a diatomic molecule, in the classical approximation, and show that the specific heat is indeed given by (4.1.51).
(b) If the specific heat is given by (4.1.51), what is the ratio $C_P/C_V = \gamma$?
(c) Compare your result with the experimental values for the diatomic gases in Table 4.1.1. Is agreement to be expected for these gases at room temperature? What about 1000°C?

Table 4.1.1 Experimental values of γ.

	H_2	N_2	CO	NO	O_2
$T = 273$ K	1.410	1.400	1.400	1.384	1.397
$T = 1273$ K	1.349	1.314	1.310	1.303	1.295

Solution on page 414

Exercise 1.13

(a) For the gas Cl_2, $\omega \simeq 10^{14}$ s^{-1} and the interatomic distance is about 2 Å. What is the range of temperatures for which the approximations we made for ζ are justified?
(b) How will the integrand in (4.1.44) look, graphically, at a temperature of 1500 K?

Solution on page 415

1.5 Diatomic molecules: vibration and rotation

The problem of the discrepancy between the theoretical and experimental values of the specific heat of a diatomic gas has been solved in the framework of quantum mechanics. Here we shall not enter into a detailed discussion on the quantum method. We shall turn to it only in order to obtain the values of the energies of the different states of the internal system, which is characterized by (4.1.41), as well as the degeneracies of the various energy levels, namely the number of different states of equal energy.

The solution is based, as we have already mentioned several times, on the fact that the internal energy of the molecule is not continuous, but assumes discrete values. In order to "feel" the excited states a high enough temperature is required, otherwise most of the molecules remain in the ground state and the internal degrees of freedom are frozen. But this is not the whole story. The internal energy of a diatomic molecule has two different sources. In addition to the molecular vibrations, discussed in the classical context in the previous section, the molecule can also rotate with respect to its center of mass. Hence it is natural to separate the kinetic part of the internal energy, Eq. (4.1.41), into a (one-dimensional) radial part and a part corresponding to the rotational (two-dimensional) motion:

$$\frac{\boldsymbol{\pi}^2}{2\mu} = \frac{\pi_\rho^2}{2\mu} + \frac{\boldsymbol{\ell}^2}{2\mu\rho^2}\,, \tag{4.1.52}$$

where π_ρ is the (relative) momentum component along ρ and $\boldsymbol{\ell}$ is the angular momentum of the molecule with respect to its center of mass:

$$\pi_\rho = \mu\dot{\rho} = \frac{\boldsymbol{\pi}\cdot\boldsymbol{\rho}}{\rho}\,, \tag{4.1.53a}$$

$$\boldsymbol{\ell} = \mu\boldsymbol{\rho}\times\dot{\boldsymbol{\rho}} = \boldsymbol{\rho}\times\boldsymbol{\pi}\,. \tag{4.1.53b}$$

▪ This is the standard procedure used in the analysis of the two-body problem in classical mechanics.

Exercise 1.14

Using definitions (4.1.53), prove Eq. (4.1.52).

Solution on page 416

The internal energy (4.1.41) becomes the sum of a vibrational part and a rotational part:

$$E_I = \frac{\pi_\rho^2}{2\mu} + \frac{K}{2}(\rho - \rho_0)^2 + \frac{\ell^2}{2\mu\rho^2}\,. \tag{4.1.54}$$

The next stage is to find the quantum energy levels of the molecule described by (4.1.54), and to calculate with their help the internal partition function $\zeta(T)$. The difficulty is in the fact that the energy in (4.1.54) is not the sum of two independent terms, since the relative distance ρ appears in the vibrational energy as well as in the rotational energy. But, since at reasonable temperatures the stretching of the molecules does not exceed 10% of ρ_0 (cf. Exercise 1.3), and anyway in applying the harmonic approximation (4.1.19) we have already restricted ourselves to small vibrations around ρ_0, we can treat the rotations as if they occurred for a molecule of constant ρ, equal to ρ_0.

■ Compare to the approximation (4.1.47), which was made for the case where the Boltzmann factor is very concentrated around $\rho = \rho_0$.

We may therefore write

$$E_I = \frac{\pi_\rho^2}{2\mu} + \frac{K}{2}(\rho - \rho_0)^2 + \frac{\ell^2}{2I}\,, \tag{4.1.55}$$

where $I = \mu\rho_0^2$ is the molecule's moment of inertia with respect to its center of mass.

Before going on, we consider the orders of magnitude involved. To this end we use several typical values of I which are given in Table 4.1.2.

Table 4.1.2 Moments of inertia of simple molecules.

Molecule	I in units of 10^{-47} kg · m^2
H_2	0.46
HCl	2.4
Cl_2	115
I_2	745

Exercise 1.15

(a) Show that $\mu\rho_0^2$ is indeed the moment of inertia of a diatomic molecule with respect to its center of mass.
(b) Calculate μ and ρ_0 for the four molecules that appear in Table 4.1.2.

Solution on page 416

Now, with the expression (4.1.55) for the internal energy of a diatomic molecule at our disposal, we can without much effort find the quantum states and their energy levels. Because the energy can be written naturally as the sum of two types of energies, i.e. a vibrational energy and a rotational energy, we shall discuss each of them separately.

Vibrational levels

The first two terms of (4.1.55) describe a one-dimensional harmonic oscillator. We have treated such an oscillator in Part III, and we already know that its states are characterized by the degree of excitation n and that its energies are quantized according to

$$E_v(n) = \left(n + \frac{1}{2}\right)\epsilon_v\,, \tag{4.1.56}$$

where ϵ_v is the spacing between the vibrational levels:

$$\epsilon_v = \hbar\omega = \hbar\sqrt{\frac{K}{\mu}}\,. \tag{4.1.57}$$

For example, in a chlorine molecule Cl_2, $\omega = 10^{14}\ \mathrm{s}^{-1}$ and hence the spacing between the vibrational levels is 10^{-20} J, which is about 0.06 eV. To these spacings there corresponds a characteristic temperature:

$$\Theta_v = \frac{\epsilon_v}{k}\,. \tag{4.1.58}$$

Above this temperature the "frozen" vibrational levels can "unfreeze." For chlorine $\Theta_v \simeq 700$ K.

Rotational states

The third term of Eq. (4.1.55) describes the energy of the rotational motion of the molecule with respect to its center of mass, in terms of the internal angular momentum of the molecule, $\boldsymbol{\ell}$. Several aspects of the quantum angular momentum have already been mentioned in Part II (Chap. 2), but this is the place to include some additional aspects. Thus, we present a brief summary of the main results of the quantum analysis of the properties of the angular momentum:

(a) The quantum angular momentum of a system assumes discrete values only, which are characterized by a nonnegative integer or half-integer J.

(b) The absolute value of the angular momentum is given by

$$|\mathbf{J}| = \sqrt{J(J+1)}\,. \tag{4.1.59}$$

$\mathbf{J}$ is dimensionless and measures the angular momentum of the system in units of $\hbar$ (see Sec. 2.1 of Part II).

(c) Each of the angular momentum's components is also quantized, assuming only discrete values. For instance,

$$J_z = -J, -J+1, \ldots, J-1, J\,. \tag{4.1.60}$$

(d) Due to the uncertainty principle it is impossible to obtain more information on the angular momentum than its magnitude $|\mathbf{J}|$ and the value of *one* of its components, which is usually chosen as J_z. Thus each state of a quantum system, for which the angular momentum is conserved, is characterized by two numbers: J and J_z. For a given J there are $2J+1$ different possible states (all the allowed values of J_z).

We shall now use all of the above in order to find the energy levels of the rotational quantum states — the rotational levels. The rotational levels are obtained from the last term on the right hand side of Eq. (4.1.55), which we call rotational energy. Since $\boldsymbol{\ell}$ is an orbital angular momentum (no spin) we can write

$$\boldsymbol{\ell} = \hbar\mathbf{J}\,. \tag{4.1.61}$$

Thus we can use (4.1.59) for the allowed values of the angular momentum squared in the notational energy

$$E_r(J) = J(J+1)\epsilon_r\,, \tag{4.1.62}$$

where ϵ_r,

$$\epsilon_r = \frac{\hbar^2}{2I}\,, \tag{4.1.63}$$

determines the difference between the rotational levels. This is the place to stress that since the origin of the angular momentum here is orbital, the allowed values of J are integers, and not half-integers, as may occur in systems in which the energy also depends upon the electronic spins.

To obtain some orders of magnitude, we again take the example of the chlorine molecule, whose moment of inertia appears in Table 4.1.2: $I = 1.15\times10^{-45}$ kg$\cdot$m^2. The value of ϵ_r is 4×10^{-24} J, which is 2.5×10^{-5} eV.

The corresponding temperature above which the rotational levels unfreeze,

$$\Theta_r = \frac{E_r(1) - E_r(0)}{k} = \frac{2\epsilon_r}{k}, \tag{4.1.64}$$

is much lower than Θ_v. For chlorine this temperature is 0.7 K.

Exercise 1.16

Calculate $\epsilon_v, \Theta_v, \epsilon_r, \Theta_r$ for HCl ($\omega = 5.7 \cdot 10^{14}\ \mathrm{s}^{-1}$).

Solution on page 417

Exercise 1.17

How many rotational levels fit into one vibrational level spacing in Cl_2 and in HCl?

Solution on page 417

Having investigated the properties of both types of internal energies of a diatomic molecule, i.e. the vibrational energy and the rotational energy, we can find the quantum energy levels of the whole molecule. The energy levels will be characterized by two numbers, n and J, and will be a sum of (4.1.56) and (4.1.62):

$$E_I(n, J) = \epsilon_v n + \epsilon_r J(J+1), \quad n, J = 0, 1, 2, \ldots, \tag{4.1.65}$$

where we have dropped the ground state vibrational energy, just as we ignored $\epsilon_0(\rho_0)$ on the way to obtaining (4.1.38).

To characterize the quantum states of a diatomic molecule, we need to specify the degree of excitation of the vibrational motion as well as that of the rotational state. The energy is determined by n and J, while a state is specified giving n, J and J_z. The vibrational state is nondegenerate, while the rotational state has a degeneracy of $2J+1$, since its energy does not depend on J_z for a given J. Hence the energy level $E_I(n, J)$ has a degeneracy of $2J + 1$.

The next step is, of course, the calculation of the internal partition function. This is, as usual, a sum of the Boltzmann factors of all the quantum states. We get

$$\zeta = \sum_{n=0}^{\infty} \sum_{J=0}^{\infty} (2J+1) \exp\{-\beta[\epsilon_v n + \epsilon_r J(J+1)]\}. \tag{4.1.66}$$

The factor $2J + 1$ is the degeneracy of the state with given J and n. It corresponds to the summation over J_z. Since the energy is a sum of two independent terms, the partition function becomes again a product of two factors, one depending only on the vibrational states and the other only on the rotational states:

$$\zeta = \zeta_v \zeta_r \,, \tag{4.1.67}$$

where

$$\zeta_v = \sum_{n=0}^{\infty} e^{-\beta\epsilon_v n} \,, \tag{4.1.68a}$$

$$\zeta_r = \sum_{J=0}^{\infty} (2J+1) e^{-\beta\epsilon_r J(J+1)} \,. \tag{4.1.68b}$$

1.6 The equipartition principle and its violation

The factorization of the internal partition function into the product of a vibrational factor and a rotational factor is not special to the quantum case. It was possible since the approximations we made rendered the internal energy of the molecule a sum of a vibrational contribution and a rotational contribution. Hence the factorization of (4.1.67) is general. The full single molecule partition function will be a product of three factors, obtained from substitution in Eq. (4.1.26):

$$z = z_{\text{c.m.}} \zeta_v \zeta_r \,. \tag{4.1.69}$$

▪ The pedant may multiply (4.1.69) by other factors corresponding to the degrees of freedom of the electrons, the nucleons, etc.

The free energy becomes a sum of terms describing the different motions: the usual (monoatomic) term $F_{\text{c.m.}}$, corresponding to the center of mass motion, and terms corresponding to the vibrational motion, the rotational motion, etc.:

$$F = F_{\text{c.m.}} - NkT(\ln\zeta_v + \ln\zeta_r) = F_{\text{c.m.}} + F_v + F_r \,. \tag{4.1.70a}$$

The energy of the gas, or the average energy per molecule, is written in the same manner:

$$\begin{aligned}\langle\epsilon\rangle &= -\frac{\partial}{\partial\beta}\ln z = -\frac{\partial}{\partial\beta}(\ln z_{\text{c.m.}} + \ln\zeta_v + \ln\zeta_r) \\ &= \langle\epsilon_{\text{c.m.}}\rangle + \langle\epsilon_v\rangle + \langle\epsilon_r\rangle \,. \end{aligned} \tag{4.1.70b}$$

The separability of the energy into independent terms lies at the very basis of the equipartition principle (see Part III, Sec. 4.3), according to which each degree of freedom appearing quadratically in the energy contributes to the energy an amount $\frac{1}{2}kT$, and thus contributes to the specific heat an amount $\frac{1}{2}k$.

The center-of-mass motion of a molecule is associated with three degrees of freedom, which are the three components of **P**, and thus $\langle \epsilon_{\text{c.m.}} \rangle = \frac{3}{2}kT$. The vibrational motion has two degrees of freedom — one is the vibrational kinetic energy $\pi_\rho^2/2\mu$, and the other is the harmonic potential term. In total, $\langle \epsilon_v \rangle = kT$.

The rotational motion of the diatomic molecule as described by $\boldsymbol{\ell}^2/2I$ has two degrees of freedom (not three), since the diatomic molecule can only rotate around two independent axes. Rotations around the axis that passes through both atoms do not contribute to the energy. In other words, the decomposition of $\boldsymbol{\pi}^2/2\mu$ in Eq. (4.1.52) was performed in such a way that the radial component was separated and taken into account in the vibrational energy, and there remain two momentum components normal to the direction of ρ, which describe the rotational motion. Hence for the rotational motion, $\langle \epsilon_r \rangle = kT$. In total we find from the equipartition principle that the average energy of a diatomic molecule is $\frac{7}{2}kT$, and that the specific heat at constant volume of a diatomic gas is $\frac{7}{2}Nk$. We have already obtained this result by direct calculation in (4.1.51).

All of the above assertions follow from the equipartition theorem, which is valid only for classical systems. Hence all of the above describes the behavior of a diatomic gas in a limited range of temperatures. The extent of this range is determined by the characteristic temperatures which have appeared in our discussion in Secs. 1.3 and 1.5 — $\Theta_e \simeq 10^4$ K, $\Theta_v \simeq 10^3$ K, $\Theta_r \simeq 10$ K — and which determine the range of excitation of the respective degrees of freedom.

Since we are not interested in the detailed structure of the atoms which come into play at temperatures above Θ_e, we shall restrict ourselves to temperatures below Θ_e. At temperatures $\Theta_v < T < \Theta_e$ all the degrees of freedom that were discussed are excited, so that the specific heat attains its "equipartition value," $\frac{7}{2}Nk$.

At temperatures below Θ_v the specific heat problem begins to emerge: the specific heat becomes smaller than its classical value. The reason is, of course, that at temperatures $\Theta_r < T < \Theta_v$ only the rotational degrees of freedom are excited (in addition, of course, to the center-of-mass degrees of freedom, which are never "frozen"). The vibrational degrees of freedom are not excited, and the vibrational contribution to the internal energy and to the specific heat is negligibly small. At these temperatures we are in the range in which the vibrational contribution to the specific heat is given by (3.2.18); it is exponentially small. This means that one k out of the $\frac{7}{2}k$ of the specific heat per molecule disappears at these temperatures. Thus C_V becomes smaller not only than the classical value of a diatomic molecule but even than the classical value of two separate atoms: the vibrational degrees of freedom are frozen.

frozen degrees of freedom

If we now lower the temperature further, to the range $T < \Theta_r$, the rotational degrees of freedom freeze as well. In the classical approximation two degrees of freedom correspond to the rotational motion, which contribute a single k to the specific heat per molecule. When $T \ll \Theta_r$, $C_V \to \frac{3}{2}Nk$, namely at these temperatures the diatomic molecules behave as structureless point particles.

We therefore have three temperature scales:

(1) Electronic — 1 eV,
(2) Vibrational — 0.1 eV,
(3) Rotational — 0.001 eV,

and correspondingly temperature regions and characteristic temperatures.

1.7 Diatomic gas — quantum calculation

We have seen that the internal partition function decomposes into a product of independent factors, which leads to the appearance of additive contributions to the free energy corresponding to each of the different factors of the partition function. We are left with the task of calculating the separate parts of the internal partition function. The vibrational part (4.1.68a) has in fact already been calculated in the previous part, and is exactly the partition function of a single harmonic oscillator Eq. (3.2.7) without the ground state energy term. It reads

$$\zeta_v = \frac{1}{1 - e^{-\Theta_v/T}}\,, \tag{4.1.71}$$

and its contribution to the specific heat is obtained by the appropriate substitutions in (3.2.16):

$$\frac{(\Delta C_V)_v}{N} = k\left(\frac{\Theta_v}{T}\right)^2 \frac{e^{\Theta_v/T}}{(e^{\Theta_v/T} - 1)^2}\,. \tag{4.1.72}$$

Note that unlike the case of the Einstein solid, where the fact that all the atoms have the same frequency is only an approximation, here all the molecules do indeed have an identical vibrational frequency.

In order to complete the discussion we have to calculate ζ_r. The summation in (4.1.68b) is hard to perform, and hence we shall treat it in two limits:

(a) Low temperatures — $T \ll \Theta_r$.

It suffices to take the first two terms in the sum, namely those with $J = 0$ and $J = 1$:

$$\zeta_r \simeq 1 + 3e^{-\Theta_r/T}\,, \tag{4.1.73}$$

and the contribution to the specific heat will be according to (4.1.33):

$$\frac{(\Delta C_V)_r}{N} = k\frac{d}{dT}\left(T^2\frac{d}{dT}\ln\zeta_r\right) \simeq 3k\left(\frac{\Theta_r}{T}\right)^2 e^{-\Theta_r/T}; \qquad (4.1.74)$$

the derivation is left to Exercise 1.18.

The right hand side of (4.1.74) tends to zero as $T \to 0$, namely the two rotational degrees of freedom freeze.

(b) High temperatures — $T \gg \Theta_r$.

It is possible to approximate the sum (4.1.68b) by an integral, since the terms in the sum change almost continuously. See, for example, the derivation of Eq. (3.4.33).

$$\zeta_r \simeq \int_0^\infty dJ(2J+1)e^{-J(J+1)\Theta_r/2T} = \frac{2T}{\Theta_r} \qquad (4.1.75)$$

and the contribution to the specific heat will be k as for two classical degrees of freedom.

Exercise 1.18

(a) Obtain the right hand side of Eq. (4.1.74).
(b) Prove (4.1.75).
(c) Prove that (4.1.75) implies a contribution of k to the specific heat per molecule.

Solution on page 418

The behavior of the specific heat as a function of the temperature is depicted in Fig. 4.1.6. Note that our explanation is still incomplete, and that there are still regions in the graph that are not "covered." Further study of this topic will take us too far afield.

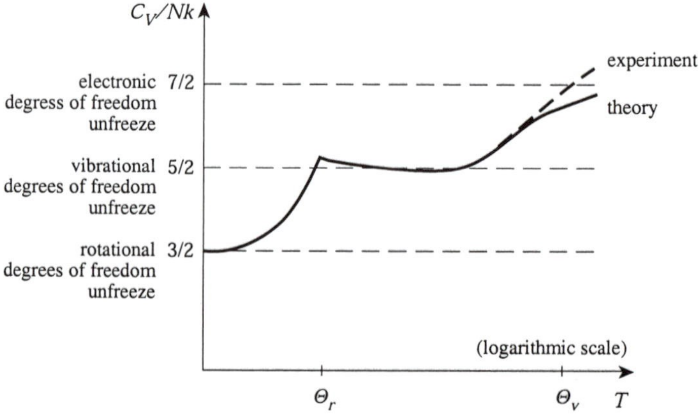

Fig. 4.1.6 The behavior of the specific heat of a diatomic gas as a function of the temperature.

In summary:

(1) The free energy of a gas of molecules with internal structure is a sum of a free energy related to the center-of-mass motion and of a free energy related to the internal motions in the molecules: $F = F_{\mathrm{c.m.}} + F_{\mathrm{I}}$.
(2) The internal motion comprises the motion of the electrons and the motion of the heavy nuclei. The motion of the nuclei may be further reduced to notations and vibrations.
(3) The treatment of the internal motions must be quantum-mechanical (if we are interested in temperatures which are not exceptionally high). There appear different scales of energy spacings. Each energy spacing — electronic, vibrational, rotational or other — determines a characteristic temperature.
(4) If the temperature is very high, the specific heat can be calculated classically. The intramolecular forces increase the specific heat above its value for a gas of unbonded (free) atoms.
(5) As the temperature decreases, each time it crosses a characteristic temperature of one of the internal motions, *the corresponding degree of freedom freezes*, and ceases to contribute to the specific heat or to the internal energy.

Chapter 2

Gases in Chemical Reactions

2.1 Conditions for chemical equilibrium

So far we have discussed gases whose basic building blocks are molecules. First (Chaps. 3 and 4 of Part III) we described the molecules as pointlike objects without any internal structure, and obtained the laws of ideal gases. In the second stage (Chap. 1 of this part) we took into account the internal structure of the molecules and the internal degrees of freedom — rotation and vibration — and mentioned the possibility of electronic excitations. We still have not considered the possibility of a molecule disintegrating into its constituent atoms, or its participating in a chemical reaction. In real gases collisions between molecules occur incessantly at different energies. In some cases, due to highly energetic collisions at least one of the molecules involved will disintegrate. On the other hand, of course, the opposite process can also occur, i.e. two free atoms which have lost their partners will meet and recombine to form a molecule. More often than not, no free atoms are left. Instead new types of molecules are created. A typical example is the process of the formation and dissociation of water molecules, as described by the reaction between molecules (not between atoms):

$$2H_2 + O_2 \rightleftharpoons 2H_2O\,.$$

In the system in which this reaction is taking place there are three types of molecules. There are molecules that disappear in the reaction, and there are others that appear. The molecules appear and disappear at constant proportions. Thus, if an oxygen molecule disappears, then twice as many hydrogen molecules must disappear, and as many water molecules must appear, etc.

Let dN_1 denote the change in the number of hydrogen molecules, dN_2 the change in the number of oxygen molecules, and dN_3 the change in the number of water molecules. The reaction must satisfy the relations

$$dN_1 = 2dN_2,\ dN_3 = -dN_1 \tag{4.2.1a}$$

or

$$\frac{dN_1}{dN_2} = 2\,, \quad \frac{dN_1}{dN_3} = -1\,. \tag{4.2.1b}$$

We can write a general formula for a chemical reaction of molecules of types $B_1, \ldots, B_M$ in the following manner:

$$b_1 B_1 + b_2 B_2 + \ldots + b_r B_r \rightleftharpoons b_{r+1} B_{r+1} + \ldots b_M B_M\,,$$

where B_i are the chemical formulae of the molecules that appear and disappear, and b_i are the smallest integers for which there are no leftovers from the reaction. In the above example,

$$B_1 = H_2\,, \quad b_1 = 2\,,$$

$$B_2 = O_2\,, \quad b_2 = 1\,,$$

$$B_3 = H_2O\,, \quad b_3 = 2\,.$$

The generalization of (4.2.1) will be a relationship between the changes in the number of molecules of the different types. If dN_i $(i = 1, \ldots, M)$ is the change in the number of molecules of type B_i, then the relationship that guarantees that the total number of *atoms*, of all types, does not change in the reaction is

$$\frac{dN_i}{dN_j} = \frac{\nu_i}{\nu_j}\,, \tag{4.2.2}$$

where

$$\nu_i = -b_i, \quad i = 1, \ldots, r\,,$$

$$\nu_i = +b_i, \quad i = r+1, \ldots, M\,.$$

In the above example,

$$\nu_1 = -2\,,$$

$$\nu_2 = -1\,,$$

$$\nu_3 = 2\,.$$

■ Note that we assume here that the atoms do not disintegrate in the collisions.

Let us suppose that the reaction takes place under conditions of constant temperature and constant volume. In this case, equilibrium will be attained when the free energy F is minimal. Thus, changes in the number of molecules will occur only if they reduce F. When the system has reached a state where the number of molecules of type B_i is such that F is minimal, no more changes will occur. The condition

$$dF = 0 \tag{4.2.3}$$

determines the relationship between the number of molecules of different types. Since we are assuming constant T and V and are only allowing the number of molecules N_i to change, dF is the change in the free energy resulting from the changes dN_i in the numbers of molecules, i.e.

$$dF = \sum_{i=1}^{M} \left(\frac{\partial F}{\partial N_i}\right)_{T,V} dN_i = \sum_{i=1}^{M} \mu_i dN_i = 0\,, \tag{4.2.4}$$

where μ_i is the chemical potential of the molecules of type B_i.

If we now substitute into Eq. (4.2.4) the relation which balances the reaction, Eq. (4.2.2), we obtain the condition for equilibrium:

$$\sum_{i=1}^{M} \nu_i \mu_i = 0\,. \tag{4.2.5}$$

Exercise 2.1

Prove Eq. (4.2.5).

Solution on page 419

The ν_i are constant integers (positive and negative). μ_i depend on the number (per unit volume) of molecules of each type. Therefore, (4.2.5) provides a relation between the densities of molecules in a state of equilibrium, and the temperature.

Exercise 2.2

Consider the following reactions:

(a) $4NH_3 + 3O_2 \rightleftharpoons 2N_2 + 6H_2O\,,$
(b) $2C_4H_{10} + 13O_2 \rightleftharpoons 8CO_2 + 10H_2O\,.$

What is the form of Eq. (4.2.5) in each case?

Solution on page 419

2.2 The law of mass action

In order to obtain the explicit form of the relationship between the densities of molecules and the temperature, we substitute into Eq. (4.2.5) the explicit form of the chemical potential, Eq. (4.1.35):

$$kT \sum_{i=1}^{M} \nu_i \left[\ln n_i - \frac{3}{2}\ln\left(\frac{2\pi M_i kT}{h^2}\right) - \ln \zeta_i\right] = 0\,, \tag{4.2.6}$$

where n_i is the density of molecules of type i, and M_i is their mass. ζ_i contains the information about the internal structure of the molecules, namely, to what extent they tend to participate in chemical reactions. From here we can immediately obtain the *law of mass action*:

law of mass action

$$n_1^{\nu_1} \cdots n_M^{\nu_M} = \frac{n_{r+1}^{|\nu_{r+1}|} \cdots n_M^{|\nu_M|}}{n_1^{|\nu_1|} \cdots n_r^{|\nu_r|}} = K(T) . \tag{4.2.7}$$

The right hand side, which is called the chemical equilibrium constant, is given by

chemical equilibrium constant

$$K(T) = \prod_{i=1}^{M} \left[\left(\frac{2\pi M_i kT}{h^2} \right)^{3/2} \zeta_i(T) \right]^{\nu_i} , \tag{4.2.8}$$

Note that for a given reaction the equilibrium constant depends only on the temperature.

Exercise 2.3

Deduce (4.2.7) and (4.2.8) from (4.2.6).

Solution on page 420

The law of mass action is a very important tool in physical chemistry (it was discovered by Guldberg and Waage in 1867). For instance, if the equilibrium constant in (4.2.7) $K(T)$ is known, then it is possible to obtain the concentrations of materials in the system at any temperature. Larger $K(T)$ implies higher product concentrations.

In the example of the formation of water from hydrogen and oxygen, we obtain from Eq. (4.2.7)

$$\frac{n_3^2}{n_1^2 n_2} = K(T) , \tag{4.2.9}$$

where n_1, n_2, n_3 are respectively the densities of the hydrogen *molecules*, oxygen *molecules* and water molecules. Equation (4.2.9) is a single equation with three unknowns. But the total amounts of hydrogen and oxygen in the system are known. If we denote the number of hydrogen *atoms* per unit volume by n_H and the number of oxygen *atoms* per unit volume by n_O, then the two additional equations will be

$$n_H = 2n_3 + 2n_1 , \tag{4.2.10a}$$

$$n_O = n_3 + 2n_2 , \tag{4.2.10b}$$

and now we have three equations with three unknowns.

Exercise 2.4

(a) Obtain the law of mass action for the first reaction in Exercise 2.2.
(b) Show that it is possible to obtain enough equations in order to find all the constituents of this reaction at equilibrium.
(c) Calculate the equilibrium constant, assuming that the internal partition functions are known.

Solution on page 420

The functions ζ_i may be determined from experiment. There exist precise measurements of the emission spectrum and absorption spectrum of various molecules. The spectrum reflects the energy differences between the different levels of the molecule which are needed for calculating the internal partition function. Actually, for the calculation only the low levels of the molecule are required, i.e. levels whose energies are not large with respect to kT. In other words, spectroscopic measurements make it possible, with the help of (4.2.8) [and (4.1.27b)], to predict the equilibrium constant of a chemical reaction!

Before proceeding, we note that in calculating the functions ζ_i for the different types of molecules participating in the reaction it is not possible to ignore the minimal value of the internal energy, as we have done in obtaining Eq. (4.1.38). The reason is that the formation or dissociation of molecules in the reaction involves energy changes. For instance, the dissociation of water into hydrogen and oxygen requires energy. Therefore a water molecule must have a lower internal energy than free hydrogen and oxygen molecules. This fact is accounted for by choosing the ground state energy in a consistent manner for all the molecules. A simple example is presented in the next section.

In order to comprehend the meaning of the law of mass action, and especially the meaning of the equilibrium constant, we shall now show that chemical equilibrium is actually determined by the free energy change in the reaction.

The starting point is Eq. (4.2.5) with the chemical potential of the molecule of type i written in the general form

$$\mu_i = \frac{\partial F}{\partial N_i} = -\frac{\partial}{\partial N_i} kT \ln Z \,. \tag{4.2.11}$$

Recall that the partition function is the product of the partition functions of all the types of molecules [Eq. (3.5.4)]:

$$Z = \prod_{i=1}^{M} \frac{(z_i)^{N_i}}{N_i!} \,. \tag{4.2.12}$$

Differentiating with respect to N_i, and using Stirling's approximation $\ln n! \simeq n \ln n - n$, we obtain the following useful expression for the chemical potential:

$$\mu_i = -kT \frac{\partial}{\partial N_i}[N_i \ln z_i - \ln(N_i!)] = -kT \ln\left(\frac{z_i}{N_i}\right) . \tag{4.2.13}$$

Substituting this expression into (4.2.5) we obtain

$$kT \sum_{i=1}^{M} \nu_i \ln\left(\frac{z_i}{N_i}\right) = 0 \tag{4.2.14}$$

or

$$-kT \sum_{i=1}^{M} \nu_i \ln N_i = -kT \sum_{i=1}^{M} \nu_i \ln z_i . \tag{4.2.15}$$

The expression $-kT \ln z_i$ may be thought of as the free energy of a single molecule of type i [cf. (3.5.10)]. Hence the right hand side of (4.2.15) is nothing but the sum of all the free energies involved in a given reaction multiplied by the number of molecules of each type. It describes, therefore, the free energy change in the reaction, and hence we shall denote it by ΔF_0. Thus we obtain

$$\sum_i \ln(N_i^{\nu_i}) = -\frac{\Delta F_0}{kT} . \tag{4.2.16}$$

Recalling that $N_i = V n_i$ and comparing with Eq. (4.2.7) we arrive at

$$N_1^{\nu_1} \cdots N_M^{\nu_M} = \exp\left(-\frac{\Delta F_0}{kT}\right) = K(T) \cdot V^{\nu_1 + \ldots + \nu_M} . \tag{4.2.17}$$

Equation (4.2.17) clarifies the connection between the equilibrium constant and the free energy change in the reaction, ΔF_0. This quantity determines the "degree of expedience" for the reactants to turn into products. Negative ΔF_0 describes a decrease in the free energy due to the reaction, and thus K will be large and at equilibrium the density of the products will be high. Conversely, positive ΔF_0 describes a reaction which leads to an increase in the free energy, and is thus disadvantageous energetically. In this case the product density is low.

Exercise 2.5

What would the law of mass action look like if we had not introduced the correction implied by Gibbs' paradox?

Solution on page 421

Another point worth considering is the system's response to temperature changes, as determined by the temperature dependence of the equilibrium constant. To this end it is convenient to express K in terms of the single molecule partition functions of all the types of participating molecules. In order to do this we return to (4.2.17).

Expressing N_i in terms of the densities and the volume we obtain (4.2.7) again, where K or $\ln K$ is given in terms of the single molecule partition functions:

$$\ln K = \sum_{i=1}^{M} \nu_i \ln\left(\frac{z_i}{V}\right) . \tag{4.2.18}$$

■ Note that in spite of the explicit appearance of the volume on the right hand side, K remains volume-independent, since z_i is proportional to the volume.

We obtain an interesting result for the temperature dependence of K. Differentiating (4.2.18) we obtain

$$\frac{d\ln K}{dT} = \sum_{i=1}^{M} \nu_i \frac{\partial \ln z_i}{\partial T} . \tag{4.2.19}$$

$\partial \ln z_i/\partial T$ is the average energy per molecule of type B_i apart from a factor kT^2 [see (3.1.3)], and hence

$$\frac{d\ln K}{dT} = \frac{\Delta E_0}{kT^2} . \tag{4.2.20}$$

ΔE_0 is the energy increase when the reaction is performed in a given direction. If ΔE_0 is positive, the reaction absorbs heat (the energy of the products is higher than the energy of the reactants). In this case K increases as the temperature rises, since the right hand side is positive, and the equilibrium will tend in the direction of the products. Hence, as the temperature is increased, the reaction proceeds in the direction in which heat is absorbed and the temperature change is canceled. This is an example of Le Chatelier's principle.

Exercise 2.6

What happens if $\Delta E_0 < 0$?

Solution on page 421

Le Chatelier's principle

Le Chatelier's principle may be formulated in a more general manner in the following form: If the external conditions change, the equilibrium state of a chemical reaction will change in a manner that decreases the external change.

2.3 Dissociation in a diatomic gas

To end this chapter, and as a "concluding exercise" for our discussion of diatomic gases, we apply the law of mass action to the calculation of the degree of dissociation of a diatomic gas, namely to the calculation of the relative part of the dissociated molecules in the gas. We suppose that the gas molecules are composed of identical atoms (like Cl_2), and the gas is at a temperature in the range $\Theta_r \ll T \ll \Theta_e$.

degree of dissociation

If we denote the atoms of the gas by A, the reaction of formation and dissociation of the molecules will be given by

$$2A \rightleftharpoons A_2 .$$

Hence at equilibrium the gas consists of free atoms at density n_1 and molecules at density n_2, so that

$$b_1 = 2,\ b_2 = 1 ,$$

$$\nu_1 = -2, \nu_2 = 1 ,$$

and by Eq. (4.2.7)

$$\frac{n_2}{n_1^2} = K(T) . \tag{4.2.21}$$

There is an additional equation, relating the total density of atoms, n_A, to the density of molecules and of free atoms. It reads

$$n_A = n_1 + 2n_2 . \tag{4.2.22}$$

Substituting (4.2.22) into (4.2.21) we obtain a quadratic equation for n_1:

$$2Kn_1^2 + n_1 - n_A = 0 , \tag{4.2.23}$$

whose positive solution is

$$n_1 = \frac{\sqrt{1+8Kn_A} - 1}{4K} . \tag{4.2.24}$$

The density of disintegrated molecules is of course $\frac{1}{2}n_1$, as each molecule is composed of two atoms. The density of molecules, were they all to remain bound, would be, for the same reason, $\frac{1}{2}n_A$. Hence the degree of dissociation will be

$$\alpha = \frac{\frac{1}{2}n_1}{\frac{1}{2}n_A} = \frac{\sqrt{1+8Kn_A} - 1}{4Kn_A} . \tag{4.2.25}$$

Note that when K is very small, $\alpha \to 1$, hence $n_1 \to n_A$, which means that most of the molecules dissociate. When K is very large, $\alpha \to 0$, hence most of the gas is diatomic.

The temperature dependence of K is obtained from Eq. (4.2.8):

$$K = \frac{h^3}{(2\pi kT)^{3/2}} \frac{M_2^{3/2}\zeta_2}{M_1^3\zeta_1^2} . \tag{4.2.26}$$

We substitute the internal partition functions of the atoms and of the molecules. Because the temperature is much lower than the electronic excitation energy, we may write $\zeta_1 = 1$. Regarding ζ_2, we first write it as a product of a rotational part and a vibrational part:

$$\zeta_2 = \zeta_r \zeta_v . \tag{4.2.27}$$

Since $T \gg \Theta_r$, we can use the expression (4.1.75):

$$\zeta_r = \frac{2T}{\Theta_r} = \frac{2IkT}{\hbar^2} , \tag{4.2.28}$$

where I is the moment of inertia of the molecule.

The vibrational part requires some thought: The initial tendency is to use (4.1.71) as it stands. But we must recall that ignoring the dependence on the ground state energy was legitimate only when the gas was composed entirely of molecules. When there are also single atoms, we must fix a common ground state for the energies of the atoms as well as the energies of the molecules. The fact that we have already written $\zeta_1 = 1$ is equivalent to having chosen the potential energy of a free atom to be zero. Hence we must add to the energies of the harmonic oscillators the minimal potential energy $U(\rho_0)$ [Eq. (4.1.19)], which is negative, as well as the quantum ground state energy $\frac{1}{2}\epsilon_v$ [Eq. (4.1.56)]. Designating their sum as the dissociation energy,

$$-\epsilon_D = U(\rho_0) + \frac{1}{2}\epsilon_v , \tag{4.2.29}$$

we can write the vibrational partition function in the form

$$\zeta_v = \sum_{n=0}^{\infty} e^{-\beta(-\epsilon_D + \epsilon_v n)} = e^{\beta\epsilon_D} \sum_{n=0}^{\infty} e^{-\beta\epsilon_v n} , \tag{4.2.30}$$

which is, of course, the partition function (4.1.68a) multiplied by a factor which translates the zero point of the energy. The result of the summation is given by (4.1.71), so that

$$\zeta_v = \frac{e^{\epsilon_D/kT}}{1 - e^{-\Theta_v/T}} \tag{4.2.31}$$

and

$$\zeta_2 = \frac{2IkT}{\hbar^2} \frac{e^{\epsilon_D/kT}}{1 - e^{-\hbar\omega/kT}}, \tag{4.2.32a}$$

where we have written $\hbar\omega$ instead of $k\Theta_v$ [Eqs. (4.1.57) and (4.1.58)].

Yet there is another factor which we have ignored so far: It is the fact that the molecules A_2 are composed of two *identical* atoms. As a result we summed over too many states in the partition function, and therefore we must correct Eq. (4.2.32a) in the spirit of Gibbs' correction, namely divide by 2. Thus the correct expression is

$$\zeta_2 = \frac{IkT}{\hbar^2} \frac{e^{\epsilon_D/kT}}{1 - e^{-\hbar\omega/kT}}. \tag{4.2.32b}$$

■ Actually one must already divide by 2 the classical internal partition function — Eq. (4.1.49) for such molecules.

Substituting everything into Eq. (4.2.26), and noting that $M_2 = 2M_1$, we obtain

$$K(T) = 4hI\sqrt{\frac{\pi}{M_1^3 kT}} \frac{e^{\epsilon_D/kT}}{1 - e^{-\hbar\omega/kT}}. \tag{4.2.33}$$

Finally, we check the behavior of the equilibrium constant in two limits. At temperatures much below Θ_v ($kT \ll \hbar\omega$) the vibrational degrees of freedom freeze and the denominator in (4.2.33) is very close to unity. Hence

$$K(T) \simeq 4hI\sqrt{\frac{\pi}{M_1^3 kT}} e^{\epsilon_D/kT}, \quad T \ll \Theta_v. \tag{4.2.34a}$$

In this range $K(T)$ decreases with T and, consequently, the degree of dissociation increases. At high temperatures ($kT \gg \hbar\omega$) the vibrational degrees of freedom unfreeze, as can be checked by the fact that (4.2.32b) yields (4.1.49), divided by 2 according to Gibbs. Since K is a directly measurable quantity, we expect that at these temperatures for which the classical approximation applies, Planck's constant will disappear from the equations, and indeed we obtain

$$K \simeq \frac{8\pi I}{\omega}\sqrt{\frac{\pi kT}{M_1^3}} e^{\epsilon_D/kT}, \qquad T \gg \Theta_v. \tag{4.2.34b}$$

This equation is valid for temperatures that are not too high, since T must be much below Θ_e and must be low enough to justify the harmonic approximation.

Chapter 3

Phonon Gas and the Debye Model

3.1 Sound waves in a crystal

In this chapter and in the next one we deal with thermodynamic properties of waves, and our final goal will be to reach Planck's distribution for black body radiation in the next chapter. But before discussing the thermodynamics of electromagnetic radiation, we return to a system we have already met in the previous part, namely the crystal of N atoms. Describing the crystal as a collection of $3N$ independent harmonic oscillators gives us a qualitative understanding of its specific heat, provided that we take into account the laws of quantum mechanics. However, we have seen that the quantitative correspondence is not satisfactory (see Fig. 3.2.4). Although the crucial step in the right direction was made by Einstein, there is still room for improvement.

We first analyze Einstein's assumption that all the harmonic oscillators have identical frequencies. The energy of the crystal according to Einstein is

$$E = \frac{1}{2}\sum_{\alpha=1}^{N}(m\dot{\mathbf{r}}_\alpha^2 + K\mathbf{r}_\alpha^2) = \sum_{\alpha}^{N}\left(\frac{\mathbf{p}_\alpha^2}{2m} + \frac{K}{2}\mathbf{r}_\alpha^2\right), \tag{4.3.1}$$

where $\mathbf{r}_\alpha$ describes the motion of atom number α relative to its equilibrium position. This expression does not take into account the fact that the potential energy of an atom in the crystal depends on the distance from its neighbors or, more precisely, on its motion relative to its neighbors. Hence in order to correct (4.3.1), we replace $\mathbf{r}_\alpha$ by the separation between atom number α and atom number β, $\mathbf{r}_\alpha - \mathbf{r}_\beta$. We thus obtain

$$E = \sum_{\alpha=1}^{N}\frac{\mathbf{p}_\alpha^2}{2m} + \frac{1}{2}\sum_{\alpha,\beta}\frac{K}{2}|\mathbf{r}_\alpha - \mathbf{r}_\beta|^2 . \tag{4.3.2}$$

Note that we could have obtained the above expression for the energy from the harmonic approximation to the energy of N atoms whose potential energy depends on the relative distance between each pair of atoms:

$$E = \sum_{\alpha=1}^{N} \frac{\mathbf{p}_\alpha^2}{2m} + \frac{1}{2}\sum_{\alpha,\beta} U(\mathbf{r}_\alpha - \mathbf{r}_\beta)\,. \tag{4.3.3}$$

Since the potential energy of a pair of atoms usually depends only on the absolute value of their relative distance from each other, we can expand U as in (4.1.19) to second order in $|\mathbf{r}_\alpha - \mathbf{r}_\beta|$ and reach (4.3.2) after dropping the minimum energy, which anyway is a shift by a constant.

▪ Compare with Eq. (4.1.1). Actually, you may think of the crystal as a sort of macroscopic molecule. The factor $\frac{1}{2}$ in front of U also appears for the same reason as in (4.1.1).

The expression (4.3.2) for the energy, unlike (4.3.1), is not a sum of single-particle energies. Thus, the calculation of the partition function may look rather difficult. But a system of N *coupled* three-dimensional oscillators is equivalent to a system of $3N$ *independent* one-dimensional oscillators. The price to be paid is that the free oscillators that are obtained are not of equal frequency; instead each oscillator has a frequency of its own. An additional complication is the fact that the independent oscillators are not related to the motions of single atoms, but to the collective motions of all the atoms of the crystal about their equilibrium positions — *vibrational modes.* The vibrational modes are just sound waves in the crystal. Each vibrational mode of frequency ω in the crystal behaves exactly as a single free harmonic oscillator of that frequency.

vibrational modes

▪ Consult a textbook on classical mechanics for a fuller account.

In order to see this consider the simple case in which the atoms in the crystal vibrate along a single direction, say, x. The equation of motion of atom α is

$$m\ddot{x}_\alpha = K(x_{\alpha+1} - x_\alpha) - K(x_\alpha - x_{\alpha-1}) = K(x_{\alpha+1} - 2x_\alpha + x_{\alpha-1})\,. \tag{4.3.4}$$

Note that this model neglects all the forces exerted upon this atom by atoms that are not its nearest neighbors.

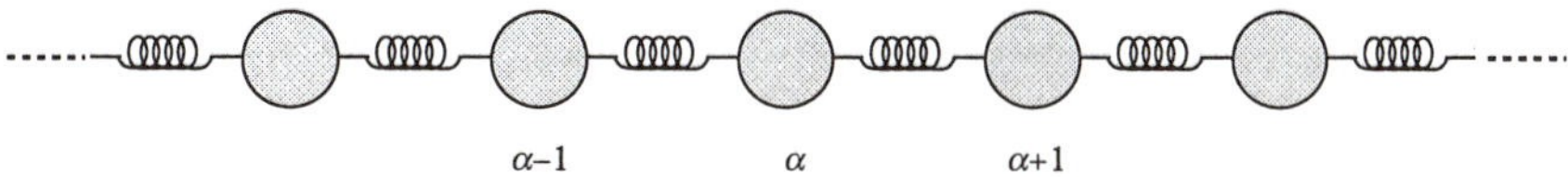

Fig. 4.3.1 One-dimensional crystal.

We stress once more that x_α is the deviation of atom α from its equilibrium position and not the distance from a fixed point that was chosen to be the origin, since if the atoms are at their equilibrium positions, they do not exert forces on each other. Here we shall only show that Eq. (4.3.4) has solutions which describe standing waves of frequencies between zero and $2\sqrt{K/m}$.

The solutions are

$$x_\alpha = a\sin(qb\alpha)\sin\omega t\,, \tag{4.3.5a}$$

where a is the amplitude, b the equilibrium distance between neighboring atoms, q is the wave number ($q = 2\pi/\lambda$) and ω is the frequency. When (4.3.5a) is substituted in (4.3.4), one obtains a relation between ω and $\mathbf{q}$:

$$\omega(q) = 2\sqrt{\frac{K}{M}}\left|\sin\left(\frac{qb}{2}\right)\right|, \quad 0 \le q \le \frac{\pi}{b}\,. \tag{4.3.5b}$$

Exercise 3.1

Show that Eqs. (4.3.5) are indeed solutions to (4.3.4).

Solution on page 422

Since the solutions are standing waves with definite frequencies, we may treat each wave (or vibrational mode) as an independent oscillator characterized by the wave number q. If the number of atoms in the chain is N, the length of the chain is $L = Nb$. The requirement that the waves be standing implies that the vibrations vanish at the two boundaries ($\alpha = 0$ and $\alpha = N$). This implies that the allowed values of q satisfy $qbN = qL = n\pi$, where n is an integer between 1 and N. Thus, q changes in steps of π/L up to the maximal value of π/b and hence takes N discrete values. Similarly, the original three-dimensional crystal will be equivalent to $3N$ independent harmonic oscillators that are characterized by a wave vector, $\mathbf{q}$. Each $\mathbf{q}$ determines the frequency, which in the simple case of a cubic crystal of identical atoms is still given by an equation similar to (4.3.5b). The total energy of the crystal (4.3.2) will thus be given by a sum of $3N$ terms, each describing the energy of a single harmonic oscillator.

The following exercise is intended to give an indication of the type of procedure that converts (4.3.2) into a sum over independent oscillators. It is a coordinate transformation into the normal modes of the system.

Exercise 3.2

The energy of the "toy crystal," in Fig. 4.3.2, is given in terms of the distances of the three "atoms," x_1, x_2, x_3, from their equilibrium positions, as

$$E = \frac{m}{2}(\dot{x}_1^2 + \dot{x}_2^2 + \dot{x}_3^2) + \frac{K}{2}[x_1^2 + (x_2 - x_1)^2 + (x_3 - x_2)^2 + x_3^2] \,.$$

Show that the change of variables

$$x_1 = \frac{1}{2}(u_1 + \sqrt{2}u_2 - u_3) \,,$$
$$x_2 = \frac{1}{\sqrt{2}}(u_1 + u_3) \,,$$
$$x_3 = \frac{1}{2}(u_1 - \sqrt{2}u_2 - u_3)$$

transforms the expression for the energy of the triatomic crystal into a sum of energies of three free oscillators. Find the frequencies of the oscillators.

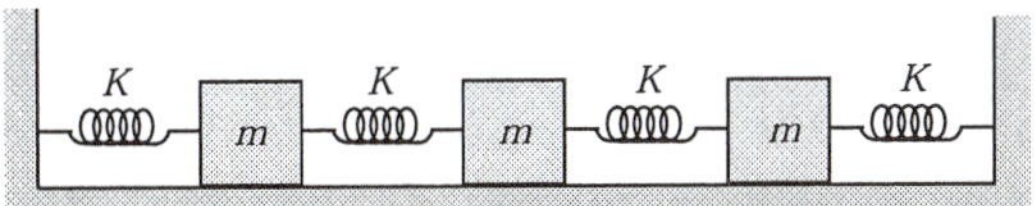

Fig. 4.3.2 A "toy crystal" of three "atoms."

Solution on page 423

3.2 Vibrational modes, phonons and enumeration of states

Since the vibrations of the crystal have to be treated quantum-mechanically, we write down the quantum expression for the energy of the $3N$ oscillators in a given microscopic state:

$$E(n_1, n_2, \ldots, n_{3N}) = \sum_{\alpha=1}^{3N} \hbar\omega_\alpha n_\alpha \,, \tag{4.3.6}$$

where we have dropped the zero point energy of the oscillators. It should be mentioned that we are now using the letter α to denote not the atoms of the crystal but the different standing waves generated in it, namely the oscillators. It is common practice to think of the excited states of these oscillators as particles, and then instead of saying that oscillator number α has a degree of excitation n_α, we say that there are n_α *phonons* of type α (or with frequency ω_α). Figure 4.3.3, for example, describes **phonon**

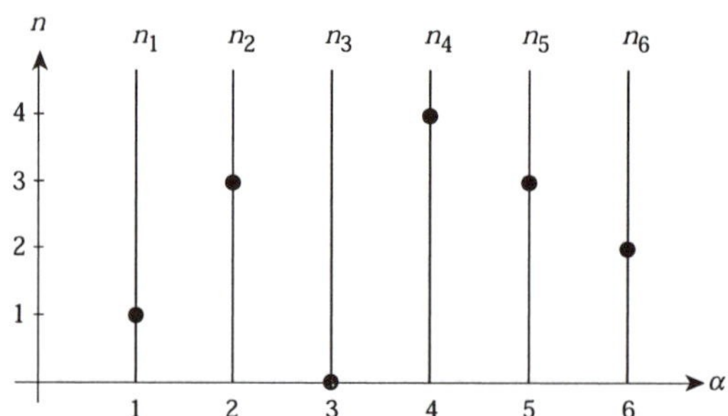

Fig. 4.3.3 Distribution of phonon numbers.

a system in which there is one phonon of type $\alpha = 1$, three phonons of type $\alpha = 2$, no phonon of type $\alpha = 3$, four phonons of type $\alpha = 4$, and so on.

From Eq. (4.3.6) it is clear that increasing the excitation number n_α by unity, or the addition of a phonon of type α, increases the energy of the system by $\hbar\omega_\alpha$. Hence it may be said that a phonon of type α has energy $\hbar\omega_\alpha$.

A given microscopic state of the crystal is thus specified by listing all the excitation numbers of all the oscillators, or listing the phonon numbers of all types that are in the crystal. In thermodynamic equilibrium it is thus possible to say that the vibrations of the crystal are, in fact, an ideal gas of phonons at temperature T. The partition function of such a system will be a product of single oscillator partition functions, but the factors in the product, z_α, will not be equal, but will change with α. The first stage in the calculation of the partition function is the calculation of z_α for a given α, by summing over all the excitation numbers n_α. Then the $3N$ different partition functions have to be multiplied together in order to obtain the partition function of the whole system. Actually it is more convenient in this case to directly calculate the *free energy*, which is a *sum* rather than a product.

We have met precisely such a calculation in Self-Assessment Exercise 2 of Part III, where we found that after summing over all excitation numbers the free energy is given by a sum over all $3N$ different oscillators:

$$F = kT \sum_{\alpha=1}^{3N} \ln(1 - e^{-\beta\hbar\omega_\alpha}), \tag{4.3.7}$$

leaving out the ground state energy.

But here there is an additional complication: While the energy is a sum of independent terms for each oscillator and for each of its components, the enumeration of the number of states must take into account the nature of the vibrational normal modes. Each three-dimensional oscillator is characterized by a wave vector $\mathbf{q}$. If the crystal is a cube of side L, the allowed values of the components of $\mathbf{q}$ are

$$q_x = \frac{\pi}{L} n_x; \quad q_y = \frac{\pi}{L} n_y, \quad q_z = \frac{\pi}{L} n_z\,, \tag{4.3.8}$$

with n_x, n_y, n_z nonnegative integers. For every vector $\mathbf{q}$, there are three possible independent oscillating modes: One in which the oscillation amplitude is longitudinal (in the direction of $\mathbf{q}$) and two in which it is transverse (perpendicular to $\mathbf{q}$). Hence, to sum over all states we must sum over all allowed wave vectors and for each wave vector over three polarizations. Yet the total number of terms must equal the total number of degrees of freedom, $3N$.

The summation over α thus transforms into a summation over these three variables:

$$F = 3kT \sum_{n_x, n_y, n_z} \ln(1 - e^{-\beta\hbar\omega(n_x, n_y, n_z)})\,, \tag{4.3.9}$$

where the factor 3 comes from the summation over the three possible polarizations, assuming that the frequencies of the standing waves, ω, do not depend on polarization.

Successive terms in the sum correspond to q's which differ by π/L. For L macroscopically large and temperatures that are not extremely low, the difference between successive terms in the sum will be infinitely small and (as in Part III, Sec. 4.4) the sum can be approximated by an integral over the region D of the positive values of the wave vectors' components:

$$F = \frac{3kTV}{\pi^3} \int_D \ln(1 - e^{-\beta\hbar\omega(\mathbf{q})}) d^3q\,. \tag{4.3.10a}$$

We have written V for L^3. The region of integration in wave vector space is depicted in Fig. 4.3.4. The points of the three-dimensional lattice represent the allowed values of $\mathbf{q}$. If the frequency of vibration ω is independent of the direction of the wave vector $\mathbf{q}$, we can replace the integration over positive q_x, q_y, q_z (1/8 of the sphere) in Eq. (4.3.10a) by an integration over the entire sphere of radius q_D, provided we correct for this by multiplying by 1/8. Hence,

$$F = \frac{3kTV}{(2\pi)^3} \int_{|\mathbf{q}| \le q_D} \ln(1 - e^{-\beta\hbar\omega(\mathbf{q})}) d^3q\,. \tag{4.3.10b}$$

This substitution represents the fact that a standing wave along the x direction contains, in fact, two waves that are moving in opposite directions with wave vectors $+q_x$ and $-q_x$ (see the remark at the end of Answer 3.1). The same applies to standing waves moving along the y and z directions.

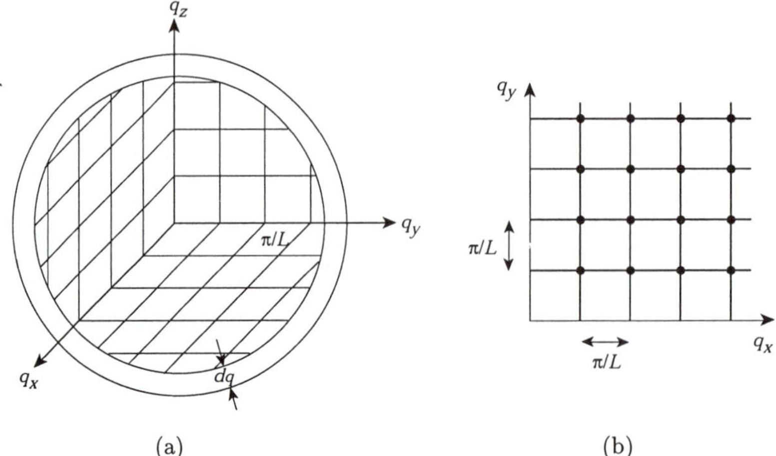

Fig. 4.3.4 The wave vectors of the vibrational modes are represented by points in wave vector space. (a) Three-dimensional representation; (b) two-dimensional section. Each of the points represents three free waves moving with all possible polarizations.

Altogether a standing wave with (a positive) wave vector $\mathbf{q}$ comprises eight traveling waves with wave vectors $(\pm q_x, \pm q_y, \pm q_z)$.

Note that the replacement of the sum over all oscillators $\sum_\alpha$, by an integral over the components of $\mathbf{q}$, which appears in (4.3.10b), implies that inside a volume d^3q in wave vector space there are $3Vd^3q/(2\pi)^3$ crystal vibrational modes. We could have arrived at this result from a slightly different direction. So far we have calculated the number of phonon states by treating the phonons as waves. But in the quantum context it is possible to consider them as particles with de Broglie momentum given by $\hbar\mathbf{q}$ (and energy $\hbar\omega$). We can now think of these phonons as particles confined to a box of volume V (the volume of the crystal). We have found (in Sec. 4.4 of Part III) that each state of a particle occupies a volume of h^3 in phase space. In addition there are three independent vibrational directions. Therefore, inside a volume dVd^3p of phase space there are $3dVd^3p/(h)^3$ states and in terms of wave vectors $3dVd^3q/(2\pi)^3$ states. Integration over the whole volume of the crystal will give us $3Vd^3q/(2\pi)^3$ states to a volume of d^3q in wave vector space.

It is highly recommended at this stage to return to Sec. 4.4 of Part III, and to compare the considerations made in the two instances.

3.3 The Debye model

In order to calculate the free energy of the crystal (4.3.10), we have to know the form of the function $\omega(q)$. Actually we have calculated it explicitly in the one-dimensional situation in Eq. (4.3.5b), but that case is too specific. The general case is three-dimensional, and the crystal structure is not always cubic and the crystal does not always consist of one type of atoms.

All these factors indicate that in the typical general case $\omega(q)$ may be a fairly complicated function. Nevertheless, Debye found that it is possible to describe the specific heat of many solids quite well, if we assume a usual free wave relationship between ω and q: $\omega = vq$, where v is the sound velocity in the crystal, provided we limit the region of integration to contain exactly $3N$ states.

The integrand in (4.3.10b) depends only on the absolute value of the wave vector, and the region of integration is a sphere of radius q_D. Thus it is possible to perform the integration as a summation over spherical shells of thickness dq (see Fig. 4.3.4a), to obtain from (4.3.10b)

$$F = \frac{3kTV}{2\pi^2} \int_0^{q_D} \ln(1 - e^{-\beta\hbar vq}) q^2 dq \,, \tag{4.3.11}$$

following the angular integration, as a result of which $d^3q = 4\pi q^2 dq$. To find the integration limit q_D, we use the fact that the number of states in a region of phase space confined by q_D should be $3N$. The number of states in an infinitesimal volume d^3q is, as we have seen, $3Vd^3q/(2\pi)^3$, and altogether we want $3N$ states inside a sphere of radius q_D. This leads to

$$q_D = \left(\frac{6\pi^2 N}{V}\right)^{1/3} \tag{4.3.12a}$$

or for the corresponding frequency, called the Debye frequency:

Debye frequency

$$\omega_D = v\left(\frac{6\pi^2 N}{V}\right)^{1/3} \,. \tag{4.3.12b}$$

Exercise 3.3

Prove Eq. (4.3.12a).

Solution on page 424

The practical aim of this whole discussion is to obtain an improved calculation of the average energy and of the specific heat of a crystal, especially in the low temperature range, where the Einstein model deviates very significantly from the experimental results (see Fig. 3.2.4).

An expression for the average energy is readily obtained from (4.3.11):

$$E = -\frac{\partial \ln Z}{\partial \beta} = \frac{3V}{2\pi^2 v^3} \int_0^{\omega_D} \frac{\hbar\omega}{e^{\beta\hbar\omega} - 1} \omega^2 d\omega \,. \tag{4.3.13}$$

Exercise 3.4

Obtain Eq. (4.3.13).

Solution on page 424

Equation (4.3.13) may be interpreted in the following manner: each phonon of frequency ω has energy $\hbar\omega$. The average number of phonons of frequency ω is the average degree of excitation found in Eq. (3.2.9): $\langle n \rangle = 1/(e^{\beta\hbar\omega} - 1)$. It is possible to identify the additional multiplicative factor as the number of microscopic states per unit frequency of a single phonon (or the density of states):

$$g(\omega) = \frac{3V}{2\pi^2 v^3}\omega^2 \,, \tag{4.3.14}$$

and then the energy density per unit frequency is the product of all these factors. We may also define an energy density per unit frequency *per unit volume*, which is

$$\rho(\omega) = \frac{3}{2\pi^2 v^3}\frac{\hbar\omega^3}{e^{\beta\hbar\omega} - 1}, \qquad \omega \le \omega_D \,. \tag{4.3.15}$$

The specific heat (at constant volume) of the crystal is obtained from Eq. (4.3.13) as

$$C = \frac{3kV}{2\pi^2 v^3}\int_0^{\omega_D} \frac{(\hbar\omega/kT)^2 e^{\hbar\omega/kT}}{(e^{\hbar\omega/kT} - 1)^2}\omega^2 d\omega \,. \tag{4.3.16}$$

This integral cannot be expressed in terms of elementary functions, and generic values must be obtained numerically. Nevertheless, at low temperatures, for which $kT \ll \hbar\omega_D$, it is possible to extend the region of integration up to infinity without changing the result of the integration, since near ω_D the integrand is vanishingly small. The physical reason for this is that at low temperatures only low frequency phonons are excited, and hence only they contribute significantly to the specific heat. We thus find that at low temperatures

$$C = Nk\frac{12\pi^4}{5}\left(\frac{kT}{\hbar\omega_D}\right)^3 \,. \tag{4.3.17}$$

Note that the exponential temperature dependence of the Einstein model has been replaced by a power dependence. Defining the Debye temperature, $\Theta_D = \hbar\omega_D/k$, we may write

$$C = Nk\frac{12\pi^4}{5}\left(\frac{T}{\Theta_D}\right)^3, \qquad T \ll \Theta_D \,. \tag{4.3.18}$$

Exercise 3.5

Prove Eq. (4.3.17).

Solution on page 425

Chapter 4

Thermodynamics of Electromagnetic Radiation

4.1 General considerations of radiation at thermal equilibrium

Inside an empty cavity of matter (Fig. 4.4.1) there usually exists electromagnetic radiation since in the material of the walls, at temperature T, there are moving charges and these radiate. Moreover, radiation in the cavity is absorbed in the walls and re-emitted, until the system of matter + radiation reaches equilibrium, characteristic of the given temperature.

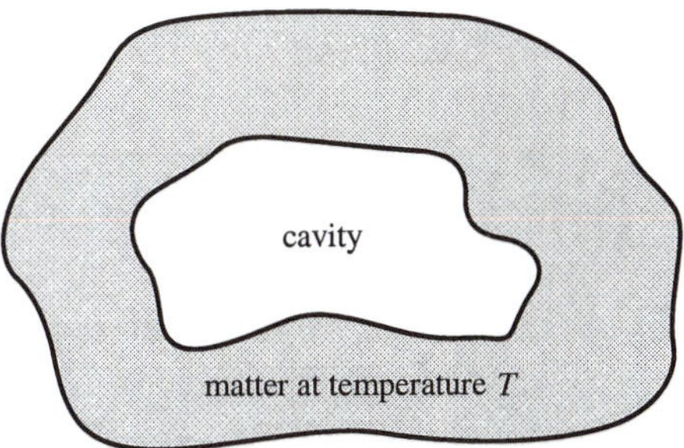

Fig. 4.4.1 Radiation inside a cavity in matter — the average energy does not depend on either the shape of the cavity or the type of matter surrounding it, but depends only on the temperature.

The radiation in the cavity is characterized by different frequencies, or different wavelengths. The amount of radiation at each frequency may be specified by the energy density around this frequency. Namely, there exists a function $\rho(\omega)$ such that $\rho(\omega)d\omega$ is the amount of energy of radiation per unit volume with frequency between ω and $d\omega$ at temperature T.

■ In what follows we shall refer to ρ also as the radiation density, even though it is actually the energy density of the radiation.

The goal of statistical mechanics is to calculate $\rho(\omega)$, which is the analog of the Maxwell–Boltzmann distribution for material particles. This calculation would have been impossible were it not for the fact that:

At equilibrium $\rho(\omega)$ is independent of the structure of the cavity or of the surrounding material.

In order to clarify this far-reaching property, we note that it is a special case of a more general independence. *A priori* we should have expected the energy density to be a function of the position inside the cavity, of the wave vector $\mathbf{q}$, or of the polarization. It is possible to show that if the radiation in the cavity is at equilibrium at a uniform temperature, then the second law of thermodynamics implies that:

(a) The distribution of the radiation in the cavity is uniform — independent of position.
(b) The distribution of the radiation in the cavity is independent of the direction of the wave vector of the radiation or of its polarization — ρ depends only on $|\mathbf{q}|$.

The frequency of electromagnetic radiation is proportional to $|\mathbf{q}|$:

$$\omega = c|\mathbf{q}| = cq\,, \tag{4.4.1}$$

where c is the speed of light. From (b) above it is inferred that ρ depends only on ω.

As a demonstration we shall bring here the argument that leads to the conclusion that $\rho(\omega)$ is independent of the shape of the cavity, its size, or the material it is made of. We shall discuss the independence of the radiation density of the position and the direction of the radiation in Sec. 4.4.

Suppose that there are two cavities A and B at the same temperature T, which differ from each other in shape and are made of different materials. We shall connect the two cavities by a small hole, such that a small amount of energy may pass from one cavity to the other (see Fig. 4.4.2).

If as a result of the passage of radiation through the hole more radiation has passed from A to B than in the opposite direction, for instance, then the radiation will not be at equilibrium with the walls of the cavity on both sides. After a while the cavities will reach a new equilibrium, and the temperatures of the walls will change. The result will be that heat will pass between two systems at the same temperature, without performing work — in contradiction to the second law of thermodynamics.

The same consideration applies to each and every frequency separately, since it is possible to insert a filter inside the hole, so that only radiation of a certain frequency ω is able to pass (the other frequencies will be

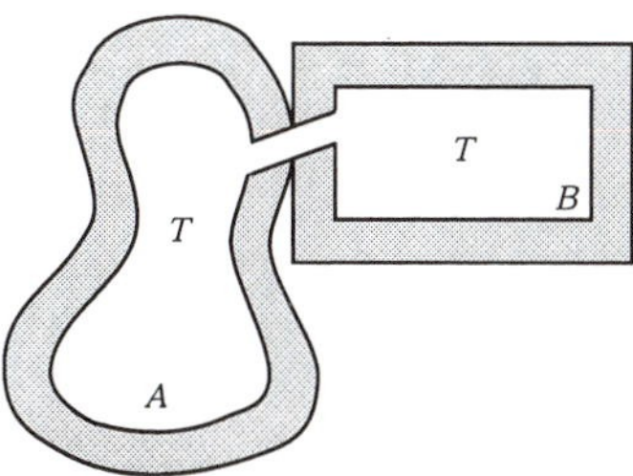

Fig. 4.4.2 Two cavities, differing from each other in shape and in the surrounding material, both at the same temperature.

reflected). Namely, the second law of thermodynamics ensures that if the two cavities are at equal temperatures, then the amount of radiation energy transferred from A to B must be equal to that transferred from B to A, at each and every frequency. But the amount of radiation energy with frequency between ω and $\omega + d\omega$, which is transferred from side to side, can be calculated exactly as for gas particles escaping from a container through a tiny aperture (see Secs. 3.2, 3.7 and Self-Assessment Exercise 4, Part I). This amount is given by

$$I(\omega)d\omega = \frac{c}{4}\rho(\omega)d\omega\,. \tag{4.4.2}$$

$I(\omega)$ is called the emissivity and represents the amount of radiation energy with frequency between ω and $\omega + d\omega$ transferred per unit time through a unit element of the aperture. This implies that the energy density $\rho(\omega)$ of one cavity is equal to that of the other cavity at every frequency. Because we have chosen two arbitrary cavities, the form of the function $\rho(\omega)$ is independent of the shape of the cavity, its size, or the material it is made of. **emissivity**

Exercise 4.1

Prove Eq. (4.4.2).

Solution on page 425

4.2 Radiation density

If indeed $\rho(\omega)$ is independent of the shape of the container and of the material it is made of, we may choose a cubic container of side L, and whose faces are made of an arbitrary material. Whatever the material, the charges within it perform thermal vibrations and emit electromagnetic radiation. The latter is emitted into the cavity between the walls and repeatedly hits them, causing the charges within them to vibrate, and so on and so forth. As a result of these exchanges an equilibrium is established between the radiation and the walls, which, as we have seen, is independent of the structure of the walls or their composition. The electromagnetic radiation in the cavity can be described as a collection of standing waves, with various frequencies, which are characterized by their wave vector $\mathbf{q}$. The frequency is determined by the wave vector according to the usual relationship, Eq. (4.4.1).

Each of these standing waves with a given wave vector is a vibration of the electromagnetic field in the cavity, which is sinusoidal in time. Hence it is possible to think of it as a vibrational mode of the electromagnetic field, namely as a harmonic oscillator with frequency $\omega = cq$, precisely in the same manner as standing sound waves in a crystal are its vibrational modes. The next step is to take into account the quantum theory. The quantization of the vibrational modes of the crystal led us to the concept of phonons, which are the excited states of these oscillators. We may also treat the vibrational modes of the electromagnetic field as quantum oscillators. The result is similar: Each oscillator α can be excited to the state n_α. In particle language this state is described as a state in which

photon

there are n_α *photons* of type α.

Each of these oscillators has energy states which are given by Eq. (3.2.1), and hence each photon carries energy of magnitude $\hbar\omega_\alpha$. However, there exists a basic difference between the phonons and the photons: The frequency of the phonons has an upper bound (ω_D in the Debye model) due to the fact that there exists a minimal wavelength determined by the distance between two neighboring atoms in the lattice or, alternatively, by the fact that the number of vibrational modes of the crystal is finite. In contrast, there is no such limit for the electromagnetic waves; hence the range of frequencies is unbounded and, correspondingly, the number of vibrational modes is infinite.

We therefore treat the electromagnetic radiation in the container as a gas of photons. At equilibrium, at temperature T, the probability of a given microscopic state specified by the collection of n_α for all α will be

$$P(n_1, n_2, \ldots) = \frac{1}{Z} \exp\left(-\beta \sum_\alpha n_\alpha \hbar\omega_\alpha\right), \tag{4.4.3}$$

where Z is the partition function:

$$\begin{aligned} Z &= \sum_{n_1=0}^{\infty} \sum_{n_2=0}^{\infty} \cdots \exp\left(-\beta \sum_\alpha n_\alpha \hbar\omega_\alpha\right) = \prod_\alpha \left(\sum_{n_\alpha} e^{-\beta n_\alpha \hbar\omega_\alpha}\right) \\ &= \prod_\alpha \left(\frac{1}{1 - e^{-\beta\hbar\omega_\alpha}}\right). \end{aligned} \tag{4.4.4}$$

This argument is identical to the one applied to phonons in the previous chapter. Thus, the free energy is

$$F = -kT \ln Z = kT \sum_{\alpha} \ln(1 - e^{-\beta\hbar\omega_\alpha}) . \tag{4.4.5}$$

But what of the summation over α?

The summation over α is a summation over all the vibrational modes of the electromagnetic field or over all possible photon states. The photon, which is the electromagnetic ray, is characterized by the wave vector $\mathbf{q}$. A second quantity characterizing the radiation is the polarization. The electric field, for example, oscillates perpendicular to $\mathbf{q}$, but there are two independent directions perpendicular to $\mathbf{q}$. Thus two different photons correspond to each $\mathbf{q}$. Note the contrast with phonons which had three possible polarizations.

Exercise 4.2

Show that a sinusoidal electromagnetic wave is a transverse wave. Use Maxwell's equations in vacuum.

Solution on page 426

The number of vibrational modes with wave vectors in a region d^3q around $\mathbf{q}$ can be found exactly as for the phonons in the previous chapter. One way is to count the number of different standing waves in a certain region in wave vector space — or, alternatively, to calculate the volume occupied by a single state in wave vector space. Reexamining Fig. 4.3.4, we note that the volume of an elementary cell is $(\pi/L)^3 = \pi^3/V$. Hence the number of standing waves in the region d^3q is Vd^3q/π^3. But the electromagnetic waves can have two polarization states, and hence the number of standing waves in d^3q is $2Vd^3q/\pi^3$. As we have already seen for phonons, it is more convenient to count *moving* wave states, for which the components of $\mathbf{q}$ may also be negative, and hence the volume d^3q contains eight times too many moving waves. Thus, we have to divide by 8 to obtain the final result, that the number of moving wave states in the region d^3q in wave vector space is $2Vd^3q/(2\pi)^3$. This result can also be obtained by a direct inspection of the phase space of the photons, precisely as we have done for the phonons.

The free energy Eq. (4.4.5) is written as an integral over $\mathbf{q}$:

$$F = \frac{2kTV}{(2\pi)^3} \int \ln(1 - e^{-\beta\hbar cq}) d^3q . \tag{4.4.6}$$

Since the "terms of the sum" (the integrand) depend only on $q = |\mathbf{q}|$ and not on the direction or the polarization, we can perform the integration over spherical layers, as illustrated in Fig. 4.3.4(a). Inside the layer the terms have identical values: There is no dependence on the angle, and dq is very small. Hence

$$\int \ln(1 - e^{-\beta\hbar cq})d^3q = 4\pi \int_0^\infty \ln(1 - e^{-\beta\hbar cq})q^2 dq\,. \tag{4.4.7}$$

The right hand side of (4.4.7) can be written, using (4.4.1), as an integral over the frequencies:

$$\frac{4\pi}{c^3}\int_0^\infty \omega^2 \ln(1 - e^{-\beta\hbar\omega})d\omega\,, \tag{4.4.8}$$

and from here we can obtain the free energy as

$$F = \frac{kTV}{\pi^2 c^3}\int_0^\infty \omega^2 \ln(1 - e^{-\beta\hbar\omega})d\omega\,. \tag{4.4.9}$$

For the average energy we find that

$$E = -\frac{\partial \ln Z}{\partial\beta} = \frac{V\hbar}{\pi^2 c^3}\int_0^\infty \frac{\omega^3 d\omega}{e^{\beta\hbar\omega} - 1}\,. \tag{4.4.10}$$

This expression for the average energy may be interpreted precisely in the manner in which we interpreted the corresponding expression in the Debye model, Eq. (4.3.13). The average degree of excitation of the quantum oscillator appears also here and the multiplying factor can be identified as the number of microscopic states per unit frequency (density of states) of a single photon:

$$g(\omega) = \frac{V}{\pi^2 c^3}\omega^2\,. \tag{4.4.11}$$

The energy density per unit volume and unit frequency stored in the electromagnetic radiation in the box will be

$$\rho(\omega) = \frac{1}{\pi^2 c^3}\cdot\frac{\hbar\omega^3}{e^{\beta\hbar\omega} - 1}\,. \tag{4.4.12}$$

■ To be compared to Eq. (4.3.15).

Dividing $\rho(\omega)$ by the energy of a photon of frequency ω, we obtain the photon *number* density per unit volume and unit frequency:

$$n(\omega) = \frac{1}{\pi^2 c^3}\frac{\omega^2}{e^{\beta\hbar\omega} - 1}\,. \tag{4.4.13}$$

4.3 Black body radiation

In order to verify that indeed we have an expression that describes well the radiation density in the container, we consider the radiation emitted through a small aperture in it. The hole has to be small, so that the amount of radiation escaping from it will be negligible, otherwise it will be impossible to maintain the state of thermodynamic equilibrium.

Such an aperture at the side of the container is an example of a *black body*. The reason for this name is that any ray incident from the outside on the hole is absorbed by the container. The probability for it to return and exit the hole is vanishingly small. Nevertheless, a black body is in no way black. In addition to being a perfect absorber, it emits radiation. As we have seen in Sec. 4.1, if the container is full of photons moving to and fro, then clearly some of them will be emitted through the hole, according to Eq. (4.4.2) and (4.4.12). A black body is therefore characterized by the power emitted by it per unit area per unit frequency (emissivity), given by **black body**

$$I(\omega) = \frac{1}{4\pi^2 c^2} \frac{\hbar\omega^3}{e^{\beta\hbar\omega} - 1} . \tag{4.4.14}$$

This equation, as well as Eq. (4.4.12), is called Planck's formula, and its discovery by Planck in 1900 marks the birth of quantum theory. Figure 4.4.3 illustrates the power distribution emitted from a black body. Note that the horizontal axis is an axis of the frequency (ν) and not angular frequency (ω).

■ Note that the units of time have disappeared from I.

The fact that Planck's constant appears in Eq. (4.4.14) immediately explains the failure of all the attempts to explain the phenomenon of black body radiation on the basis of classical physics. Nevertheless, (4.4.12) and (4.4.14) have a classical limit which is obtained when $\beta\hbar\omega \ll 1$.

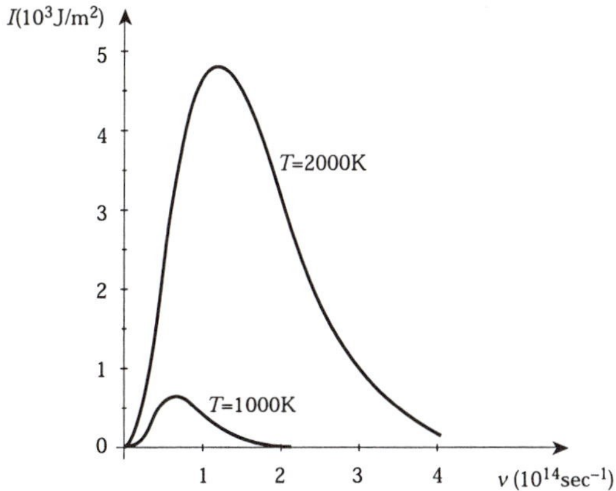

Fig. 4.4.3 The emissivity, or the emission rate of radiation per unit area of the black body, according to Planck's law.

Exercise 4.3

(a) Prove that in the classical limit the energy distribution has the form

$$\rho(\omega) = \frac{kT\omega^2}{\pi^2 c^3}\,.$$

(b) To what physical conditions does this limit correspond?

Solution on page 427

The expression obtained in the last exercise is called the Rayleigh–Jeans law and is as far as classical physics could go. The original classical calculation was not performed as a limit of the quantum result, but in the following manner: Electromagnetic radiation is equivalent to an ensemble of harmonic oscillators, and according to the equipartition law each one contributes an average energy of kT. Hence the energy per unit frequency in the cavity is obtained by multiplying the density of states (4.4.11) by kT. In order to obtain the energy per unit volume we have to divide by V.

The Rayleigh–Jeans formula is yet another indication for the limitations of the equipartition law.

From Eq. (4.4.14) it is possible to obtain two important characteristics of black body radiation.

(a) Wien's law: The emissivity of a black body has a maximum at a frequency $\nu_{\max}$ that increases with temperature according to

$$\frac{\nu_{\max}}{T} = 5.88 \times 10^{10}\ \mathrm{s^{-1}K^{-1}}\,. \qquad (4.4.15)$$

See for example the location of the maxima in Fig. 4.4.3.

(b) The Stefan–Boltzmann law: The total power emitted per unit area of a black body (namely a sum over all frequencies) is proportional to the fourth power of the temperature:

$$I = \sigma T^4, \quad \sigma = \frac{2\pi^5 k^4}{15h^3c^2} = 5.67 \times 10^{-8}\ \mathrm{Js^{-1}m^{-2}K^{-4}}\,. \qquad (4.4.16)$$

Both of these laws were discovered experimentally many years before Planck found the theoretical explanation for the properties of black body radiation.

Exercise 4.4

(a) Using Planck's formula, Eq. (4.4.14), prove (4.4.15).
(b) Prove (4.4.16).

Solution on page 427

Another interesting quantity is the energy density per unit volume $u(T)$ in the black body radiation, which we can readily obtain from the expression for the total energy of the radiation in the container, i.e. Eq. (4.4.10). A procedure analogous to the one used for obtaining the Stefan–Boltzmann law (Exercise 4.4) leads to a direct proportionality to the fourth power of the temperature:

energy density

$$u(T) = \frac{4\sigma}{c} T^4 . \tag{4.4.17}$$

The proportionality constant $4\sigma/c$ is 7.57×10^{-16} $\mathrm{Jm^{-3}K^{-4}}$.

To end this section let us check the thermodynamic properties of the radiation gas confined to a container.

Exercise 4.5

(a) Calculate the entropy and the specific heat of the radiation.

(b) Calculate the radiation's chemical potential. Explain your result in terms of the number of photons.

Solution on page 429

Exercise 4.6

(a) Calculate the pressure of the radiation and its relation to the energy density. Compare with the relationship obtained in the kinetic theory in Part I, Eq. (1.1.7).

(b) The difficulty, or the ease, of compressing a system is measured by the compressibility, which is defined as

compressibility

$$K = -V \left(\frac{\partial P}{\partial V} \right)_T .$$

Calculate K for the radiation gas. Compare your result to the one obtained for a gas of particles.

Solution on page 430

Exercise 4.7

(a) At what temperature will the pressure of the radiation gas be equal to one atmosphere ($1 \text{ atm} = 1.013 \times 10^5 \text{ N/m}^2$)?

(b) At what temperature will the pressure of the radiation be equal to 10^{-10} atm?

(c) The temperature at the sun's center is 10^7 K. What is the pressure of the radiation? Compare it to the pressure of the gas at the center of the sun, which is 10^{11} atm.

(d) Calculate the pressure of the sun's radiation on the earth's surface, given that the power of the radiation reaching earth from the sun is 1400 W/m^2.

Solution on page 431

As a final note we show in Fig. 4.4.4 the astounding correspondence between Planck's formula and the experimental results for the energy density, performed by Coblentz in 1916. Observe that the horizontal axis is the *wavelength* and that ρ is the energy per unit volume per unit wavelength, and hence its dimensions are $[E]/[L]^4$. Regarding the relationship between $\rho(\lambda)$ and $\rho(\nu)$, see Self-Assessment Exercise 6.

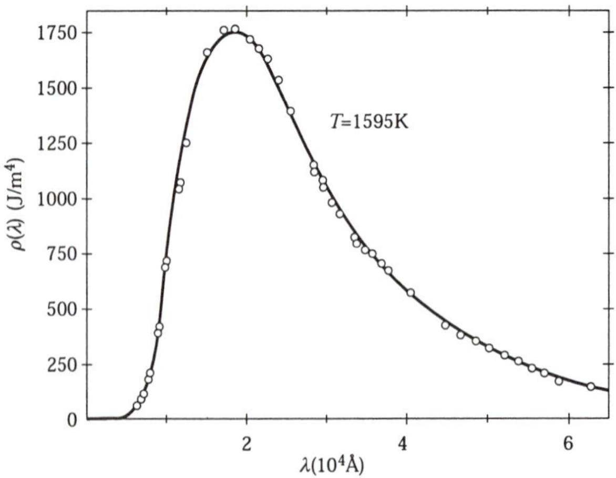

Fig. 4.4.4 Black body radiation — experimental results vs. Planck's formula.

4.4 Absorption and emission of radiation — Kirchhoff's law

The relatively simple considerations that we applied to a volume of space with radiation in equilibrium, at a given temperature, allow us to reach some surprising conclusions concerning the behavior of bodies that absorb and emit radiation at nonequilibrium conditions.

First, using physical considerations alone, without any calculations, it is possible to show that inside a cavity containing radiation at equilibrium at temperature T, the radiation density is independent of position.

■ See also Sec. 4.1, where we have shown that the radiation density is independent of the shape of the cavity or of the surrounding material.

Suppose that at frequency ω, $\rho(\omega)$ is different at two locations inside the cavity. In these locations we shall place two identical bodies whose temperature is T. They are coated by a material which is transparent only to radiation with frequency ω (a filter). The body around which the radiation density is higher will absorb more energy than the one around which the radiation density is lower. As a consequence heat will be transferred between the two bodies whose temperatures are equal, in violation of the second law of thermodynamics.

In a similar manner it is possible to show that the radiation density is independent of the direction of $\mathbf{q}$, or of the polarization of the radiation. The proofs are left to Self-Assessment Exercise 5.

Any body, not necessarily a black body, can be characterized by its emissive power function $L(\mathbf{q}, T)$, expressing the intensity of radiation emitted by the body per unit area per unit time directed along $\mathbf{q}$ with frequency $\omega = c|\mathbf{q}|$:

emissive power

$$\Delta E_{\text{em}} = L(\mathbf{q}, T) \cos\theta \Delta\omega\Delta\Omega\,, \tag{4.4.18}$$

where θ is the angle between $\mathbf{q}$ and the normal to the surface and $d\Omega$ is the solid angle into which the radiation is emitted. The factor $\cos\theta$ projects the unit area of the body in the direction perpendicular to q (Fig. 4.4.5). Note that we assume that L does not vary from point to point over the surface.

L depends only on the properties of the radiating body and on its temperature. This does not mean that the radiation must have the same temperature as the body. In fact this is a case in which the radiation is not at equilibrium with the body but is emitted by it in a continuous manner as, for instance, the radiation emitted by the filament of a light bulb.

Similarly, it is possible to characterize a body by its absorbing power $A(\mathbf{q}, T)$, which is also dependent upon the properties of the body and on its temperature. $A(\mathbf{q}, T)$ gives the fraction of the intensity of the radi-

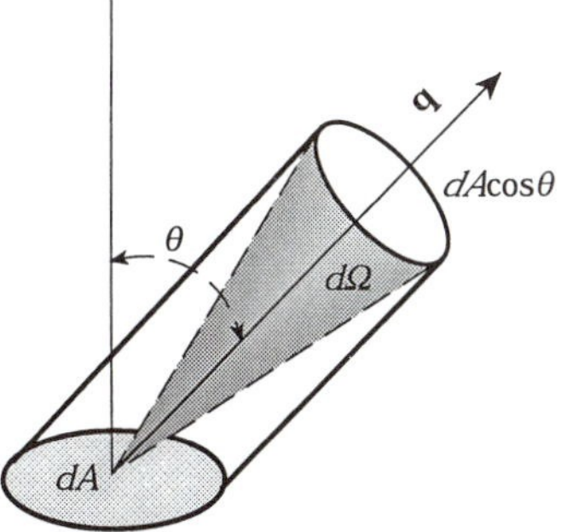

Fig. 4.4.5 Geometry of radiation emission: θ is the angle between the normal to the area element and $\mathbf{q}$. ϕ is measured in the plane of dA. $d\Omega = \sin\theta d\theta d\phi$ is the solid angle corresponding to the infinitesimal elements $d\theta$ and $d\phi$. An area dA is seen to be of magnitude $dA\cos\theta$ if viewed at an angle θ.

ation, with wave vector $\mathbf{q}$ and frequency $\omega(\mathbf{q})$, incident on the body at temperature T, that is absorbed by it. Note that $A(\mathbf{q}, T)$ is a dimensionless quantity. L has the dimensions of energy per unit area.

Kirchhoff's law

Kirchhoff's law, which we now prove, states that the ratio of L and A depends only on ω and T:

$$\frac{L(\mathbf{q}, T)}{A(\mathbf{q}, T)} = \frac{c}{4\pi}\rho(\omega, T), \tag{4.4.19}$$

where $\rho(\omega, T)$ is the radiation density in equilibrium inside a cavity at temperature T, namely (4.4.12). This means that L/A is actually independent of the properties of the body!

■ In order to emphasize the temperature dependence of the radiation density, we write here $\rho(\omega, T)$.

We conclude the treatment of the properties of radiation by presenting the arguments that prove Kirchhoff's law:

First, we calculate the intensity of radiation with wave vector $\mathbf{q}$ which is incident per unit time on the unit surface area of a body that is placed inside a cavity, in which there is radiation in equilibrium at temperature T. To do this, we repeat an argument identical to the one in Sec. 3.7 of Part I; however, here the density of the particles is spatially uniform and they all have velocity c (see also Solution 4.1). The amount of energy arriving from a region $d\Omega$ around the direction $\mathbf{q}$, and which is incident upon the unit area ΔA in unit time Δt, will be the same expression obtained in Solution 4.1 but without the angular integration:

$$\Delta E_{\text{in}} = c\rho(\omega, T)\Delta\omega\frac{\Delta\Omega}{4\pi}\cos\theta\,. \tag{4.4.20}$$

Only a fraction $A(\mathbf{q}, T)$ of this radiation is absorbed. The intensity absorbed is $A(\mathbf{q}, T) \cdot \Delta E_{\text{in}}$. The intensity absorbed per unit area per unit time, from radiation directed along $\mathbf{q}$ with frequency $\omega = c|\mathbf{q}|$, is

$$\Delta E_{\text{abs}} = A(\mathbf{q}, T)c\rho(\omega, T)\frac{\cos\theta}{4\pi}\Delta\omega\Delta\Omega\,. \tag{4.4.21}$$

A body that is in equilibrium at temperature T will remain at equilibrium (at the same temperature) if placed in a cavity in which there is radiation in equilibrium at the same temperature T.

Hence the energy flux absorbed by the body while it is in the cavity must equal the flux emitted by the body. Moreover, the balance must be maintained in full detail. Namely, the amount of energy absorbed at every frequency and from every direction per unit time, must be identical to the amount emitted per unit time in the same direction and with the same frequency.

Exercise 4.8

Show that the balance must be detailed.

Solution on page 432

The energy absorbed for a certain $\mathbf{q}$ is simply given by (4.4.21) and the energy emitted with the same $\mathbf{q}$ is given by (4.4.18). The fact that the balance is detailed implies that $\Delta E_{\text{abs}} = \Delta E_{\text{em}}$, and hence

$$L(\mathbf{q}, T) = \frac{c}{4\pi} A(\mathbf{q}, T)\rho(\omega, T),$$

which is Kirchhoff's law. We re-emphasize that even though we have obtained it making use of an equilibrium state, the law is also valid out of equilibrium.

Exercise 4.9

(a) Show that the emissive power of a black body is

$$L_B(\mathbf{q}, T) = \frac{c}{4\pi}\rho(\omega, T). \tag{4.4.22}$$

(b) Explain the relationship between the emissive power L and the emissivity I.

Solution on page 432

In addition to the fact that the ratio of L and A of a certain body is independent of the body's properties, we can further deduce from (4.4.19) the surprising fact that a material that is a better absorber of radiation is also a better emitter. Thus, a black body is a better emitter than any other body! The reason why a black body usually appears to be black is that at the temperatures at which we encounter black bodies the wavelength of most of the emitted radiation is too large for the radiation to be seen.

Exercise 4.10

Calculate the frequencies for which the emissive power of a black body is maximal at 100 K, 1000 K, 10,000 K, 100,000 K and the values of the maximum emissive power for these temperatures.

Solution on page 433

4.5 Role of black body radiation in modern physics

The phenomenon of thermal radiation has had great importance in the evolution of modern physics. In addition it has a wide range of practical applications. Its most important contribution has been that it pointed out the inadequacy of classical physics and brought Max Planck to propose in 1900 the idea of quantization of electromagnetic radiation — more precisely, the quantization of the energy of a harmonic oscillator in units of $\hbar\omega$.

As we have seen in the previous sections, this assumption is the basis for obtaining the spectral distribution of black body radiation as given in Eq. (4.4.14), and hence the year 1900 is considered to be the year of the birth of quantum theory. However, in spite of the success of Planck's explanation, initially its physical significance and its generality were not totally clear. Only after Einstein had shown in 1905 that the same assumption is sufficient for explaining the photoelectric effect and in 1907 that it also leads to an explanation of the temperature dependence of the heat capacities of solids (see Part III), were the importance and generality of Planck's idea of quantization recognized.

Planck's formula has many practical applications related to the measurement of the temperatures of hot bodies which are hard to approach (extremely hot bodies such as steel furnaces, distant bodies such as celestial objects, etc.). By measuring the body's spectrum of radiation and comparing it to Planck's law it is possible to determine its temperature. In this way it is possible to measure, for instance, the heat production of an atomic reactor and to reach conclusions on its output, from measurements made from a high-flying surveillance aircraft.

Astronomers often compare the spectrum of radiation emitted by distant celestial objects to Planck's formula in order to measure their temperatures and to verify if the radiation emitted by them is black body radiation. Since usually the spectrum of the radiation does not fit Planck's law, they conclude that this radiation is not thermal radiation (or that it is not all thermal radiation). The study of the spectrum is bound, in this case, to shed light on the processes that are taking place in the distant object. In other words, the thermal radiation in this case amounts to a "background noise"and the detailed knowledge of its behavior allows subtraction of the background from the signal and to remain with its relevant part. These astronomical measurements are currently possible in a wide spectral range, from radio-frequency waves up to X-rays and γ-rays.

Another interesting phenomenon that was discovered as a result of astronomical measurements of electromagnetic radiation is the "cosmic background radiation," also known as the "three-degree radiation." In 1965 Penzias and Wilson discovered electromagnetic radiation with wavelengths ranging from several millimeters to several centimeters and whose spectrum corresponded to black body radiation at a temperature of 2.7 K. This radiation is incident from all directions of space at uniform intensity, from which it was concluded that it probably fills the universe uniformly, just as if the universe were residing inside a huge "oven" whose walls were at a temperature of 2.7 K.

The currently accepted explanation for the origin of this radiation is that it was created at the time of the big bang, which is assumed to have occurred about 10^{10} years ago. The whole universe was created then in a huge explosion of a compressed and very hot ball of particles and radiation. Since then the universe has been in a perpetual state of expansion, and its energy density has been decreasing. The temperature of the radiation filling the universe has decreased, therefore, to the value observed today, about 3 K. The whole phenomenon fits other available evidence concerning the big bang and the expansion of the universe, and is considered an important verification of this cosmological theory.

In April 1992 it was reported that the satellite COBE (Cosmic Background Explorer), which was launched to carry out precise measurements of the cosmic background radiation, had detected a slight nonuniformity in it, manifested as variations of order 15 μK around the average temperature of 2.7 K. These fluctuations imply that not long after the big bang there began to appear deviations from the state of uniform density in the universe. These deviations are probably responsible for the large scale structure of the universe today.

Another area in which black body radiation plays a central role is the physics of black holes. A black hole is a region in space from which no **black hole** material or radiation can escape; in contrast, there is no restriction on entering it and thus it serves as a perfect trap. The existence of such regions is predicted by Einstein's general theory of relativity around stars which were so heavy (above about five solar masses) that they eventually collapsed under their own gravitational attraction, to scales that were so small that the escape velocity from them became larger than the velocity of light. This prevents any possibility of escape from the entire region for which the escape velocity is larger than c. A nonrelativistic calculation shows that if M is the mass of the star, then the black hole extends over the region for which $c < \sqrt{2GM/r}$ or $r < 2GM/c^2$. An exact relativistic calculation yields the same result. Namely, the radius of a black hole is determined by its mass as $2GM/c^2$. Surprisingly, Hawking discovered in 1974 that if the laws of quantum theory are also taken into account then the black hole is no blacker than a black body: It emits particles and electromagnetic radiation with an energy or frequency distribution of a black body at a temperature which is also determined by the mass:

$T = hc^3/16\pi^2 kGM$. For a black hole of five solar masses this temperature is about 10^{-8} K and hence the phenomenon is rather marginal. But there may also exist much smaller black holes in the universe (that were created not as a result of the collapse of stars but in other ways). Their temperatures could be much higher. Such a small black hole is unstable as a result of the Hawking effect: The more energy it loses by radiating, the more its mass decreases, then its temperature increases and according to the Stefan–Boltzmann law (4.4.16) its radiated power increases (in spite of the decrease of the radius and the surface area) and so on, till it is totally dissipated in one last energy burst.

Appendix

Calculation of Some Integrals

In the course of this part we needed integrals of the form

$$I_\ell = \int_0^\infty \frac{x^\ell dx}{e^x - 1}, \tag{4.A.1}$$

where ℓ is an integer. In order to calculate them we start with integrals that represent the factorial function:

$$\int_0^\infty x e^{-x} dx = 1,$$

$$\int_0^\infty x^2 e^{-x} dx = -\int_0^\infty x^2 \frac{d}{dx}(e^{-x}) dx = 2\int_0^\infty x e^{-x} dx = 2,$$

$$\int_0^\infty x^3 e^{-x} dx = -\int_0^\infty x^3 \frac{d}{dx}(e^{-x}) dx = 3\int_0^\infty x^2 e^{-x} dx = 6,$$

$$\int_0^\infty x^n e^{-x} dx = \ell \int_0^\infty x^{\ell-1} e^{-x}\, dx,$$

and hence

$$\int_0^\infty x^\ell e^{-x} dx = \ell!. \tag{4.A.2}$$

Returning to the original integral (4.A.1), we note that the expression $1/(e^x - 1)$ is a geometric sum:

$$\frac{1}{e^x - 1} = \frac{e^{-x}}{1 - e^{-x}} = e^{-x} + e^{-2x} + e^{-3x} + \ldots = \sum_{m=1}^\infty e^{-mx}.$$

Substituting in the integral (4.A.1), we obtain

$$\int_0^\infty \frac{x^\ell}{e^x - 1} dx = \sum_{m=1}^\infty \int_0^\infty x^\ell e^{-mx}\, dx$$

$$= \int_0^\infty y^\ell e^{-y} dy \sum_{m=1}^\infty \frac{1}{m^{\ell+1}} = \ell! \sum_{m=1}^\infty \frac{1}{m^{\ell+1}}, \tag{4.A.3}$$

where we have performed the change of variables $y = mx$ and used (4.A.2). The sum is a known function in mathematical literature and is called the Reimann zeta function:

$$\zeta(\ell) = \sum_{m=1}^{\infty} \frac{1}{m^{\ell}}\,. \tag{4.A.4}$$

■ Actually this function is defined by (4.A.4) also for nonintegral values of ℓ and even complex ones.

The zeta function has been studied extensively and its values appear in numerical tables. The following table contains useful values.

ℓ	2	3	4	5	6	7	8
$\zeta(\ell)$	$\frac{\pi^2}{6} \simeq 1.645$	1.202	$\frac{\pi^4}{90} \simeq 1.082$	1.037	$\frac{\pi^6}{945} \simeq 1.017$	1.008	$\frac{\pi^8}{9450} \simeq 1.004$

We have therefore obtained

$$\int_0^{\infty} \frac{x^{\ell}dx}{e^x - 1} = \ell!\zeta(\ell + 1)\,. \tag{4.A.5}$$

Self-assessment exercises

Exercise 1 *Solution on page 434*

The hydrogen atom has a spectral line whose wavelength is 21 cm (this line is of extreme importance in astrophysics). The ground state is nondegenerate, and the first excited level, from which the line is emitted, is triply degenerate. The next level is found 10 eV above the ground state.

(a) What is the excitation energy of the first excited level?
(b) What is ζ at a temperature of 0.1 K?

In atomic chlorine the first spectral line corresponds to a photon with an energy of 0.11 eV. The ground level is four times degenerate, and the first excited level is doubly degenerate. The second level is higher than the first level by 1 eV.

(c) What is the characteristic temperature of the first excited level?
(d) Calculate ζ at very low temperatures, with respect to the characteristic temperature of the second level.
(e) Calculate the contribution of the atomic levels to the specific heat. Sketch it in detail. Have we already seen a specific heat with such a structure? Why is there nevertheless a difference?

Exercise 2 *Solution on page 436*

Consider a given monoatomic gas. The first three electronic levels are at a distance of 1 eV from one another. In addition, each atom has a magnetic moment, whose projection along the direction of the field can take five values: $\mu = -2\mu_0, -\mu_0, 0, \mu_0, 2\mu_0$, where $\mu_0 = 2 \times 10^{-20}$ erg/gauss. Calculate the changes in the chemical potential and in the specific heat resulting from this structure in a magnetic field of 10^6 gauss, and illustrate their temperature dependence graphically.

Exercise 3 *Solution on page 439*

For a reaction described by

$$2\mathrm{NaOH} + \mathrm{H_2SO_4} \rightleftharpoons \mathrm{Na_2SO_4} + 2\mathrm{H_2O}\,,$$

(a) Write the law of mass action.
(b) Write the additional equations.
(c) Is the system solvable? Explain.

Exercise 4 *Solution on page 440*

Calculate the equilibrium constant in the gaseous reaction:

$$\mathrm{Cl_2} + \mathrm{H_2} \rightleftharpoons 2\mathrm{HCl}\,.$$

The vibrational angular frequencies are: For chlorine $10^{14}\ \mathrm{s}^{-1}$, for hydrogen $8.3 \times 10^{14}\ \mathrm{s}^{-1}$ and for HCl $5.7 \times 10^{14}\ \mathrm{s}^{-1}$. The moments of inertia are given in Table 4.1.2.

Exercise 5 *Solution on page 442*

At the beginning of Sec. 4.4 we showed that the radiation density in a cavity in equilibrium at temperature T is independent of position. In a similar way:

(a) Prove that $\rho(\mathbf{q}, T)$ is independent of the direction of $\mathbf{q}$.
(b) Prove that $\rho(\mathbf{q}, T)$ is independent of the radiation's polarization.
(c) Is it possible to prove that ρ is independent of ω?

Exercise 6 *Solution on page 443*

(a) Prove that the radiation emitted per unit surface area per unit time by a black body, for wavelengths ranging between λ and $\lambda + d\lambda$, is

$$\tilde{I}(\lambda)d\lambda = \frac{2\pi hc^2}{\lambda^5} \frac{d\lambda}{e^{\beta hc/\lambda} - 1}\,.$$

(b) Calculate the wavelength $\lambda_{\max}$ for which the emitted energy is maximal.
(c) The intensity of the radiation emitted by the sun has a maximum at around $\lambda = 5 \times 10^{-7}$ m. What is the temperature of the sun's surface?

Exercise 7 *Solution on page 445*

(a) What is the average number of photons per unit volume in a cavity at temperature T? How many photons per cm^3 does the cosmic background radiation contain?
(b) What is the ratio between the average number of photons and the average number of gas particles at the same pressure, volume and temperature as that of the three-degree radiation?
(c) Is the ratio you found in (b) valid at any temperature?

(d) Calculate the specific heat at constant volume per photon. Compare it to the specific heat per molecule of an ideal gas.
(e) A student calculated C_V/N for the photon gas discussed in (d) in the following manner: Using the result of Exercise 4.6 (a) he wrote $E = 3PV$, and using Eq. (xi) in the solution of (b) of the present exercise he wrote

$$E = 3PV = \frac{3NkT}{1.1106} = 2.7NkT\,.$$

The specific heat at constant volume is the derivative of E with respect to T, and hence

$$\frac{C_V}{N} = \frac{1}{N}\left(\frac{\partial E}{\partial T}\right)_{N,V} = 2.7k\,.$$

This result differs, of course, from that obtained in (d). Where is the error?

Exercise 8 *Solution on page 447*

Bodies whose color is black are bodies that absorb most of the radiation incident upon them, and those are also what we technically defined as black bodies (see Secs. 4.3 and 4.4). But there we stated that a black body emits radiation better than any other kind of body.

(a) How would you settle the contradiction?
(b) In what conditions will a black body appear to be yellow?
(c) How is it possible, at regular conditions, to convince the sceptic that indeed a black body is not all that black?

Exercise 9 *Solution on page 447*

The law of mass action is useful not only for chemical reactions but also for reactions of any other kind, such as reactions between nuclei or between elementary particles at high energies. A typical process is the creation and annihilation of electron–positron pairs, which we may think of as a chemical reaction:

$$e^+ + e^- \rightleftharpoons \gamma\,.$$

Such a gas of electrons, positrons and photons is actually hard to come by, but it could be created at extreme temperature conditions such as those not long after the big bang or inside stars.

Calculate the density of electrons and positrons in such a gas at nonrelativistic temperatures ($kT \ll mc^2$, where m is the electron mass). Assume that the whole gas is electrically neutral.

Solutions to exercises in the text

Solution 1.1 *Exercise on page 342*

(a) To obtain (4.1.6) we write

$$\mathbf{p}_\alpha = m_\alpha \dot{\mathbf{r}}_\alpha = m_\alpha(\dot{\mathbf{R}}_\alpha + \dot{\boldsymbol{\rho}}_\alpha) = \frac{m_\alpha}{M}(M\dot{\mathbf{R}}) + m_\alpha \dot{\boldsymbol{\rho}}_\alpha = \frac{m_\alpha}{M}\mathbf{P} + \boldsymbol{\pi}_\alpha \,,$$

where we have used (4.1.3), (4.1.4) and (4.1.5).

(b) In order to obtain (4.1.7) we rewrite (4.1.4) in the form

$$\boldsymbol{\rho}_\alpha = \mathbf{r}_\alpha - \mathbf{R}$$

and then multiply it by m_α and sum over α:

$$\sum_{\alpha=1}^{q} m_\alpha \boldsymbol{\rho}_\alpha = \sum_{\alpha=1}^{q} m_\alpha \mathbf{r}_\alpha - \sum_{\alpha=1}^{q} m_\alpha \mathbf{R} = \sum_{\alpha=1}^{q} m_\alpha \mathbf{r}_\alpha - M\mathbf{R} \,.$$

Using (4.1.2), we obtain (4.1.7a):

$$\sum_{\alpha=1}^{q} m_\alpha \boldsymbol{\rho}_\alpha = 0 \,.$$

Differentiating the last equation, we obtain (4.1.7b):

$$0 = \frac{d}{dt}\sum_{\alpha=1}^{q} m_\alpha \boldsymbol{\rho}_\alpha = \sum_{\alpha=1}^{q} m_\alpha \dot{\boldsymbol{\rho}}_\alpha = \sum_{\alpha=1}^{q} \boldsymbol{\pi}_\alpha \,.$$

(c) To prove (4.1.8) we use (4.1.6) and write

$$\begin{aligned}\sum_{\alpha=1}^{q} \frac{\mathbf{p}_\alpha^2}{2m_\alpha} &= \sum_{\alpha=1}^{q} \frac{1}{2m_\alpha}\left(\frac{m_\alpha}{M}\mathbf{P} + \boldsymbol{\pi}_\alpha\right)^2 \\ &= \frac{\mathbf{P}^2}{2M^2}\sum_{\alpha=1}^{q} m_\alpha + \frac{\mathbf{P}}{M}\cdot\sum_{\alpha=1}^{q} \boldsymbol{\pi}_\alpha + \sum_{\alpha=1}^{q} \frac{\boldsymbol{\pi}_\alpha^2}{2m_\alpha} \,.\end{aligned}$$

The middle term is zero by (4.1.7b), and thus we obtain (4.1.8):

$$\sum_{\alpha=1}^{q} \frac{\mathbf{p}_\alpha^2}{2m_\alpha} = \frac{\mathbf{P}^2}{2M} + \sum_{\alpha}^{q} \frac{\boldsymbol{\pi}_\alpha^2}{2m_\alpha} \,.$$

Solution 1.2 *Exercise on page 344*

Multiplying (4.1.12b) by m_1 and (4.1.12a) by m_2 and subtracting we have

$$m_2\boldsymbol{\pi}_1 - m_1\boldsymbol{\pi}_2 = m_2\mathbf{p}_1 - m_1\mathbf{p}_2\,.$$

Equation (4.1.14b) implies that the left hand side of the last equation is $M\boldsymbol{\pi}_1$. We can write its right hand side as

$$m_2\mathbf{p}_1 - m_1\mathbf{p}_2 = m_2 m_1(\dot{\mathbf{r}}_1 - \dot{\mathbf{r}}_2) = m_2 m_1(\dot{\boldsymbol{\rho}}_1 - \dot{\boldsymbol{\rho}}_2)$$

and from here

$$\boldsymbol{\pi}_1 = \mu(\dot{\boldsymbol{\rho}}_1 - \dot{\boldsymbol{\rho}}_2)$$

and with the help of (4.1.14b) we can also obtain $\boldsymbol{\pi}_2$ as $-\boldsymbol{\pi}_1$.

Solution 1.3 *Exercise on page 345*

The potential between the two atoms is

$$U(\rho) = \epsilon\left[\left(\frac{a}{\rho}\right)^2 - 2\frac{a}{\rho}\right], \tag{i}$$

where ρ is the distance between the two atoms, and a and ϵ are parameters that determine the equilibrium point and the depth of the potential.

(a) When the distance between the atoms is the equilibrium distance, the potential energy is minimal. We thus differentiate (i) and equate the derivative to zero:

$$\frac{dU}{d\rho} = \epsilon\left(-2\frac{a^2}{\rho^3} + 2\frac{a}{\rho^2}\right) = 0 \tag{ii}$$

$$\Downarrow$$

$$\rho = a\,,$$

namely the equilibrium distance is a. In order to verify that $\rho = a$ is indeed the minimum point, one can differentiate U a second time and verify that at $\rho = a$, $U'' > 0$. See (c) below.

(b) The binding energy is obtained by substituting the equilibrium distance $\rho = a$ into (i):

$$U(\rho = a) = -\epsilon\,. \tag{iii}$$

(c) The harmonic approximation to $U(\rho)$ will be

$$U(\rho) = U(a) + \frac{1}{2}U''(a)(\rho - a)^2\,. \tag{iv}$$

The linear term does not appear because at the equilibrium distance $U'(\rho_0) = U'(a) = 0$.

The value of the second derivative at the equilibrium distance

$$U''(a) = \epsilon\left(2\cdot 3\frac{a^2}{\rho^4} - 2\cdot 2\frac{a}{\rho^3}\right)_{\rho=a} = \frac{2\epsilon}{a^2}\,. \qquad \text{(v)}$$

Equation (iv) will therefore take the form

$$U(\rho) = -\epsilon + \frac{\epsilon}{a^2}(\rho - a)^2\,. \qquad \text{(vi)}$$

The "spring constant" is twice the coefficient of the square deviation from equilibrium:

$$K = 2\frac{\epsilon}{a^2} = \frac{2\times 2.5\times 1.6\times 10^{-19}}{(2\times 10^{-10})^2} = 20\ \text{N/m}\,. \qquad \text{(vii)}$$

(d) The energy needed to increase the distance between the atoms from a to $1.05a$ may be calculated in the harmonic approximation:

$$\Delta U = \frac{\epsilon}{a^2}(0.05a)^2 = \frac{1}{400}\epsilon\,. \qquad \text{(viii)}$$

If $\epsilon = 2.5$ eV, $\Delta U = 0.25\times 10^{-3}$ eV and the corresponding temperature is of order

$$T = \frac{\Delta U}{k} = \frac{0.25\times 10^{-3}\times 1.6\times 10^{-19}}{1.38\times 10^{-23}} \approx 70\ \text{K}\,. \qquad \text{(ix)}$$

(e) Similar arguments show that ρ increases by 10% at room temperature.

Solution 1.4 *Exercise on page 345*

From Eq. (4.1.11) we obtain

$$\begin{cases} M\mathbf{R} = m_1\mathbf{r}_1 + m_2\mathbf{r}_2\,, \\ \boldsymbol{\rho}_{12} \ = \mathbf{r}_1 - \mathbf{r}_2\,, \end{cases}$$

and in order to express $\mathbf{r}_1$ and $\mathbf{r}_2$ in terms of $\mathbf{R}$ and $\boldsymbol{\rho}_{12}$ we solve the two equations with two unknowns. We multiply the second equation by m_2 and to add it to the first equation:

$$M\mathbf{R} + m_2\boldsymbol{\rho}_{12} = m_1\mathbf{r}_1 + m_2\mathbf{r}_1\,,$$

and from here

$$\mathbf{r}_1 = \mathbf{R} + \frac{m_2}{M}\boldsymbol{\rho}_{12}\,.$$

Then we multiply the second equation by m_1 and subtract it from the first equation we obtain

$$M\mathbf{R} - m_1\boldsymbol{\rho}_{12} = m_2\mathbf{r}_2 + m_1\mathbf{r}_2$$

and from here

$$\mathbf{r}_2 = \mathbf{R} - \frac{m_1}{M}\boldsymbol{\rho}_{12} \,.$$

In a similar manner we obtain from Eqs. (4.1.12) and (4.1.13), the momentum variables,

$$\begin{cases} \mathbf{P} &= \mathbf{p}_1 + \mathbf{p}_2 \,, \\ M\boldsymbol{\pi}_{12} &= m_2\mathbf{p}_1 - m_1\mathbf{p}_2 \,. \end{cases}$$

Actually, there is no need to solve these equations. It is possible to simply use the previous solution, substituting $\mathbf{p}_2$ for $\mathbf{r}_1$, $-\mathbf{p}_1$ for $\mathbf{r}_2$, $\mathbf{P}$ for $\boldsymbol{\rho}_{12}$ and $-\boldsymbol{\pi}_{12}$ for $\mathbf{R}$. We then obtain

$$\mathbf{p}_2 = -\boldsymbol{\pi}_{12} + \frac{m_2}{M}\mathbf{P} \,,$$

$$\mathbf{p}_1 = \boldsymbol{\pi}_{12} + \frac{m_1}{M}\mathbf{P} \,.$$

Solution 1.5 *Exercise on page 345*

The energy of a triatomic molecule will be, according to (4.1.9)

$$E_{\text{mol}} = \frac{\mathbf{P}^2}{2M} + \frac{\boldsymbol{\pi}_1^2}{2m_1} + \frac{\boldsymbol{\pi}_2^2}{2m_2} + \frac{\boldsymbol{\pi}_3^2}{2m_3} + U_{12}(\boldsymbol{\rho}_{12}) + U_{23}(\boldsymbol{\rho}_{23}) + U_{31}(\boldsymbol{\rho}_{31}) \,, \quad \text{(i)}$$

where we have denoted the potential between atoms α and β by $U_{\alpha\beta}$, and

$$\boldsymbol{\rho}_{\alpha\beta} = \boldsymbol{\rho}_\alpha - \boldsymbol{\rho}_\beta \,. \quad \text{(ii)}$$

We now note that E_{mol} is expressed in terms of too many variables, but there is a dependence between them:

$$\boldsymbol{\pi}_1 + \boldsymbol{\pi}_2 + \boldsymbol{\pi}_3 = 0 \,, \quad \text{(iii)}$$

$$\boldsymbol{\rho}_{12} + \boldsymbol{\rho}_{23} + \boldsymbol{\rho}_{31} = 0 \,. \quad \text{(iv)}$$

It is possible to transform to relative coordinates and to write the kinetic energy as a sum of $\mathbf{P}^2/2M$ with the two corresponding kinetic energy terms, for example the relative motion between atoms 1 and 2 and the motion of atom number 3 with respect to the center of mass of 1 and 2. In contrast, the potential energies will not separate into a sum of three independent terms since $\boldsymbol{\rho}_{12}$, for example, depends on the other two, and since the potentials are not linear, $U_{12}(\rho_{12}) = U_{12}(-\boldsymbol{\rho}_{23} - \boldsymbol{\rho}_{31})$ does separate into a sum.

Solution 1.6 *Exercise on page 349*

The internal energy E is given by (3.1.3):

$$E = -\frac{\partial \ln Z}{\partial \beta}$$

or by the thermodynamic relation [see (2.0.25) or (3.1.28)]:

$$E = F + TS\,.$$

Since F and S have already been given explicitly in Eqs. (4.1.28) and (4.1.30), we prefer to calculate E from the thermodynamic relation:

$$\begin{aligned} E = &-NkT\left[\ln\frac{V}{N} + \frac{3}{2}\ln\left(\frac{2\pi MkT}{h^2}\right) + 1 + \ln\zeta\right] \\ &+NkT\left[\ln\frac{V}{N} + \frac{3}{2}\ln\left(\frac{2\pi MkT}{h^2}\right) + \frac{5}{2}\right] + NkT\frac{d}{dT}(T\ln\zeta)\,. \end{aligned}$$

Differentiating terms, collecting and dividing by N, we obtain (4.1.32).

Solution 1.7 *Exercise on page 349*

The specific heat at constant volume is given by the derivative of the internal energy with respect to the temperature. Using (4.1.32) we obtain

$$C_V = \left(\frac{\partial E}{\partial T}\right)_{N,V} = \frac{3}{2}Nk + Nk\frac{d}{dT}\left(T^2\frac{d}{dT}\ln\zeta\right),$$

which is exactly (4.1.33).

The specific heat at constant pressure is given by the derivative with respect to T of the enthalpy:

$$C_P = \left(\frac{\partial \mathsf{H}}{\partial T}\right)_{N,P},$$

where

$$\mathsf{H} = E + PV\,.$$

From (4.1.29) we obtain

$$\mathsf{H} = E + NkT$$

and hence

$$C_P = \left(\frac{\partial \mathsf{H}}{\partial T}\right)_{N,P} = Nk + \frac{\partial E}{\partial T} = Nk + C_V$$

$$\Downarrow$$

$$C_P - C_V = Nk\,.$$

Solution 1.8 *Exercise on page 350*

In order to calculate the corrections, implied by the internal structure of the molecule, to the entropy, the internal energy and the specific heat, we have to calculate the partition function $\zeta(T)$. Because $\epsilon_1 \ll \epsilon_2, \epsilon_3, \ldots$ it suffices, at temperatures that are not too high, to take only the first two levels. Assuming that there is a single state at energy ϵ_1 (no degeneracy),

$$\zeta(T) = 1 + e^{-\beta\epsilon_1} . \tag{i}$$

The correction to the entropy per molecule is given by (4.1.31). Writing for short ϵ instead of ϵ_1 we obtain

$$\Delta s = \frac{d}{dT}(kT \ln \zeta) = \frac{d}{dT}[kT \ln(1 + e^{-\beta\epsilon})] \tag{ii}$$

$$\Downarrow$$

$$\Delta s = k \left[\ln(1 + e^{-\beta\epsilon}) + \frac{\beta\epsilon}{1 + e^{\beta\epsilon}} \right] . \tag{iii}$$

The following figure illustrates graphically the temperature dependence of Δs:

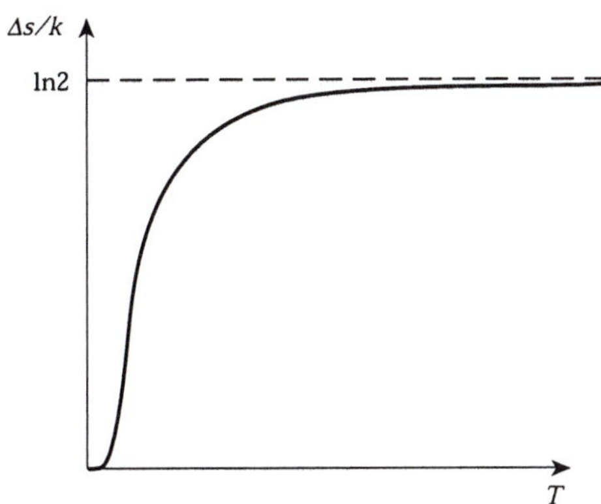

If the temperature is very low ($kT \ll \epsilon$), then $\zeta \to 1$ and from (ii) it is clear that $\Delta s \to 0$. The approach to zero is given by the power expansion of (iii):

$$\Delta s = k(e^{-\beta\epsilon} + \beta\epsilon e^{-\beta\epsilon} + \cdots) \approx \frac{\epsilon}{kT} e^{-\epsilon/kT} . \tag{iv}$$

The correction to the internal energy is given by (4.1.32)

$$\frac{\Delta E}{N} = kT^2 \frac{d}{dT} \ln \zeta = \frac{\epsilon}{e^{\beta\epsilon} + 1} . \tag{v}$$

The following figure illustrates the temperature dependence of $\Delta E/N$:

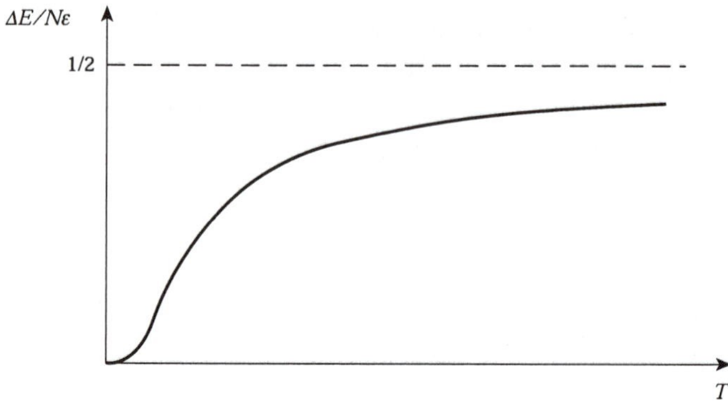

At low temperatures the correction to the internal energy per molecule will behave in the following manner:

$$\frac{\Delta E}{N} \approx \epsilon e^{-\beta\epsilon} = \epsilon e^{-\epsilon/kT},$$

and tends to zero when $T \to 0$.

Finally, the correction to the specific heat per molecule is obtained from (4.1.33):

$$\frac{\Delta C_V}{N} = k\frac{d}{dT}\left(T^2\frac{d}{dT}\ln\zeta\right) = \frac{d}{dT}\frac{\Delta E}{N} = \frac{k(\beta\epsilon)^2 e^{\beta\epsilon}}{(1+e^{\beta\epsilon})^2} = \frac{k(\beta\epsilon)^2}{4\cosh^2(\beta\epsilon/2)},$$

which at low temperatures behaves in the following manner:

$$\frac{\Delta C_V}{N} \sim k(\beta\epsilon)^2 e^{-\beta\epsilon} = k\left(\frac{\epsilon}{kT}\right)^2 e^{-\epsilon/kT}.$$

A graphical illustration of $\frac{\Delta C_V}{N}$ is given in the following figure:

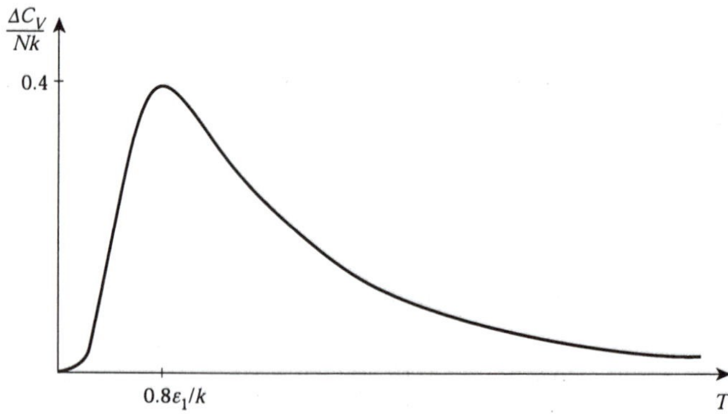

Note the similarity between these results and the corresponding quantities of the paramagnet (Part II, Sec. 5.3). The reason is, of course, that we considered molecules with only two molecular states, as for the single spin of the paramagnet.

Solution 1.9 *Exercise on page 351*

The internal partition function, when $kT \ll \epsilon_1 - \epsilon_0$, is obtained from (4.1.36):

$$\zeta(T) \approx g_0 e^{-\beta\epsilon_0} \tag{i}$$

or

$$\ln\zeta = \ln g_0 - \beta\epsilon_0 \,. \tag{ii}$$

Substituting (ii) into Eq. (4.1.31) we obtain

$$\Delta s = \frac{d}{dT}(kT\ln\zeta) = \frac{d}{dT}(kT\ln g_0 - \epsilon_0) = k\ln g_0 \,.$$

When $g_0 = 1$, namely when the ground level is nondegenerate, $\Delta S = 0$.

Since C_V is derived from S, also $\Delta C_V = 0$, as is obtained by substitution of (i) into Eq. (4.1.33):

$$\Delta C_V = Nk\frac{d}{dT}\left(T^2\frac{d}{dT}\left(\frac{-\epsilon_0}{kT}\right)\right) = 0 \,.$$

We have therefore found that, for the conditions of the exercise, S and C_V of a monoatomic gas are identical to their values for a gas of particles devoid of internal structure.

The change in the chemical potential is obtained from (4.1.35), and is

$$\Delta\mu = -kT\ln\zeta = \epsilon_0 \,,$$

which is equivalent to a constant addition to all the energies of the problem.

Solution 1.10 *Exercise on page 351*

If the ground level is doubly degenerate, we substitute into Eq. (i) of the preceding solution $g_0 = 2$, and obtain

$$\zeta(T) = 2e^{-\beta\epsilon_0} \,.$$

The extra entropy per molecule is obtained using (4.1.31):

$$\Delta s = \frac{d}{dT}(kT\ln\zeta) = k\ln 2 \,,$$

which is independent of the temperature, so again there is no change in the specific heat.

The change in the chemical potential is obtained from (4.1.35):

$$\Delta\mu = kT \ln \zeta = -kT \ln 2 + \epsilon_0 \,.$$

Solution 1.11 *Exercise on page 354*

We change the variables to $x = \rho - \rho_0$, to obtain

$$\begin{aligned}\int_{-\infty}^{\infty} \rho^2 e^{-\beta K(\rho-\rho_0)^2/2} d\rho &= \int_{-\infty}^{\infty} (\rho_0 + x)^2 e^{-\beta K x^2/2} dx \\ &= \rho_0^2 \int_{-\infty}^{\infty} e^{-\beta K x^2/2} dx + \int_{-\infty}^{\infty} x^2 e^{-\beta K x^2/2} dx \,.\end{aligned}$$

The mixed term (with $x\rho_0$) vanishes upon integration due to the anti-symmetry of the integrand. The first term gives

$$\rho_0^2 \sqrt{\frac{2\pi}{\beta K}} \,.$$

See e.g. Exercise (1.13), Part I. The second is proportional to $(\beta K)^{-3/2}$. See e.g. Exercise (1.14), Part I. It is negligible relative to the first, because $\rho_0^2 \gg 1/\beta K$.

Solution 1.12 *Exercise on page 355*

(a) The internal partition function in the classical approximation is given by Eq. (4.1.49). To calculate the internal energy we substitute (4.1.49) into (4.1.32) and obtain

$$\begin{aligned}\langle \epsilon \rangle &= \frac{3}{2}kT + kT^2 \frac{d}{dT} \ln \zeta = \frac{3}{2}kT + kT^2 \frac{d}{dT}(2 \ln T) \\ &= \frac{3}{2}kT + 2kT = \frac{7}{2}kT \,.\end{aligned}$$

The specific heat at constant volume is

$$C_V = \frac{\partial E}{\partial T} = \frac{7}{2}Nk \,,$$

as in (4.1.51).

(b) From (4.1.34)

$$C_P - C_V = Nk$$

and from (4.1.51)

$$C_V = \frac{7}{2}Nk \,,$$

hence

$$\gamma = \frac{C_P}{C_V} = \frac{Nk + C_V}{C_V} = \frac{9}{7} = 1.286\,.$$

(c) Room temperature is not in the range of temperatures for which our approximations are valid and indeed γ is quite far from 1.286. On the other hand, at $T = 1273$ K we expect a better agreement, and indeed the values of γ decrease and for oxygen we obtain a value that differs from the classical value by only 0.6%.

Solution 1.13 *Exercise on page 356*

(a) In order for the classical approximation to be justified we required that $\hbar\omega \ll kT \ll K\rho_0^2$. To a good approximation μ of Cl_2 molecules is half the mass of the chlorine nucleus, since the mass of the electrons is negligible compared to the mass of the nucleus. The mass of the most common isotope of chlorine is 35 atomic mass units (1 amu = 1.66×10^{-27} kg), namely to a good approximation

$$\mu = \frac{1}{2} \times 35 \times 1.66 \times 10^{-27} \text{ kg} \approx 2.9 \times 10^{-26} \text{ kg}\,.$$

Hence

$$K\rho_0^2 = \mu\omega^2\rho_0^2 = 2.9 \times 10^{-26} \times 10^{28} \times 4 \times 10^{-20} = 1.16 \times 10^{-17} \text{ J}\,,$$
$$\hbar\omega = 1.05 \times 10^{-34} \times 10^{14} = 1.05 \times 10^{-20} \text{ J}\,.$$

And the corresponding temperature range is

$$750 \text{ K} \ll T \ll 8.5 \times 10^5 \text{ K}\,.$$

(b) A graphical illustration of the integrand in Eq. (4.1.44) is given in the following figure. The bell approximation is justified, since except for a very narrow region of ρ, the integrand is negligible everywhere.

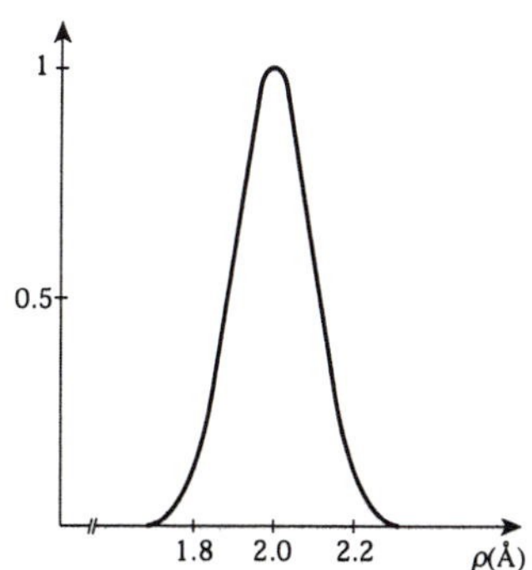

Solution 1.14 *Exercise on page 357*

To prove (4.1.52) we write $\boldsymbol{\ell}$ using (4.1.53b):

$$\boldsymbol{\ell}^2 = (\boldsymbol{\rho} \times \boldsymbol{\pi})^2 = \rho^2\pi^2 - (\boldsymbol{\rho} \cdot \boldsymbol{\pi})^2 . \tag{i}$$

It is possible to prove (i) using identities relating vector products with scalar products or directly:

$$\boldsymbol{\ell}^2 = (\rho\pi \sin\theta)^2 = \rho^2\pi^2(1 - \cos^2\theta) = \rho^2\pi^2 - (\boldsymbol{\rho} \cdot \boldsymbol{\pi})^2 , \tag{ii}$$

where θ is the angle between the directions of $\boldsymbol{\rho}$ and $\boldsymbol{\pi}$. Dividing Eq. (i) by ρ^2 and using (4.1.53a), we obtain

$$\frac{\boldsymbol{\ell}^2}{\rho^2} = \pi^2 - \left(\frac{\boldsymbol{\pi} \cdot \boldsymbol{\rho}}{\rho}\right)^2 = \pi^2 - \pi_\rho^2 , \tag{iii}$$

Dividing Eq. (iii) by 2μ we arrive at (4.1.52).

Solution 1.15 *Exercise on page 358*

(a) The moment of inertia with respect to the center of mass of the molecule is

$$I = m_1\rho_1^2 + m_2\rho_2^2 . \tag{i}$$

Using (4.1.11) and the relation between the relative coordinates and those of the laboratory frame (see Exercise 1.4), we obtain

$$\boldsymbol{\rho}_1 = \mathbf{r}_1 - \mathbf{R} = \frac{m_2}{M}\boldsymbol{\rho} , \tag{ii}$$

$$\boldsymbol{\rho}_2 = \mathbf{r}_2 - \mathbf{R} = -\frac{m_1}{M}\boldsymbol{\rho} , \tag{iii}$$

where we have written $\boldsymbol{\rho}_{12} = \boldsymbol{\rho}$. Substituting (ii) and (iii) into (i), we find

$$I = m_1\left(\frac{m_2}{M}\right)^2 \rho^2 + m_2\left(\frac{m_1}{M}\right)^2 \rho^2 = \mu\rho^2 ,$$

and if $\rho = \rho_0$, the moment of inertia is equal to $\mu\rho_0^2$.

(b) The values of μ and ρ_0 for the molecules in Table 4.1.2 are summarized in the following table:

Molecule	μ (amu)	μ (10^{-27} kg)	I (10^{-47} kg $\cdot$ m^2)	ρ_0 (10^{-10} m)
H_2	1/2	0.83	0.46	0.7
HCl	35/36	1.60	2.40	1.2
Cl_2	35/2	29	115	2.0
I_2	127/2	105	745	2.7

Solution 1.16 *Exercise on page 360*

For HCl

$$I = 2.4 \times 10^{-47} \text{ kg} \cdot \text{m}^2 ,$$

$$\omega = 5.7 \times 10^{14} \text{ s}^{-1} ,$$

$$\epsilon_v = \hbar\omega = (1.05 \times 10^{-34})(5.7 \times 10^{14}) = 6 \times 10^{-20} \text{ J} = 0.37 \text{ eV} ,$$

$$\Theta_v = \frac{\epsilon_v}{k} = 4300 \text{ K} ,$$

$$\epsilon_r = \frac{\hbar^2}{2I} = \frac{(1.05 \times 10^{-34})^2}{2(2.4 \times 10^{-47})} = 2.3 \times 10^{-22} \text{ J} = 1.44 \times 10^{-3} \text{ eV} ,$$

$$\Theta_r = \frac{2\epsilon_r}{k} = 33 \text{ K} .$$

Solution 1.17 *Exercise on page 360*

The spacing between the vibrational levels is constant and for Cl_2 is equal to 0.06 eV, as we have already calculated. In order to find out how many rotational levels "fit" into this spacing we have to find the value of J for which $E_r(J) = \epsilon_v$, or to solve

$$J(J+1)\epsilon_r = \epsilon_v .$$

We have also calculated ϵ_v for Cl_2, and obtained 2.5×10^{-5} eV. Hence

$$J(J+1) = \frac{\epsilon_v}{\epsilon_r} = \frac{0.06}{2.5 \times 10^{-5}} = 2400 ,$$

and the positive solution of this equation is $J \cong 48.5$.

That is, one has to go up 49 rotational levels in order to cross the first vibrational spacing.

For HCl we use the results we obtained in the solution to the preceding exercise for ϵ_v and ϵ_r.

$$J(J+1) = \frac{0.37}{1.44 \times 10^{-3}} = 257$$

and

$$J \approx 16 .$$

Namely, one has to go up 16 rotational levels of HCl in order to reach the first vibrational level.

Solution 1.18 *Exercise on page 364*

(a)

$$T^2 \frac{d}{dT} \ln \zeta_r = T^2 \frac{d}{dT} \ln(1 + 3e^{-\Theta_r/T}) = T^2 \frac{3e^{-\Theta_r/T}}{1 + 3e^{-\Theta_r/T}} \frac{\Theta_r}{T^2}$$
$$= \frac{3\Theta_r}{3 + e^{\Theta_r/T}} ,$$

and differentiating once more, we obtain

$$\frac{(\Delta C_V)_r}{N} = 3k\Theta_r \frac{d}{dT} \frac{1}{3 + e^{\Theta_r/T}} = 3k \left(\frac{\Theta_r}{T}\right)^2 \frac{e^{\Theta_r/T}}{(3 + e^{\Theta_r/T})^2}$$
$$\approx 3k \left(\frac{\Theta_r}{T}\right)^2 e^{-\Theta_r/T} ,$$

where we have reached the last expression by neglecting the factor of 3 in the denominator compared to the exponential, which is dominant for $T \to 0$.

(b) We shall again write the continuum approximation to the partition function at high temperatures:

$$\zeta_r \approx \int_0^\infty dJ(2J+1)e^{-J(J+1)\Theta_r/2T} .$$

We now perform a change of variables:

$$x = J(J+1) ,$$

so that

$$dx = (2J+1)dJ .$$

Hence the integral becomes

$$\zeta_r = \int_0^\infty dx e^{-x\Theta_r/2T} = \frac{2T}{\Theta_r} .$$

(c) According to (4.1.33),

$$\frac{(\Delta C_V)_r}{N} = k \frac{d}{dT} \left(T^2 \frac{d}{dT} \ln \frac{2T}{\Theta_r}\right) = k .$$

Solution 2.1 *Exercise on page 368*

We choose to express all of the differentials in Eq. (4.2.2) in terms of dN_1, so that

$$dN_2 = \frac{\nu_2}{\nu_1} dN_1,\ dN_3 = \frac{\nu_3}{\nu_1} dN_1, \ldots, dN_M = \frac{\nu_M}{\nu_1} dN_1 \,.$$

Substituting all these expressions into (4.2.4) we obtain

$$\mu_1 dN_1 + \mu_2 \frac{\nu_2}{\nu_1} dN_1 + \mu_3 \frac{\nu_3}{\nu_1} dN_1 + \cdots + \mu_M \frac{\nu_M}{\nu_1} dN_1 = 0 \,,$$

and after dividing by dN_1 and multiplying by ν_1 we obtain (4.2.5).

Solution 2.2 *Exercise on page 368*

(a)

$$4NH_3 + 3O_2 \rightleftharpoons 2N_2 + 6H_2O \,,$$

$$\begin{array}{lll} B_1 = NH_3 \,, & b_1 = 4 \,, & \nu_1 = -4 \,, \\ B_2 = O_2 \,, & b_2 = 3 \,, & \nu_2 = -3 \,, \\ B_3 = N_2 \,, & b_3 = 2 \,, & \nu_3 = 2 \,, \\ B_4 = H_2O \,, & b_4 = 6 \,, & \nu_4 = 6 \,. \end{array}$$

And the form of Eq. (4.2.5) becomes

$$-4\mu_1 - 3\mu_2 + 2\mu_3 + 6\mu_4 = 0 \,.$$

(b)

$$2C_4H_{10} + 13O_2 \rightleftharpoons 8CO_2 + 10H_2O \,,$$

$$\begin{array}{lll} B_1 = C_4H_{10}, & b_1 = 2 \,, & \nu_1 = -2, \\ B_2 = O_2, & b_2 = 13 \,, & \nu_2 = -13, \\ B_3 = CO_2, & b_3 = 8 \,, & \nu_3 = 8, \\ B_4 = H_2O, & b_4 = 10 \,, & \nu_4 = 10, \end{array}$$

and Eq. (4.2.5) reads

$$-2\mu_1 - 13\mu_2 + 8\mu_3 + 10\mu_4 = 0 \,.$$

Solution 2.3 *Exercise on page 369*

We substitute the expression for the chemical potential (4.1.35) into the equilibrium equation of the reaction and obtain (4.2.6):

$$0 = \sum_{i=1}^{M} \nu_i \mu_i = \sum_{i=1}^{M} \nu_i kT \left[\ln n_i - \frac{3}{2} \ln \left(\frac{2\pi M_i kT}{h^2} \right) - \ln \zeta_i \right]$$

$$\Downarrow$$

$$\sum_{i=1}^{M} \nu_i \ln n_i = \sum_{i=1}^{M} \left[\frac{3}{2} \nu_i \ln \left(\frac{2\pi M_i kT}{h^2} \right) + \nu_i \ln \zeta_i \right]$$

$$\Downarrow$$

$$\sum_{i=1}^{M} \ln (n_i)^{\nu_i} = \sum_{i=1}^{M} \ln \left[\left(\frac{2\pi M_i kT}{h^2} \right)^{3/2} \zeta_i \right]^{\nu_i}$$

$$\Downarrow$$

$$\ln \left[\prod_{i=1}^{M} (n_i)^{\nu_i} \right] = \ln \left\{ \prod_{i=1}^{M} \left[\left(\frac{2\pi M_i kT}{h^2} \right)^{3/2} \zeta_i \right]^{\nu_i} \right\}$$

$$\Downarrow$$

$$\prod_{i=1}^{M} (n_i)^{\nu_i} = \prod_{i=1}^{M} \left[\left(\frac{2\pi M_i kT}{h^2} \right)^{3/2} \zeta_i \right]^{\nu_i},$$

and denoting the right hand side by $K(T)$, we obtain both (4.2.7) and (4.2.8).

Solution 2.4 *Exercise on page 370*

(a) The law of mass action for the reaction

$$4NH_3 + 3O_2 \rightleftharpoons 2N_2 + 6H_2O$$

is obtained using the table we constructed in the solution to Exercise 2.2. We use n_1 up to n_4 to denote the densities of the constituents of the reaction, respectively from left to right. The law of mass action reads

$$\frac{n_3^2 n_4^6}{n_1^4 n_2^3} = K(T). \tag{i}$$

A single relationship is obtained between the four densities for a given $K(T)$.

(b) The reaction conserves the number of atoms of each element. Denoting by $n_{\rm N}$, $n_{\rm H}$ and $n_{\rm O}$ the densities of nitrogen, hydrogen and oxygen atoms, the reaction equation implies

$$n_{\rm N} = n_1 + 2n_3\,,$$
$$n_{\rm H} = 3n_1 + 2n_4\,,$$
$$n_{\rm O} = 2n_2 + n_4$$

namely a total of four equations with four unknowns.

(c) We shall find $K(T)$ using Eq. (4.2.8) with $M = 4$:

$$K(T) = \frac{M_3^3 M_4^9}{M_1^6 M_2^{9/2}} \left(\frac{2\pi kT}{h^2}\right)^{3/2} \frac{\zeta_3^2\zeta_4^6}{\zeta_1^4\zeta_2^3}\,,$$

where we have used the fact that here

$$\sum_{i=1}^{4} \nu_i = 1\,.$$

Solution 2.5 *Exercise on page 371*

Without the Gibbs correction the chemical potential of the ith constituent is given by Eq. (3.5.1), with the last term coming from the internal partition function:

$$\mu_i = -kT\left[\ln V + \frac{3}{2}\ln\left(\frac{2\pi M_i kT}{h^2}\right)\right] - kT\ln\zeta_i\,.$$

The condition for equilibrium, Eq. (4.2.5), takes the form

$$\sum_{i=1}^{M}\nu_i\mu_i = -kT\ln V\sum_{i=1}^{M}\nu_i - \frac{3}{2}kT\sum_{i=1}^{M}\nu_i\ln\left(\frac{2\pi M_i kT}{h^2}\right) - kT\sum_{i=1}^{M}\nu_i\ln\zeta_i = 0\,,$$

or

$$\left(\frac{1}{V}\right)^{\sum\nu_i} = \left[\left(\frac{2\pi M_1 kT}{h^2}\right)^{3/2}\zeta_1\right]^{\nu_1} \cdot \ldots \cdot \left[\left(\frac{2\pi M_M kT}{h^2}\right)^{3/2}\zeta_M\right]^{\nu_M}\,,$$

which is a meaningless relation between the temperature and the volume.

Solution 2.6 *Exercise on page 372*

When $\Delta E_0 < 0$,

$$\frac{d\ln K}{dT} < 0\,,$$

namely K decreases with increasing temperature.

The reaction releases heat, and hence an increase in temperature will lead to the reverse reaction, which absorbs heat and tends to cancel the change in temperature. Thus, equilibrium will tend towards the original composition.

Solution 3.1 *Exercise on page 378*

(a) We write (4.3.4) in the form

$$\ddot{x}_\alpha = \Omega^2(x_{\alpha+1} - 2x_\alpha + x_{\alpha-1}),$$

with $\Omega^2 = K/M$, and substitute (4.3.5a) in order to check if it is a solution. Differentiating twice with respect to t is equivalent to multiplying by $-\omega^2$, and thus we find that (4.3.4) is satisfied provided:

$$-\omega^2 a \sin(qb\alpha)\sin\omega t = \Omega^2 a[\sin qb(\alpha+1) - 2\sin qb\alpha + \sin qb(\alpha-1)]\sin\omega t.$$

The expression in the square brackets can be written, using identities between trigonometric functions, in the form

$$\begin{aligned} 2\sin(qb\alpha)\cos(qb) - 2\sin(qb\alpha) &= -2\sin(qb\alpha)(1-\cos qb) \\ &= -4\sin(qb\alpha)\sin^2\left(\frac{qb}{2}\right), \end{aligned}$$

■ $\sin x + \sin y = 2\sin\frac{x+y}{2}\cos\frac{x-y}{2}$.

and the right hand side becomes

$$-4\Omega^2 a \sin(qb\alpha)\sin^2\left(\frac{qb}{2}\right)\sin\omega t.$$

Comparing this expression to the left hand side we find that (4.3.5a) is a solution provided that

$$\omega^2 = 4\Omega^2\sin^2\left(\frac{qb}{2}\right),$$

from which $\omega(q)$ is obtained. The range of q values in Eq. (4.3.5b) can be obtained by studying the form of the solution (4.3.5a). Since the function $\sin(qb\alpha)$ is periodic and α is an integer, the function with a wave number $q + 2\pi/b$ is identical to that with q. Moreover, a solution with $q + \pi/b$ is equivalent to a solution with $q - \pi/b$. It is therefore clear that we will not get any new solutions from wave numbers outside the range $0 \le q \le \pi/b$.

Solution 3.2 *Exercise on page 379*

First, we shall perform the change of variables in the kinetic energy part

$$E_k = \frac{m}{2}\left[\frac{1}{4}(\dot{u}_1 + \sqrt{2}\dot{u}_2 - \dot{u}_3)^2 + \frac{1}{2}(\dot{u}_1 + \dot{u}_3)^2 + \frac{1}{4}(\dot{u}_1 - \sqrt{2}\dot{u}_2 - \dot{u}_3)^2\right]$$
$$= \frac{m}{2}(\dot{u}_1^2 + \dot{u}_2^2 + \dot{u}_3^2)\,, \tag{i}$$

where all of the mixed terms cancel out.

The potential part can be written as the sum

$$E_p = K(x_1^2 + x_2^2 + x_3^2) - K(x_1x_2 + x_2x_3)\,. \tag{ii}$$

The first term assumes a form similar to that of the kinetic energy, since it involves exactly the same coefficients:

$$K(x_1^2 + x_2^2 + x_3^2) = K(u_1^2 + u_2^2 + u_3^2)\,. \tag{iii}$$

The second part of the potential also reduces to a sum of squares:

$$K(x_1x_2 + x_2x_3) = K(x_1 + x_3)x_2 = \frac{K}{\sqrt{2}}(u_1 - u_3)(u_1 + u_3) = \frac{K}{\sqrt{2}}(u_1^2 - u_3^2)\,, \tag{iv}$$

so that overall

$$E_p = \left(1 + \frac{1}{\sqrt{2}}\right)Ku_1^2 + Ku_2^2 + \left(1 - \frac{1}{\sqrt{2}}\right)Ku_3^2\,. \tag{v}$$

The total energy of the three coupled oscillators transforms, therefore, to an expression that looks exactly like the sum of the energies of three free oscillators, where each of the new variables is a linear combination of the original variables.

The energy can be written as

$$E = \epsilon_1 + \epsilon_2 + \epsilon_3\,, \tag{vi}$$

with

$$\epsilon_1 = \frac{m}{2}\dot{u}_1^2 + \left(1 + \frac{1}{\sqrt{2}}\right)Ku_1^2\,, \tag{vii}$$

$$\epsilon_2 = \frac{m}{2}\dot{u}_2^2 + Ku_2^2\,, \tag{viii}$$

$$\epsilon_3 = \frac{m}{2}\dot{u}_3^2 + \left(1 - \frac{1}{\sqrt{2}}\right)Ku_3^2\,. \tag{ix}$$

Each of these free oscillators has a different frequency. Denoting $\Omega^2 = K/m$ we obtain

$$\omega_1^2 = (2 + \sqrt{2})\Omega^2,\ \omega_2^2 = 2\Omega^2,\ \omega_3^2 = (2 - \sqrt{2})\Omega^2\,. \tag{x}$$

Comment

The change of variables performed in this solution may seem to be a *deus ex machina*, but this is not so. This is a simple example of the general way of dealing with harmonic vibrations of systems of N particles. The general result is that Eq. (4.3.1) is written as a sum over single oscillator energies ϵ_α, where

$$\epsilon_\alpha = \frac{m}{2}(\dot{u}_\alpha^2 + \omega_\alpha^2 u_\alpha^2)\,.$$

Solution 3.3 *Exercise on page 383*

Since the total number of states inside a sphere of radius q_D is $3N$, we obtain the condition

$$\frac{3V}{(2\pi)^3}\int d^3q = 3N\,.$$

Integrating over the angles in wave vector space, we obtain

$$\frac{3V}{2\pi^2}\int_0^{q_D} q^2 dq = 3N\,,$$

and after integrating over q,

$$\frac{Vq_D^3}{2\pi^2} = 3N \Rightarrow q_D = \left(\frac{6\pi^2 N}{V}\right)^{1/3}\,.$$

Solution 3.4 *Exercise on page 383*

Since $F = -kT\ln Z$ and F is given in Eq. (4.3.11), we can write

$$\ln Z = -\frac{F}{kT} = -\beta F\,,$$

so that

$$E = -\frac{\partial \ln Z}{\partial \beta} = \frac{\partial(\beta F)}{\partial \beta} = \frac{3V}{2\pi^2}\frac{\partial}{\partial \beta}\int_0^{q_D} \ln(1 - e^{-\beta\hbar v q})q^2 dq\,,$$

where in the last step have substituted (4.3.11). We change the variable of integration to ω, differentiate with respect to β and obtain

$$E = \frac{3V}{2\pi^2 v^3}\frac{\partial}{\partial \beta}\int_0^{\omega_D} \ln(1 - e^{-\beta\hbar\omega})\omega^2 d\omega = \frac{3V}{2\pi^2 v^3}\int_0^{\omega_D} \frac{e^{-\beta\hbar\omega}\hbar\omega}{1 - e^{-\beta\hbar\omega}}\omega^2 d\omega\,.$$

Multiplying the numerator and the denominator in the integrand by $e^{\beta\hbar\omega}$, we obtain (4.3.13).

Note that we could differentiate under the integration sign because the limit of the integration, ω_D is independent of β.

Solution 3.5 *Exercise on page 384*

To calculate the specific heat, namely the integral in Eq. (4.3.16), we change variable to $x = \hbar\omega/kT$. We have to replace the upper integration limit by

$$x_D = \frac{\hbar\omega_D}{kT},$$

so that

$$C = \frac{3kV}{2\pi^2}\left(\frac{kT}{\hbar v}\right)^3 \int_0^{x_D} \frac{x^4 e^x}{(e^x - 1)^2} dx\,.$$

Using (4.3.12b) we write

$$\left(\frac{kT}{\hbar v}\right)^3 = \left(\frac{kT}{\hbar\omega_D}\right)^3 \cdot \frac{6\pi^2 N}{V}$$

and then

$$C = 9Nk\left(\frac{kT}{\hbar\omega_D}\right)^3 \int_0^{x_D} \frac{x^4 e^x}{(e^x - 1)^2} dx\,.$$

This is where the low temperature approximation enters. When T is very small, $x_D \gg 1$, and then we can approximate the integral by

$$I \equiv \int_0^{\infty} \frac{x^4 e^x}{(e^x - 1)^2} dx\,.$$

The value of this integral can be looked up in tables. See also the appendix at the end of this part. We write

$$I = -\int_0^{\infty} x^4 \frac{d}{dx}\left[\frac{1}{e^x - 1}\right] dx = 4\int_0^{\infty} \frac{x^3 dx}{e^x - 1},$$

and using (4.A.5) we find the value of the integral to be $I = 4\pi^4/15$. Then by substituting in the expression for C, we obtain (4.3.17).

Solution 4.1 *Exercise on page 387*

The present problem is a simplified version of the argument made in Sec. 3.7 (the appendix) of Part I. There are three simplifications on our side:

(a) The distribution of radiation is position independent.
(b) All the photons have the same speed — c.
(c) The mean free path is infinite.

We can use Fig. 1.3.7. The amount of radiation energy with frequency between ω and $\omega + d\omega$ in a volume element $dV (= r^2 dr \sin\theta d\theta d\phi)$ is $\rho(\omega) d\omega dV$. The radiation propagates from dV in all directions at a speed c. The part $\Delta A \cos\theta / 4\pi r^2$ (the ratio between the solid angle spanned by the area element and the whole sphere) of the radiation is directed towards the surface element which concerns us.

The energy of the radiation between ω and $\omega + d\omega$, which will cross the unit area during unit time Δt, will be the sum of all the contributions coming from all the volume elements which are found in half a sphere of radius $c\Delta t$. Namely,

$$\begin{aligned}\Delta E &= \int_0^{c\Delta t} dr \int_0^{\pi/2} d\theta \int_0^{2\pi} d\phi r^2 \sin\theta \rho(\omega)\Delta\omega \frac{\Delta A \cos\theta}{4\pi r^2} \\ &= \frac{1}{2}\Delta A \Delta\omega \rho(\omega) \int_0^{c\Delta t} dr \int_0^{\pi/2} \sin\theta \cos\theta d\theta\,,\end{aligned}$$

where we have performed the integration over ϕ. We are thus left with elementary integrals over r and θ whose result is $\frac{1}{2}c\Delta t$. Hence

$$\Delta E = \frac{c}{4}\rho(\omega)\Delta\omega\Delta t\Delta A$$

and the amount of energy per unit time per unit area and per unit frequency is given by (4.4.2).

Solution 4.2 *Exercise on page 389*

A sinusoidal electromagnetic wave is an electric field and a magnetic field of the form

$$\begin{aligned}\mathbf{E} &= \mathsf{E}\sin(\mathbf{q}\cdot\mathbf{r} - \omega t)\,, \\ \mathbf{B} &= \mathsf{B}\sin(\mathbf{q}\cdot\mathbf{r} - \omega t)\,,\end{aligned}$$

where E and B describe the wave's amplitude and the direction of its polarization. We shall now use the two Maxwell equations

$$\begin{aligned}\nabla\cdot\mathbf{E} &= 0\,, \\ \nabla\cdot\mathbf{B} &= 0\,.\end{aligned}$$

The calculation of the divergence is the same for the two fields. For example,

$$\begin{aligned}\nabla\cdot\mathbf{E} &= \frac{\partial E_x}{\partial x} + \frac{\partial E_y}{\partial y} + \frac{\partial E_z}{\partial z} = \left(\mathsf{E}_x\frac{\partial}{\partial x} + \mathsf{E}_y\frac{\partial}{\partial y} + \mathsf{E}_z\frac{\partial}{\partial z}\right)\sin(\mathbf{q}\cdot\mathbf{r} - \omega t) \\ &= \mathbf{q}\cdot\mathsf{E}\cos(\mathbf{q}\cdot\mathbf{r} - \omega t)\,,\end{aligned}$$

and since the divergence vanishes at every point at all times,

$$\mathbf{q}\cdot\mathsf{E} = 0\,,$$

namely E is perpendicular to $\mathbf{q}$, and hence $\mathbf{E}$ is also perpendicular to $\mathbf{q}$.

In exactly the same manner it is possible to show that **B** is perpendicular to **q**, which shows that this is a transverse wave.

Solution 4.3 *Exercise on page 392*

(a) When

$$\beta\hbar\omega \ll 1\,, \tag{i}$$

we can expand the exponential function in the denominator of (4.4.12):

$$e^{\beta\hbar\omega} \approx 1 + \beta\hbar\omega\,, \tag{ii}$$

and then

$$\rho(\omega) \approx \frac{1}{\pi^2 c^3}\frac{\hbar\omega^3}{\beta\hbar\omega} = \frac{kT\omega^2}{\pi^2 c^3}\,. \tag{iii}$$

(b) In the Einstein model, condition (i) can be interpreted as a condition for high temperatures (see Sec. 2.3 of Part III). This is so since in the Einstein model all the oscillators have the same frequency ω. In contrast, for a black body, all the frequencies from zero up to infinity appear. Hence there is no temperature that satisfies condition (i) for *all* the oscillators. On the other hand the condition is satisfied by the low frequency modes, once the temperature is given. The Rayleigh–Jeans law therefore describes the Planck distribution in the low frequency range.

Solution 4.4 *Exercise on page 392*

(a) First we write the frequency (and not the angular frequency) distribution by changing variables in (4.4.14) and using the fact that $I(\nu)d\nu = I(\omega)d\omega$:

$$I(\nu)d\nu = \frac{2\pi h}{c^2}\frac{\nu^3}{e^{\beta h\nu} - 1}d\nu\,. \tag{i}$$

In order to find the frequency at which the emissivity is maximal we differentiate with respect to ν and obtain

$$\frac{dI}{d\nu} = \frac{2\pi h\nu^2}{c^2}\frac{3e^{\beta h\nu} - \beta h\nu e^{\beta h\nu} - 3}{(e^{\beta h\nu} - 1)^2}\,. \tag{ii}$$

The sign of the derivative is determined by the numerator of the second fraction in (ii), which we write after the change of variables $x = \beta h\nu$ in the form

$$A(x) = 3e^x - xe^x - 3\,. \tag{iii}$$

This function is positive between $x = 0$ and $x = x_{\max}$ and negative for $x > x_{\max}$ (check this). It vanishes at the point $x = x_{\max}$ which is a solution of the equation $A(x) = 0$. This is a transcendental equation to be solved numerically. Equation (iii) can be written in the form

$$x = f(x)\,, \tag{iv}$$

with $f(x) \equiv 3(1 - e^{-x})$. It can be iterated according to

$$x_{n+1} = f(x_n) \tag{v}$$

until $|x_{n+1} - x_n|$ becomes less than a desired precision.

We choose arbitrarily $x_0 = 2.5$ and iterate according to (v). The results are summarized in the table below.

n	0	1	2	3	4
x	2.500000	2.753745	2.808933	2.819192	2.821038
		5	6	7	8
		2.821368	2.821427	2.821437	2.821439

Hence x converges to the value $x_{\max} = 2.8214$ and from here

$$\frac{h\nu_{\max}}{kT} = 2.8214\,. \tag{vi}$$

Substituting the values of Planck's constant and Boltzmann's constant we obtain (4.4.15).

(b) In order to find the total power emitted from a unit surface area of a black body, we have to integrate over Eq. (i):

$$I = \int_0^\infty I(\nu)d\nu = \frac{2\pi}{c^2h^3\beta^4}\int_0^\infty \frac{x^3}{e^x - 1}dx\,,$$

after the change of variable to $x = \beta h\nu$.

In the solution to Exercise 3.5, we found that this integral is equal to $\pi^4/15$. Hence

$$I = \frac{2\pi^5(kT)^4}{15c^2h^3} = \sigma T^4\,.$$

Note that by integrating the Rayleigh–Jeans distribution we obtain a diverging result that implies an emission of infinite power. This is the well-known *ultraviolet catastrophe.*

Solution 4.5 *Exercise on page 393*

(a) The free energy, for the case of electromagnetic radiation, is given by (4.4.9):

$$F = \frac{kTV}{\pi^2 c^3} \int_0^\infty \omega^2 \ln(1 - e^{-\beta\hbar\omega}) d\omega \,. \tag{i}$$

With $x = \beta\hbar\omega = \hbar\omega/kT$ it becomes

$$F = \frac{(kT)^4 V}{\pi^2 c^3 \hbar^3} \int_0^\infty x^2 \ln(1 - e^{-x}) dx \,. \tag{ii}$$

The integral in (ii) is a numerical constant whose value is found after integration by parts using the integral in the preceding exercise:

$$\int_0^\infty x^2 \ln(1 - e^{-x}) dx = -\frac{1}{3} \int_0^\infty \frac{x^3}{1 - e^{-x}} dx = -\frac{\pi^4}{45} \,, \tag{iii}$$

and from here

$$F(T, V) = -\frac{\pi^2 V (kT)^4}{45(\hbar c)^3} = -\frac{4\sigma}{3c} V T^4 \,, \tag{iv}$$

$$S = -\left(\frac{\partial F}{\partial T}\right)_V = \frac{16\sigma}{3c} V T^3 \,, \tag{v}$$

$$C_V = T \left(\frac{\partial S}{\partial T}\right)_V = \frac{16\sigma}{c} V T^3 \,. \tag{vi}$$

(b) The chemical potential is

$$\mu = \left(\frac{\partial F}{\partial N}\right)_{T,N} = 0 \,. \tag{vii}$$

The reason is that the free energy is independent of the number of photons in the container, which is not a constant (conserved) quantity. Photons are created and annihilated freely as a result of absorption and emission by the walls of the container. The meaningful quantity in this context is the average number of photons [see Eq. (4.4.13) and Self-Assessment Exercise 7 (a)]:

$$\langle N \rangle = \frac{V}{\pi^2 c^3} \int_0^\infty \frac{\omega^2}{e^{\beta\hbar\omega} - 1} d\omega = \left(\frac{kT}{\hbar c}\right)^3 \frac{V}{\pi^2} \int_0^\infty \frac{x^2}{e^x - 1} dx \,. \tag{viii}$$

The integral is a dimensionless number, and hence $\langle N \rangle$ is proportional to VT^3, as are S and C_V.

Solution 4.6 *Exercise on page 393*

(a) In the solution to the preceding exercise we have calculated the free energy of electromagnetic radiation:

$$F = -\frac{4\sigma}{3c} V T^4 \,. \tag{i}$$

The radiation pressure is therefore

$$P = -\left(\frac{\partial F}{\partial V}\right)_T = \frac{4\sigma}{3c} T^4 \,. \tag{ii}$$

Using the expression for the energy density Eq. (4.4.17),

$$u = \frac{4\sigma}{c} T^4 \,, \tag{iii}$$

we can rewrite (ii) in the form

$$P = \frac{1}{3} u \tag{iv}$$

and obtain the equation of state

$$PV = \frac{1}{3} E \,, \tag{v}$$

which is exactly Eq. (1.1.7), obtained in Part I for particles moving at the speed of light and satisfying $\epsilon = pc$.

(b) The compressibility is

$$K = -V \left(\frac{\partial P}{\partial V}\right)_T = 0 \tag{vi}$$

since P, at constant T, is independent of the volume [see Eq. (ii)]. In an ideal gas, on the contrary,

$$P = \frac{NkT}{V} \,, \tag{vii}$$

and from here

$$K = -V \left(\frac{\partial P}{\partial V}\right)_T = \frac{NkT}{V} = P \,. \tag{viii}$$

The big difference is due to the fact that when the volume of the gas decreases at constant temperature, the number of particles remains constant, and the pressure increases. In contrast, when the volume of radiation decreases at constant T, the number of photons decreases [see Eq. (viii) in the solution of the preceding exercise], and the pressure remains constant.

Solution 4.7 *Exercise on page 393*

In Solution 4.6 we obtained the pressure of radiation:

$$P = \frac{4\sigma}{3c} T^4 , \tag{i}$$

hence

$$T = \left(\frac{3cP}{4\sigma} \right)^{1/4} . \tag{ii}$$

(a) For $P = 1$ atm,

$$T = \left(\frac{3(3 \times 10^8)(1.013 \times 10^5)}{4(5.67 \times 10^{-8})} \right)^{1/4} = 1.4 \times 10^5 \text{ K} .$$

(b) For $P = 10^{-10}$ atm,

$$T = \left(\frac{3(3 \times 10^8)(10^{-10} \times 1.013 \times 10^5)}{4(5.67 \times 10^{-8})} \right)^{1/4} = 450 \text{ K} .$$

(c) The pressure of radiation at the center of the sun ($T = 10^7$ K) is

$$P = \frac{4(5.67 \times 10^{-8})(10^7)^4}{3(3 \times 10^8)} = 2.5 \times 10^{12} \text{ N/m}^2 = 2.5 \times 10^7 \text{ atm} .$$

Since the pressure of the *gas* there is 10^{11} atm, the pressure of radiation in the sun is still negligible in the balance for mechanical equilibrium.

(d) From Eq. (iv) in Solution 4.6,

$$P = \frac{1}{3} u , \tag{i}$$

and by comparing (4.4.16) to (4.4.17) we find that

$$u = \frac{4}{c} I , \tag{ii}$$

and hence

$$P = \frac{4I}{3c} . \tag{iii}$$

The radiation pressure on Earth's surface is

$$P = \frac{4 \times 1400}{3 \times 3 \times 10^8} = 6.22 \times 10^{-6} \text{ N/m}^2 = 6 \times 10^{-11} \text{ atm} .$$

Solution 4.8 *Exercise on page 397*

Were the detailed balance not to hold separately for each frequency, we would be able to find a frequency in which the body absorbs more radiation than it emits. As a result, the energy density inside the container would decrease at this frequency. The radiation distribution would deviate from Planck's distribution and as a result thermodynamic equilibrium would be violated, even though initially the temperature of the body was equal to the temperature of the radiation. A similar situation would have occurred were detailed balance not to hold for each direction of q separately.

Since it is impossible for the system, once it has reached thermodynamic equilibrium, to leave it, our assumption regarding the detailed balance must be wrong.

Solution 4.9 *Exercise on page 397*

(a) A black body has been defined as a body that absorbs all the radiation incident upon it. The absorbing power of body A was defined as the ratio between the intensity of radiation absorbed by the body and the intensity of radiation incident upon it. Hence, for a black body $A = 1$ and (4.4.19) implies that

$$L_B = \frac{c}{4\pi}\rho \, .$$

(b) The emissivity has been defined as the total power emitted per unit area per unit frequency, whereas the emissive power is the power emitted by a unit area *in a certain direction* (determined by the direction of the wave vector) per unit solid angle per unit frequency. The relation between the two quantities is, therefore,

$$dI = L \cos\theta d\Omega \, .$$

For a black body, L is independent of the angle and is given by (4.4.22). Thus,

$$I(\omega) = \frac{c}{4\pi}\rho(\omega) \int \cos\theta d\Omega = \frac{c}{4}\rho(\omega) \, ,$$

which is Eq. (4.4.2). We have performed the integration only in the directions that describe emission, namely $0 \leq \theta \leq \pi/2$.

Solution 4.10 *Exercise on page 397*

We shall write the emissive power as a function of the frequency ν using Eq. (4.4.22):

$$L_B(\mathbf{q}, T) = \frac{c}{4\pi}\rho(\omega, T) = \frac{1}{4\pi^3 c^2}\frac{\hbar\omega^3}{e^{\beta\hbar\omega} - 1} = \frac{1}{\pi c^2}\cdot\frac{h\nu^3}{e^{\beta h\nu} - 1}.$$

The maximum of L_B satisfies Wien's law, (4.4.15), which we derived from the emissivity I, and hence we can calculate ν_{max} at each of the temperatures from Wien's law and substitute in the function L_B. It is somewhat simpler to first substitute (4.4.15) into L_B and express $L_{B_{\text{max}}}$ as a function of the temperature only:

$$L_{B_{\text{max}}} = L_B(\nu = \nu_{\text{max}}) = \frac{(kT)^3}{\pi c^2 h^2}\cdot\frac{x_{\text{max}}^3}{e^{x_{\text{max}}} - 1}.$$

If we substitute the physical constants (and $x_{\text{max}} = 2.8214$) and if T is measured in K and L_B in J m^{-2}, then

$$L_{B_{\text{max}}} = 3.018 \times 10^{-20} T^3 .$$

The numerical results are summarized in the following table:

T	(K)	100	1,000	10,000	100,000
ν_{max}	(Hz)	5.88×10^{12}	5.88×10^{13}	5.88×10^{14}	5.88×10^{15}
$L_{B_{\text{max}}}$	(J m^{-2})	3.018×10^{-14}	3.018×10^{-11}	3.018×10^{-8}	3.018×10^{-5}

Solutions to self-assessment exercises

Solution 1 *Exercise on page 403*

(a) The scheme of the low levels that interest us is

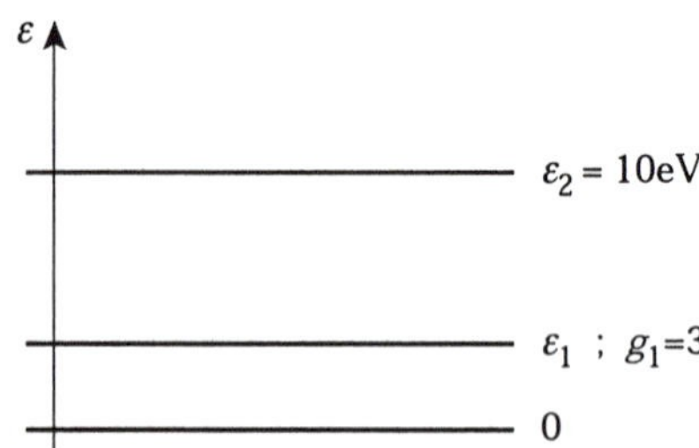

The wavelength of the emitted line in the transition from ϵ_1 to the ground state is 21 cm. Hence

$$\epsilon_1 = \hbar\omega = \frac{hc}{\lambda} = \frac{(6.626 \times 10^{-34})(3 \times 10^8)}{0.21}$$
$$\approx 9.5 \times 10^{-25}\ \mathrm{J} \approx 5.9 \times 10^{-6}\ \mathrm{eV}\,.$$

(b) Because $\epsilon_2 = 10$ eV,

$$\beta(\epsilon_2 - \epsilon_1) \approx \beta\epsilon_2 \approx \frac{1.6 \times 10^{-18}}{(1.38 \times 10^{-23})T} \approx \frac{1.2 \times 10^5}{T}\,,$$

so that for $T \ll 10^5$ K we may use the approximation (4.1.36). Taking $\epsilon_0 = 0$, we have

$$\zeta = 1 + g_1 e^{-\beta\epsilon_1}\,.$$

Specifically for 0.1 K,

$$\zeta(0.1\ \mathrm{K}) = 1 + 3e^{-\beta\epsilon_1} = 1 + 3\exp\left(-\frac{\epsilon_1}{kT}\right)$$
$$= 1 + 3\exp\left(-\frac{9.5 \times 10^{-25}}{1.38 \times 10^{-24}}\right) \approx 2.5\,.$$

(c) The characteristic temperature of the first excited level is

$$\Theta_1 = \frac{\epsilon_1}{k} \approx \frac{0.11 \times 1.6 \times 10^{-19}}{1.38 \times 10^{-23}} \approx 1,250 \text{ K}.$$

(d) The level scheme is:

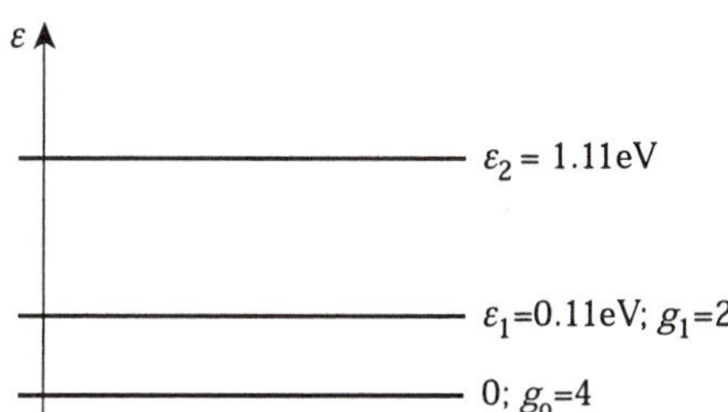

The characteristic temperature of the second excited level is

$$\Theta_2 = \frac{\epsilon_2}{k} \approx \frac{1.11 \times 1.6 \times 10^{-19}}{1.38 \times 10^{-23}} \approx 1.3 \times 10^4 \text{ K}.$$

Namely, we are looking for ζ for $T \ll 1.3 \times 10^4$ K, where the term containing ϵ_2 is negligible. We thus have

$$\zeta(T) = g_0 + g_1 e^{-\beta\epsilon_1} = 4 + 2e^{-\Theta_1/T}.$$

(e) The contribution of the atomic levels to the specific heat is derived from ζ using Eq. (4.1.33):

$$\frac{\Delta C_V}{N} = k\frac{d}{dT}\left(T^2\frac{d}{dT}\ln\zeta\right).$$

Substituting $\zeta(T)$ we obtain

$$T^2\frac{d}{dT}\ln\zeta = T^2\frac{d}{dT}\ln(2 + e^{-\Theta_1/T}) = \frac{\Theta_1 e^{-\Theta_1/T}}{2 + e^{-\Theta_1/T}} = \frac{\Theta_1}{2e^{\Theta_1/T} + 1}.$$

Substituting this in the expression for C_V yields

$$\frac{C_V}{N} = \frac{3}{2}k + k\frac{d}{dT}\frac{\Theta_1}{2e^{\Theta_1/T} + 1} = \frac{3}{2}k + \frac{2k(\Theta_1/T)^2 e^{\Theta_1/T}}{(2e^{\Theta_1/T} + 1)^2}.$$

The graph describes the specific heat as a function of temperature, where the temperature is measured in units of Θ_1. Only the contribution due to the internal structure, above the ideal gas value, is depicted. We have seen a specific heat of similar structure for the paramagnet (see Exercise 1.8 in this part and Fig. 2.5.2).

The similarity is due to the fact that the atoms of the gas have two (active) electronic energy levels, as does the paramagnetic moment. But in contrast the two levels have different degeneracies, and thus the specific heat does not look exactly like that of the paramagnet. In addition, in the gas there are degrees of freedom related to translational motion of the molecules. As a result, the specific heat of the gas will not vanish when $T \to 0$, nor when $T \to \infty$.

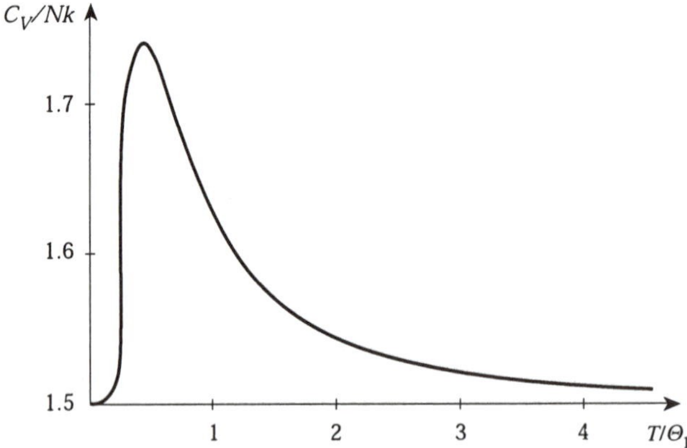

Only in a narrow range of temperatures will a maximum appear whose origin is similar to that in the paramagnetic case. If T is less than the energy spacing between the two levels, it is impossible to excite the atoms to the second level, and the heat must be absorbed by translation. When the temperature equals the spacing, the system can absorb energy in a much more efficient manner than by translation — the atoms pass from the ground level to the excited one. And later on, when there are no more molecules to excite, translation is again the only option.

Solution 2 *Exercise on page 403*

The electronic levels: the ground level is at $\epsilon_0 = 0$, the first excited level at $\epsilon_1 = 1$ eV, and the second excited state $\epsilon_2 = 2$ eV.

The corresponding characteristic electronic temperature

$$\Theta_e = \frac{1.6 \times 10^{-19}}{1.38 \times 10^{-23}} \approx 11,600 \text{ K}. \qquad \text{(i)}$$

The distances between the magnetic levels are of order $\mu_0 H$, and they define a characteristic magnetic temperature:

$$\Theta_M = \frac{\mu_0 H}{k} = \frac{(2 \times 10^{-27} \text{ J gauss}^{-1}) \times 10^6 \text{ gauss}}{1.38 \times 10^{-23} \text{ J K}^{-1}} \approx 145 \text{ K} \ll \Theta_e. \qquad \text{(ii)}$$

The contributions to the partition function can be written as

$$\zeta_e = 1 + e^{-\epsilon_1/kT} + e^{-\epsilon_2/kT} = 1 + e^{-\Theta_e/T} + e^{-2\Theta_e/T} , \qquad \text{(iii)}$$

$$\begin{aligned}\zeta_M &= \sum_{\sigma=-2}^{2} e^{-\sigma\mu_0 H/kT} = 1 + e^{\Theta_M/T} + e^{-\Theta_M/T} + e^{2\Theta_M/T} + e^{-2\Theta_M/T} \\ &= 1 + 2\cosh\left(\frac{\Theta_M}{T}\right) + 2\cosh\left(\frac{2\Theta_M}{T}\right) . \qquad \text{(iv)}\end{aligned}$$

The change in the chemical potential will be calculated using (4.1.35):

$$\begin{aligned}\Delta\mu &= -kT\ln\zeta = -kT(\ln\zeta_e + \ln\zeta_M) \\ &= -kT\left\{ \ln\left[1 + \exp\left(-\frac{\Theta_e}{T}\right) + \exp\left(-\frac{2\Theta_e}{T}\right)\right]\right. \\ &\quad \left. + \ln\left[1 + 2\cosh\left(\frac{\Theta_M}{T}\right) + 2\cosh\left(\frac{2\Theta_M}{T}\right)\right]\right\} . \qquad \text{(v)}\end{aligned}$$

At very low temperatures ($T \ll \Theta_M$) the first ln is negligible, whereas the second one behaves as

$$\ln\left[\cosh\left(\frac{2\Theta_M}{T}\right)\right] \approx \frac{2\Theta_M}{T} ,$$

so that there

$$\Delta\mu \approx -2k\Theta_M .$$

At very high temperatures ($T \gg \Theta_M$) both terms in the braces are constant, and hence $\Delta\mu$ decreases linearly with the temperature. The following figure illustrates $\Delta\mu$ as a function of T. The temperature scale is logarithmic.

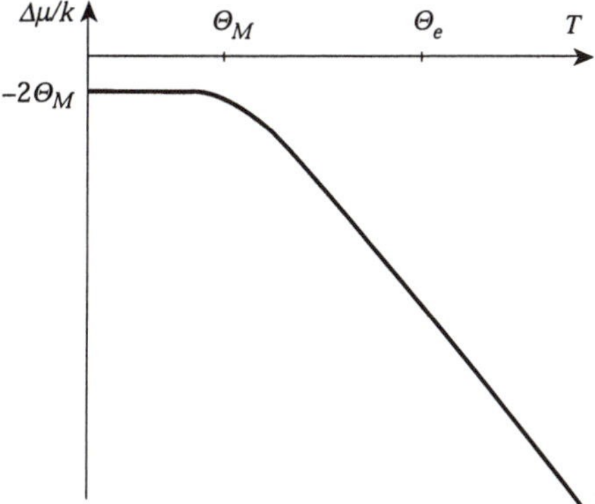

The change in the specific heat (per atom) will be calculated using (4.1.33):

$$\frac{\Delta c_V}{k} = \frac{\Delta C_V}{Nk} = \frac{d}{dT}\left[T^2\left(\frac{d}{dT}\ln\zeta\right)\right] . \qquad \text{(vi)}$$

Even without calculating the derivatives we can obtain an idea of the nature of the dependence of Δc_V upon T. Since $\ln\zeta$ is a sum of two terms, the specific heat Δc_V will also be a sum of a magnetic part and a part originating in the atomic levels. The specific heat originating from the magnetic moment "unfreezes" at temperatures around Θ_M. Hence it behaves as the specific heat of a paramagnet whose general features are depicted in Fig. 2.5.2 (in Part II). The specific heat originating in the atomic levels unfreezes at temperatures around Θ_e, and looks like the one illustrated in the solution of the preceding exercise, which also resembles that of a paramagnet. Overall we therefore obtain a specific heat with two high peaks: one at $T \approx \Theta_M$ and the other at $T \approx \Theta_e$, as illustrated in the following figure:

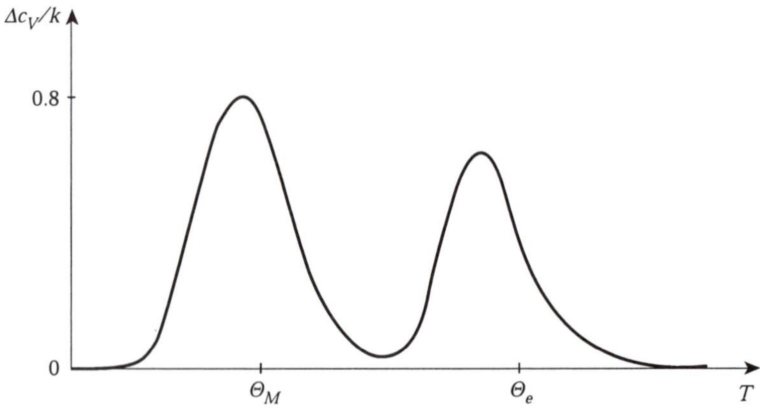

Note that here as well the temperature scale is logarithmic.

In order to obtain the values of Δc_V around the points of the maxima ($T \approx \Theta_M, \Theta_e$), we have to calculate explicitly each of the contributions. We can obtain the first directly from Eq. (vi), but it is much simpler to use Exercise 5.7(d) in Part II:

$$\frac{(\Delta c_V)_M}{k} = (\beta\mu_0 H)^2 \left[\frac{1}{4\sinh^2\left(\frac{1}{2}\beta\mu_0 H\right)} - \frac{25}{4\sinh^2\left(\frac{5}{2}\beta\mu_0 H\right)}\right], \tag{vii}$$

where we have substituted $J = 2$ (for which $2J+1 = 5$) and $\mu_0 = g\mu_B$. Equation (vii) can be written using Θ_M:

$$\frac{(\Delta c_V)_M}{k} = \left(\frac{\Theta_M}{2T}\right)^2 \left[\frac{1}{\sinh^2(\Theta_M/2T)} - \frac{25}{\sinh^2(5\Theta_M/2T)}\right] \tag{viii}$$

and for $T = \Theta_M$:

$$\frac{(\Delta c_V)_M}{k} = 0.75\,. \tag{ix}$$

A more detailed study, which we shall not reproduce here, reveals that the maximum of $(\Delta c_V)_M$ is not attained at exactly $T = \Theta_M$ but at $T \approx 0.75\Theta_M$, and that $(\Delta c_V)_M/k \approx 0.8$.

The second contribution to the specific heat can again be calculated directly from Eq. (vi). However, it is again preferable to use the known results for the paramagnet from Exercise 5.7 of Part II, since the atomic excitations behave here exactly like a paramagnet with spin $J = 1$. Replacing $\beta g \mu_B H$ by Θ_e/T we write

$$\frac{(\Delta c_V)_e}{k} = \left(\frac{\Theta_e}{2T}\right)^2 \left[\frac{1}{\sinh^2(\Theta_e/2T)} - \frac{9}{\sinh^2(3\Theta_e/2T)}\right] \tag{x}$$

and then for $T = \Theta_e$

$$\frac{(\Delta c_V)_e}{k} = 0.42\,. \tag{xi}$$

Again the maximum of $(\Delta c_V)_e$ is not attained exactly at $T = \Theta_e$ but at $T = 0.53\Theta_e$, and $(\Delta c_V)_e/k = 0.64$ at the maximum.

The reason why the specific heat contributed by the atomic excitations is smaller than that originating from the magnetic moment is that the number of atomic levels is smaller than the number of the magnetic levels.

Solution 3 *Exercise on page 403*

(a) The reaction:

$$2\mathrm{NaOH} + \mathrm{H_2SO_4} \rightleftharpoons \mathrm{Na_2SO_4} + 2\mathrm{H_2O}\,, \tag{i}$$

$$\begin{array}{llll} B_1 = \mathrm{NaOH}, & B_2 = \mathrm{H_2SO_4}, & B_3 = \mathrm{Na_2SO_4}, & B_4 = \mathrm{H_2O}, \\ b_1 = 2, & b_2 = 1, & b_3 = 1, & b_4 = 2, \\ \nu_1 = -2, & \nu_2 = -1, & \nu_3 = 1, & \nu_4 = 2\,. \end{array}$$

The law of mass action reads

$$\frac{n_3 n_4^2}{n_1^2 n_2} = K(T)\,. \tag{ii}$$

(b) The additional equations are

$$n_{\mathrm{Na}} = n_1 + 2n_3\,, \tag{iii}$$

$$n_{\mathrm{O}} = n_1 + 4n_2 + 4n_3 + n_4\,, \tag{iv}$$

$$n_{\mathrm{H}} = n_1 + 2n_2 + 2n_4\,, \tag{v}$$

$$n_{\mathrm{S}} = n_2 + n_3\,. \tag{vi}$$

There are four additional equations, since four types of atoms appear in the reaction. Overall there will be five equations for determining the densities. Hence it appears that if the atomic densities are given arbitrary values, it will be impossible to find a solution for all five equations.

(c) The solution of the mystery is the fact that the four equations (iii)–(vi) are not independent, as you will find if you try to solve them and to calculate the four densities n_1, n_2, n_3, n_4 in terms of the atomic densities. The dependence between the equations is also reflected in the fact that

$$n_{\mathrm{H}} + n_{\mathrm{Na}} + 6n_{\mathrm{S}} - 2n_{\mathrm{O}} = 0 .$$

Namely, any three of the densities determine the fourth density. This is of course a special property of reaction (i). As a result, we have only three independent accompanying equations and together with Eq. (ii) we can obtain a single solution for the densities of the molecules.

Solution 4 *Exercise on page 404*

In the reaction

$$\mathrm{Cl_2} + \mathrm{H_2} \rightleftharpoons 2\mathrm{HCl} ,$$

$$\begin{aligned} &B_1 = \mathrm{Cl_2}, \quad &B_2 = \mathrm{H_2}, \quad &B_3 = \mathrm{HCl}, \\ &b_1 = 1, &b_2 = 1, &b_3 = 2, \\ &\nu_1 = -1, &\nu_2 = -1, &\nu_3 = 2 . \end{aligned}$$

We calculate K using Eq. (4.2.8):

$$\begin{aligned} K(T) &= \prod_{i=1}^{3} \left[\left(\frac{2\pi M_i kT}{h^2} \right)^{3/2} \zeta_i(T) \right]^{\nu_i} \\ &= \left(\frac{2\pi kT}{h^2} \right)^{3/2 \sum_i \nu_i} \prod_{i=1}^{3} [M_i^{3/2} \zeta_i(T)]^{\nu_i} = \frac{M\zeta_3^2}{\zeta_1 \zeta_2} , \end{aligned} \tag{i}$$

where

$$M = \left(\frac{M_3}{M_1 M_2} \right)^{3/2} . \tag{ii}$$

Note that since $\sum_i \nu_i = 0$ the temperature dependence of K enters only through the internal partition functions.

In the approximation of our calculation, it is possible to write for each of the diatomic gases

$$\zeta = \zeta_r \cdot \zeta_v , \tag{iii}$$

so that

$$K(T) = M \frac{\zeta_{3r}^2}{\zeta_{1r}\zeta_{2r}} \cdot \frac{\zeta_{3v}^2}{\zeta_{1v}\zeta_{2v}} . \tag{iv}$$

ζ_v cannot be obtained from (4.1.68a), because we have to take into account the different dissociation energies (binding energies) of each of the molecules. Thus we take the corrected equation (4.2.31):

$$\zeta_v = \frac{e^{\epsilon_D/kT}}{1 - e^{-\Theta_v/T}} \,. \tag{v}$$

From here we calculate the second (vibrational) term in the equilibrium constant (iv):

$$\frac{\zeta_{3v}^2}{\zeta_{1v}\zeta_{2v}} = \frac{e^{\Delta\epsilon/kT}(1 - e^{-\Theta_{1v}/T})(1 - e^{-\Theta_{2v}/T})}{(1 - e^{-\Theta_{3v}/T})^2} \,, \tag{vi}$$

where $\Delta\epsilon$ is the difference between the dissociation energies of the reactants and the products in the reaction — in other words, the energy released in the reaction:

$$\Delta\epsilon = 2\epsilon_{3D} - \epsilon_{1D} - \epsilon_{2D} \,. \tag{vii}$$

The first factor in Eq. (iii), ζ_r, is obtained from (4.1.68b) as

$$\zeta_r = \sum_{J=0}^{\infty}(2J+1)\exp[-\beta\epsilon_r J(J+1)] = \sum_{J=0}^{\infty}(2J+1)\exp\left[-\frac{J(J+1)\Theta_r}{2T}\right] \,, \tag{viii}$$

where we have also used in Eq. (4.1.64).

But, the first (rotational) term in the equilibrium constant cannot be calculated exactly, since, as we have seen, the summation in (viii) is difficult to perform. It can instead be studied in the ranges of high and low temperatures. In order to get an idea of the temperature scales, we calculate Θ_r and Θ_v. The results appear in the following table:

	ω (10^{14} s^{-1})	I (10^{-47} kg $\cdot$ m^2)	$\Theta_v = \hbar\omega/k$ (K)	$\Theta_r = \hbar^2/kI$ (K)
Cl_2	1.0	115	725	0.7
H_2	8.3	0.46	6014	175.0
HCl	5.7	2.4	4130	33.5

From this data it is clear that due to the low rotational temperature of chlorine, we can use the approximation (4.1.73) only for very low temperatures (below 0.1 K, $T \ll \Theta_{1r}$). On the other hand, at temperatures above 1000 K ($T \gg \Theta_{2r}$) we can already write, using (4.1.75),

$$\frac{\zeta_{3r}^2}{\zeta_{1r}\zeta_{2r}} = \frac{\Theta_{1r}\Theta_{2r}}{\Theta_{3r}^2} \,, \tag{ix}$$

so that at these temperatures

$$K(T) = M\frac{\Theta_{1r}\Theta_{2r}}{\Theta_{3r}^2}\frac{e^{\Delta\epsilon/kT}(1-e^{-\Theta_{1v}/T})(1-e^{\Theta_{2v}/T})}{(1-e^{-\Theta_{3v}/T})^2}. \tag{x}$$

Solution 5 *Exercise on page 404*

(a) We assume that $\rho(\mathbf{q},T)$ depends on the direction of $\mathbf{q}$; then it is possible to find two directions for which

$$\rho(\mathbf{q}_1,T) < \rho(\mathbf{q}_2,T),$$

where

$$|\mathbf{q}_1| = |\mathbf{q}_2| = \frac{\omega}{c}.$$

We take two identical bodies which are originally at equilibrium with the cavity at temperature T. We coat each of them with reflecting mirrors on all sides but one. The side without a mirror we cover with a filter that transmits radiation with frequency ω only and reflects all other frequencies. We place the two bodies so that the filter of body No. 1 is perpendicular to $\mathbf{q}_1$ and the filter of body No. 2 perpendicular to $\mathbf{q}_2$, making sure that only radiation directed along $\mathbf{q}_1$ be incident on body No. 1 and only radiation directed along $\mathbf{q}_2$ be incident on body No. 2. We can achieve this by placing a collimator made of two slits in front of each filter along the corresponding direction. In addition we ensure that the bodies be positioned with the same orientation with respect to their respective collimators. The system is illustrated in the following figure.

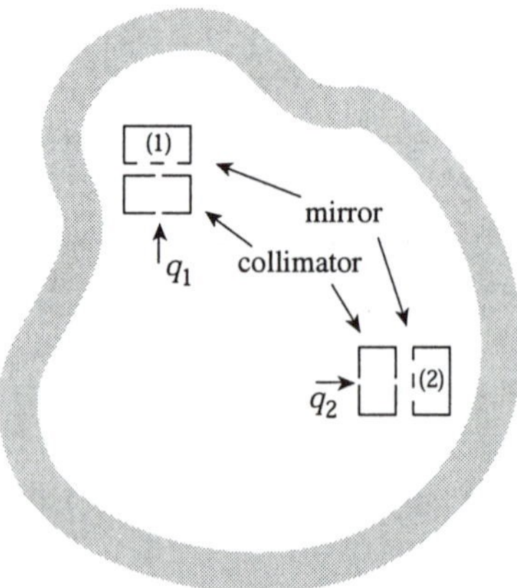

Now, since the two bodies absorb radiation with the same frequency, ω, and the radiation incident on each is perpendicular to the filter-coated side, the two absorb the same fraction of the radiation incident upon them. Hence if $\rho(\mathbf{q}_1, T) < \rho(\mathbf{q}_2, T)$, then body No. 2 absorbs more energy than body No. 1. As a result, its temperature increases and heat will be transferred from one region to another without any work performed, in contradiction to the second law.

Note that we have ignored the radiation emitted by the bodies. Emission of radiation actually occurs, but since the bodies are identical and are at the same temperature, they emit radiation of the same intensity. Thus our conclusion, that body No. 2 will absorb more energy than body No. 1, remains valid.

(b) We now consider a radiation cavity at temperature T, in which ρ depends on the polarization of the radiation. Namely, we assume that the density of radiation with polarization $\mathbf{n}_2$ is greater than the density of radiation with polarization $\mathbf{n}_1$. We place two identical bodies at temperature T, filter-coated as before. Instead of collimators, we shall place a polarizing filter with polarization $\mathbf{n}_1$ in front of one of them and a filter with polarization $\mathbf{n}_2$ in front of the other. In addition we rotate the bodies so that they be oriented in the same way with respect to the axis of their polarizers. Body No. 2 will heat up more than body No. 1, since it will have more radiation incident upon it, in contradiction to the second law.

(c) So far we have managed to prove, using the second law of thermodynamics, that ρ is independent of position, the radiation's direction of propagation, and its polarization. We try to prove in a similar manner that ρ is independent of ω. Namely, we shall attempt to show that a dependence of ρ upon ω leads to a transfer of heat between regions with the same temperature, without performing work.

We assume that the density of radiation with ω_2 is higher than that with ω_1. We take two bodies which were at equilibrium with the radiation in the cavity at temperature T. We coat one with a filter which only transmits radiation with frequency ω_1, and coat the other with a filter that only transmits radiation with frequency ω_2. Using a line of argument similar to the one used above, we can seemingly conclude that body No. 2 will now heat up, as there is more energy at ω_2. However, the absorbing power and the emittance both depend on the frequency, so that we may not reach such a conclusion. The fact that the energy density depends on the frequency does not, therefore, contradict the second law.

Solution 6 *Exercise on page 404*

(a) The emissivity per unit angular frequency of a black body is given by Eq. (4.4.14):

$$I(\omega) = \frac{1}{4\pi^2 c^2} \frac{\hbar\omega^3}{e^{\beta\hbar\omega} - 1} .$$

In order to obtain the emissivity per unit wavelength, $\tilde{I}(\lambda)$, it is not enough to "naively" substitute $\omega = 2\pi c/\lambda$. We must also ensure that the same power (per unit surface area) be emitted in the frequency range $d\omega$ around ω, as in the range $d\lambda$ around the corresponding λ. In mathematical terms,

$$I(\omega)d\omega = \tilde{I}(\lambda)d\lambda \,.$$

This implies that in order to obtain $\tilde{I}(\lambda)$ we should not only express ω in terms of λ in the function $I(\omega)$, but also multiply it by the derivative $d\omega/d\lambda$. Actually, it is the absolute value of the derivative that must be taken, since the energy distribution must remain positive.

Thus

$$\tilde{I}(\lambda) = I\left(\frac{2\pi c}{\lambda}\right)\left|\frac{d\omega}{d\lambda}\right| = \frac{1}{4\pi^2c^2}\cdot\frac{\hbar(2\pi c/\lambda)^3}{e^{\beta\hbar 2\pi c/\lambda}-1}\cdot\frac{2\pi c}{\lambda^2} = \frac{2\pi hc^2}{\lambda^5}\frac{1}{e^{\beta hc/\lambda}-1}\,.$$

(b) In order to find λ_{max} we differentiate with respect to λ, like in Exercise 4.4, and obtain

$$\frac{d\tilde{I}}{d\lambda} = \frac{2\pi hc^2}{\lambda^6}\frac{(\beta hc/\lambda)e^{\beta hc/\lambda} - 5e^{\beta hc/\lambda} + 5}{(e^{\beta hc/\lambda}-1)^2}\,.$$

The sign of the derivative is determined by the numerator of the second fraction, which, after the change of variables $x = \beta hc/\lambda$, we rewrite in the form

$$B(x) = 5 + xe^x - 5e^x \,.$$

This function is positive from $x = 0$ up to $x = x_{\text{max}}$, and is negative for $x > x_{\text{max}}$. It vanishes once for $x = x_{\text{max}}$. Hence, in general, $\tilde{I}(\lambda)$ has a form similar to that of $I(\omega)$ (see Fig. 4.4.4).

In order to find the maximum, namely x_{max}, we again use the iterative method of Exercise 4.4. We write the equation $B(x) = 0$ in the form

$$x = f(x) \equiv 5(1 - e^{-x})\,.$$

The result of the iterations is summarized in the following table:

n	0	1	2	3	4	5
x	4.500000	4.944455	4.964386	4.965089	4.965113	4.965114

We see that x converges to the value $x_{\text{max}} = 4.9651$, so that

$$\frac{hc}{kT\lambda_{\text{max}}} = 4.9651\,,$$

or

$$\lambda_{\max} T = 2.898 \times 10^{-3}\ \mathrm{mK} = 0.2898\ \mathrm{cmK}\,.$$

This relationship is also referred to as Wien's law, like Eq. (4.4.15). Note that $\lambda_{\max}\nu_{\max} = c$ *is not satisfied.*

(c) The temperature on the surface of the sun is, according to (b) above,

$$T \approx \frac{0.3}{5 \times 10^{-5}} \approx 6,000\ \mathrm{K}\,.$$

Solution 7 *Exercise on page 404*

(a) We treat the radiation as the excited states of quantum harmonic oscillators with frequencies ω_α. The energy levels of such an oscillator are integral multiples of $\hbar\omega_\alpha$. Inside the cavity, in which there are oscillators of all frequencies, there will be n_α photons with energy $\hbar\omega_\alpha$, which contribute $n_\alpha\hbar\omega_\alpha$ to the total energy.

The average number of photons with energy $\hbar\omega_\alpha$ will be calculated in exactly the same way in which we calculated the number of phonons with energy ω (Exercise 2.3a of Part III):

$$\langle n_\alpha \rangle = \frac{\sum_{n_\alpha=0}^{\infty} n_\alpha e^{-\beta\hbar\omega_\alpha n_\alpha}}{\sum_{n_\alpha=0}^{\infty} e^{-\beta\hbar\omega_\alpha n_\alpha}} = \frac{\sum_{n_\alpha=0}^{\infty} n_\alpha e^{-x n_\alpha}}{\sum_{n_\alpha=0}^{\infty} e^{-x n_\alpha}}\,, \tag{i}$$

where we have denoted $x = \beta\hbar\omega_\alpha$. The expression on the right hand is calculated by noting that

$$\langle n_\alpha \rangle = -\frac{d}{dx} \ln \sum_{n_\alpha=0}^{\infty} e^{-x n_\alpha} = -\frac{d}{dx} \ln \frac{1}{1-e^{-x}} = \frac{e^{-x}}{1-e^{-x}} = \frac{1}{e^x - 1}\,, \tag{ii}$$

or

$$\langle n_\alpha \rangle = \frac{1}{e^{\beta\hbar\omega_\alpha} - 1}\,. \tag{iii}$$

The average number of photons per unit volume is a sum over all frequencies — over all the vibrational modes, namely

$$\langle n \rangle = \frac{\langle N \rangle}{V} = \frac{1}{V}\sum_\alpha \langle n_\alpha \rangle = \frac{1}{V}\sum_\alpha \frac{1}{e^{\beta\hbar\omega_\alpha} - 1}\,, \tag{iv}$$

where N is the total number of photons in the cavity.

Because the frequencies are very dense, it is possible to pass from the sum to an integral. To this end we multiply $\langle n_\alpha \rangle$ by the number of states in the region d^3q:

$$d\mathcal{N} = \frac{2V d^3q}{(2\pi)^3} = \frac{V q^2 dq}{\pi^2} = \frac{V}{\pi^2 c^3}\omega^2 d\omega\,, \tag{v}$$

and then

$$\langle n \rangle = \int_0^\infty \frac{\omega^2}{\pi^2 c^3} \frac{d\omega}{e^{\beta\hbar\omega} - 1}, \tag{vi}$$

which is the integral of Eq. (4.4.13). After changing variables to $x = \beta\hbar\omega$ we obtain

$$\langle n \rangle = \left(\frac{kT}{\hbar c}\right)^3 \frac{1}{\pi^2} \int_0^\infty \frac{x^2 dx}{e^x - 1} = \frac{2.404}{\pi^2} \left(\frac{kT}{\hbar c}\right)^3, \tag{vii}$$

where the integral has been calculated using Eq. (4.A.5) with $\ell = 2$.

Equation (vii) can also be written in the form

$$\langle n \rangle = \frac{144.24}{\pi^4} \frac{\sigma T^3}{kc}. \tag{viii}$$

Substituting $T = 3\text{K}$ we obtain about 550 photons per cm^3.

(b) In order to find the relationship between the density of photons and the pressure, we calculate the equation of state. We have

$$P = -\left(\frac{\partial F}{\partial V}\right)_{N,T}. \tag{ix}$$

Actually we have already calculated this in Solution 4.6 and obtained

$$P = \frac{4\sigma}{3c} T^4. \tag{x}$$

Dividing this equation by Eq. (viii) from (a) above, we obtain

$$\frac{N_p}{V} = \frac{108.18}{\pi^4} \frac{P}{kT} = 1.1106 \frac{P}{kT}, \tag{xi}$$

where we have denoted the average number of photons by N_p.

The number of molecules in an ideal gas satisfies

$$\frac{N_m}{V} = \frac{P}{kT}, \tag{xii}$$

so that

$$\frac{N_p}{N_m} \approx 1.11. \tag{xiii}$$

(c) Suppose that we want to emulate an ideal gas whose pressure is 1 atm by a photon gas. From Eq. (x) it is clear that there exists a single temperature for which this is possible: It is 1.4×10^5 K. Once we have fixed the temperature, the density of molecules is fixed by the equation of state (xii), and the density of photons is fixed by the equation of state (xi) to be 1.11 times larger. This ratio is satisfied, therefore, only for the special temperature for which the pressure of the radiation equals the pressure of the gas.

(d) We have calculated the specific heat of a photon gas in Exercise 4.5, and obtained

$$C_V = \frac{16\sigma}{c} V T^3 \,. \tag{xiv}$$

Taking N from Eq. (viii), we find that the specific heat per photon is

$$\frac{C_V}{N_p} = \frac{\pi^4 k}{9.015} = 10.8k \tag{xv}$$

and is independent of temperature, as in the case of an ideal gas.

Each photon contributes to the specific heat about $11k$, compared to $1.5k$ per molecule in an ideal gas. Nevertheless, we must emphasize that the ratio C_V/N is to be interpreted differently in each case, since in the case of a gas of photons, their number increases with temperature. Hence not only does each photon contribute to the specific heat more than a molecule of an ideal gas, but the number of photons changes as well.

(e) This question remains unsolved. Make sure that you know the answer.

Solution 8 *Exercise on page 405*

(a) There exists an ambiguity in the language. In order for a body to be called black in everyday language, it should be a body that does not reflect the radiation incident upon it and does not emit radiation in the visible range. Thus an object can be black in terms of everyday language if it absorbs all the radiation in the visible range and emits most of the radiation in the invisible range [see (c) below].

(b) The black body (in the technical sense) will appear yellow if we heat it to a temperature at which the energy emission is maximal in the yellow wavelengths. As we found in the solution of Exercise 6, this temperature is fixed by Wien's law and is around 6,000 K. An example of such an object is the sun.

(c) In order to get convinced that the black body is not black, it must be observed through an instrument that is sensitive to the wavelengths in which the radiation emission is concentrated. At room temperature this wavelength is (according to Wien's law) about 10^{-3} cm, which is in the infrared region.

Solution 9 *Exercise on page 405*

We cannot use Eqs. (4.2.7) and (4.2.8) directly, since they were obtained for a Boltzmann gas, whereas here we must treat at least the photons differently, since they obey Planck's distribution. However, we can begin from the basic condition for chemical equilibrium, Eq. (4.2.5), which gives in this case

$$\mu_+ + \mu_- - \mu_\gamma = 0 \,, \tag{i}$$

where $\mu_\pm$ are, respectively, the chemical potential of $e^\pm$ and μ_γ is the chemical potential of the photons. From Exercise 4.5 we have $\mu_\gamma = 0$ and hence

$$\mu_+ + \mu_- = 0\,. \tag{ii}$$

The chemical potential of a relativistic electron gas can be obtained from Eq. (4.2.13), where the partition function of a single electron is

$$z = \frac{2V}{h^3}\int d^3p \exp\left[-\beta\sqrt{(mc^2)^2 + (cp)^2}\right]\,. \tag{iii}$$

The factor 2 originates from the two spin states of the electron. The Boltzmann factor has been generalized to contain the relativistic expression for the energy.

As we shall see in the next part, electrons do not obey Boltzmann statistics. Boltzmann statistics applies only in the limit in which the electron density is very low and our calculation is to be interpreted in this light. The low density is guaranteed by the fact that the temperature of the gas is assumed to be nonrelativistic ($kT \ll mc^2$) so that pair creation is quite rare. In Part V we discuss such a gas in the opposite limit, $kT \gg mc^2$.

The integral appearing in Eq. (iii) cannot be calculated explicitly. However, for nonrelativistic particles for which $mc^2 \gg cp$ it is possible to expand the energy and to obtain

$$\epsilon = mc^2 + \frac{p^2}{2m}\,, \tag{iv}$$

which is the nonrelativistic expression for the energy plus the rest energy mc^2. We can now calculate the integral and obtain the usual nonrelativistic expression, Eq. (3.4.37), with a correction originating from the spin and the rest mass:

$$z = \frac{8\pi V}{h^3}\int_0^\infty dpp^2 \exp\left[-\beta\left(mc^2 + \frac{p^2}{2m}\right)\right] = \frac{2e^{-\beta mc^2}V(2\pi mkT)^{3/2}}{h^3}\,, \tag{v}$$

and using Eq. (4.2.13) the chemical potential becomes

$$\mu = kT\ln\left[\frac{n}{2}\left(\frac{h^2}{2\pi mkT}\right)^{3/2}\right] + mc^2\,. \tag{vi}$$

This is the chemical potential of an ideal gas with a correction which we may think of as a shift of the energy scale. The factor $\frac{1}{2}$ inside the logarithmic function originates from the spin states of the electron. Note that there is no room here for the internal partition function, since the electron (at least as far as we know today) is a particle devoid of internal structure. Even if it had such structure, it does not come into play at the energies discussed.

Using Eq. (vi) for the chemical potential of the electrons and the positrons in Eq. (ii), we obtain

$$kT \ln \left[\frac{n_+ n_-}{4} \left(\frac{h^2}{2\pi mkT} \right)^3 \right] + 2mc^2 = 0 \,, \qquad \text{(vii)}$$

where n_+, n_- are the densities of positrons and electrons, respectively.

From here we obtain the law of mass action, which will contain only the densities of the electrons and the positrons:

$$n_+ n_- = 4 \left(\frac{2\pi mkT}{h^2} \right)^3 e^{-2mc^2/kT} \,. \qquad \text{(viii)}$$

If the gas is electrically neutral, then $n_+ = n_-$ so that

$$n_+ = n_- = 2 \left(\frac{2\pi mkT}{h^2} \right)^{3/2} e^{-mc^2/kT} \,. \qquad \text{(ix)}$$

Note that the rest energy mc^2 determines the scale of temperatures at which pair creation becomes significant. For instance, at a temperature for which $kT/mc^2 = 0.1$ ($T \approx 6 \times 10^9$ K), which is found at the edge of the approximation's region of validity,

$$n_\pm = 1.6 \times 10^{24} \text{ cm}^{-3} \,.$$

The density of photons at such a temperature, Eq. (viii) in the solution to Exercise 7, is much higher. We find

$$n_\gamma = 4.2 \times 10^{30} \text{ cm}^{-3} \,.$$

Part V

Of Fermions and Bosons

Introduction

In the preceding parts we analyzed the thermodynamic properties of systems in the canonical ensemble. In this part we will get acquainted with a generalization of the canonical ensemble to systems that can exchange particles with their surroundings. This generalization, the grand canonical ensemble, which we present in Chap. 1, is of double importance. On the one hand it contributes to a deeper understanding of the statistical physics of classical particles. On the other hand, the new ensemble is the natural tool for the development of statistical mechanics of systems of identical particles in quantum conditions. This is not because a system of (nonrelativistic) quantum-mechanical particles does not conserve the number of particles. It is merely of great technical convenience and is analogous to the application of the canonical ensemble to isolated systems, which have a constant energy.

Using the grand canonical ensemble we develop, in Chap. 2, the tools for the statistical mechanics of systems of identical quantum particles. Some of the elements have been laid down in Part IV, in the discussion of the Debye model and black body radiation, and upon this we shall expand and build the structure. The importance of the applications of the statistical physics of identical particles cannot be overstated. We will get acquainted with some of them in different degrees of detail. The main application we shall meet in Chap. 3 concerns electrons, which obey the Pauli principle and therefore are subject to the Fermi–Dirac statistics. They are responsible for the properties of metals and especially the various conductivity phenomena. Chapter 4 is devoted to the quantum statistical mechanics of bosons, particles subject to the Bose–Einstein statistics, and to the associated phenomenon of superfluidity. It concludes with the spectacular manifestation of electrical conductivity — superconductivity.

Chapter 1

Grand Canonical Ensemble

1.1 Definitions and motivation

In the preceding parts we have seen that the canonical ensemble (and the free energy) is a most useful description of a system, since in most cases the temperature of the thermodynamic system (and not its energy) is the controlled quantity. Moreover, even when the system is isolated, it is more convenient to think of it as being in contact with a heat bath, to calculate its free energy and from it to derive the properties of the isolated system. See, for instance, how this was accomplished for the paramagnet in Sec. 1.4 of Part III.

grand canonical ensemble

We now introduce the *grand canonical ensemble*, which allows us to relax the constraint of a fixed number of particles in the system. This would have been the natural thing to do in dealing with chemical reactions, or with systems that can exchange particles with a reservoir. But here our motivation for introducing this ensemble, as mentioned in the introduction, is purely technical, so we move directly to the formal aspects.

The grand canonical ensemble is defined as a set of systems all with the same type of degrees of freedom and the same energy function. Each system is defined by a state α which is specified by the number of particles it contains, N, and by the microscopic state i of its N particles. The probability of a state is a generalization of the Gibbs probability, Eq. (3.1.35). It is given by

$$P_\alpha \sim e^{-\beta(E_i - \mu N)} . \tag{5.1.1}$$

grand partition function

and the *grand partition function* is defined as

$$\mathcal{Z}(T, V, \mu) = \sum_\alpha e^{-\beta(E_i - \mu N)} . \tag{5.1.2}$$

$\mathcal{Z}$ is again the normalization of the probabilities P_α. The summation consists of two parts: a sum over the particle number N and for each particle number, over all microscopic states i of a system with that number of particles. This second sum is the same as in a canonical ensemble with N particles and it gives the canonical partition function for N particles. Thus

$$\mathcal{Z}(T,V,\mu)=\sum_N e^{\beta\mu N}\sum_i e^{-\beta E_i}=\sum_N e^{\beta\mu N} Z(T,V,N)=\sum_N e^{\beta[\mu N-F(T,V,N)]}\,, \tag{5.1.3}$$

where the appearance of the free energy is due to the relation $Z = e^{-\beta F}$ [Eq. (3.1.30)].

Averages of physical quantities (observables) are computed as expectations with the probability distribution (5.1.1), normalized by $\mathcal{Z}$, summing over all values of N and for each value of N, over all microscopic states i. If A_α, or $A(N,i)$, is an observable giving a value for each state in the ensemble, a value that depends on the number of particles N as well as on the microscopic state i of the system, then the average of A in the ensemble is

$$\langle A\rangle = \sum_\alpha A_\alpha P_\alpha = \mathcal{Z}^{-1}\sum_{N=0}^{\infty}\sum_i A(N,i)e^{-\beta(E_i-\mu N)}\,. \tag{5.1.4}$$

Exercise 1.1

Prove that $\mathcal{Z}\to 1$ in the limit where $\mu\to-\infty$.

Solution on page 514

To justify the choice of this probability distribution we follow the logic of Chap. 1 of Part III, showing that these probabilities lead to average quantities which can be identified with thermodynamic quantities satisfying the thermodynamic laws.

1.2 Connection to thermodynamics

The identification of the thermodynamic work follows closely the discussion in Sec. 1.2 of Part III. For any state i of N particles $\delta W_i = -\partial E_i/\partial X$ and the thermodynamic work is the average of this quantity. Namely,

$$\delta W = -\mathcal{Z}^{-1}\sum_{N,i} e^{\beta(\mu N-E_i)}\frac{\partial E_i}{\partial X}dX = \frac{1}{\beta}\frac{\partial \ln \mathcal{Z}}{\partial X}dX\,. \tag{5.1.5}$$

Thus, for a volume change $\delta W = PdV$ and then

$$P = \frac{1}{\beta}\frac{\partial \ln\mathcal{Z}}{\partial V} = \frac{\partial}{\partial V}(kT\ln\mathcal{Z})\,. \tag{5.1.6}$$

$\mathcal{Z}$ depends on V, on T (or β) and also on μ. Hence we have

$$N = \frac{\partial}{\partial \mu}(kT \ln \mathcal{Z}), \tag{5.1.7}$$

where N on the left hand side is the average number of particles.

Exercise 1.2

Prove (5.1.7).

Solution on page 514

Comparing (5.1.6) and (5.1.7) to the rightmost Eq. (2.0.31), we expect that it be possible to identify the potential Ω with $-kT \ln \mathcal{Z}$.

Consider the function

$$\Omega \equiv -kT \ln \mathcal{Z}, \tag{5.1.8}$$

which is our candidate for a generalization of the relation between the free energy and the canonical partition function, $F = -kT \ln Z$. Calculating the derivative of Ω with respect to T we have

$$\left(\frac{\partial \Omega}{\partial T}\right)_{V,\mu} = -k \ln \mathcal{Z} - kT \frac{d\beta}{dT}\left(\frac{\partial \ln \mathcal{Z}}{\partial \beta}\right)_{V,\mu} = \frac{\Omega}{T} - \frac{E}{T} + \frac{\mu}{T}N, \tag{5.1.9}$$

Grand potential

where E and N are the averages of the energy and the particle number. Now if Ω is identified as the grand potential, then its temperature derivative, at constant V and μ, should be the entropy. See for example Eq. (2.0.31). Indeed, Eq. (5.1.9) corresponds to the thermodynamic relation Eq. (2.0.29), expressing the fact that Ω is the potential providing the thermodynamic information in terms of V, T and μ.

Exercise 1.3

Complete the derivation of Eq. (5.1.9).

Solution on page 515

Exercise 1.4

Another way of obtaining the identification $\Omega = -kT \ln \mathcal{Z}$ is to use the expression for the entropy in terms of the probabilities of the microscopic states [Eq. (2.6.7)]:

$$S = -k \sum_{\alpha} P_\alpha \ln P_\alpha .$$

Perform this identification.

Solution on page 515

Given the potential Ω we derive all the familiar thermodynamic quantities via Eq. (2.0.31)

$$\frac{\partial \Omega}{\partial T} = -S, \quad \frac{\partial \Omega}{\partial V} = -P, \quad \frac{\partial \Omega}{\partial \mu} = -N, \tag{5.1.10}$$

where we keep in mind that the independent variables are T, V and μ, and each derivative is calculated with the two unspecified variables kept constant. We can also write

$$d\Omega = -SdT - PdV - Nd\mu. \tag{5.1.11}$$

But there is a slight complication, as we shall see in detail in chapters 3 and 4: The appearance of μ as a variable, while computationally very convenient, is not natural. Thermodynamic properties of systems are eventually measured with a given number of particles, or density. However, in the grand canonical framework quantities like pressure (equation of state), or the specific heat, are given as functions of T, V and μ. Thus, we must first use the rightmost of Eqs. (5.1.10), to eliminate μ in terms of T, V and N, or rather in terms of T and $n = N/V$.

Chapter 2

Statistical Mechanics of Identical Quantum Particles

2.1 Classification of states — occupation numbers

quantum statistics

The passage to the statistical mechanics of systems of identical quantum particles is achieved in two steps. The first step was already presented in Sec. 4.4 of Part III. It concerns the appearance of quantum states. The second is *quantum statistics.* In the first step one takes into consideration the fact that quantum particles cannot be described by a full specification of their coordinates and momenta, because of the uncertainty principle. Instead, particles are described by wave functions which satisfy boundary conditions imposed by the container to which they are confined. If the particles are noninteracting, then they are independent of each other and each particle is described by its own wave function.

As we have seen in Sec. 4.4 of Part III, the wave function of a particle in a rectangular box is a standing (de Broglie) wave in each of the three perpendicular directions. Such a state is specified by three integers n, p, q, which determine the allowed (quantized) wavelengths, or wave numbers, in the three perpendicular directions and the corresponding three components of the quantized de Broglie momentum. (See Part IV Sec. 3.2.) The set of three integers, specifying the wave function of the single particle, we denote schematically by a single index k. The energy of a particle in the state k is ϵ_k and in the case of a particle in a box, ϵ_k is given by Eq. (3.4.27). This scheme can be extended to any quantum system of noninteracting particles, with k interpreted as a label of the single particle states.

A state of a system of many noninteracting particles can be specified by listing which particle is in which of the accessible single particle states, as we have done in Sec. 4.4 of Part III. The energy of such a many-particle state can be written as the sum of the single particle energies, i.e.

$$E = \sum_{j=1}^{N} \epsilon_{k_j}\,, \tag{5.2.1}$$

where k_j is the state of particle j.

In each microscopic state we can identify the number of particles in a given single particle state k. This number, which is called the *occupation number*, of state k, is denoted by n_k. Clearly,

occupation number

$$\sum_k n_k = N\,, \tag{5.2.2}$$

where N is the total number of particles.

Exercise 2.1

What is the largest value that n_k can have (for each k) in a system of N particles?

Solution on page 515

The energy of the N-particle state is written as

$$E = \sum_k n_k \epsilon_k\,. \tag{5.2.3}$$

At this stage, a large number of microscopic states share the same set of occupation numbers. We return to this issue below. Note also that the number of sets k is cubically infinite, hence for a given N most occupation numbers are zero.

Exercise 2.2

How many individual microscopic states have the same set of occupation numbers n_k?

Solution on page 516

Exercise 2.3

Justify Eq. (5.2.3).

Solution on page 516

The canonical partition function is, as usual, the sum

$$Z = \sum_{\text{states}} e^{-\beta E}\,.$$

As long as the only constraint is the total number of particles, the partition sum can be carried out as in Sec. 4.4 of Part III, i.e. by summing over all triplets of integers corresponding to the index k for each particle. This sum, over N vectors of integers, becomes a product of identical sums. For a box of macroscopic size, if the temperature is not extremely low, each of these sums is very well approximated by an integral, which is essentially the classical result. Temperatures at which the sum deviates significantly from the integral are so low as to be beyond any experimental interest. See for example Exercise 4.10. In conclusion, the quantization of the single particle wave functions is not a quantum effect of great impact, until we get to discussing Bose–Einstein condensation in Chap. 4, below. Hence we turn to the second step.

2.2 Quantum statistics — many-particle states

The principal effect of quantum mechanics on the thermodynamic properties of systems of identical particles is brought about by the quantum-mechanical constraints on the identification of allowed, distinguishable microscopic states of the system. Such constraints follow from symmetry properties that must be obeyed by wave functions of many identical particles. Here we pass directly to the implications for the microscopic states. The discussion of the underlying wave functions is a subject for a more advanced course. It turns out that all possible particles divide into two types:

fermion

Pauli principle

(1) Fermi–Dirac particles (or fermions) — no single particle state can be occupied by more than one particle. This goes by the name of the Pauli principle. Hence the states are characterized by occupation numbers that can be either 0 or 1. To any set of (0,1) occupation numbers corresponds a single microscopic state of the system. Fermions are found to possess half-integer spin (in units of $\hbar$).

boson

(2) Bose–Einstein particles (or bosons) — single particle states can be occupied by any number of particles, but for any distribution of occupation numbers there is a single microscopic state of the system. Bosons possess integer spin (in units of $\hbar$).

Since atoms are composed of different combinations of protons, neutrons and electrons, all of which are fermions, there are two possibilities: atoms that are composed of an odd number of fermions have half-integral spins, and are fermions, while atoms made up of an even number of fermions have integral spin, and are bosons. Thus, for example, the (neutral) atoms H^1, He^4 and Li^7 are bosons whereas H^2, He^3 and Li^6 are fermions.

Exercise 2.4

(a) Why is there no need to take into account the orbital angular momentum of the electrons in order to determine if an atom is a boson or a fermion?
(b) Show that if a neutral atom contains an odd number of neutrons then it is a fermion, and if it contains an even number of neutrons then it is a boson.

Solution on page 516

Before proceeding let us clarify the relation between two ways of specifying microscopic states of noninteracting identical particles. Both ways use the single particle states denoted symbolically by $\{k\}$. The first way has been employed since Part II to describe states of distinguishable particles. Accordingly, the microscopic states are specified by the N numbers $\{k_j\}$ with $j = 1, \ldots, N$. It leads to a partition function that is the Nth power of the single particle partition function. This way is incompatible with the indistiguishability of identical particles and leads to the Gibbs paradox. The error is corrected introducing the factor $1/N!$ in the partition function, as we have done in Chap. 5 of Part III.

The alternative way of specifying microscopic states uses occupation numbers $\{n_k\}$, as in the previous section. The partition sum would then be over all sets of occupation numbers for all possible single particle states, $\{k\}$, respecting the constraint Eq. (5.2.2). This sum can be carried out, to reproduce the classical result for the partition function, provided one keeps in mind that the number of N-particle states that correspond to a given distribution of n_k's (Exercise 2.2) is

$$\frac{N!}{n_1! n_2! \cdots n_k! \cdots}.$$

■ This may be considered as an unofficial exercise for the enterprising student.

On the other hand, in quantum statistics a state with a given set of $\{n_k\}$ corresponds to a *single* N-particle quantum state. This makes computing the canonical partition sum for systems of particles impossible.

It is important to contrast this case with that of the analyses of the Debye model and black body radiation in Part IV. There the *degree of excitation* played exactly the role of the present occupation number. But there were no underlying particles to those "occupation numbers." The reason we could perform the sum over states in the case of the phonons or photons was that their total number was not fixed. Phonons and photons can be created and annihilated freely and so the sum over their occupation numbers became simple. It is precisely for this reason that we employ the grand canonical ensemble in the calculation of the partition function of quantum many particle systems.

2.3 Thermodynamics of fermions and bosons

The partition function of an ideal gas of identical particles is a sum over all the values of n_k, as follows:

$$Z = \sum_{n_1, n_2, \ldots, n_k \ldots} e^{-\beta E(n_1, n_2, \ldots, n_k, \ldots)} . \tag{5.2.4}$$

If we neglect the forces between the particles, the total energy of a state is a sum of the single particle energies and is given by Eq. (5.2.3). Thus we have

$$Z = \sum_{n_1, n_2, \ldots} \exp[-\beta(n_1\epsilon_1 + n_2\epsilon_2 + \ldots + n_k\epsilon_k + \ldots)] . \tag{5.2.5}$$

We would have liked to write this sum as a product of independent sums over each of the occupation numbers separately, as we did for the Einstein solid (Sec. 2.2 of Part III). But here we are faced with a problem that did not arise in the previous cases, namely the values of the occupation numbers are not independent of each other since $\sum_k n_k = N$ [Eq. (5.2.2)] is fixed. It is the total number of particles in the system. Thus

$$Z \neq \left(\sum_{n_1} e^{-\beta\epsilon_1 n_1}\right)\left(\sum_{n_2} e^{-\beta\epsilon_2 n_2}\right) \cdots \left(\sum_{n_k} e^{-\beta\epsilon_k n_k}\right) \cdots$$

Exercise 2.5

Explain why we succeeded in factorizing the partition function of the paramagnet and the Einstein solid while here this step fails.

Solution on page 516

Since what is blocking the factorization of the canonical partition function, Eq. (5.2.5), is the constraint (5.2.2), implied by the fixed number of particles, we turn to the grand canonical ensemble. The constraint is removed at the expense of introducing the chemical potential μ. We calculate the grand canonical partition function, in which all possible values of N enter.

We turn to Eq. (5.1.2), where a microscopic state α is characterized by all the occupation numbers $\{n_k\}$. The set of $\{n_k\}$ determines, via Eqs. (5.2.2) and (5.2.3), the number of particles as well as the energy of the state. Hence,

$$\mathcal{Z}(T,V,\mu)=\sum_{\alpha}\exp[\beta(\mu N-E_i)]=\sum_{\{n_k\}}\exp\left[\beta\sum_{k=1}^{\infty}n_k(\mu-\epsilon_k)\right]. \quad (5.2.6)$$

Now there is no constraint on the occupation numbers. They take on independent values, provided those are consistent with the particles' statistics. Hence it is possible to write $\mathcal{Z}$ as a product of "single state" partition functions:

$$\mathcal{Z}=\prod_k\sum_{n_k}e^{\beta(\mu-\epsilon_k)n_k}. \quad (5.2.7)$$

Exercise 2.6

Prove Eq. (5.2.7).

Solution on page 517

All that is left is to calculate the summations in the product (5.2.7) for a given k:

$$\mathcal{Z}_k=\sum_{n_k}e^{\beta(\mu-\epsilon_k)n_k}. \quad (5.2.8)$$

Quantum statistics dictates only two options, either $n_k=0{,}1$ for a gas of fermions or $n_k=0,1,\ldots,\infty$ for bosons.

In the case of fermions there are only two terms on the right hand side of (5.2.8). Thus,

$$\mathcal{Z}_k^{(F)}=\sum_{n_k=0}^{1}e^{\beta(\mu-\epsilon_k)n_k}=1+e^{\beta(\mu-\epsilon_k)}, \quad (5.2.9a)$$

while for bosons the right hand side is an infinite geometric series:

$$\mathcal{Z}_k^{(B)}=\sum_{n_k=0}^{\infty}e^{\beta(\mu-\epsilon_k)n_k}=(1-e^{\beta(\mu-\epsilon_k)})^{-1}. \quad (5.2.9b)$$

Note that the summation in (5.2.9b) converges only when μ is lower than all the energy levels of the system, including the ground level. If the ground level is $\epsilon_1=0$ the chemical potential must be negative. Conversely, the summation in (5.2.9a) consists of only two terms, and thus there is no such constraint on the chemical potential of a gas of fermions.

We continue by calculating the thermodynamic potential $\Omega=-kT\ln\mathcal{Z}$ [Eq. (5.1.8)], which for fermions and bosons, respectively, reads

$$\Omega^{(F)}=-kT\sum_k\ln(1+e^{\beta(\mu-\epsilon_k)}), \quad (5.2.10a)$$

$$\Omega^{(B)}=kT\sum_k\ln(1-e^{\beta(\mu-\epsilon_k)}). \quad (5.2.10b)$$

Recall that the summation is over all a single particle states denoted schematically by k.

From these expressions for Ω, and given the energy, ϵ_k, of the single particle state k, we can proceed to calculate all the thermodynamic properties of the system, as functions of T, V and μ.

2.4 Average occupation numbers

In the grand canonical ensemble the number of particles, N, is not fixed, but the probability distribution of the different states is such that the number actually fluctuates very little around an average number that is determined by T, V and μ. This average is obtained from Eq. (5.1.7). Substituting for Ω the expressions we obtained in Eqs. (5.2.10), we obtain, for fermions and bosons respectively,

$$N^{(F)} = -\left(\frac{\partial\Omega^{(F)}}{\partial\mu}\right)_{V,T} = \sum_k \frac{1}{e^{\beta(\epsilon_k-\mu)}+1}, \tag{5.2.11a}$$

$$N^{(B)} = -\left(\frac{\partial\Omega^{(B)}}{\partial\mu}\right)_{V,T} = \sum_k \frac{1}{e^{\beta(\epsilon_k-\mu)}-1}. \tag{5.2.11b}$$

■ Note the inversion of the sign in the exponent compared to (5.2.10).

$N^{(F)}$ and $N^{(B)}$ are the *average* number of particles in a state of thermodynamic equilibrium, and should have been denoted by $\langle N\rangle$. But where the meaning is clear we drop the braces, as we have been doing all along.

Since N is a sum over all the single particle states, each term of the sum, corresponding to a given single particle state k, is the average number of particles in that state. For a state k with energy ϵ_k we find that there are *on average*

$$\langle n_k^{(F)}\rangle = (e^{\beta(\epsilon_k-\mu)}+1)^{-1} \tag{5.2.12a}$$

fermions and

$$\langle n_k^{(B)}\rangle = (e^{\beta(\epsilon_k-\mu)}-1)^{-1} \tag{5.2.12b}$$

bosons.

Equation (5.2.12b) should look familiar. In Part III we found that the average degree of excitation of a harmonic oscillator is given by exactly such an expression with $\mu = 0$ [see Eq. (3.2.9)]. The connection between the two becomes clear if the excited states of the harmonic oscillators are considered as particles — phonons with energy $\hbar\omega$, as explained in Chap. 3 of Part IV. The average number of phonons with frequency ω in the crystal is the average degree of excitation of the oscillator that was discussed in Part III.

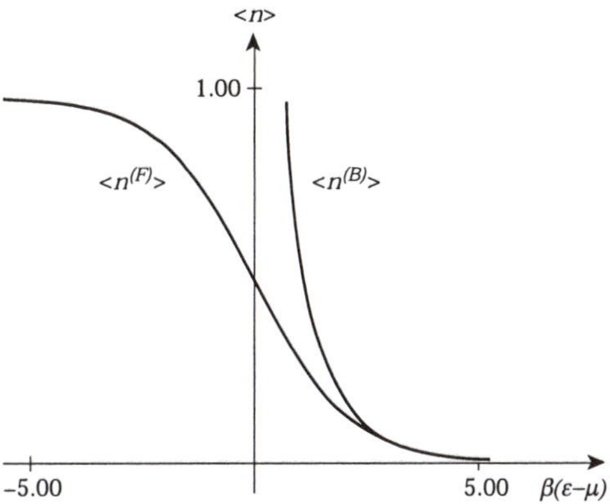

Fig. 5.2.1. The average occupation numbers of a single particle state as a function of $\beta(\epsilon - \mu)$. The index k has been omitted.

Considering more closely the these average occupation numbers, we observe that the average occupation number for fermions is consistent with the Pauli principle: If the variable $n_k^{(F)}$ can take the value 0 or 1 only, then its (thermal) average must be less than 1. And, indeed, since the exponential function is always positive, the denominator in (5.2.12a) is larger than 1 and hence the average $\langle n_k^{(F)} \rangle < 1$. Bosons do not satisfy the Pauli principle, and at low T tend to accumulate at the low energy levels. Only the thermal fluctuations stop them from all ending up in the ground level. Equation (5.2.12b) reveals that indeed $\langle n_k^{(B)} \rangle$ increases with decreasing ϵ_k and diverges for $\epsilon_k \to \mu$. The implications of this divergence will be discussed later on.

Fermi–Dirac distribution

Bose–Einstein distribution

Figure 5.2.1 describes the energy dependence of the two average occupation numbers. These are the Fermi–Dirac distribution and the Bose–Einstein distribution. Plotting the average occupation numbers as a function of $\beta(\epsilon - \mu)$ allows us to consider the graphs as expressing the dependence of $\langle n_k^{(F)} \rangle$ and $\langle n_k^{(B)} \rangle$ on ϵ, at constant T, or, alternatively, as the dependence on the temperature (β) at fixed single particle energy.

For large values of $\beta(\epsilon - \mu)$ the two graphs in Fig. 5.2.1 merge, since the exponential dominates the denominator. This is where both distributions tend toward the classical approximation — the Maxwell–Boltzmann distribution, for which $\langle n_k \rangle \sim e^{-\beta\epsilon_k}$.

Thus, we have found that a consistent discussion of systems of identical particles, together with the fact that only two kinds of occupation numbers are allowed, yields the Boltzmann distribution only as an approximation. The fundamental distributions (5.2.12) are different from the Boltzmann distribution and the results obtained from them are in many cases significantly different from the results obtained using the Boltzmann distribution.

Chapter 3

Electrical Conductivity in Metals

3.1 The Drude model

An important application of the Fermi–Dirac distribution is the clarification of the electrical conductivity and other physical properties of metals. The comprehensive discussion of this subject is a matter for an entire course. We will limit ourselves to a presentation of the basic conclusions obtained in simplified conditions.

For many years electrical conductivity was considered a mystery, and only the discovery of the electron by J. J. Thomson in 1897 pointed to a possible mechanism for electrical conductivity in metals: The electric current is a current of electrons. Because the application of even the smallest potential on a metal wire gives rise to a significant electric current, metals must contain a huge (macroscopic) number of electrons that are moving essentially freely. This was the starting point for the first attempt to explain the conductivity in metals, made by Drude in 1900. Macroscopic numbers of electrons immediately bring to mind a statistical treatment, and indeed to this day this is the accepted approach to analyzing this problem. Since the only tools available to Drude, at the turn of the 20th century, were the kinetic theory and the Boltzmann distribution, he applied them to the electrons in metals. To this end he asserted that the conduction electrons in a metal behave as an ideal gas. In contemporary terms this assertion would be based on the following assumptions:

- When many atoms of a metal create a crystalline structure, the external electrons (valence electrons) of the atoms detach from them and move freely in the metal. The atoms that are left behind become positive ions. The rigid lattice of the crystalline metal is made up of these ions.
- Although the electrons are charged, the interactions between them are negligible and they do not affect one another's motion. This assumption will be justified in Sec. 4.7.
- The electrons are in a state of thermodynamic equilibrium as a result of collisions with the ions in the metal. Between collisions the electrons

move in a straight line and at constant speed and the forces applied to them by the ions are neglected.

■ Note the similarity to photons of a black body, where equilibrium is not attained by the mutual interactions but by interactions with the walls of the container. Between collisions the photons move as free particles.

Exercise 3.1

Calculate the number of conduction electrons per cubic centimeter of sodium and potassium.

Solution on page 517

Based on these assumptions and applying the kinetic theory, we derive Ohm's law: Consider a segment of a metal wire along which there exists a uniform electric field. The wire contains a very large number of electrons that are moving about with different speeds in different directions. See e.g. Fig. 5.3.1. The velocity of an electron between two collisions is determined by Newton's second law:

$$m\frac{d\mathbf{v}}{dt} = -e\mathbf{E}\,, \tag{5.3.1}$$

i.e.

$$\mathbf{v} = \mathbf{v}_0 - \frac{e}{m}\mathbf{E}t\,, \tag{5.3.2}$$

where $\mathbf{v}_0$ is its velocity immediately after the last collision. t, therefore, is the time elapsed from the last collision.

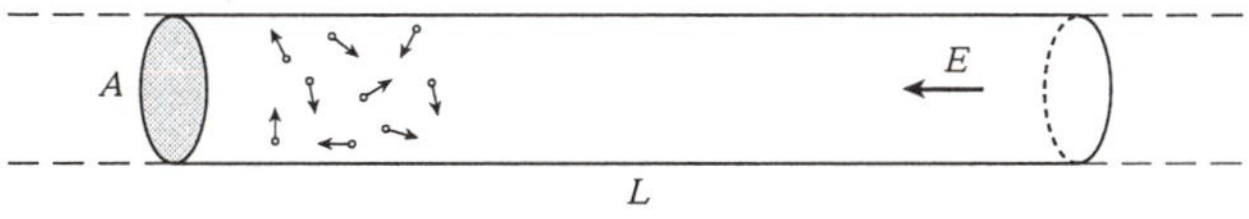

Fig. 5.3.1. Conduction in a metal wire.

Note that an electric current exists only if the number of electrons moving to the right (against the field) is different from the number of electrons moving to the left. The relevant quantity is the average velocity, $\langle\mathbf{v}\rangle$, of all the (free) electrons at a given moment. We have to average Eq. (5.3.2) over all the free electrons in the wire. Each of them has undergone its last collision at a different time and has come out of it with a different velocity. We thus obtain

$$\langle\mathbf{v}\rangle = -\frac{e\tau}{m}\mathbf{E}\,, \tag{5.3.3}$$

where τ is the mean free time of the electrons.

Exercise 3.2

Justify Eq. (5.3.3).

Solution on page 517

This is a typical transport problem, with a drift velocity proportional to the force. The mobility coefficient is τ/m, as in Part I, Sec. 3.4. The final stage is the calculation of the electric current density, namely the current that crosses a unit area per unit time. All the electrons with velocity $\mathbf{v}$, whose number per unit volume is $n(\mathbf{v})$, contribute a factor of $-en(\mathbf{v})\mathbf{v}$ to the current density. Hence the current density due to all the electrons in the wire is a sum over all the velocities:

$$\mathbf{J} = -e\sum_{\mathbf{v}} n(\mathbf{v})\mathbf{v} = -ne\langle\mathbf{v}\rangle\,, \tag{5.3.4}$$

where n is the volume density of all the electrons in the wire (calculated in Exercise 3.1) and $\langle\mathbf{v}\rangle$ is their average velocity. The negative sign corresponds to the negative charge of the electrons, and will cancel out against the negative sign in Eq. (5.3.3). Substituting Eq. (5.3.3) into Eq. (5.3.4) yields the local form of Ohm's law:

Ohm's law

$$\mathbf{J} = \sigma\mathbf{E}\,, \tag{5.3.5a}$$

where the electrical conductivity σ is given by

electrical conductivity

$$\sigma = \frac{ne^2\tau}{m}\,. \tag{5.3.5b}$$

The inverse of the electrical conductivity is called the electrical resistivity: $\rho = 1/\sigma$. Note the similarity between ρ and the "friction coefficient" μ of Sec. 2.4 of Part I. Both are inversely proportional to the mobility and in fact both describe the resistance to the motion of the particles due to the many collisions they undergo during their motion.

electrical resistivity

Exercise 3.3

Show that when the electric field is uniform along the wire's length, Eq. (5.3.5a) is equivalent to Ohm's law, $I = V/R$.

Solution on page 518

Exercise 3.4

In Part I we discussed the isothermal atmosphere, which is analogous to the conducting wire in that a constant force acts on the particles. Explain why a constant current of particles, proportional to the acceleration of gravity, does not arise in the isothermal atmosphere.

Solution on page 518

An additional success of Drude's model was in providing an explanation for the experimental law that was discovered in the middle of the 19th century by Wiedemann and Franz, relating the electrical conductivity of metals to their thermal conductivity, $\bar{K}$ (for $\bar{K}$ see Part I, Sec. 3.6). According to the Wiedemann–Franz law the ratio of the thermal conductivity and the electrical conductivity is proportional to the temperature with a proportionality coefficient that is approximately independent of the material. This law applies to a large number of metals. Several examples appear in the following table:

Wiedemann–Franz law

Table 5.3.1. Thermal conductivity, resistivity $\rho = 1/\sigma$ and the proportionality constant in the Wiedemann–Franz law. Sodium and potassium are liquids at a temperature of 373 K.

		Li	Be	Na	Mg	Al	K	Cu	Ag	Sb	Au
	$\bar{K}$(100 W/mK)	0.71	2.3	1.38	1.5	2.38	1.0	3.85	4.18	0.18	3.1
273 K	$\rho(10^{-8}\ \Omega\text{m})$	8.55	2.8	4.2	3.9	2.45	6.1	1.56	1.51	39	2.04
	$\bar{K}/\sigma T(10^{-8}\ \text{W}\Omega/\text{K}^2)$	2.22	2.36	2.12	2.14	2.14	2.23	2.20	2.31	2.57	2.32
	$\bar{K}$(100 W/mK)	0.73	1.7	—	1.5	2.30	—	3.82	4.17	0.17	3.1
373 K	$\rho(10^{-8}\ \Omega\text{m})$	12.4	5.3	—	5.6	3.55	—	2.24	2.13	59	2.84
	$\bar{K}/\sigma T(10^{-8}\ \text{W}\Omega/\text{K}^2)$	2.43	2.42	—	2.25	2.19	—	2.29	2.38	2.69	2.36

To obtain an explanation for this law Drude assumed that the thermal conductivity of a metal originates principally from the gas of free electrons which can transfer heat with greater ease than the ions which are bound to their positions. Support for this assumption is found in the large difference between the thermal conductivities of metals and insulators. Since the difference between the two types lies in the density of free electrons, which in metals is many orders of magnitude greater than in insulators, it is natural to assume that the larger thermal conductivity is due to the same phenomenon which gives rise to a higher electrical conductivity. Hence we can also calculate the thermal conductivity of a metal by the kinetic theory, as we have done in Sec. 3.6 of Part I, and use the result (1.3.51):

$$\bar{K} = \frac{1}{3} n\bar{v}\ell c = \frac{1}{3} n\bar{v}^2 \tau c\,, \tag{5.3.6}$$

where $\bar{v}$ is the thermal velocity and c is the specific heat per electron. If we now divide (5.3.6) by (5.3.5b) we obtain

$$\frac{\bar{K}}{\sigma} = \frac{m\bar{v}^2 c}{3e^2}\,. \tag{5.3.7}$$

So far all the arguments have been completely general, independent of the form of the velocity distribution of the electrons. If we assume, according to the Drude model, that the distribution is Boltzmannian, we may write $3kT$ for $m\bar{v}^2$, and instead of the specific heat per electron we can use the ideal gas expression $\frac{3}{2}k$. Thus

$$\frac{\bar{K}}{\sigma} = \frac{3k^2 T}{2e^2}\,, \tag{5.3.8}$$

and this is exactly the form of the Wiedemann–Franz law, where the proportionality constant is expressed in terms of the fundamental physical constants e and k.

3.2 A critique of the Drude model

In order to estimate the extent of the Drude model's success we must check its quantitative accord with experiment. First we confront with experiment the relation (5.3.8). We calculate the proportionality coefficient and obtain

$$\frac{3k^2}{2e^2} = 1.11 \times 10^{-8}\ \mathrm{W\Omega/K^2}\,. \tag{5.3.9}$$

This is about one half of the typical value in Table 5.3.1, but it can be considered a success in light of the rough assumptions made. But what actually casts doubt on the validity of the calculation is the problem of the specific heat. In deriving Eq. (5.3.8) we took $3k/2$ for the specific heat per electron. In metals the number of electrons is at least equal to the number of ions. The contribution of the electrons to the specific heat is expected to be very significant. But the experimental specific heat of solids does not surpass the Dulong–Petit value, which is 3R per mole and does not approach 4.5R per mole — the sum of the contribution of the crystal and the expected electronic contribution. In other words, free electrons do not obey the equipartition law. This fact puts in doubt the validity of our calculation, since it turns out that experimentally there is no support for the validity of the equipartition law for electrons in a metal.

Another difficulty shows up upon a deeper examination of Ohm's law (5.3.5). Actually we cannot calculate from first principles the electrical conductivity and compare it to experimental results, because we do not know the mean free time, τ, which depends on the details of the microscopic structure of the metal. However, we can use known values of the resistivity ρ to calculate τ, or the mean free path ℓ_D, according to the Drude model:

$$\ell_D \approx \frac{\sqrt{3mkT}}{ne^2\rho}\,. \tag{5.3.10}$$

Exercise 3.5

(a) Show that the right hand side of (5.3.10) has the dimensions of length.
(b) Prove (5.3.10).

Solution on page 518

Table 5.3.2. Values of ℓ_D for several metals at three different temperatures. The valence v, the atomic mass A and the density d are required for the calculation. The weak temperature dependence of the density is not taken into account and the densities appearing are at room temperature.

		Li	Be	Na	Mg	Al	K	Cu	Ag	Sb	Au
v		1	2	1	2	3	1	1	1	5	1
A		6.941	9.012	22.99	24.305	26.982	39.098	63.55	107.87	121.75	196.97
$d(\mathrm{gr/cm^3})$		0.53	1.85	0.97	1.74	2.70	0.86	8.96	10.50	6.62	19.3
77 K	$\rho(10^{-8}\ \Omega\mathrm{m})$	1.04	—	0.8	0.62	0.3	1.38	0.2	0.3	8	0.5
	$\ell_D(\text{Å})$	44.0	—	104	39.4	38.8	115	124	120	8.03	71.3
273 K	$\rho(10^{-8}\ \Omega\mathrm{m})$	8.55	2.8	4.2	3.9	2.45	6.1	1.56	1.51	39	2.04
	$\ell_D(\text{Å})$	10.1	5.7	37.2	11.8	8.9	49.0	29.9	44.8	0.62	33.0
373 K	$\rho(10^{-8}\ \Omega\mathrm{m})$	12.4	5.3	—	5.6	3.55	—	2.24	2.13	59	2.84
	$\ell_D(\text{Å})$	8.1	3.6	—	9.6	7.2	—	24.4	37.1	0.48	27.7

Table 5.3.2 gives the values of the mean free paths for several metals according to the Drude model, using Eq. (5.3.10).

Exercise 3.6

Obtain the values of ℓ_D that appear in Table 5.3.2.

Solution on page 519

From Table 5.3.2 it appears that around room temperature the values of the mean free path are reasonable, since the interatomic separation in all the metals listed in the table is 3–5 Å. A mean free path of this magnitude corresponds well to the assumption that the resistivity originates from electrons colliding with the ions of the crystal. However, the values of ℓ_D at low temperature, 77 K, pose a problem. The interatomic distances are almost unaffected by the decrease in temperature, but the mean free path is 10–20 times larger. This ratio of the two quantities raises the question: How can an electron pass without hindrance over 10–20 ions in the crystal?

These two difficulties, the problem of the specific heat and the problem of the mean free path, along with other difficulties which we have not mentioned here, remained unresolved for 25 years until the appearance of quantum theory and the discovery of the Pauli principle and the ensuing conclusions regarding the statistical properties of electrons. We discuss these in the coming sections.

3.3 The Sommerfeld model

The discovery of the Pauli principle and the Fermi–Dirac distribution, given in Eq. (5.2.12a), provided Sommerfeld with the tools for explaining the electrical conductivity as well as the thermal conductivity of metals which were left unexplained within the framework of classical physics. Sommerfeld replaced the Maxwell–Boltzmann distribution of the Drude model by a velocity distribution of electrons that obey Fermi–Dirac statistics. This is similar to what Planck had done, replacing the classical distribution function with his own to explain the properties of black body radiation.

We start by calculating the thermodynamic potential Ω for electrons in a metal. We already have the general expression for this potential as a sum over single particle states, i.e. Eq. (5.2.10a), and what remains is to identify the single particle states and to calculate the sum. The single particle states are states of free electrons confined to move inside the metal. Their energy is $p^2/2m$. They are quantum particles, so that in principle we should perform the summation over all the quantum states of a particle in a three-dimensional box, as in Sec. 4.4 of Part III (see Sec. 3.2 of Part IV as well). For the temperatures of interest, in a macroscopic volume V, the summation can be replaced by an integration. Schematically we make the replacement

$$\sum_k \to V \int \frac{d^3p}{h^3}, \tag{5.3.11}$$

where $\mathbf{p}$ is the de Broglie momentum. But we should keep in mind that the electron has spin $\frac{1}{2}$, and thus a complete specification of its state must also include the spin: $+\frac{1}{2}$ or $-\frac{1}{2}$.

Since to any $\mathbf{p}$, specifying a single particle state, correspond two spin states, and since the single particle energy is independent of the spin direction, the sum over the spins contributes an overall factor of 2. The thermodynamic potential is, therefore,

$$\Omega = -2kTV \int \ln \left\{ 1 + \exp \left[\beta \left(\mu - \frac{p^2}{2m} \right) \right] \right\} \frac{d^3p}{h^3}, \tag{5.3.12}$$

and carrying out the integration over all the directions of momentum we obtain

$$\Omega = -\frac{8\pi VkT}{h^3} \int_0^\infty \ln \left\{ 1 + \exp \left[\beta \left(\mu - \frac{p^2}{2m} \right) \right] \right\} p^2 dp. \tag{5.3.13}$$

From this expression for the thermodynamic potential we can obtain the average number of electrons using Eq. (5.1.11):

$$N = \frac{8\pi V}{h^3} \int_0^\infty \frac{p^2 dp}{\exp \left[\beta \left(\frac{p^2}{2m} - \mu \right) \right] + 1}. \tag{5.3.14}$$

The same result could have been obtained from Eq. (5.2.10a).

In principle, given the expression for $\Omega(T, V, \mu)$, Eq. (5.3.13), the thermodynamic properties of the system follow by applying Eqs. (5.1.10). In practice, things are more complicated, since usually it is the number (or density) of the electrons and not the chemical potential that is the controlled variable. Hence we must re-express the chemical potential in terms of the number of particles (as well as the temperature and the volume), and this is done using Eq. (5.3.14), which expresses this relation implicitly. The integral must be calculated as best we can. Then inverting the resulting relation $N(T, V, \mu)$, or better still $N/V = n(T, \mu)$, one obtains $\mu(T, n)$, which replaces the dependence on μ by a dependence on n in all the thermodynamic quantities.

■ The two-dimensional analog of Eq. (5.3.14) is an integral which can be calculated explicitly, and it is possible to find the exact relationship between μ and n. See Self-Assessment Exercise 3.

Next we calculate the probability density function in velocity space for the electrons. It is given by

$$f(\mathbf{v})d^3v = \frac{2m^3}{nh^3} \frac{1}{\exp\left[\beta\left(\frac{mv^2}{2} - \mu\right)\right] + 1} d^3v\,, \tag{5.3.15}$$

where n is the number of electrons per unit volume. This function is the Fermi–Dirac analog of the Maxwell–Boltzmann velocity distribution function (1.1.49).

Exercise 3.7

(a) Derive (5.3.15).
(b) Is $f(\mathbf{v})$ normalized? Explain.

Solution on page 520

Using the Fermi–Dirac velocity distribution function we can calculate the average energy of an electron in the metal:

$$\langle\epsilon\rangle = \int \frac{m}{2} v^2 f(\mathbf{v}) d\tau = \frac{4\pi m^4}{nh^3} \int_0^\infty \frac{v^4 dv}{\exp\left[\beta\left(\frac{mv^2}{2} - \mu\right)\right] + 1}\,. \tag{5.3.16}$$

■ An alternative way of obtaining the same result is found in Self-Assessment Exercise 2(b).

Equation (5.3.16) may also be written as an integral over all the energies after changing variables to $\epsilon = \frac{1}{2}mv^2$, and then defining an energy distribution function:

$$f_\epsilon(\epsilon) = \frac{4\pi(2m)^{3/2}\epsilon^{1/2}}{nh^3} \cdot \frac{1}{e^{\beta(\epsilon-\mu)} + 1}\,, \tag{5.3.17a}$$

so that

$$\langle \epsilon \rangle = \int_0^\infty \epsilon f_\epsilon(\epsilon) d\epsilon\,. \tag{5.3.17b}$$

Exercise 3.8

(a) Prove (5.3.17).
(b) Verify that $f_\epsilon(\epsilon)$ is normalized.

Solution on page 474

As implied by the form in which (5.3.17a) is written, the energy distribution function of the electrons can be interpreted as a product of two factors:

The first is called the degeneracy, or the density of states per unit energy. It is the number of quantum states in the energy range between ϵ and $\epsilon + d\epsilon$:

$$g(\epsilon) = \frac{4\pi V(2m)^{3/2}\epsilon^{1/2}}{h^3}\,. \tag{5.3.18}$$

This factor is quite general, and is independent of the statistical nature of the particles.

The second is the average number of electrons with energy ϵ as in Eq. (5.2.12a), and we write it once more, now without the braces,

$$n_\epsilon(\epsilon) = \{\exp[\beta(\epsilon - \mu) + 1]\}^{-1}\,. \tag{5.3.19}$$

The information that the electrons are fermions is contained in (5.3.19). In order to transform the electron numbers into probabilities, $\langle n_\epsilon \rangle$ must further be divided by N.

3.4 Electrons at high and low temperatures

To proceed we calculate the various integrals approximately. In each range of parameters we introduce approximations valid for that range. This is how we proceeded also in calculating the internal partition function of a diatomic gas (Secs. 1.4 and 1.7 of Part IV). We will study the behavior of the electron gas at high temperatures and at low temperatures. Hence we need to know how the distribution functions of the velocities or of the energies behave in these limits. But here a difficulty arises: The dependence on the temperature (or on β) in (5.3.19) is more complicated than may seem. The chemical potential μ, once replaced by the density n, becomes dependent on the temperature (as well as on n).

To clarify how the chemical potential depends on the temperature, we use Eq. (5.3.14). Dividing both sides by V and changing variables to

$\epsilon = p^2/2m$ we have for the density

$$n = \frac{4\pi(2m)^{3/2}}{h^3} \int_0^\infty \frac{\epsilon^{1/2} d\epsilon}{e^{\beta(\epsilon-\mu)} + 1} \,. \tag{5.3.20}$$

In the limit in which the electron density is very low, i.e. $n \to 0$, the value of the integral must become very small. The right hand side is a monotonic increasing function of $\beta\mu$, since the integrand is at every ε. Hence, the integral decreases as $\beta\mu$ decreases, and for it to become very small, we must have $\beta\mu \to -\infty$. Consequently, for low electron density, it is possible to approximate $n_\epsilon(\epsilon)$ by the Boltzmann factor:

$$n_\epsilon(\epsilon) = Ce^{-\beta\epsilon}, \quad C = e^{\beta\mu}\,, \tag{5.3.21}$$

where we have dropped the braces again. The same classical behavior is obtained also in the high temperature limit. If as $\beta \to 0$, $\beta\mu$ approaches a finite limit, the integrand in (5.3.20) becomes independent of ϵ and the integral diverges. For a given value of n this leads to a contradiction, unless, again, $\beta\mu \to -\infty$ in the limit of high T. One arrives again at the classical Boltzmann distribution.

■ Observe that $\beta\mu \to -\infty$ as $T \to \infty$ is indeed the property of the chemical potential of an ideal gas — Eq. (3.5.9).

At low temperatures the behavior is quite different. When $T \to 0$, or $\beta \to \infty$, the factor $e^{\beta(\epsilon-\mu)}$ depends on the energy ϵ in the following manner: If $\epsilon < \mu$ the exponent is negative, and $e^{\beta(\epsilon-\mu)} \to 0$. If $\epsilon > \mu$ the exponent is positive, and $e^{\beta(\epsilon-\mu)} \to \infty$. Thus, in the limit $T \to 0$,

$$n_\epsilon(\epsilon) = \begin{cases} 1, & \epsilon < \mu\,, \\ 0, & \epsilon > \mu\,. \end{cases} \tag{5.3.22}$$

In this limit the integral in Eq. (5.3.20) is elementary, and the following relation is obtained between the chemical potential at absolute zero, μ_0, and the density of electrons:

$$\mu_0 = \frac{h^2}{2m}\left(\frac{3n}{8\pi}\right)^{2/3} \equiv \epsilon_F\,. \tag{5.3.23}$$

The chemical potential at absolute zero is called the Fermi energy and is denoted by ϵ_F. **Fermi energy**

Exercise 3.9

Derive (5.3.23).

Solution on page 521

The meaning of ϵ_F is clear from Eq. (5.3.22). Even at absolute zero, the electrons must have kinetic energy. This is the crucial difference with respect to the Maxwell–Boltzmann distribution for which, as the temperature decreases, the number of low energy particles increases and at $T = 0$ all the particles are concentrated at the ground level $\epsilon = 0$. This can also be deduced from the equipartition law: $\langle\epsilon\rangle \sim kT$. The reason for the different behavior of the electrons is, of course, the Pauli principle, which forbids the accumulation of many electrons in the same state. Hence the electrons have no alternative but to stack one above the other, in energy, and to fill out all the allowed low energy states. The last among them, with the highest energy, are at ϵ_F. Such a state of a gas is called a *degenerate Fermi gas*.

degenerate Fermi gas

Thus, we could have obtained the Fermi energy simply as an answer to the following question: What energy will be reached by N electrons occupying the N lowest single particle states?

Exercise 3.10

Show that the N electrons fill out in momentum space a sphere of radius p_F which satisfies

$$\frac{p_F^2}{2m} = \epsilon_F\,.$$

Solution on page 521

Fermi sphere

Fermi momentum

Fermi velocity

The surface of the sphere described in Exercise 3.10 is called the Fermi sphere and p_F is called the *Fermi momentum.* Hence we will say that at $T \to 0$ the electrons fill out the Fermi sphere, i.e. all the states up to the Fermi momentum or up to the Fermi energy. The "top" electrons, those with Fermi energy, move at the Fermi velocity, $v_F = (p_F/m)$.

In order to get an idea of the orders of magnitude involved, we present in Table 5.3.3 the values of the Fermi energy and the Fermi velocity of several metals.

Table 5.3.3. Values of the Fermi energy, Fermi velocity and Fermi temperature for several metals. For the last row, the Fermi wavelength, see Exercise 3.18.

	Li	Be	Na	Mg	Al	K	Cu	Ag	Sb	Au
ϵ_F(eV)	4.7	14.3	3.2	7.1	11.7	2.1	7.0	5.5	10.9	5.5
$v_F(10^6\ \text{m s}^{-1})$	1.3	2.25	1.1	1.6	2.0	0.85	1.6	1.4	2.0	1.4
$T_F(10^4$ K)	5.5	16.6	3.8	8.2	13.6	2.5	8.2	6.4	12.7	6.4
λ_F(Å)										

Exercise 3.11

Obtain some of the values of ϵ_F and v_F that appear in Table 5.3.3.

Solution on page 522

Note the energy values reached by the electrons at temperature $T = 0$. For electrons in a classical ideal gas to attain kinetic energies of this order, their temperature must obey

$$kT_F = \epsilon_F \,. \tag{5.3.24}$$

This temperature, which is different for each material, is called the *Fermi temperature* and appears in the third row of Table 5.3.3. It reaches tens of thousands of degrees. As we shall see later on, the Fermi temperature determines the scale for high versus low temperatures of electrons. For $T \gg T_F$ the electrons behave as a classical gas, whereas for $T \ll T_F$ the system becomes degenerate.

Fermi temperature

■ Ordinary metals melt way before reaching T_F.

Exercise 3.12

Show that the *average* kinetic energy of an electron at $T = 0$ is $\frac{3}{5}\epsilon_F$.

Solution on page 523

The Fermi energy plays a double role: One as the maximal (kinetic) energy state occupied by electrons at $T = 0$; the other is the value of the chemical potential at $T = 0$. The relation between the two is clear. The chemical potential describes the increase in the system's free energy at constant volume and temperature upon addition of a single particle. At $T = 0$ the free energy is equal to the internal energy, and hence the chemical potential is the excess internal energy due to the addition of the particle. Since at $T = 0$ all the levels below ϵ_F are occupied, a particle that is added to the system must have energy ϵ_F. This is therefore also the value of the chemical potential at $T = 0$.

Exercise 3.13

(a) How much energy is added to an electron gas at $T = 0$ if an electron with kinetic energy ϵ larger than ϵ_F is added?

(b) How much energy does an electron gas at $T = 0$ lose if we remove from it an electron with kinetic energy ϵ less than ϵ_F?

Solution on page 523

We have found that as $T \to 0$ the chemical potential tends to a positive constant, ϵ_F, and that at $T \to \infty$ it decreases to $-\infty$. Hence it is reasonable to expect that μ decreases as a function of temperature. We will verify this in the next section.

3.5 Metals at room temperature

Now that we have found that an electron gas behaves at high temperatures as a classical Boltzmann gas and at low temperatures as a degenerate Fermi gas, we will apply it to realistic systems, i.e. to metals. We first clarify whether one of the two limits is valid for metals at room temperature. Actually, the difficulties of the Drude model, discussed in Sec. 3.2, indicate that an electron gas does not behave as a classical gas already at room temperature. We will thus check at what temperature does the degenerate gas approximation become valid; in particular, up to what temperature does $\mu = \epsilon_F$ still hold approximately.

To this end we return to Eq. (5.3.20), which implicitly determines the dependence of the chemical potential upon T and n. In terms of the Fermi energy we rewrite Eq. (5.3.20) in the form

$$\frac{2}{3}\epsilon_F^{3/2} = \int_0^\infty \frac{\epsilon^{1/2} d\epsilon}{e^{\beta(\epsilon-\mu)}+1} \tag{5.3.25}$$

(see also Exercise 3.12). We change the integration variable by $\epsilon = x\epsilon_F$ (x is a dimensionless variable that measures the energy in units of ϵ_F). Equation (5.3.25) becomes

$$\frac{2}{3} = \int_0^\infty \frac{x^{1/2} dx}{\exp\left[\frac{\epsilon_F}{kT}\left(x - \frac{\mu}{\epsilon_F}\right)\right]+1} \,. \tag{5.3.26}$$

From Eq. (5.3.23) we know that ϵ_F depends only on the density and not on the temperature. Hence, in the low temperature limit,

$$\frac{kT}{\epsilon_F} \ll 1 \,, \tag{5.3.27a}$$

or, in terms of the Fermi temperature, (5.3.24),

$$T \ll T_F \,. \tag{5.3.27b}$$

As long as (5.3.27b) holds, we see that for x satisfying $x > \mu/\epsilon_F$ the integrand vanishes and for $x < \mu/\epsilon_F$ the exponential term vanishes and the denominator is approximately 1. Hence at temperatures that are much lower than the Fermi temperature it is possible to write

$$\frac{2}{3} = \int_0^{\mu/\epsilon_F} x^{1/2} dx \,, \tag{5.3.28}$$

which leads to $\mu = \epsilon_F$. We found, therefore, that the Fermi temperature does indeed set the scale with respect to which the temperature T is considered low. Because the Fermi temperature of metals is well above

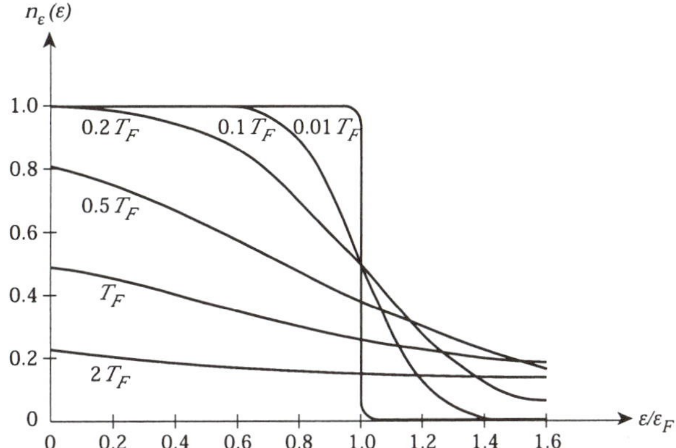

Fig. 5.3.2. The Fermi–Dirac occupation function at different temperatures as a function of the dimensionless variable ϵ/ϵ_F.

10^4 K (see Table 5.3.3), it is possible to treat an electron gas at room temperature, or even well above it, as a degenerate Fermi gas.

To illustrate the situation, we depict in Fig. 5.3.2 the Fermi–Dirac occupation function $n_\epsilon(\epsilon)$ at different temperatures. The graphs take into account the temperature dependence of μ, implicit in Eq. (5.3.20). The figure makes it clear that at temperatures up to several hundredths of T_F the degenerate gas approximation is excellent, since the curve with $T = 0.01T_F$ is very close to the step function.

From this figure it is also possible to learn about the temperature dependence of the chemical potential. From Eq. (5.3.19) we see that at $\epsilon = \mu$, $n_\epsilon = \frac{1}{2}$. Thus, the chemical potential, at any temperature, can be read from the corresponding graph as the energy for which $n_\epsilon = \frac{1}{2}$. Since the intersection of the graph with the line $n_\epsilon = \frac{1}{2}$ moves to the left as the temperature increases, we conclude that the chemical potential decreases with temperature. We already anticipated such a behavior at the end of the preceding section.

Exercise 3.14

Calculate the temperature at which the chemical potential of an electron gas vanishes.

Solution on page 523

3.6 Thermodynamics of the Sommerfeld model

We first apply what we have learned about the properties of the Fermi-Dirac distribution to the calculation of the specific heat: While the Boltzmann distribution predicts that each electron will contribute $\frac{3}{2}kT$ to the specific heat of the metal, there is no experimental evidence for such a contribution; actually the Debye model explains the specific heat of metals very nicely, based on the vibrations of the ions alone. We thus calculate the specific heat of an electron gas with the hope that its contribution at room temperature be indeed negligible.

The specific heat, C_V, of an electron gas is obtained by differentiating $E(T,V,N)$ with respect to T. The total energy of the gas is $N\langle\epsilon\rangle$, where $\langle\epsilon\rangle$ is given by Eq. (5.3.17b). Actually we will also need averages of powers of the energy higher than 1, and thus we write

$$\langle\epsilon^l\rangle = \frac{3}{2\epsilon_F^{3/2}}\int_0^\infty \frac{\epsilon^{l+1/2}d\epsilon}{e^{\beta(\epsilon-\mu)}+1}, \tag{5.3.29}$$

where we have used the results of Exercise 3.12 to simplify the form of the energy distribution function. The integral on the right hand side cannot be calculated explicitly. But, fortunately, the temperature range of interest will be always much smaller than the Fermi temperature, and hence we may expand the integral in Eq. (5.3.29) in powers of the temperature. Such an expansion is made possible by the fact that at low temperatures the Fermi–Dirac function is very similar to a step function (see Fig. 5.3.2) so that the main contribution to the integral is

$$\int_0^\mu \epsilon^{l+1/2}d\epsilon = \frac{2}{2l+3}\mu^{l+3/2}.$$

This is the lowest order term in a series and the corrections will contain higher powers of the temperature. The first two terms will suffice for our needs:

$$\langle\epsilon^l\rangle \approx \frac{3}{2l+3}\frac{\mu^{l+3/2}}{\epsilon_F^{3/2}} + \left(l+\frac{1}{2}\right)\frac{\pi^2}{4}\cdot\frac{\mu^{l-1/2}(kT)^2}{\epsilon_F^{3/2}}. \tag{5.3.30}$$

The proof of (5.3.30) is lengthy. It is contained in the solution to the next exercise and may be skipped altogether on first reading.

Exercise 3.15

Derive Eq. (5.3.30).

Hint: Integrate by parts, and note that the derivative of the Fermi–Dirac function differs significantly from zero only in a narrow range around $\epsilon = \mu$.

Solution on page 524

We now proceed with the calculation of the specific heat. We use Eq. (5.3.30) with $l = 1$. But the temperature derivative of E or $\langle\epsilon\rangle$ must be calculated at constant N and V rather than constant μ and V, and thus we have to express μ in terms of N (as well as T and V). We do this using Eq. (5.3.30) with $l = 0$, which is also equivalent to the normalization condition for f_ϵ resulting from (5.3.20) or (5.3.25). For $l = 0$, to second order in T, we have

$$1 \approx \frac{\mu^{3/2}}{\epsilon_F^{3/2}} + \frac{\pi^2}{8}\frac{(kT)^2}{\epsilon_F^{3/2}\mu^{1/2}}. \tag{5.3.31}$$

This is an equation for μ in terms of the electron density contained in ϵ_F. It does not have a simple solution, but since μ is very close to ϵ_F we can replace $\mu^{1/2}$ by $\epsilon_F^{1/2}$ in the second term on the right hand side, which is already small due to the explicit factor T^2. Hence we find an expression for the dependence of the chemical potential on the temperature up to the second order:

$$\mu(T, V, N) \approx \epsilon_F \left[1 - \frac{\pi^2}{12}\left(\frac{kT}{\epsilon_F}\right)^2\right]. \tag{5.3.32}$$

■ Note that μ decreases with T, as expected.

Using this expression, we calculate the temperature dependence of the average energy up to second order in T. It reads

$$\langle\epsilon\rangle = \frac{3}{5}\epsilon_F \left[1 + \frac{5\pi^2}{12}\left(\frac{kT}{\epsilon_F}\right)^2\right]. \tag{5.3.33}$$

As $T \to 0$ the result of Exercise 3.12 is obtained. Moreover, unlike the ideal gas, the average energy depends on the density.

Exercise 3.16

(a) Show that (5.3.32) is a solution of (5.3.31) to order T^2.
(b) Derive (5.3.33).

Solution on page 526

We see, therefore, that the energy of an electron gas increases quadratically with temperature. By differentiating (5.3.33) with respect to T we find that the specific heat at low temperatures is linear in the temperature:

$$C_V \approx \frac{\pi^2}{2} Nk\frac{T}{T_F}. \tag{5.3.34}$$

As already mentioned, this expression is valid also at room temperature, since the typical values of the Fermi temperature are of order 10^4 K.

Hence, the contribution of the electrons to the specific heat of metals at room temperature is not $\frac{3}{2}k$ per electron, but about $0.05k$ per electron, which is the behavior we were hoping to find. It should further be noted that the contribution of the electrons to the specific heat of metals is not negligible at every temperature. As we saw in Part IV, the contribution of the phonons to the specific heat at low temperatures is proportional to T^3 [Eq. (4.3.18)], and hence at low enough temperatures the contribution of the phonons becomes much smaller than that of the electrons, and the specific heat will behave according to Eq. (5.3.34). Such a behavior has in fact been observed experimentally and provided additional support for the Sommerfeld model.

Exercise 3.17

Lithium has a Debye temperature of 400 K. Find at what temperature the phonon and electron contributions to the specific heat are equal.

Solution on page 527

We now return to the Wiedemann–Franz law to test the implications of the Sommerfeld model for the ratio of the thermal conductivity to the electrical conductivity. The result obtained from the Drude model for this ratio was reasonable and we should hope that the result of the Sommerfeld model for the specific heat would not damage this agreement.

The thermal conductivity of the metal is given by Eq. (5.3.6). However, we have to reinterpret the quantities appearing in it. c is the specific heat per electron, whose correct value is given by Eq. (5.3.34). τ is the mean free time, and is related to the electrical conductivity via Eq. (5.3.5b). We are left with the average velocity $\bar{v}$. This velocity cannot be the average velocity of the Maxwell–Boltzmann distribution, which at room temperature is of the order 10^5 m s^{-1}, but a velocity of the order of the Fermi velocity, which is 10 times larger. In the ratio of the thermal conductivity to the electrical conductivity, given in Eq. (5.3.7), the specific heat per electron decreases by a factor of about 30, and the average velocity increases by a factor of about 10 and its square by a factor of 100. These two changes, which more or less cancel each other out, leave the Wiedemann–Franz law valid even though the pictures of the electrical and thermal conductivities have changed drastically.

To perform a quantitative study of the ratio $\bar{K}/\sigma$ one should clarify which velocity to substitute for $\bar{v}$. We note that not all free electrons in the metal can participate in the thermal conduction but only those with energy very near the Fermi energy. Electrons with energy significantly smaller than the Fermi energy ($0.5\epsilon_F$, for instance) cannot change their kinetic energy by collisions, since the states are occupied up to the Fermi energy. The Pauli principle implies that for an electron to change its momentum it must find an empty state to go into. Thus to change its momentum, the deep electron must "jump," as a result of a collision, to the neighborhood of the Fermi energy. For the latter to occur it has to collide with an electron with kinetic energy of at least $1.5\epsilon_F$, but such an electron is very rare, as follows from Fig. 5.3.2, for example.

The "effective" electrons that can absorb and transfer energy are, therefore, those found near the Fermi energy, at a range of energies of width kT. Hence, the electrons that participate in the thermal conduction, which involves the transfer of kinetic energy, are with energies very near to the Fermi energy. They all have the Fermi velocity, namely $\bar{v} = v_F$. If we now substitute this into (5.3.7) along with (5.3.34), we obtain

$$\frac{\bar{K}}{\sigma} = \frac{mv_F^2}{3e^2} \cdot \frac{\pi^2}{2} \cdot \frac{kT}{T_F} = \frac{\pi^2 k^2}{3e^2} T \,, \tag{5.3.35}$$

and this is once more the Wiedemann–Franz law with a different coefficient.

Note that the temperature dependence has remained the same in spite of all the transformations. But there is a difference: In the Drude model the linear dependence originates from the fact that the average velocity is proportional to $T^{1/2}$; in contrast, in the Sommerfeld model the average velocity does not depend on the temperature (for $T/T_F \ll 1$) but as a compensation the *specific heat* is proportional to the temperature!

When we calculate the coefficient, we obtain

$$\frac{\pi^2 k^2}{3e^2} = 2.44 \times 10^{-8}\ \mathrm{W\Omega/K^2} \,, \tag{5.3.36}$$

which agrees well with the experimental values that appear in Table 5.3.1.

Another difficulty of the Drude model was the problem of the mean free path, which attains puzzling values of 10–20 interatomic distances. We thus check what values are obtained in the Sommerfeld model. To this end we write the relation between the mean free path and the resistivity in the form (see Exercise 3.5)

$$\ell = \frac{m\bar{v}}{ne^2\rho} \,, \tag{5.3.37}$$

where $\bar{v}$ is the average velocity of the electrons.

The Drude model assumes that $\bar{v}$ is of order of the thermal velocity of a Boltzmann electron gas, which at room temperature is of order 10^5 m s^{-1}. In the Sommerfeld model we use the Fermi–Dirac distribution to obtain $\bar{v}$, leading to a velocity of the order of the Fermi velocity. Since the Fermi velocity is of order 10^6 m s^{-1}, the mean free path becomes much larger. Already at room temperature the mean free path is 10 times larger than the values in the Drude model, and at lower temperatures it can reach 1000 Å.

It appears, therefore, that the problem of the mean free path, which in the Drude model was limited to fairly low temperatures (see Table 5.3.2), worsened in the Sommerfeld model, in which it is present already at room temperature and above. To understand how it is possible that the electrons in a metal almost do not feel the existence of the ions, we first note that so far we have not taken into account the fact that the electrons have wavelike properties. The de Broglie wavelength of a typical electron in the metal, namely the wavelength of an electron with Fermi energy, called the Fermi wavelength, λ_F, is about 2–10 Å. See Exercise 3.18.

Fermi wavelength

Exercise 3.18

Insert the values of λ_F missing from Table 5.3.3.

Solution on page 527

This value of the Fermi wavelength implies that the electrons in the metal cannot collide with the ions as if they (the electrons) were pointlike particles. Moreover, it seems that the fact that the wavelength is some 10 times larger than the atomic sizes, supports the intuition that the electrons do not "see" the ions, just as waves with wavelength λ are of no use for observing objects of a size less than λ.

At this point a consistent quantum-mechanical treatment of the dynamics of the electrons that are moving in a potential due to the periodic ionic lattice of the metal is called for. Here we leave this subject. We shall only remark that a quantum analysis of the motion of the electrons in a periodic potential was actually carried out in the 1930's by F. Bloch. Surprisingly, the result of that analysis is that the electrical resistivity of a perfectly periodic crystal is zero! Hence the origin of the electrical resistivity cannot be the collisions between the electrons and the ions and the mean free path that we calculated does not correspond to the scattering of the electrons by the ions. It turns out, however, that other scattering mechanisms exist: electrons are scattered by the deviations from perfect periodicity of the crystal. Such deviations exist in any real metal in the form of phonons (at finite temperature), missing ions, impurities of foreign materials and other defects. It is the scattering off such defects that gives rise to electrical resistivity.

Chapter 4

Boson Gas

4.1 Bose–Einstein distribution

We turn to the properties of systems of bosons, whose (average) number in a state with energy ϵ is given by the Bose–Einstein distribution:

Bose–Einstein distribution

$$\langle n_\epsilon(\epsilon)\rangle = \{\exp[\beta(\epsilon-\mu)]-1\}^{-1}. \tag{5.4.1}$$

Actually we have already seen in Parts III and IV that it is possible to treat the phonons in a crystal and the photons of a black body as bosons. Since phonons and photons are created and annihilated freely, their chemical potential is zero. In contrast, a gas of identical atoms with integral spin is a system of bosons with a given number of particles, so its chemical potential will not be zero. We discuss, therefore, a gas of N bosons in a container of volume V at temperature T. As we saw in Chaps. 1 and 2, it is best to use the grand canonical ensemble in which the chemical potential μ is a controlled variable instead of the number of particles, as a technically natural way of dealing with constraints imposed by quantum statistics.

The thermodynamic potential for a gas of bosons is given by Eq. (5.2.10b) as a sum over single particle states and may be written as an integral over all possible particle momenta, just as we did for the fermion gas, Eqs. (5.3.12) and (5.3.13). If we restrict our discussion to bosons with spin 0, we obtain the Bose–Einstein analog of Eq. (5.3.13) as

$$\Omega = \frac{4\pi VkT}{h^3}\int_0^\infty \ln\left\{1-\exp\left[\beta\left(\mu-\frac{p^2}{2m}\right)\right]\right\}p^2dp\,, \tag{5.4.2}$$

and differentiating this with respect to μ we obtain the average number of bosons:

$$N = \frac{4\pi V}{h^3}\int_0^\infty \frac{p^2dp}{\exp\left[\beta\left(\frac{p^2}{2m}-\mu\right)\right]-1}\,, \tag{5.4.3}$$

as a function of T, V and μ. For bosons with nonzero spin J, these expressions have to be multiplied by $2J+1$. As for the fermion gas, Eq. (5.4.3) is used to express μ in terms of N, or better still, in terms of the average density n. We write the energy distribution function in the form

$$f_\epsilon(\epsilon) = \frac{1}{N} g(\epsilon) n_\epsilon(\epsilon) , \tag{5.4.4}$$

where n_ϵ is given now by (5.4.1) (without the braces) and $g(\epsilon)$ describes the number of quantum states in the range between ϵ and $\epsilon + d\epsilon$:

$$g(\epsilon) = \frac{2\pi V (2m)^{3/2} \epsilon^{1/2}}{h^3} = V \bar{g}(\epsilon) , \tag{5.4.5}$$

where $\bar{g}$ is introduced to make the volume dependence explicit. Note that between this function and the density of electronic states [Eq. (5.3.18)] there is a factor of 2 originating in the spin. The derivation of these expressions is the same as for fermions.

The average density takes on a form analogous to Eq. (5.3.20):

$$n = \frac{N}{V} = \frac{2\pi (2m)^{3/2}}{h^3} \int_0^\infty \frac{\epsilon^{1/2} d\epsilon}{e^{\beta(\epsilon - \mu)} - 1} , \tag{5.4.6}$$

and just as for fermions, the Boltzmann limit is obtained at low densities ($n \to 0$) or high temperatures ($T \to \infty$) and in these limits $\beta\mu \to -\infty$.

4.2 Chemical potential at low temperatures

We are interested in the behavior of the boson gas at low temperatures. Since the Pauli principle does not apply to bosons, they tend to concentrate at the lowest energy levels and only the thermal fluctuations prevent them from accumulating all in the ground level (see Fig. 5.2.1). At low temperatures, where the thermal fluctuations are small, we may expect a concentration of a macroscopic number of bosons in the ground level $\epsilon = 0$. Substituting into the Bose–Einstein distribution Eq. (5.4.1), $\epsilon = 0$, we find that at low temperatures, if a number of the order of N bosons are in the ground level, then

$$N \approx \frac{1}{e^{-\beta\mu} - 1} , \tag{5.4.7}$$

or

$$\mu \approx \frac{-kT}{N} . \tag{5.4.8}$$

Exercise 4.1

Prove (5.4.8). What are typical values of the chemical potential at low temperatures?

Solution on page 527

At first sight such a behavior is quite reasonable, because it is consistent with the tendency of the chemical potential to decrease at high temperatures and is also consistent with the requirement that the chemical potential should be negative. However, a more discerning inspection reveals that all is not well with our understanding of the Bose–Einstein distribution: The occupation of the the ground level, which at low temperatures is expected to be very large, does not appear at all in the expression for the density of particles, Eq. (5.4.6). This equation gives the density of particles as a sum of contributions from all energies. But the contribution of the ground level, namely the region near $\epsilon = 0$, to the integral is zero.

This difficulty leads to an apparent paradox, which is a consequence of the relation between the density of particles and the chemical potential, Eq. (5.4.6). The integral on the right hand side is an increasing function of μ, since the integrand at every value of ϵ is an increasing function of μ. But the integral cannot become larger than its value at $\mu = 0$, since μ cannot become positive. Consequently, the right hand side of Eq. (5.4.6) has an upper bound, which is its value at $\mu = 0$. This would imply that the density, n, at a given temperature, of our noninteracting boson gas cannot rise above a certain maximum value, determined by the maximum of the right hand side. For densities above that value, there is no solution for μ. This value is

$$n_* = \frac{2.612(2\pi mkT)^{3/2}}{h^3} . \tag{5.4.9}$$

It corresponds to a maximal number of particles, in a given volume V, $N_* = Vn_*$. Note that the maximal density decreases when T decreases.

Exercise 4.2

(a) Show that n is an increasing function of μ.
(b) Derive Eq. (5.4.9).

Solution on page 528

This conclusion is, of course, unreasonable, since it is impossible for there to be a restriction on the number of noninteracting bosons in a given volume. Even fermions which obey the Pauli principle do not resist the addition of fermions; they only force the "new" fermions to occupy higher energy levels. Supposing that the system is prepared at a density and temperature for which there exists a solution for μ, as T is lowered the solution disappears. Moreover, at $T = 0$ the equation will allow only zero density, which is clearly absurd.

What went wrong?

4.3 Bose–Einstein condensation

In order to clarify the reason for the inconsistency we return to basics: The calculation of the thermodynamic potential Ω and its derivatives. Equations (5.4.2) and (5.4.5) were obtained from (5.2.10b) by replacing the summation over the discrete quantum states k by an integration over a continuous momentum. But in doing so we lost the contribution of the ground level, just because $g(0) = 0$.

If the temperature is not too low (or the density not too high) the occupation of the ground level does not differ greatly from that of the other levels. In such conditions, each of the terms in the sum (5.2.10b) or (5.2.11b) is of order 1, while the sums are at least of order N. Hence the error introduced in neglecting the ground level is insignificant.

But at low temperatures the chemical potential tends to zero as $1/N$, Eq. (5.4.8), and the term corresponding to the ground level in the various sums must be examined more closely. In the sum for Ω, Eq. (5.2.10b), this term is

$$\Omega_0 = kT\ln(1 - e^{\beta\mu}) \approx kT\ln(1 - e^{-a/N}) \approx kT\ln(a/N)\,,$$

where we wrote [Eq. (5.4.8)] $\beta\mu \approx -a/N$ with a of order 1. This term can still be neglected, since $\ln N$ is negligible compared to N as $N\to\infty$. In contrast, at low T, the ground level term in the sum for N in Eq. (5.2.11b) is

$$N_0 = \frac{1}{e^{-\beta\mu} - 1} \approx \frac{1}{e^{a/N} - 1} \approx \frac{N}{a}\,.$$

This term is of the same order of magnitude as the entire sum, despite the fact that it originates from the (negligible) ground level term in Ω. Neglecting it leads to the contradiction we encountered. Thus we split the expression for Ω into two parts:

$$\Omega = \Omega_0 + kTV\int_0^\infty \bar{g}(\epsilon)\ln(1 - e^{\beta(\mu-\epsilon)})d\epsilon\,, \tag{5.4.10}$$

where $\bar{g}$ [Eq. (5.4.5)] has been introduced to make the extensive character of the second term explicit.

By the same reasoning we rewrite Eq. (5.4.3), divided by N as

$$n = n_0 + \int_0^\infty \frac{\bar{g}(\epsilon)d\epsilon}{e^{\beta(\mu-\epsilon)} - 1}\,, \tag{5.4.11a}$$

with

$$n_0 = \frac{1}{V}\frac{1}{e^{-\beta\mu} - 1}\,. \tag{5.4.11b}$$

Here the role of $\bar{g}$ in the second term of (5.4.11a) is to render it of order 1 (independent of the volume), as the first term.

As long as the density n is low compared to n_* (or the temperature high enough), μ is not very small, and the density of particles in the ground level, n_0, is negligible compared to the total density n. n is then essentially due to the second term of (5.4.10), which we may designate by n_e. The addition of bosons causes an increase in the chemical potential, which becomes less negative and at the same time the integral (i.e. n_e) tends to its maximal value n_*. When n becomes larger than n_*, there is no more room in the excited states and the remaining bosons occupy the ground level. n_* is thus the density of bosons occupying the excited states when $\mu = 0$, when also $n_e = n_*$. The fact that all the additional bosons occupy the ground level is a reflection of the vanishing of the chemical potential: The addition of a particle does not add energy to the system.

A boson gas in such a state is called a *degenerate Bose gas*, and the phenomenon of the aggregation of bosons in the ground level is called *Bose–Einstein condensation.* Note the contrast with the degenerate Fermi gas in which any additional particle piles up at the top of the distribution. Conversely, in a degenerate Bose gas any additional particle goes straight to the bottom.

degenerate Bose gas

Bose–Einstein condensation

The same phenomenon occurs while a Bose gas is cooled: At a sufficiently high temperature the overwhelming majority of bosons occupy the excited levels and only a negligible minority occupy the ground level. A decrease in temperature (at constant N, V) causes the chemical potential to adjust itself and to increase (to become less negative). Since the chemical potential is bounded by zero, there appears a critical temperature T_c, which is the temperature at which the chemical potential (first) vanishes. In order to find this temperature we can use Eq. (5.4.9) with $n_* = n$, which yields

$$T_c = \frac{h^2}{2\pi mk}\left(\frac{n}{2.612}\right)^{2/3} . \tag{5.4.12}$$

Exercise 4.3

(a) Show that the chemical potential increases as the temperature decreases, namely that μ increases with β.

(b) Obtain Eq. (5.4.12).

(c) Show that the maximal density of particles occupying the excited states can be written in the form

$$n_* = n\left(\frac{T}{T_c}\right)^{3/2} .$$

(d) Explain the relation between n_e and n_* when $T < T_c$ and when $T > T_c$.

Solution on page 529

Any further cooling below T_c leaves the chemical potential at $\mu = 0$. As implied by Eq. (5.4.9), or by Exercise 4.3, the density of particles occupying the excited states decreases with decreasing temperature. Because the number of particles is constant, the difference is concentrated in the ground level, and as the temperature decreases, the density of bosons in the ground level n_0 increases:

$$n_0 = n - n_* = n\left[1 - \left(\frac{T}{T_c}\right)^{3/2}\right]. \tag{5.4.13}$$

Figure 5.4.1 illustrates the temperature dependence of n_0 and n_e. At temperatures $T < T_c$ Bose–Einstein condensation takes place.

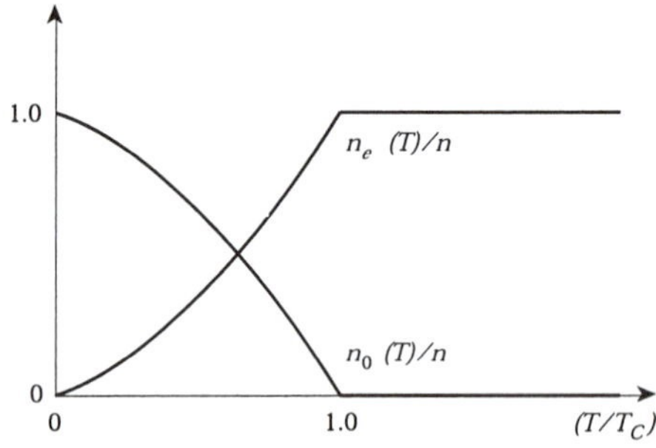

Fig. 5.4.1. The density of bosons in the ground state n_0, and the density in the excited states n_e as a function of temperature.

4.4 Superfluidity

After having found that an ideal boson gas behaves at low temperatures very differently from a Boltzmann gas, the following question arises: In what physical conditions does this behavior manifest itself?

The first quantity we consider is the critical temperature, T_c. In order to obtain an estimate of T_c we calculate the critical temperature of 10^{22} atoms in a volume of 1 cm^3. Assuming that the mass is that of a hydrogen atom, 1.7×10^{-24} gr, we obtain a critical temperature of 7 K. For heavier atoms we obtain even lower critical temperatures.

This may seem a very disappointing result, because at such low temperatures almost all substances are already either solid or liquid (at atmospheric pressure). The lowest freezing points are of nitrogen, which freezes at 63 K; neon, at 25 K; and hydrogen, at 14 K. Helium (He^4) is exceptional, in that it liquefies at 4.2 K and does not freeze even at $T = 0$. This has delayed the discovery of superfluidity in regular gases until very recently.

Nevertheless, we may place some hope in liquid helium, or specifically in He^4, whose atoms are bosons (the atoms of He^3 are fermions). The fact that liquid He^4 does not freeze even at the lowest temperatures implies that the interatomic forces in it are very weak. Its low density — 0.14 gr/cm^3 — and its exceptionally low viscosity — 40 μP — indicate that its properties are closer to those of a dense gas than of a liquid. At room temperature water has a viscosity of 0.01 P, and typical gases such as nitrogen and helium have a viscosity of 2×10^{-4} P. Since the viscosity is proportional to $T^{1/2}$ (see Exercise 3.14 of Part I), a typical viscosity for gases at 4 K would have been (if there existed such gases) of order 10^{-5} P. The viscosity of *gaseous* helium in this region is indeed about 20 μP. Moreover, since the experiments of Keesom and Kapitza of the twenties and thirties, it has been known that liquid He^4 drastically changes it properties at a temperature of 2.17 K. The change is so significant that each of the phases has been given a name of its own: He I is the name of the liquid above 2.17 K, and He II, below 2.17 K.

The most pronounced property of He II is its ability to flow through capillaries without any friction. The measurement of the viscosity of He II in such flows suggests a value not in excess of 10^{-11} of that of He I. This is the property of *superfluidity* of He II. Another of its prominent properties is an extraordinarily large thermal conductivity of order 10^4–10^5 W/mK. **superfluidity**

■ Compare with the thermal conductivities of metals appearing in Table 5.3.1.

These two properties manifest themselves in a series of spectacular phenomena, such as the fact that the liquid can "crawl" on the sides of an empty vessel immersed in a bath of He II and fill it up (Fig. 5.4.2a) or crawl out of it when the vessel is taken out of the bath (Fig. 5.4.2b), or the fact that He II boils without bubbles.

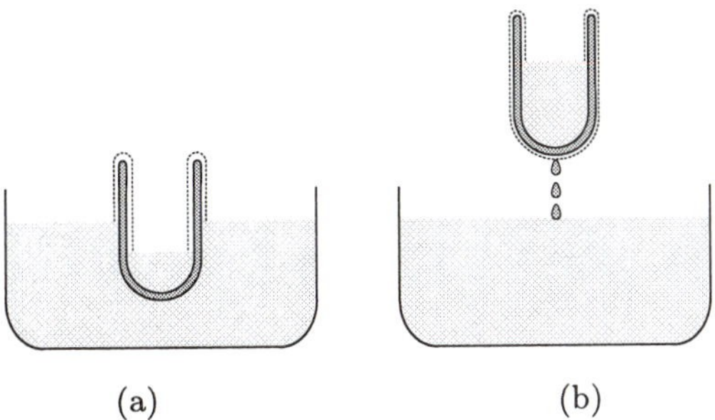

Fig. 5.4.2. (a) An empty vessel immersed in a bath of He II is filled up by the liquid crawling on its sides until the heights become equal. (b) If the vessel is taken out of the bath, the liquid crawls out of it and drips back into the bath.

All of the above led F. London in 1938 to interpret the phase transition from He I to He II as a Bose–Einstein condensation. According to this explanation, at temperatures below the transition temperature a macroscopic number of helium atoms (a finite density) occupy the ground level and a macroscopic number of atoms are in the excited levels. These two populations behave as two different liquids — one "normal," He I-like, and the other superfluid — that coexist in the same vessel. London identified the atoms of the normal liquid with the atoms occupying the excited states, whereas the superfluid atoms were identified with the atoms occupying the ground level.

The existence of two liquids is actually confirmed experimentally, and a good example of this is the fact that different methods of measuring the viscosity yield different results. One way is to measure the rate of liquid flow in a capillary. This rate is inversely proportional to the viscosity. Measurements of this kind yield the extremely small viscosity values mentioned above and have led to the term "superfluidity." On the other hand, measurements of the viscosity of He II by the method of the rotating cylinder, which is illustrated in Fig. 5.4.3, yield values very close to those of He I.

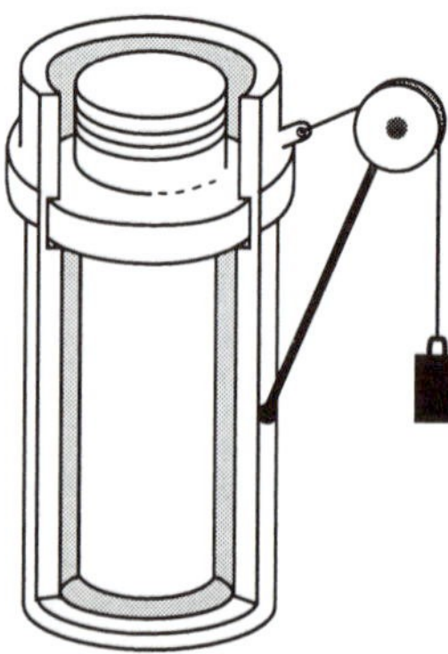

Fig. 5.4.3. Viscosity measurement by the method of the rotating cylinder. The inner cylinder rotates due to the pull of the weight and creates a velocity gradient in the liquid filling the space between the inner and outer cylinders.

The reason for this is that the viscosity of the "normal" component of He II prevents it from participating in the flow through capillaries, and thus that flow is totally superfluid. In contrast, in the rotating cylinder the whole liquid participates. The superfluid component does not contribute, of course, but the macroscopic number of "normal" atoms which participate in this flow give rise to a viscosity like that of He I.

■ Note that we do not explain here why Bose–Einstein condensation implies a zero viscosity. We will return to this later.

The superfluid component is also responsible for the phenomenon of the crawling of liquid helium. A thin layer of helium atoms, originating from the helium vapor around the vessel, accumulates on its inner and outer surfaces and gives rise to a continuous connection between the inside of the vessel and the bath. Since the superfluid component flows without viscosity there is no resistance to its atoms' tendency to decrease their energy by moving along the layer from the higher liquid surface towards the lower liquid surface, as gasoline is pumped out of the tank of a car and into an external container with the help of a rubber pipe. In our case the pipe is created spontaneously by the helium layer.

The huge thermal conductivity is also a reflection of the fact that He II consists of two different components. The heating of a certain region A with respect to a nearby region B leads to a difference in the concentration of the atoms of the superfluid component, since their number decreases with increasing temperature, according to Eq. (5.4.13). Hence the concentration of the superfluid component in region A is lowered, and as a result a current of the superfluid component is set up from B to A, in order to equate the concentrations. This current causes an increase in the total density and the total pressure of the helium in region A, which in turn gives rise to a current of the normal component from A to B. In a steady state in which a temperature gradient is created between A and B ($T_A > T_B$), there will appear, therefore, a current of the superfluid component from B to A and a current of the normal component from A to B.

Since the atoms of the superfluid component are all in the ground level, they cannot carry heat. On the other hand, the normal component carries heat, so that the temperature difference we created is accompanied by a flow of heat from the hot region to the cold region. This mechanism of heat conduction is very efficient. It leads to the fact that He II boils without bubbles: Any local temperature rise near the boiling point does not lead to the evaporation of the region and to the appearance of a bubble, but is instead immediately carried to the edge of the liquid, causing its surface to evaporate.

Thus, we have described qualitatively the manner in which Bose–Einstein condensation is related to the phenomenon of superfluidity in He II. The description of liquid helium as an ideal boson gas must be considered no more than a rough approximation, on which we will improve in Sec. 4.6. But before doing that we shall see that several properties of liquid helium can nevertheless be captured within this description.

4.5 Bose–Einstein condensation in helium

The first point we discuss is the critical temperature. If the phase transition from He I to He II is indeed a Bose–Einstein condensation, then the critical temperature given by (5.4.12) must correspond to the transition temperature 2.17 K, at least approximately (since, after all, liquid helium is not an ideal gas). And, indeed, substituting the appropriate quantities for helium we obtain a critical temperature of about 3 K, which is close enough to the transition temperature to be considered as supporting our explanation.

Exercise 4.4

(a) Calculate the critical temperature for Bose condensation of He^4.

(b) Calculate the critical temperature for Bose condensation of diatomic hydrogen H_2 if the density of liquid hydrogen is 0.06 gr/cm^3. Would you expect superfluidity in liquid hydrogen as well?

Solution on page 530

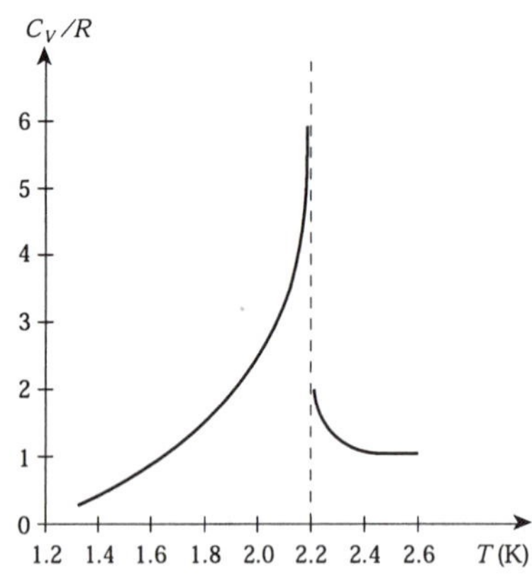

Fig. 5.4.4. The molar specific heat of liquid He^4. C_V is measured in units of the gas constant $R = 8.3$ J K^{-1}.

The next issue is the specific heat. Measurements of the specific heat around the transition temperature 2.17 K reveal a sharp increase at this temperature, which is evidence for a sharp change in the properties of the liquid. We shall presently see that the specific heat of the Bose gas increases with T for $T < T_c$ and decrease for $T > T_c$. At $T = T_c$ there is a cusp (see Fig. 5.4.5). The quantitative difference between the experimental and theoretical curves reflects the fact that the description of liquid helium as an ideal gas is too coarse. To obtain better agreement, the interactions between the atoms must be taken into account, which we will not do here.

The calculation of the specific heat, C_V, of a boson gas below the critical temperature is quite simple, since the chemical potential is constant ($\mu = 0$) and the energy is due entirely to the atoms in the excited states:

$$E = \frac{2.012V(2\pi m)^{3/2}(kT)^{5/2}}{h^3} . \tag{5.4.14}$$

Exercise 4.5

(a) Derive Eq. (5.4.14).

(b) Show that the average energy per particle (when $T < T_c$) is

$$\frac{E}{N} = 0.770kT_c\left(\frac{T}{T_c}\right)^{5/2} .$$

Solution on page 530

Differentiating the energy with respect to T we obtain the specific heat for $T < T_c$:

$$C_V = \frac{5.030V(2\pi mkT)^{3/2}}{h^3} \cdot k = 1.925Nk\left(\frac{T}{T_c}\right)^{3/2} . \qquad (5.4.15)$$

The specific heat of a boson gas below the critical temperature thus behaves like $T^{3/2}$ compared to that of a degenerate fermion gas which is linear in T. It reaches a maximum of $1.925k$ per particle, compared to $\frac{3}{2}k$ for a classical gas. We further emphasize that the specific heat and the energy do not depend on the number of particles! The reason for this is that below T_c all the additional particles accumulate in the ground level, and do not contribute to the energy.

We now calculate the specific heat above the critical temperature. To this end we need the temperature dependence of the energy, but now the chemical potential is also temperature-dependent. Hence we first obtain the temperature dependence of the chemical potential. We do this, approximately, for temperatures near T_c and hence for values of μ near zero, as we did for the fermion gas near $T = 0$. The details of the calculation are, of course, different.

For $T > T_c$ the occupation of the ground level is negligible, and hence it is possible to use Eq. (5.4.6), which is an implicit relation between μ, T and n. To calculate the integral on the right hand side of (5.4.6), as an expansion for small values of μ, we write the difference between the maximal density in the excited states, n_*, and n:

$$n_* - n = \frac{2\pi(2m)^{3/2}}{h^3}\int_0^\infty \left[\frac{1}{e^{\beta\epsilon}-1} - \frac{1}{e^{\beta(\epsilon-\mu)}-1}\right]\epsilon^{1/2}d\epsilon\,. \qquad (5.4.16)$$

Though n_* is known, Eq. (5.4.9), the advantage of this form is that the integral on the right hand side of (5.4.16) can now be calculated when μ is close to zero. Its magnitude is $\pi kT\sqrt{-\mu}$. (See solution to Exercise 4.6.)

■ Recall that $\mu < 0$.

Thus

$$\mu \approx -\frac{h^6}{32\pi^4 m^3}\left(\frac{n_* - n}{kT}\right)^2 = -\left\{2.612\left[1 - \left(\frac{T_c}{T}\right)^{3/2}\right]\right\}^2 \frac{kT}{4\pi} . \qquad (5.4.17)$$

Note that since n_* increases with T, μ becomes more negative with increasing temperature above T_c.

Exercise 4.6

Derive Eq. (5.4.17).

Solution on page 531

Next we calculate the energy E at temperatures near T_c (but above it). We calculate the difference between the energy E_*, for $\mu = 0$, and the actual E, in analogy with the calculation of $n_* - n$. E_* is given by the right hand side of Eq. (5.4.14). We have

$$E_* - E = \frac{2\pi V(2m)^{3/2}}{h^3} \int_0^\infty \left[\frac{1}{e^{\beta\epsilon} - 1} - \frac{1}{e^{\beta(\epsilon-\mu)} - 1}\right] \epsilon^{3/2} d\epsilon\,, \qquad (5.4.18)$$

and for small μ we find that

$$E \approx E_* + \frac{3}{2} N_* \mu\,. \qquad (5.4.19)$$

Since μ is negative, the energy is actually smaller than the maximal value E_*, which it can attain at temperature T.

Exercise 4.7

Derive (5.4.19).

Solution on page 532

The temperature dependence of the energy is quite complicated even in this approximation: The temperature dependence of E_* is given by (5.4.14), the temperature dependence of n_* by (5.4.9), and the temperature dependence of μ by (5.4.17), and in all

$$E \approx \left\{0.770 - 0.814\left[1 - \left(\frac{T_c}{T}\right)^{3/2}\right]^2\right\} \left(\frac{T}{T_c}\right)^{3/2} NkT\,. \qquad (5.4.20)$$

Differentiating, we find the specific heat at temperatures near T_c:

$$C_V \approx \left[1.629 + 0.407\left(\frac{T_c}{T}\right)^{3/2} - 0.111\left(\frac{T}{T_c}\right)^{3/2}\right] Nk\,. \qquad (5.4.21)$$

This is a function that attains a maximum value of $1.925Nk$ at $T = T_c$ and decreases with T. Far from $T = T_c$, Eq. (5.4.21) is no longer valid.

It is clear that when $T \gg T_c$ the boson gas behaves like an ideal gas, and hence $C_V \to \frac{3}{2}Nk$. Figure 5.4.5 illustrates the behavior of the specific heat as a function of temperature. The similarity to Fig. 5.4.4 is not complete, but it is quite significant.

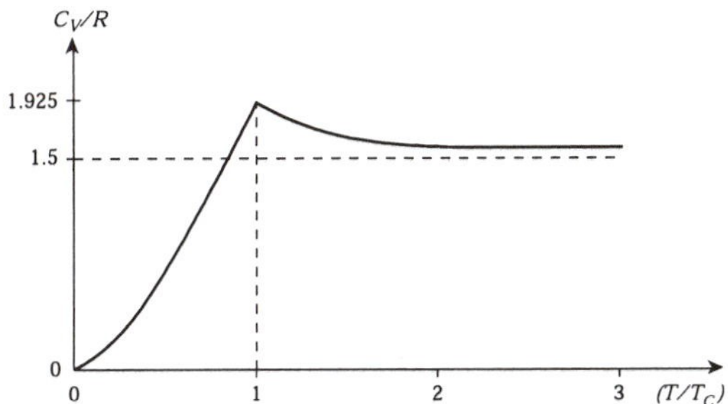

Fig. 5.4.5. The specific heat of a boson gas as a function of temperature.

4.6 Viscosity of a superfluid

In describing the properties of liquid helium below 2.17 K, we mentioned that when it behaves as an inviscid fluid, it is the superfluid component, namely the atoms occupying the ground level $\epsilon = 0$, that is responsible for all the observed spectacular phenomena. But we did not explain why a liquid whose atoms are all in the ground state is inviscid. The first explanation of this was given in the 1940's by Landau.

We start by considering the motion of a body in a liquid under the influence of an external force. The viscosity of the liquid is expressed as a frictional force acting on the moving body, originating in the collisions between the body and the molecules of the liquid (see Secs. 2.4 and 3.6 of Part I). We have seen that it is possible to give a satisfactory account of the viscosity of gases considering their molecules as free particles colliding from time to time with the body.

This assumption cannot underlie a quantitative analysis of the viscosity of a liquid. The molecules of a liquid cannot be considered free particles: The special properties of the liquid are due to forces between its molecules. Hence it is not possible to describe the collisions of the body with the molecules of the liquid as a sequence of independent collisions with free particles. A collision with one particle necessarily involves the nearby molecules. Another aspect of this is the fact that the total energy of the liquid cannot be considered as a sum of single molecule energies, since there is a significant contribution due to the intermolecular potential energy. It is, therefore, impossible to define single particle states of the molecules of a liquid. Hence a body moving through a liquid and losing energy does not transfer it to single molecules but to the liquid as a whole. The energy states of a liquid are states of the collective motion of all the molecules.

In this sense there is a great similarity between the liquid and the solid crystal discussed in Chap. 3 of Part IV. There we have seen that the description of a crystal as N coupled three-dimensional harmonic oscillators can be replaced by a description in terms of $3N$ independent one-dimensional harmonic oscillators. Each of these oscillators has a well-defined frequency, and it describes a collective vibration of the molecules of the crystal, namely a sound wave. This is also the case in a liquid: A body moving through it will excite sound waves, which are collective motions of the liquid. The energy of these excitations originates in energy losses of the moving body. The advantage of the crystal over the liquid is its periodic structure, which allows a precise calculation of the dispersion relation, namely the relation between the frequency of the sound waves and their wave number. In a liquid it is not possible to calculate the dispersion relation in a simple manner and it has to be found from experiment.

Another ingredient is that liquid helium is a quantum liquid. The sound waves excited in it should be treated as excited states of quantum harmonic oscillators where each oscillator (phonon) is characterized by its wave vector, $\mathbf{q}$, and its angular frequency, ω. We thus describe liquid helium at very low temperatures, as collective excitations above a quantum-mechanical ground state, in which the energy of the liquid is minimal. These excitations, which are themselves free bosons, are at thermodynamic equilibrium, and all the information regarding the properties of the liquid is contained in the dispersion relation $\omega(\mathbf{q})$ or $\epsilon(\mathbf{q})$ of these excitations. Thus, a body moving through liquid helium at $T = 0$ does not excite single atoms from the ground level but rather excites quanta such as phonons, for example. These excitations carry momentum and energy.

If the body moving through the liquid excites a phonon with momentum $\hbar\mathbf{q}$ and energy $\epsilon(\mathbf{q})$, it is at the expense of the body's motion. Momentum and energy conservation imply that

$$\begin{cases} \mathbf{p} = \mathbf{p}' + \hbar\mathbf{q}\,, \\ \dfrac{p^2}{2M} = \dfrac{p'^2}{2M} + \epsilon(q)\,, \end{cases} \tag{5.4.22}$$

where M is the mass of the body, $\mathbf{p}$ is its initial momentum and $\mathbf{p}'$ is its momentum after the excitation of the phonon. It turns out, as we proceed, that the two equations in (5.4.22) can be satisfied simultaneously only in a narrow range of $\mathbf{p}$ or, equivalently, only in a narrow range of velocities of the body. This means that in the range of velocities in which the two equations in (5.4.22) are not satisfied, the body moving through the liquid cannot give rise to an excitation. Hence it cannot lose energy and will move through the liquid without friction. This is equivalent to zero viscosity, or superfluidity. In order to see when the two equations in (5.4.22) are satisfied simultaneously, we substitute $\mathbf{p}' = \mathbf{p} - \hbar\mathbf{q}$ in the second equation (5.4.22) and obtain

$$\hbar \mathbf{q} \cdot \mathbf{v} = \frac{(\hbar q)^2}{2M} + \epsilon(q)\,, \tag{5.4.23}$$

where $\mathbf{v}(= \mathbf{p}/M)$ is the initial velocity of the body.

Exercise 4.8

(a) Prove Eq. (5.4.23).
(b) Show that if the body has a macroscopic mass, then to a good approximation the condition for phonons to be excited is

$$\hbar \mathbf{q} \cdot \mathbf{v} = \epsilon(q)\,.$$

Use 240 m s^{-1} for the speed of sound in liquid He.

Solution on page 533

Therefore, in order to give rise to an excitation with momentum $\hbar\mathbf{q}$ the velocity of the body must satisfy (5.4.23). The right hand side of this equation is positive. Hence, the angle θ between $\mathbf{v}$ and $\mathbf{q}$ must be acute. Furthermore, the most "efficient" state is that in which $\theta = 0$, namely the excited phonon, propagates in the same direction as the body. In this case the velocity of the body attains the minimum value which still allows the excitation of a phonon with momentum $\hbar q$. All other angles require a larger velocity. The minimal velocity required for the excitation of a phonon with momentum $\hbar q$ thus depends on q as follows:

$$v_{\min} = \frac{\epsilon(q)}{\hbar q}\,. \tag{5.4.24}$$

The question now is: What is the form of $\epsilon(q)$? $\epsilon(q)$ could describe ordinary sound waves propagating at a speed v_s. Measurements of the propagation of sound waves in liquid helium yield $v_s = 237$ m s^{-1}. For sound waves,

$$\epsilon(q) = \hbar\omega(q) = \hbar v_s |\mathbf{q}|\,, \tag{5.4.25}$$

and if we substitute this into Eq. (5.4.24) we find that the minimal speed required for the excitation of a phonon of any wavelength is the sound velocity in helium, v_s. Hence any body which moves through liquid helium at $T = 0$ with a speed lower than v_s, will move without viscosity. But this conclusion does not fit the experimental results, which give a much smaller critical speed.

The reason for the discrepancy is that the dispersion relation is not simply that of phonons, Eq. (5.4.25), but is more complicated. The dispersion relation has been measured experimentally by scattering slow neutrons from liquid helium. It is depicted in Fig. 5.4.6. The measurement

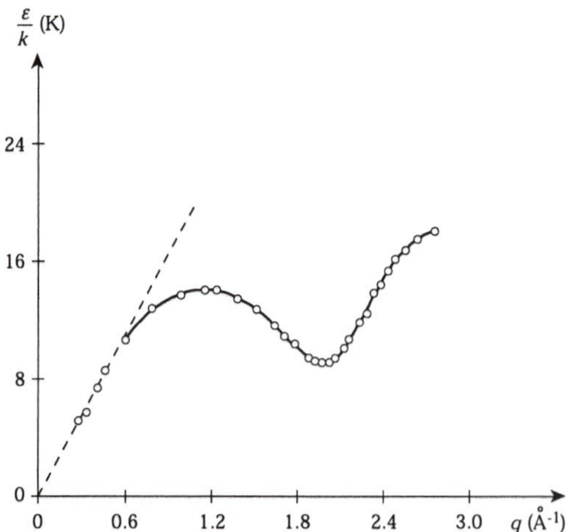

Fig. 5.4.6. The dispersion relation $\epsilon(q)$ for excitations in liquid helium at a temperature of 1.12 K, measured using neutrons with wavelength 4.04 Å. The wave number is given in units of Å^{-1} and the energy in units of K. For instance, $\epsilon/k = 32$ K corresponds to an energy of 4.4×10^{-22} J or 2.8×10^{-3} eV. The minimum of ϵ is attained at $q = 1.94$ Å^{-1}, $\epsilon/k = 8.67$ K. *D. G. Henshaw & A. D. B. Woods, Phys. Rev.* **121**, *1260 (1961).*

is based on the fact that slow neutrons give rise to a single excitation in each scattering event, so that by measuring their energy and momentum loss, Eq. (5.4.22), it is possible to deduce $\epsilon(q)$.

At small wave numbers (large wavelengths) the dispersion relation looks indeed like that of ordinary sound waves, namely like (5.4.25), and is represented by the dashed line in the figure. But at larger wave numbers the trend changes and $\epsilon(q)$ begins to decrease, attains a minimum at $q = 1.94$ Å^{-1}, $\epsilon/k = 8.67$ K, and then increases again.

In order to understand how such a dispersion relation affects the minimum speed required to create an excitation in the liquid, we shall use a graphical approach: From Eq. (5.4.24) we conclude that the minimum speed required in order to create an excitation with wave number q is the slope of the straight line connecting the point on the dispersion curve corresponding to the wave number q and the origin — see Fig. 5.4.7. A different minimal speed is required for each q and, as is made clear by the figure, due to the deviation of the dispersion relation from the straight line, an increase of q decreases v_{min}.

Hence v_{min} has a minimal value which is obtained when the straight line is tangent to the dispersion curve. In this state the body moves through the liquid at the critical speed v_c, and it can only produce excitations corresponding to the tangent point which is (almost exactly) the minimum point of the dispersion curve. If the body's speed is less than v_c, the body will produce no excitations, since momentum and energy conservation cannot be simultaneously satisfied. If the body's

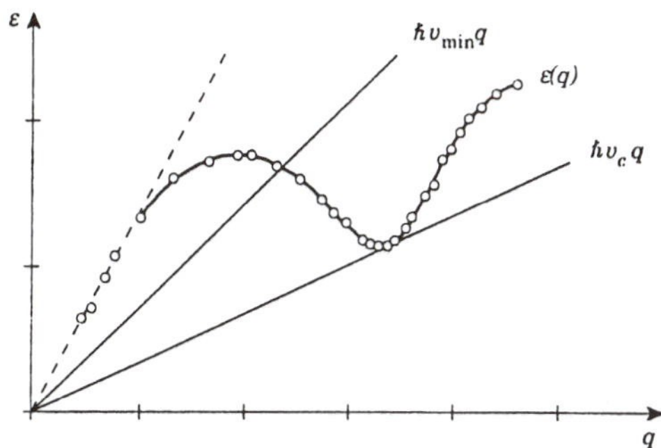

Fig. 5.4.7. A graphical solution of Eq. (5.4.24).

speed is larger than v_c there can be additional excitations at other values of q.

To come back to the case in which it is liquid helium that flows across a stationary body, we observe that the creation of excitations in the liquid can only depend on the relative velocity between the liquid and the body. Hence we obtain the same criterion for the critical speed for superfluidity, this time for the flow velocity of liquid helium past a body at rest. Note that our considerations apply to a body of arbitrary shape, across which the liquid is flowing. Such a body can also be a capillary tube. In a coordinate frame moving with the liquid (in which the liquid is at rest), we can use Eqs. (5.4.22) and proceed in the same way which led to Eq. (5.4.24). Then we return to the coordinate frame in which the liquid is flowing and the capillary is at rest and obtain the condition for the critical speed for superfluid flow:

$$v_c = \min\left[\frac{\epsilon(q)}{\hbar q}\right] . \tag{5.4.26}$$

As we have seen, the graphical meaning of this condition is, finding the slope of the line which passes through the origin and is tangent to the dispersion curve. This condition is called Landau's condition.

Landau's condition

Exercise 4.9

(a) Calculate the speed of sound in He II from the graph in Fig. 5.4.6.
(b) Calculate Landau's critical speed.

Solution on page 533

A critical speed of about 60 m s^{-1}, as resulted in the previous exercise, is attained in very special circumstances. Under ordinary circumstances a much lower speed is obtained, on the order of 1 cm/s. In light of our discussion, this implies that there exist excitations which are of a different type, and are not included in the dispersion relation in Fig. 5.4.6. These excitations are also created by the motion of the body through He II, and are the cause for the low critical speed which is observed. It turns out that these excitations are vortices. It is possible to excite them in a controlled manner and to study their properties in experiments performed in He II, in a rotating vessel. But we leave the subject at this point.

Before concluding we emphasize that the phenomenon of superfluidity can be understood only by taking into account the interatomic forces between the helium atoms, since their existence is responsible for the existence of the collective excitations in the liquid. In the absence of interatomic forces the liquid would behave as a gas of free particles and its excitations would be the excitations of the free atoms themselves. The dispersion relation of a free particle is the usual $\epsilon(q) = (\hbar q)^2/2m$, and applying the Landau condition (5.4.26) to it, we would find $v_c = 0$, since the minimum of the function $\epsilon(q)/q$ for free particles satisfies

$$\min\left[\frac{\epsilon(q)}{q}\right] = \min\left[\frac{\hbar^2 q}{2m}\right] = 0\,. \tag{5.4.27}$$

Alternatively, it is possible to see that the tangent to the dispersion curve $\epsilon = (\hbar q)^2/2m$, which passes through the origin, is the q axis itself, whose slope is zero. This means that, in the absence of forces between the atoms, the creation of excitations in liquid helium at $T = 0$ is possible at any flow speed, and thus the system will not behave as a superfluid.

The discussion so far has regarded superfluidity at absolute zero. As noted already, at $T > 0$ it is possible to treat liquid helium as a collection of excitations above the ground state. Hence a body moving through liquid helium at $T > 0$ will lose energy in one of two ways. The first is by creating excitations from the ground state, exactly as at $T = 0$, and this can occur only above the critical speed. The second is by colliding with existing excitations in the liquid due to the thermal fluctuations. These latter excitations disappear at $T \to 0$, and have thus been ignored till now.

The thermal excitations behave as an ideal gas of phonons (with its particular dispersion relation) which propagate against a background of helium atoms in the ground state of the liquid. The motion of a body through the gas, which involves collisions with the phonons, cannot occur without energy loss. In this picture the origin of the viscosity of liquid helium at temperatures above absolute zero are the excitations themselves, not the helium atoms occupying the single particle excited states. The latter lost their meaning the moment we took into account the interatomic forces.

The two-component description of He II takes, therefore, a different meaning. The superfluid component is composed of the atoms themselves, whereas the normal component is composed of the phonon gas, which propagates against the background of the ground state of the atoms.

In superfluid flow at $T > 0$, such as the flow of liquid helium through a capillary out of a container, helium atoms flow through the tube without carrying excitations along with them. Most of the phonons remain behind inside the container, since their viscosity does not allow them to pass through the tube.

Exercise 4.10

Would you expect the specific heat of He^4 at very low temperatures to be proportional to $T^{3/2}$, like in a condensed boson gas?

Solution on page 534

4.7 Fermi liquid and superconductivity

Helium has two stable isotopes: He^4 atoms, with spin 0, are bosons and He^3 atoms, with spin 1/2, are fermions. Liquid He^4 exhibits, as we have seen, quantum behavior on a macroscopic scale. The following question, therefore, arises: Will He^3 also exhibit at low temperatures a behavior determined by the fermionic nature of its atoms?

First we mention that it is possible to obtain macroscopic amounts of He^3, which enable the measurement of the physical properties of He^3. It turns out that they are significantly different from those of He^4. He^3 liquefies at atmospheric pressure at a temperature of 3.2 K (compared to 4.2 K in He^4), and like He^4 it does not freeze at $T = 0$. The density of He^3 is 0.07 gr/cm^3, which is half of that of He^4, and its viscosity around 1 K is 25 μP, about half of that of He^4. To a first approximation it is possible to think of liquid He^3 as an ideal fermion gas, as we did for liquid He^4 in Sec. 4.5, which we treated as an ideal boson gas. The first question is: What is the degree of degeneracy of liquid He^3? In order to answer this we calculate its Fermi temperature, and find that it is quite low — 4.5 K!

Exercise 4.11

Calculate the Fermi temperature of He^3.

Solution on page 534

From the discussion of the Fermi system in Chap. 3 we know that T must be much below T_F for the gas of fermions to be degenerate. Hence, at temperatures at which He^4 behaves as a degenerate boson gas, He^3 is far from being degenerate. The temperatures at which the fermionic nature of He^3 will begin to appear are expected to be at least 10 times lower (see Fig. 5.3.2). A phenomenon characteristic of a degenerate Fermi gas that is actually exhibited by He^3 at temperatures below 0.5 K, is the rapid increase of the mean free time between collisions with decreasing temperature. The reason is again the Pauli principle, which limits, when $T \ll T_F$, the number of atoms that can participate in collisions to those found in an energy range kT around the Fermi level. Recall that we used an identical argument to obtain the ratio of the thermal to the electrical conductivity in the Sommerfeld model, in Sec. 3.6. An estimate of the relative number of atoms in the range $\Delta\epsilon = kT$ around the Fermi level is obtained from the energy distribution function $f_\epsilon(\epsilon)$ (see Solution 3.12):

$$f_\epsilon(\epsilon_F)kT \approx \frac{kT}{\epsilon_F}. \tag{5.4.28}$$

In a collision two such atoms must be involved, and the probability (per unit length) for a He^3 atom to collide decreases by a factor of $(kT/\epsilon_F)^2$. Since the probability per unit length is given by 1 over the mean free path [see Sec. 3.2 of Part I and especially Eq. (1.3.14)], it follows that the mean free path in a degenerate fermion gas increases by a factor of $(\epsilon_F/kT)^2$, compared to that calculated in Part I for Boltzmann particles. Hence,

$$\ell \approx \frac{1}{n\sigma}\left(\frac{\epsilon_F}{kT}\right)^2, \tag{5.4.29a}$$

where n is the density of fermions (He^3 atoms or electrons, etc.) and σ is their cross section, which we took in Part I to be $4\pi a^2$. All numerical factors of order unity have been suppressed.

The mean free time is obtained by dividing ℓ by the typical velocity of the particles that participate in collisions, namely the Fermi velocity:

$$\tau \approx \frac{1}{n\sigma v_F}\left(\frac{\epsilon_F}{kT}\right)^2. \tag{5.4.29b}$$

This sharp increase in the mean free path with decreasing temperature affects the transport coefficients, such as the viscosity and the thermal conductivity, which are proportional to ℓ or τ. The viscosity, for instance, will be given by Eq. (1.3.47) with $\bar{v} = v_F$:

$$\eta \approx \frac{1}{3} m v_F \ell n. \tag{5.4.30}$$

When the density n is constant, so are v_F and ϵ_F, and by (5.4.29a)

$$\eta \propto \frac{1}{T^2}. \tag{5.4.31}$$

Exercise 4.12

(a) How will the viscosity of a degenerate fermion gas vary with the mass of the fermions?
(b) How will the viscosity vary with density, at constant T?

Solution on page 534

Such variation of the viscosity with temperature has actually been measured experimentally and indicates that it is possible, at least approximately, to describe liquid He^3 as a fermion gas, and that the approximation improves as the temperature decreases.

A very long mean free path is not only characteristic of He^3 but appears also in metals, as we have already mentioned. It is also consistent with the fact that the electric forces between the conduction electrons in metals are neglected in the Drude model. The justification for Drude's assumption is, of course, the good agreement with experiment of the results obtained by applying the kinetic theory to the conduction electrons. Later on we will see that this analogy between the properties of liquid He^3 and the properties of electrical conductivity does not end here but persists at lower temperatures and promises several surprises.

To better understand the properties of liquid He^3 we have to take into account the interatomic forces, which our description of an ideal fermion gas has ignored. We must, therefore, clarify what are the fundamental excitations of the liquid above its ground state, as we did for He^4. These excitations, which describe collective motions of He^3 atoms, will themselves be fermions this time, namely they will satisfy the Pauli principle. The properties of the liquid will depend on their dispersion relation $\epsilon(q)$, which is determined by the forces between the atoms. Such a liquid has been given the name of "Fermi liquid." Using this description Landau succeeded in 1956 in explaining the low temperature properties of liquid He^3 down to a few mK (10^{-3} K). **Fermi liquid**

At the beginning of the seventies it was discovered that He^3 changes its properties drastically at a temperature of about 1 mK, and that this change is accompanied by a sharp change in the specific heat which is very similar to the one of He^4 at 2.17 K. The similarity is not only superficial — it turned out that indeed He^3 behaves below the transition temperature of 1 mK as a superfluid! To understand how superfluidity, a phenomenon characteristic of a boson liquid, is made possible in He^3 one postulates a mechanism that correlates the motion of pairs of He^3 atoms. The pair forms a sort of diatomic molecule which obeys Bose–Einstein statistics, and a macroscopic number of such pairs is equivalent to a boson liquid. If the pairs were stable, a liquid of such pairs would undergo a Bose–Einstein condensation at a critical temperature of a Bose gas with a mass double that of He^3, which is about 1 K. It would behave as a superfluid below that temperature. But He^3 becomes superfluid only below 1 mK.

Exercise 4.13

Calculate the critical temperature of Bose–Einstein condensation for pairs of He^3 atoms.

Solution on page 535

One concludes, therefore, that the temperature of 1 mK, below which ordinary He^3 becomes superfluid, cannot be interpreted as the critical temperature for Bose–Einstein condensation and the analog of He I does not exist in He^3. The reason is that above this temperature there are no bosons in the liquid. This transition temperature is the temperature below which it becomes "advantageous" for the He^3 atoms to move in pairs. The mechanism, which is indirect and complicated, gives rise to an effective attraction between atoms with energies near the Fermi energy, due to an interaction of the magnetic moment of each atom with its surroundings, and the back reaction of the surroundings on another atom. The forces induced in this way are very weak, and the resulting pairs are loose and extended; they can survive only at very low temperature, where almost all thermal fluctuations are eliminated.

For the sake of completeness we mention that actually there exist two phases of superfluid He^3. They differ in their magnetic properties, and are called He^3A and He^3B. He^3A appears at pressures between 21 and 34 atm for temperatures ranging between 2.2 mK and 2.8 mK. He^3B appears for the entire pressure range between 0 and 34 atm (above this pressure He^3 solidifies). At zero pressure He^3B appears at a temperature of 1 mK, and increasing the pressure increases the transition temperature up to 2.5 mK.

Electrons are also fermions, hence one would expect the phenomenon of superfluidity in He^3 to have an electronic analog. The electronic analog of the superfluidity of He^3, known as superconductivity, was discovered by Kamerlingh–Onnes in 1911.

As implied by its name, superconductivity is the ability of materials to maintain a current without resistivity. In fact, current can flow in superconducting rings for months or even years with no observable loss. Superconductivity is much more common than superfluidity. About 30 metallic elements and a countless number of compounds and alloys are known to be superconductors. Superconductivity appears at temperatures below a critical temperature that depends on the specific material. Until 1987 superconductivity was considered a low temperature phenomenon. Its discovery in 1911 was in mercury and the critical temperature measured was about 4 K. All the superconductors that followed had critical temperatures of similar magnitude, except for a few that reached 20 K. In 1987 this barrier was broken, and materials with critical temperatures above 100 K have been discovered.

A full theoretical explanation for superconductivity, which was given in 1957 by Bardeen, Cooper and Schrieffer (BCS theory), is quite complicated and deserves an entire course. A mere "shadow" of this explanation may be brought here: Superconductivity is interpreted as the superfluidity of the electronic Fermi liquid. Just as in He^3 below the transition temperature, there exists in superconductors a mechanism that creates a net attraction between pairs of electrons with energies close to the Fermi energy. The electric charge of an electron induces a charge density in its surroundings, and the latter exerts a force on another electron. In this way electron pairs (Cooper pairs) are created. The electrons in such a pair move in a correlated manner, even if the distance between them is large and there are many other electrons between them, some of which belong to other Cooper pairs. Due to this correlated movement which exists in the ground state, it is difficult to create excitations in the system and hence the electron pairs can move without friction, like in a boson superfluid. Since a Cooper pair has a charge, of $2e$, the motion of the pairs is an electric current, and their superfluid flow is an electric current without resistivity, namely superconductivity.

Cooper pairs

Superconductors exhibit not only a spectacular electrical behavior, they also have some extraordinary magnetic properties. A good example is perfect diamagnetism, called the Meissner effect. Diamagnetism is the induction of magnetization opposing the external magnetic field. It implies a negative susceptibility:

dia-magnetism

$$\mathsf{M} = \chi\mathbf{H}, \qquad \chi < 0\,, \tag{5.4.32}$$

where M is the magnetization density [see Chap. 1 and Sec. 5.3 of Part II; note that χ in Eq. (2.5.15) is defined in terms of the total magnetization and not in terms of the magnetization density].

Perfect diamagnetism in a superconductor is the fact that the magnetization totally cancels the magnetic field inside the material, i.e.

$$\mathbf{B} = 0, \quad \text{or} \quad \mathsf{M} = -\frac{1}{4\pi}\mathbf{H}\,. \tag{5.4.33}$$

See Eq. (2.1.3). The magnetization is induced by permanent screening currents on the surface of the superconductor which appear immediately upon the application of the magnetic field. This phenomenon is limited to low magnetic fields. Above a critical value, H_c, of the field, which depends on the material and on the temperature, the magnetic field penetrates the bulk material. At that point superconductivity disappears.

Another unusual phenomenon is observed when cooling takes place in the presence of a magnetic field, from a temperature $T > T_c$ to a temperature $T < T_c$. Above T_c the field penetrates the material, as usual. When the external field is turned off at a temperature $T < T_c$, the induced magnetization is trapped in the material and a macroscopic magnetic field is felt outside the superconductor.

One can go on to describe many additional fascinating aspects of superconductivity, but for us this is a good point to end our discussion of superconductivity, and indeed the entire course.

Appendix

Calculation of Some Integrals

In the course of this part we needed integrals of the form

$$J_\nu = \int_0^\infty \frac{x^{\nu-1}}{e^x+1}dx\,, \tag{5.A.1}$$

where ν is an integer or half-integer. Note that the expression $1/(e^x+1)$ is the sum of the geometric series with alternating signs:

$$\frac{1}{e^x+1} = \frac{e^{-x}}{1+e^{-x}} = e^{-x} - e^{-2x} + e^{-3x} - \cdots = \sum_{m=1}^{\infty}(-1)^{m+1}e^{-mx}\,, \tag{5.A.2}$$

and substituting into (5.A.1) we obtain in a similar manner to (4.A.3)

$$\begin{aligned}\int_0^\infty \frac{x^{\nu-1}}{e^x+1}dx &= \sum_{m=1}^{\infty}(-1)^{m+1}\int_0^\infty x^{\nu-1}e^{-mx}dx \\ &= \int_0^\infty y^{\nu-1}e^{-y}dy \sum_{m=1}^{\infty}\frac{(-1)^{m+1}}{m^\nu}\,.\end{aligned} \tag{5.A.3}$$

We are thus left with the calculation of the sum and the integral appearing on the right hand side of (5.A.3), and both are quite similar to the ones calculated in the appendix to Part IV.

We first calculate the integral for integer values of ν. It is identical to the one we calculated in (4.A.2). Denote the integral by $\Gamma(\nu)$,

$$\Gamma(\nu) = \int_0^\infty x^{\nu-1}e^{-x}dx\,. \tag{5.A.4}$$

Integrating by parts for arbitrary ν we obtain

$$\int_0^\infty x^{\nu-1}e^{-x}dx = (\nu-1)\int_0^\infty x^{\nu-2}e^{-x}dx\,, \tag{5.A.5}$$

or

$$\Gamma(\nu) = (\nu-1)\Gamma(\nu-1)\,. \tag{5.A.6}$$

Using the fact that $\Gamma(2) = 1$ and iterating (5.A.6) we regain for integer ν Eq. (4.A.2):

$$\Gamma(\nu) = (\nu - 1)!\,. \tag{5.A.7}$$

For half-integer ν we can repeat the iteration since the recursion relation (5.A.6) still holds. But we need the value of $\Gamma(1/2)$. This integral is calculated via the change of variable $x = z^2$:

$$\Gamma\left(\frac{1}{2}\right) = \int_0^\infty x^{-1/2} e^{-x} dx = 2\int_0^\infty \exp(-z^2) dz = \sqrt{\pi}\,, \tag{5.A.8}$$

where the last integral has been calculated in Part I.

Thus, for example, we have

$$\Gamma\left(\frac{3}{2}\right) = \frac{1}{2}\sqrt{\pi},\ \Gamma\left(\frac{5}{2}\right) = \frac{3}{4}\sqrt{\pi},\ \Gamma\left(\frac{7}{2}\right) = \frac{15}{8}\sqrt{\pi}\,. \tag{5.A.9}$$

We are left with the calculation of the infinite sum in (5.A.3):

$$S = \sum_{m=1}^{\infty} \frac{(-1)^{m+1}}{m^\nu} = 1 - \frac{1}{2^\nu} + \frac{1}{3^\nu} - \frac{1}{4^\nu} + \frac{1}{5^\nu} - \frac{1}{6^\nu} + \cdots, \tag{5.A.10}$$

which we now express in terms of the zeta function, defined for every $\nu > 1$ by Eq. (4.A.4). Separating the even and odd terms in the sum we find that $S = S_1 - S_2$, where

$$S_1 = 1 + \frac{1}{3^\nu} + \frac{1}{5^\nu} + \cdots, \tag{5.A.11a}$$

$$S_2 = \frac{1}{2^\nu} + \frac{1}{4^\nu} + \frac{1}{6^\nu} + \cdots = \frac{1}{2^\nu}\left(1 + \frac{1}{2^\nu} + \frac{1}{3^\nu} + \cdots\right) = 2^{-\nu}\zeta(\nu)\,. \tag{5.A.11b}$$

S_1 can also be expressed in terms of $\zeta(\nu)$:

$$S_1 = \zeta(\nu) - S_2 = (1 - 2^{-\nu})\zeta(\nu)\,,$$

and then

$$S = S_1 - S_2 = \zeta(\nu) - 2S_2 = (1 - 2^{1-\nu})\zeta(\nu)\,. \tag{5.A.12}$$

We have thus found that for $\nu > 1$

$$\int_0^\infty \frac{x^{\nu-1}}{e^x + 1} dx = (1 - 2^{1-\nu})\Gamma(\nu)\zeta(\nu)\,, \tag{5.A.13}$$

and for the sake of completeness we rewrite Eq. (4.A.5) in a form that emphasizes the similarities and the differences:

$$\int_0^\infty \frac{x^{\nu-1}}{e^x - 1} dx = \Gamma(\nu)\zeta(\nu)\,. \tag{5.A.14}$$

Actually both these equations are valid not only for integer and half-integer values of ν but for any real ν satisfying $\nu > 1$.

Finally, we list some useful values of $\zeta(\nu)$ and $\Gamma(\nu)$:

ν	$\frac{1}{2}$	1	$\frac{3}{2}$	2	$\frac{5}{2}$	3	$\frac{7}{2}$	4
$\zeta(\nu)$	-1.460	—	2.612	$\frac{\pi^2}{6}$	1.341	1.202	1.127	$\frac{\pi^4}{90}$
$\Gamma(\nu)$	$\sqrt{\pi}$	1	$\frac{1}{2}\sqrt{\pi}$	1	$\frac{3}{4}\sqrt{\pi}$	2	$\frac{15}{8}\sqrt{\pi}$	6

Self-assessment exercises

Exercise 1 *Solution on page 536*

A classical (Boltzmann) ideal gas is enclosed in a container of volume V, and can exchange energy and particles with a bath of temperature T and chemical potential μ.

(a) Prove that the average number of particles in the container $\langle N\rangle$ is related to the thermodynamic potential Ω by

$$\langle N\rangle = -\beta\Omega(T, V, \mu)\,.$$

(b) Prove that the probability of finding exactly N particles inside the container is given by

$$P_N = e^{-\langle N\rangle}\frac{\langle N\rangle^N}{N!}\,,$$

where $\langle N\rangle = -\beta\Omega$. This is the Poisson distribution.

Exercise 2 *Solution on page 537*

(a) In Sec. 2.4 it is stated that the chemical potential of a phonon gas is zero, and that hence it is possible to obtain the average number of phonons in a crystal with frequency ω by substituting $\mu = 0$ into Eq. (5.2.11b). Explain how it is possible that $\mu = 0$, if the free energy of an Einstein solid explicitly depends on N [Eq. (3.2.12)] and also the free energy in the Debye model, as given in Eq. (4.3.11), depends on it through q_D [Eq. (4.3.12a)].

(b) Show directly from the partition function that the average energy of a gas of fermions or bosons can be written in terms of its average occupation numbers:

$$E = \sum_k \langle n_k\rangle\epsilon_k\,.$$

Exercise 3 *Solution on page 538*

Calculate the chemical potential of a two dimensional fermion gas as a function of the temperature T and the density of particles per unit area $n = N/A$. (Assume that each fermion has a single spin state.)

Exercise 4 *Solution on page 539*

(a) Calculate the pressure of a degenerate electron gas, and find the relation between the pressure and the energy density, E/V.
(b) Calculate the pressure of the electron gas in aluminum.

Exercise 5 *Solution on page 540*

(a) Calculate the energy density of an extremely relativistic gas of degenerate fermions with spin 1/2. Compare to the nonrelativistic case and determine under what conditions this description of an extremely relativistic degenerate fermion gas is valid.
(b) Calculate the pressure of the gas in (a) and find the relation between the pressure and the energy density.

Exercise 6 *Solution on page 542*

(a) Calculate the pressure of a boson gas below the condensation temperature T_c, and explain why it does not depend on the volume.
(b) Explain why there is no need to correct the expression for the thermodynamic potential Ω of a degenerate fermion gas by adding a contribution of the ground state, as required for a degenerate boson gas.

Exercise 7 *Solution on page 543*

Check if the phenomenon of Bose–Einstein condensation occurs in a two-dimensional boson gas.

Exercise 8 *Solution on page 544*

Show that the energy density of a gas of massless fermions with spin 1/2 and zero chemical potential at temperature T, is 7/8 that of black body radiation at the same temperature.

■ This may serve as an approximation to the gas of cosmic neutrino particles.

Exercise 9 *Solution on page 545*

Solve Self-Assessment Exercise 9 of Part IV, for extremely relativistic temperatures: $kT \gg mc^2$.

Solutions to exercises in the text

Solution 1.1 *Exercise on page 455*

We write the grand canonical partition function (5.1.2) as a sum of two contributions: One from a state denoted by $\alpha = 0$, in which there are no particles in the system, and the other from all the other states. In the state $\alpha = 0$

$$N = 0, \quad E = 0\,. \tag{i}$$

Hence

$$\mathcal{Z} = 1 + \sum_{N>0} \sum_i e^{\beta(\mu N - E_i(N))}\,. \tag{ii}$$

Because all the numbers N are positive, if $\mu \to -\infty$, all the terms in the sum must vanish since

$$\lim_{\mu\to-\infty} e^{\beta\mu N} = 0\,. \tag{iii}$$

Thus, only the 1 remains on the right hand side of (ii) and indeed

$$\lim_{\mu\to-\infty} \mathcal{Z} = 1\,. \tag{iv}$$

Solution 1.2 *Exercise on page 456*

The average number of particles is calculated using the grand canonical probabilities:

$$\begin{aligned} N &= \mathcal{Z}^{-1} \sum_{N=0}^{\infty} \sum_i N e^{\beta(\mu N - E_i)} = \mathcal{Z}^{-1} \frac{1}{\beta} \frac{\partial}{\partial \mu} \sum_\alpha e^{\beta(\mu N - E_i)} \\ &= \frac{1}{\beta} \frac{\partial}{\partial \mu} \ln \mathcal{Z} = \frac{\partial}{\partial \mu} (kT \ln \mathcal{Z})\,, \end{aligned}$$

which is Eq. (5.1.7).

Solution 1.3 *Exercise on page 456*

To complete the derivation of Eq. (5.1.9) we must compute

$$kT\frac{d\beta}{dT}\left(\frac{\partial \ln \mathcal{Z}}{\partial \beta}\right)_{V,\mu}.$$

Since

$$\frac{d\beta}{dT} = -\frac{1}{kT^2},$$

we can write

$$kT\frac{d\beta}{dT}\left(\frac{\partial \ln \mathcal{Z}}{\partial \beta}\right)_{V,\mu} = -\frac{1}{T}\frac{1}{\mathcal{Z}}\left(\frac{\partial \mathcal{Z}}{\partial \beta}\right)_{V,\mu}.$$

Substituting the partition function from (5.1.2) and taking the derivative with respect to β, we can write

$$\frac{1}{\mathcal{Z}}\left(\frac{\partial \mathcal{Z}}{\partial \beta}\right)_{V,\mu} = \frac{1}{\mathcal{Z}}\sum_{N=0}^{\infty}\sum_i(\mu N - E_i)e^{\beta(\mu N - E_i)} = \mu\langle N\rangle - \langle E\rangle,$$

where use has been made of the definition of the average values, Eq. (5.1.4). This leads to Eq. (5.1.9), once the brackets have been dropped.

Solution 1.4 *Exercise on page 456*

Substituting the grand canonical probabilities only into the logarithms of $\sum_\alpha P_\alpha \ln P_\alpha$ gives

$$S = -k\sum_\alpha P_\alpha[\beta(\mu N - E_i) - \ln \mathcal{Z}].$$

After performing the summation, we obtain

$$S = -k\beta(\mu\langle N\rangle - \langle E\rangle) + k\ln \mathcal{Z} = -\frac{\mu\langle N\rangle - \langle E\rangle}{T} + k\ln \mathcal{Z}.$$

Multiplying by T, dropping the average braces and rearranging, it becomes

$$TS + \mu N - E = kT\ln \mathcal{Z},$$

and using (2.0.29), we find that

$$\Omega = -kT\ln \mathcal{Z}.$$

Solution 2.1 *Exercise on page 459*

Since n_k is the number of particles in state k, it is impossible for it to be larger than N. In an extreme case all the particles may be in the same state k_0 and then $n_{k_0} = N$, and all other n_k are zero.

Solution 2.2 *Exercise on page 459*

The number of states with the same set of n_k is the number of ways the N particles can be distributed in groups of n_k each. It is the combinatorial factor that expresses the fact that all N particles can be interchanged, but interchanges of particles within each group do not produce new states. Hence the number is

$$\frac{N!}{n_1!n_2!\cdots n_k!\cdots}$$

Solution 2.3 *Exercise on page 459*

The n_1 particles in the state with energy ϵ_1 contribute $n_1\epsilon_1$ to the total energy. In the same way, for each k we obtain a contribution of $n_k\epsilon_k$, and overall we obtain (5.2.3).

Solution 2.4 *Exercise on page 461*

(a) The orbital angular momentum of the electrons is always an integer. Hence the addition of the orbital angular momentum cannot change an integer spin into a half-integer spin, and vice versa.

(b) In a neutral atom the number of electrons is equal to the number of protons, and hence their total spin is always an integer. There are still the neutrons. Since each has half-integer spin, an odd number of neutrons contribute a half-integer to the total spin and an even number of the neutrons will contribute integer spin, so that the neutrons determine the statistical behavior of the atom.

Solution 2.5 *Exercise on page 462*

The difference stems from the fact that the paramagnet and the Einstein solid are *not gases.* Consequently each particle has a constant position distinguishing it from its neighbors. Hence, even though the spins of the paramagnet are identical, and so are the oscillators of the Einstein solid, the method of characterizing the macroscopic states is *not* by using occupation numbers but by specifying the quantum state of each of the spins or the oscillators. For the paramagnet, there is one state in which all the spins are pointing along the field and N *different* states in which a single spin points opposite to the field and all the others point along the field. In contrast, if we think of the spins as identical particles which are indistinguishable, then we have to count all of these N possibilities as a single quantum state.

This difference in points of view allows us to write the partition function of the paramagnet and the Einstein solid as a product of single particle partition functions, but does not allow us to do it for the partition function of a gas of identical particles.

Solution 2.6 *Exercise on page 463*

We write (5.2.6) in more detail as

$$\sum_{n_1,\ldots,n_k,\ldots} \exp\{\beta[(\mu-\epsilon_1)n_1+\ldots+(\mu-\epsilon_k)n_k+\ldots]\}\,.$$

Since the summation is over all possible values of the occupation numbers in an independent manner, it is possible to replace the sum of products by a product of sums:

$$\mathcal{Z}\ (T,V,\mu)=\left(\sum_{n_1} e^{\beta(\mu-\epsilon_1)n_1}\right)\cdot\ldots\cdot\left(\sum_{n_k} e^{\beta(\mu-\epsilon_k)n_k}\right)\cdot\ldots\,,$$

and this is exactly Eq. (5.2.7).

Solution 3.1 *Exercise on page 467*

The valence of sodium and potassium is 1, namely each of the atoms has a single outer electron. Hence the density of conduction electrons is equal to the number of atoms per unit volume. Sodium has an atomic weight of 23, namely 23 grams contain an Avogadro number of atoms. Its density is $0.97\mathrm{g/cm}^3$, and hence

$$n_c(\mathrm{Na})=\frac{6.02\times10^{23}\times0.97}{23}=2.5\times10^{22}\ \mathrm{cm}^{-3}=2.5\times10^{28}\ \mathrm{m}^{-3}\,.$$

The atomic weight of potassium is 39 and its density is $0.86\ \mathrm{g/cm}^3$, hence

$$n_c(\mathrm{K})=1.3\times10^{22}\ \mathrm{cm}^{-3}=1.3\times10^{28}\ \mathrm{m}^{-3}\,.$$

Solution 3.2 *Exercise on page 468*

In averaging (5.3.2) we have to calculate the sum of averages:

$$\langle\mathbf{v}\rangle.=\langle\mathbf{v}_0\rangle-\frac{e}{m}\mathbf{E}\langle t\rangle\,.$$

$\langle\mathbf{v}_0\rangle$ is the average of all the electron velocities immediately after their last collision. Because the motions of the electrons are independent of each other, and their collisions with the ions are random, $\langle\mathbf{v}_0\rangle$ must vanish.

In the second term we must calculate the average time elapsed since the last collision. This is exactly the mean free time τ, of Part I (Sec. 3.2). This leads to Eq. (5.3.3).

Solution 3.3 *Exercise on page 468*

When the electric field is uniform along the wire, the potential difference between the edges of a segment of length L is $V = E \cdot L$. Since the field is uniform, Eq. (5.3.5a) implies that the current density is also uniform, and the current crossing through the entire section, of area A, is $I = JA$. By substituting into (5.3.5a) we obtain

$$\frac{I}{A} = \frac{\sigma V}{L} \Rightarrow I = \left(\frac{\sigma A}{L}\right) V .$$

We can identify the prefactor of the potential as the conductance, and thus the resistance is

$$R = \frac{L}{\sigma A} = \frac{\rho L}{A} .$$

ρ is the resistivity of the material.

Solution 3.4 *Exercise on page 468*

Suppose we "mix" the isothermal atmosphere until a uniform density is obtained and then allow it to evolve spontaneously. In this case a downward current will be created, in exactly the same way that an electron current results when an electric field is applied. In contrast to the electric case, the molecular current cannot continue indefinitely, since the atmosphere has a bottom. As a result more molecules accumulate near the bottom, giving rise to a density gradient, which in turn creates an upward diffusive current. This drives the system towards equilibrium where the two currents must be equal, as we have seen in Part I (Sec. 3.5).

In contrast, the electrons arriving at the edge of the segment of the wire can cross it and continue along the electrical circuit, eventually returning from the other side of the wire by way of the voltage source. Hence no density gradients appear along the wire, and no diffusive currents.

Solution 3.5 *Exercise on page 470*

(a) The numerator on the right hand side has the dimensions of momentum:

$$[\sqrt{mkT}] = [M][v] = \mathrm{kg \cdot m\ s^{-1}} = \mathrm{J \cdot s\ m^{-1}} .$$

Resistivity has the dimensions of resistance × length, or

$$[\rho] = \Omega \cdot \mathrm{m} = \frac{\mathrm{V \cdot m}}{\mathrm{A}} = \frac{\mathrm{J \cdot m}}{\mathrm{C \cdot A}} = \frac{\mathrm{J \cdot m \cdot s}}{\mathrm{C^2}} .$$

The whole denominator has the dimensions of

$$[ne^2\rho] = \mathrm{J \cdot s\ m^{-2}} .$$

Hence, the dimensions of the right hand side are

$$\frac{\mathrm{J \cdot s\ m^{-1}}}{\mathrm{J \cdot s\ m^{-2}}} = \mathrm{m}\,.$$

(b) The mean free path is given approximately by the product of the average speed between collisions, $\bar{v}$, and the mean free time. The average speed according to the Drude model is the thermal speed obtained from the Boltzmann distribution:

$$\frac{1}{2}m\bar{v}^2 \approx \frac{3}{2}kT\,,$$

and using (5.3.5b),

$$\ell \approx \bar{v}\tau \approx \frac{m\bar{v}}{ne^2\rho} = \frac{\sqrt{3mkT}}{ne^2\rho}\,.$$

Solution 3.6 *Exercise on page 471*

We use Eq. (5.3.10) to calculate ℓ_D. One must first calculate the density of conduction electrons, n. Since an amount of A grams contains an Avogadro number, N_0, of atoms, the density of atoms per unit volume is $N_0 d/A$, where the mass density d is expressed in $\mathrm{g/cm^3}$. Since each atom contributes v conduction electrons,

$$n = \frac{vN_0 d}{A}\,,$$

where n is in units of $\mathrm{cm^{-3}}$. Hence, substituting in Eq. (5.3.10), we obtain

$$\ell_D \approx \frac{A\sqrt{3mkT}}{vN_0 de^2\rho}\,.$$

For Li at $T = 77$ K we substitute

$$A = 6.941 \times 10^{-3}\ \mathrm{kg},\ m = 9.11 \times 10^{-31}\ \mathrm{kg},\ k = 1.38 \times 10^{-23}\ \mathrm{J\ K^{-1}}, v = 1\,,$$

$$N_0 = 6.02 \times 10^{23},\ d = 530\ \mathrm{kg\ m^{-3}},\ e = 1.6 \times 10^{-19}\ \mathrm{C}, \rho = 1.04 \times 10^{-8}\ \Omega\mathrm{m}\,,$$

and obtain

$$\ell_D = 4.40 \times 10^{-9}\ \mathrm{m} = 44.0\ \text{Å}\,.$$

Note that we have expressed all the quantities in SI units in order to refrain from expressing the electron charge and the resistivity in electrostatic units.

The calculation of the remaining quantities in the table is carried out in exactly the same way, and there is no point in bringing it here. Do check for yourself several more values.

Solution 3.7 *Exercise on page 473*

(a) Reconstituting the angular factor of 4π to the integral in (5.3.14) or calculating N directly from (5.3.12) using Eq. (5.1.11), we obtain

$$N = \frac{2V}{h^3}\int \frac{d^3p}{\exp\left[\beta\left(\frac{p^2}{2m}-\mu\right)\right]+1}\,.$$

Since the number of electrons is an integral over all momenta, the number of electrons with momentum in the region d^3p around $\mathbf{p}$ is given by the integrand [compare with the transition from Eq. (5.2.11) to Eq. (5.2.12)]. The *probability* for a given electron to have momentum in the region d^3p around $\mathbf{p}$ is the number of electrons in this region divided by the total number of electrons, N:

$$f_p(\mathbf{p})d^3p = \frac{2}{nh^3}\frac{d^3p}{\exp\left[\beta\left(\frac{p^2}{2m}-\mu\right)\right]+1}\,,$$

where $n = N/V$. Substituting $\mathbf{p} = m\mathbf{v}$ we obtain (5.3.15).

(b) The normalization of $f(\mathbf{v})$ is not automatic. It implies a relation between T, V, N and μ, e.g. Eq. (5.3.14). This is an expression of the fact that the independent variables we started with were T, V, μ. Once we have the thermodynamic potential and the average N is expressed in terms of T, V, μ, it is possible to invert the relation and, in principle, to express μ in terms of T, V and N. If we think of Eq. (5.3.14) as an equation from which it is possible to obtain such a relation, then it also serves as a normalization condition,

$$\frac{8\pi}{nh^3}\int_0^\infty \frac{p^2dp}{\exp\left[\beta\left(\frac{p^2}{2m}-\mu\right)\right]+1} = 1\,,$$

and after the variable change $\mathbf{p} = m\mathbf{v}$ we find that $f(\mathbf{v})$ is also normalized.

Solution 3.8 *Exercise on page 474*

(a) The variable change $\epsilon = mv^2/2$ leads to $d\epsilon = mvdv$. Hence we obtain from (5.3.16)

$$\langle\epsilon\rangle = \frac{4\pi}{nh^3}\int_0^\infty \frac{(mv(\epsilon))^3d\epsilon}{e^{\beta(\epsilon-\mu)}+1} = \frac{4\pi(2m)^{3/2}}{nh^3}\int_0^\infty \frac{\epsilon^{3/2}d\epsilon}{e^{\beta(\epsilon-\mu)}+1}\,.$$

To bring the integral to the standard form of an average,

$$\int_0^\infty \epsilon f_\epsilon(\epsilon)d\epsilon\,,$$

we define the energy distribution function according to Eq. (5.3.17a).

(b) In order to verify that $f_\epsilon(\epsilon)$ is normalized we calculate its integral, performing the variable change $\epsilon = p^2/2m$:

$$\int_0^\infty f_\epsilon(\epsilon)d\epsilon = \frac{4\pi(2m)^{3/2}}{nh^3}\int_0^\infty \frac{\epsilon^{1/2}d\epsilon}{e^{\beta(\epsilon-\mu)}+1}$$
$$= \frac{8\pi}{nh^3}\int_0^\infty \frac{p^2dp}{\exp\left[\beta\left(\frac{p^2}{2m}-\mu\right)\right]+1},$$

and the expression we obtained is indeed equal to 1 if Eq. (5.3.14) is satisfied.

Solution 3.9 *Exercise on page 475*

In the limit $T \to 0$ the integrand in (5.3.20) becomes $\epsilon^{1/2}$ for $\epsilon < \mu_0$ and 0 for $\epsilon > \mu_0$. Hence, the region of integration is $0 \le \epsilon < \mu_0$, so that

$$n = \frac{4\pi(2m)^{3/2}}{h^3}\int_0^{\mu_0} \epsilon^{1/2}d\epsilon = \frac{8\pi(2m\mu_0)^{3/2}}{3h^3},$$

and Eq. (5.3.23) is obtained.

Solution 3.10 *Exercise on page 476*

The energy of an electron is given by

$$\epsilon = \frac{1}{2m}(p_x^2 + p_y^2 + p_z^2),$$

and each of the momentum components is quantized. Since the energy depends only on the magnitude of $\mathbf{p}$, the state with the lowest energy of the N electrons will be that in which the electrons are closest to the origin (of momentum space). In this state the electrons occupy all the states inside a sphere (the Fermi sphere) whose radius, p_F, is determined by their number.

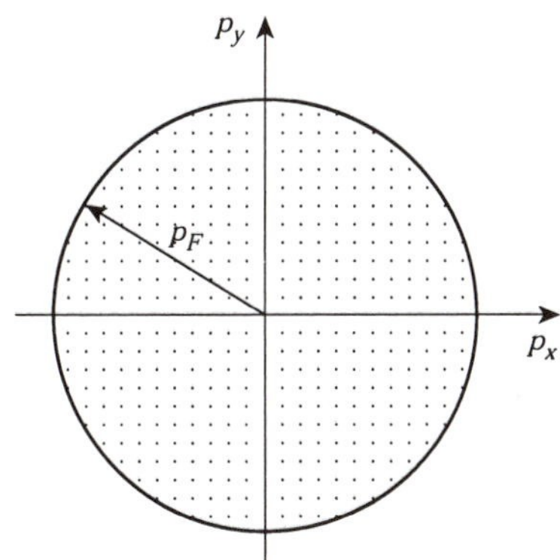

A two-dimensional projection of the three-dimensional case is illustrated in the figure. The volume of this sphere is $\frac{4}{3}\pi p_F^3$, and the energy of the extremal states with momentum p_F is $p_F^2/2m$. If we show that this energy is equal to ϵ_F, as defined in Eq. (5.3.23), we arrive at the required result. To this end we note that since a single state occupies a volume of h^3 in phase space, the volume of a single state in momentum space (see figure) is h^3/V. The number of states contained in the Fermi sphere is given by the ratio of its volume and the volume of a single state. But this is still to be multiplied by 2 in order to take into account the two spin states corresponding to each momentum state. Hence

$$2\frac{\frac{4}{3}\pi p_F^3}{h^3/V} = N\,,$$

which leads to the Fermi momentum,

$$p_F = \left(\frac{3n}{8\pi}\right)^{1/3} h\,,$$

and the relationship to the Fermi energy

$$\epsilon_F = \frac{p_F^2}{2m}$$

is indeed satisfied.

Solution 3.11 *Exercise on page 477*

To calculate ϵ_F using Eq. (5.3.23) we need the density of conduction electrons n, which was calculated in Solution 3.6 to be

$$n = \frac{vN_0 d}{A}\,.$$

v is the number of conduction electrons contributed by each atom in the metal, N_0 is Avogadro's number, d is the mass density of the metal and A is the atomic weight. We thus obtain

$$\epsilon_F = \frac{h^2}{2m}\left(\frac{3vN_0 d}{8\pi A}\right)^{2/3}\,.$$

The values of v, A and d are given in Table 5.3.2 in cgs units, so we also express h, m and n in these units. We find for Li

$$\epsilon_F = \frac{(6.626\times 10^{-27})^2}{2\times 9.11\times 10^{-28}}\left(\frac{3\times 1\times 6.02\times 10^{23}\times 0.53}{8\pi\times 6.941}\right)^{2/3}$$

$$= 7.5\times 10^{-12}\ \text{erg} = 4.7\ \text{eV}\,.$$

■ Note that in calculating the chemical potential at $T = 0$ we have used values of the density at room temperature, introducing an error of 1–2%.

The Fermi velocity is

$$v_F = \sqrt{\frac{2\epsilon_F}{m}} = \sqrt{\frac{2 \times 7.5 \times 10^{-12}}{9.11 \times 10^{-28}}} = 1.3 \times 10^8 \text{ cm/s} = 1.3 \times 10^6 \text{ m/s}.$$

The calculation of ϵ_F and v for the other metals is performed in the same way. Do check out several more values for yourself.

Solution 3.12 *Exercise on page 477*

We can calculate $\langle\epsilon\rangle$ directly using (5.3.17). It is possible to write $f_\epsilon(\epsilon)$ in a simpler form in terms of the Fermi energy:

$$f_\epsilon(\epsilon) = \frac{3}{2\epsilon_F^{3/2}} \cdot \frac{\epsilon^{1/2}}{e^{\beta(\epsilon-\mu)} + 1}.$$

Using the approximation (5.3.22), we obtain

$$\langle\epsilon\rangle = \frac{3}{2\epsilon_F^{3/2}} \int_0^{\epsilon_F} \epsilon^{3/2} d\epsilon = \frac{3}{2\epsilon_F^{3/2}} \cdot \frac{2}{5}\epsilon_F^{5/2} = \frac{3}{5}\epsilon_F.$$

Solution 3.13 *Exercise on page 477*

(a) At $T = 0$ all the lowest energy levels are filled with electrons, and hence there can be no electron with energy higher than ϵ_F. An electron that is artificially inserted with an energy that is too high will thermalize and lose its excess energy, $\epsilon - \epsilon_F$, which will eventually reach the heat bath that maintains the electron gas at $T = 0$. The energy added to the gas will thus be equal to ϵ_F.

(b) As in (a), it is impossible at $T = 0$ for there to be a "hole" at energy ϵ below ϵ_F. Hence, in order to maintain the temperature at $T = 0$, the hole will be filled by an electron from above which will create a new hole which will be filled by an electron above it, and so on. The net result is the filling-out of the hole by an electron dropping down from the Fermi level to fill the hole. The excess energy, $\epsilon_F - \epsilon$, eventually reaches the heat bath. Thus, the gas loses energy ϵ_F.

Solution 3.14 *Exercise on page 479*

Substitute $\mu = 0$ into (5.3.25) to obtain

$$\frac{2}{3}\epsilon_F^{3/2} = \int_0^\infty \frac{\epsilon^{1/2}}{e^{\beta\epsilon} + 1} d\epsilon.$$

Perform the variable change $x = \beta\epsilon$ to obtain

$$\frac{2}{3}(\beta\epsilon_F)^{3/2} = \int_0^\infty \frac{x^{1/2}dx}{e^x + 1},$$

and this equation determines the required temperature in terms of ϵ_F. The integral is read in Eq. (5.A.13) of the appendix with $\nu = 3/2$. Its value is

$$\left(1 - \frac{1}{\sqrt{2}}\right)\Gamma\left(\frac{3}{2}\right)\zeta\left(\frac{3}{2}\right) = 0.678,$$

and hence

$$\beta\epsilon_F = 1.01,$$

or

$$T = 0.989T_F.$$

Solution 3.15 *Exercise on page 480*

At low temperatures the Fermi–Dirac occupation function (5.3.19) looks like a step function, and its derivative is significantly different from zero only in a narrow range around $\epsilon = \mu$ (see figure). Note that we do not assume that $\mu = \epsilon_F$.

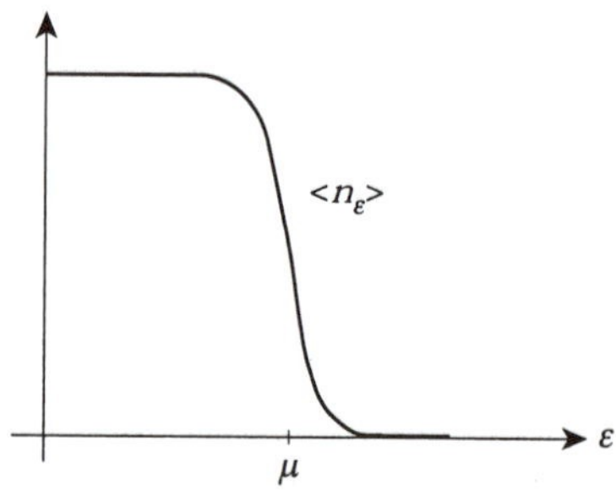

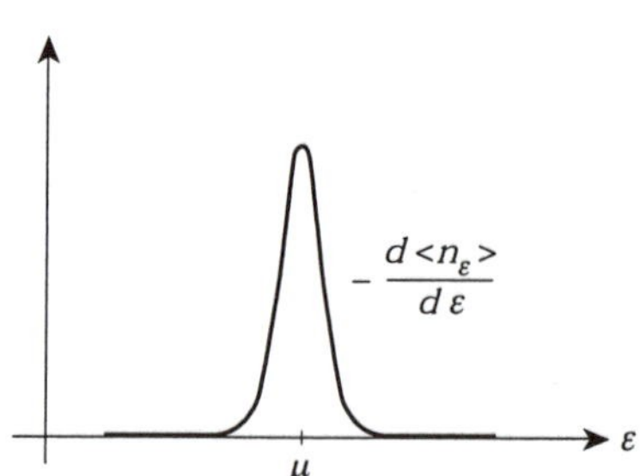

In order to use this fact we integrate (5.3.29) by parts so that instead of n_ϵ its derivative will appear. The integral on the right hand side of (5.3.29) becomes

$$I_l = \frac{2}{2l+3}\int_0^\infty n_\epsilon \left[\frac{d}{d\epsilon}\epsilon^{l+3/2}\right] d\epsilon = -\frac{2}{2l+3}\int_0^\infty \frac{dn_\epsilon}{d\epsilon}\epsilon^{l+3/2}d\epsilon, \qquad \text{(i)}$$

where we have used the fact that n_ϵ vanishes fast for $\epsilon \to \infty$.

Clearly, the principal contribution to the integral in (i) comes from the region around $\epsilon = \mu$. Hence we expand the function $\epsilon^{l+3/2}$ around this point and take the first terms. A Taylor expansion around $\epsilon = \mu$ yields, to the second order,

$$\epsilon^{l+3/2} = \mu^{l+3/2} + \left(l + \frac{3}{2}\right)\mu^{l+1/2}(\epsilon - \mu)$$
$$+\frac{1}{2}\left(l + \frac{3}{2}\right)\left(l + \frac{1}{2}\right)\mu^{l-1/2}(\epsilon - \mu)^2 + \ldots \qquad \text{(ii)}$$

Substituting this into (i) we obtain to the second order

$$I_l \approx -\frac{2}{2l+3}\mu^{l+3/2}\int_{-\infty}^{\infty}\frac{dn_\epsilon}{d\epsilon}d\epsilon - \mu^{l+1/2}\int_{-\infty}^{\infty}\frac{dn_\epsilon}{d\epsilon}(\epsilon - \mu)d\epsilon$$
$$-\frac{1}{2}\left(l + \frac{1}{2}\right)\mu^{l-1/2}\int_{-\infty}^{\infty}\frac{dn_\epsilon}{d\epsilon}(\epsilon - \mu)^2 d\epsilon\,. \qquad \text{(iii)}$$

In addition to substituting (iii), for mathematical convenience we extended the range of integration to negative values of ϵ, justified by the fact that $dn_\epsilon/d\epsilon$ vanishes for $\epsilon \ll \mu$. The next step is to perform the variable change $x = \beta(\epsilon - \mu)$, which renders the function $dn_\epsilon/d\epsilon$ symmetric around $x = 0$, and the "step" in n_ϵ is translated to $x = 0$. Explicitly

$$n_\epsilon(x) = \frac{1}{e^x + 1}\,, \qquad \text{(iv)}$$

$$\frac{dn_\epsilon}{dx} = -\frac{e^x}{(e^x+1)^2} = -\frac{1}{4\cosh^2(x/2)}\,. \qquad \text{(v)}$$

Equation (iii) is rewritten in the form

$$I_l \approx -\frac{2}{2l+3}\mu^{l+3/2}\int_{-\infty}^{\infty}\frac{dn_\epsilon}{dx}dx - \mu^{l+1/2}(kT)\int_{-\infty}^{\infty}\frac{dn_\epsilon}{dx}x\,dx$$
$$-\frac{1}{2}\left(l + \frac{1}{2}\right)\mu^{l-1/2}(kT)^2\int_{-\infty}^{\infty}\frac{dn_\epsilon}{dx}x^2 dx\,. \qquad \text{(vi)}$$

We now note the following:

- The integral in the first term in (vi) is actually the difference between the values of n_ϵ at $x \to \infty$ and at $x \to -\infty$, which is -1. This is the principal contribution to I_l.
- The integrand in the second term is an odd function, whereas the region of integration is symmetric with respect to the origin. Hence the second term vanishes.
- Making use of the fact that the integrand in the third term is an even function, and integrating by parts, yield

$$\int_{-\infty}^{\infty}\frac{dn_\epsilon}{dx}x^2 dx = 2\int_0^{\infty}\frac{dn_\epsilon}{dx}x^2 dx$$
$$= -4\int_0^{\infty}n_\epsilon x\,dx = -4\int_0^{\infty}\frac{x}{e^x+1}dx = -\frac{\pi^2}{3}\,.$$

The value of the last integral is found from the appendix.

Substituting in (vi) we obtain

$$I_l \approx \frac{2}{2l+3}\mu^{l+3/2} + \left(l+\frac{1}{2}\right)\frac{\pi^2}{6}\mu^{l-1/2}(kT)^2 , \tag{vii}$$

which, when substituted in (5.3.29), gives (5.3.30).

We summarize the stages of the calculation of I_l:

(1) Integration by parts.
(2) The expansion of $\epsilon^{l+1/2}$ around $\epsilon = \mu$ up to the second order.
(3) Extension of the range of integration to $-\infty < \epsilon < \infty$.
(4) The variable change $x = \beta(\epsilon - \mu)$.
(5) Calculation of the first integral in (vi).
(6) Noting that the second integral vanishes (parity).
(7) Calculation of the third integral with the help of parity properties and the appendix.

Solution 3.16 *Exercise on page 481*

(a) Equation (5.3.31) implies that $(\mu/\epsilon_F)^{3/2}$ is somewhat smaller than 1:

$$\left(\frac{\mu}{\epsilon_F}\right)^{3/2} \approx 1 - \frac{\pi^2}{8}\frac{(kT)^2}{\epsilon_F^{3/2}\mu^{1/2}} .$$

In order to obtain the first correction in the temperature we substitute $\mu = \epsilon_F$ on the right hand side. Any correction to μ in the denominator will produce terms of higher order than T^2. Inverting the power in the above equation and expanding according to $(1-x)^p \simeq 1 - px$, we have

$$\frac{\mu}{\epsilon_F} \approx \left[1 - \frac{\pi^2}{8}\left(\frac{kT}{\epsilon_F}\right)^2\right]^{2/3} \approx 1 - \frac{\pi^2}{12}\left(\frac{kT}{\epsilon_F}\right)^2 .$$

This leads to Eq. (5.3.32).

(b) Substituting $n = 1$ in (5.3.30) we obtain

$$\langle\epsilon\rangle \approx \frac{3}{5}\frac{\mu^{5/2}}{\epsilon_F^{3/2}} + \frac{3\pi^2}{8}\frac{\mu^{1/2}(kT)^2}{\epsilon_F^{3/2}} .$$

In order to obtain the T^2 correction we insert Eq. (5.3.32) for μ. In the second term on the right hand side it suffices to substitute $\mu = \epsilon_F$, since any correction to μ will produce terms of order higher than T^2. In the first term we must include in $\mu^{5/2}$ also the term with T^2 and then expand in T, i.e.

$$\mu^{5/2} \approx \epsilon_F^{5/2}\left[1 - \frac{5\pi^2}{24}\left(\frac{kT}{\epsilon_F}\right)^2\right] .$$

Overall we obtain

$$\langle\epsilon\rangle \approx \frac{3}{5}\epsilon_F\left[1-\frac{5\pi^2}{24}\left(\frac{kT}{\epsilon_F}\right)^2\right]+\frac{3\pi^2}{8}\frac{(kT)^2}{\epsilon_F}=\frac{3}{5}\epsilon_F+\frac{\pi^2}{4}\frac{(kT)^2}{\epsilon_F}\,,$$

which is Eq. (5.3.33).

Solution 3.17 *Exercise on page 482*

The phonon contribution to the specific heat is given by (4.3.18) as

$$C_{\text{ph}} = Nk\cdot\frac{12\pi^4}{5}\left(\frac{T}{\Theta_D}\right)^3,$$

while that of the electrons is

$$C_{\text{el}} = Nk\cdot\frac{\pi^2}{2}\frac{T}{T_F}\,.$$

Equating the two contributions we obtain

$$T=\sqrt{\frac{5}{24\pi^2}\frac{\Theta_D^3}{T_F}}\,.$$

The Fermi temperature of lithium is 5.5×10^4K (Table 5.3.3), and the equality is obtained at a temperature of 5 K.

Solution 3.18 *Exercise on page 484*

The Fermi wavelength is the de Broglie wavelength corresponding to an electron with Fermi energy:

$$\lambda_F=\frac{h}{p_F}=\frac{h}{\sqrt{2m\epsilon_F}}\,.$$

We first calculate λ_F for $\epsilon_F = 1$eV:

$$\lambda_F(1\text{ eV})=\frac{6.626\times10^{-34}}{\sqrt{2\times9.11\times10^{-31}\times1.6\times10^{-19}}}=12.27\text{ Å}\,,$$

and then, if ϵ_F is in units of eV, λ_F is obtained in Å, by

$$\lambda_F=\frac{12.27}{\sqrt{\epsilon_F}}\,.$$

Thus, for example for Li, we obtain $\lambda_F = 5.66$ Å. Do calculate the other values yourself and fill out the table.

Solution 4.1 *Exercise on page 486*

From (5.4.7) we obtain

$$e^{-\beta\mu}-1\approx\frac{1}{N}\,,$$

or

$$\beta\mu \approx -\ln\left(1+\frac{1}{N}\right),$$

and since N is macroscopic, we can expand the right hand side and obtain

$$\beta\mu \approx -\frac{1}{N},$$

and this is Eq. (5.4.8).

Substituting $T = 1$ K, $N = 10^{22}$ yields

$$\mu \approx -1.4 \times 10^{-45}\ \mathrm{J} = -8.6 \times 10^{-27}\ \mathrm{eV}.$$

Solution 4.2 *Exercise on page 487*

(a) To show that n increases with μ we calculate the derivative from (5.4.6):

$$\frac{\partial n}{\partial \mu} = \frac{2\pi(2m)^{3/2}\beta}{h^3}\int_0^\infty \frac{e^{\beta(\epsilon-\mu)}\cdot \epsilon^{1/2}d\epsilon}{[e^{\beta(\epsilon-\mu)}-1]^2}.$$

The integrand on the right hand side is everywhere positive, and hence

$$\frac{\partial n}{\partial \mu} > 0.$$

The maximum of n at $\mu = 0$ is attained at the edge of the region of allowed values of μ. We further remark that the fact that n increases with μ is general, and is a manifestation of the fact that μ increases with n at constant T, expressing the fact that particles tend to flow from a high density region to a low density region. See, for example, Part III, Sec. 1.5.

(b) Substituting $\mu = 0$ into (5.4.6) we obtain

$$n_* = \frac{2\pi(2m)^{3/2}}{h^3}\int_0^\infty \frac{\epsilon^{1/2}d\epsilon}{e^{\beta\epsilon}-1} = \frac{2\pi(2mkT)^{3/2}}{h^3}\int_0^\infty \frac{x^{1/2}dx}{e^x-1},$$

where $x = \beta\epsilon$. The integral can be calculated using Eq. (5.A.14) of the appendix, to give $1.306\sqrt{\pi}$. From here

$$n_* = \frac{2.612(2\pi mkT)^{3/2}}{h^3},$$

which is the result we sought.

Solution 4.3 *Exercise on page 489*

(a) To obtain the dependence of μ on β we write Eq. (5.4.6) with $x = \beta\epsilon$:

$$n = \frac{2\pi}{h^3}\left(\frac{2m}{\beta}\right)^{3/2}\int_0^\infty \frac{x^{1/2}dx}{e^{x-\beta\mu}-1}, \tag{i}$$

or

$$\frac{nh^3}{2\pi}\left(\frac{\beta}{2m}\right)^{3/2} = \int_0^\infty \frac{x^{1/2}dx}{e^{x-\beta\mu}-1}. \tag{ii}$$

When n is constant, the left hand side increases with β and so the integral on the right hand side must also increase with β. The right hand side is a monotonic increasing function of $\beta\mu$, since the integrand is increasing at every point. Hence, $\beta\mu$ must become less negative as β increases. Since β itself increases, μ must increase with β and become less negative.

(b) Substituting $\mu = 0$ into Eq. (i) we obtain the same integral as in calculating n_* in Exercise 4.2, and thus

$$n = \frac{2.612(2\pi mkT)^{3/2}}{h^3}, \tag{iii}$$

and from here Eq. (5.4.12) is immediately obtained.

Note the physical distinction between the two cases. In the previous exercise the density changed as a result of a change in the chemical potential. Here the density is constant, and it is the change in temperature that causes the change in the chemical potential.

(c) Using Eq. (5.4.12) it is possible to express n in terms of T_c:

$$n = \frac{2.612(2\pi mkT_c)^{3/2}}{h^3}, \tag{iv}$$

and from Eq. (5.4.9)

$$n_* = \frac{2.612V(2\pi mkT)^{3/2}}{h^3}. \tag{v}$$

By dividing the two we obtain

$$\frac{n_*}{n} = \left(\frac{T}{T_c}\right)^{3/2}. \tag{vi}$$

(d) n_e is the density of bosons occupying the excited states, and cannot be larger than the maximum n_*. When $T < T_c$, $n_* < n$ and therefore $n_e = n_*$ and all other $(n - n_*)$ bosons condensate in the ground state. When $T > T_c$, $n_* > n$ and only a tiny fraction of the bosons occupy the ground state and thus $n_e = n$.

Solution 4.4 *Exercise on page 494*

(a) In order to calculate T_c for helium we have to find its atomic density. This can be obtained from its mass density, d, which is 0.14 g/cm^3, and the atomic mass, which is four atomic mass units. Hence

$$T_c = \frac{h^2}{2\pi mk}\left(\frac{d}{2.612m}\right)^{2/3} = \frac{h^2}{2\pi km^{5/3}}\left(\frac{d}{2.612}\right)^{2/3}$$
$$= \frac{(6.626\times 10^{-27})^2}{2\pi\times 1.38\times 10^{-16}\times(4\times 1.66\times 10^{-24})^{5/3}}\left(\frac{0.14}{2.612}\right)^{2/3} = 3.1\text{ K}.$$

(b) Instead of substituting all the numerical values again, we note that the mass of a hydrogen molecule is half that of a helium atom and the mass density of hydrogen is 3/7 that of helium. Hence, to obtain the critical temperature of liquid hydrogen we have to multiply the critical temperature of He^4 by $2^{5/3}\times(3/7)^{2/3}$, to obtain 5.6 K.

Since hydrogen liquefies around 20 K and freezes at 14 K, this temperature range is too far from the value of 5.6 K, and we do not expect superfluidity in liquid hydrogen.

Solution 4.5 *Exercise on page 494*

(a) The total energy of a boson gas can be calculated using the energy distribution function, (5.4.4). We write

$$E = N\langle\epsilon\rangle = N\int_0^\infty \epsilon f(\epsilon)d\epsilon = \frac{2\pi V(2m)^{3/2}}{h^3}\int_0^\infty \frac{\epsilon^{3/2}d\epsilon}{e^{\beta(\epsilon-\mu)}-1}.$$

Below the critical temperature $\mu = 0$, and with $x = \beta\epsilon$, we obtain

$$E = \frac{2\pi V(2m)^{3/2}(kT)^{5/2}}{h^3}\int_0^\infty \frac{x^{3/2}dx}{e^x-1}.$$

The integral on the right hand side is calculated using the appendix [Eq. (5.A.14)], and its value is

$$\Gamma\left(\frac{5}{2}\right)\zeta\left(\frac{5}{2}\right) = 1.006\sqrt{\pi}.$$

Substituting, we obtain

$$E = \frac{2.012V(2\pi m)^{3/2}(kT)^{5/2}}{h^3}.$$

(b) First we calculate the average energy per particle in the excited levels ($\epsilon > 0$) by dividing Eq. (5.4.14) by $N_e = N_* = Vn_*$ [with n_* given by Eq. (5.4.9)]:

$$\frac{E}{N_*} = 0.770kT\,.$$

Next, we express the relation between the number of excited particles and the total number of particles using Solution 4.3(c) for the corresponding densities, to obtain

$$E = 0.770N\left(\frac{T}{T_c}\right)^{3/2} kT\,,$$

which leads directly to the average energy per particle, E/N.

Solution 4.6 *Exercise on page 496*

Writing the integrand in Eq. (5.4.16) as one fraction we obtain

$$n_* - n = \frac{2\pi(2m)^{3/2}(e^{-\beta\mu} - 1)}{h^3}\int_0^\infty \frac{e^{\beta\epsilon}\epsilon^{1/2}d\epsilon}{(e^{\beta\epsilon} - 1)(e^{\beta(\epsilon-\mu)} - 1)}\,.$$

Since μ is small, the major contribution to the integral comes from regions of small ϵ. We expand the exponentials in the integrand to linear order, and obtain

$$n_* - n \approx -\mu kT \cdot \frac{2\pi(2m)^{3/2}}{h^3}\int_0^\infty \frac{d\epsilon}{\epsilon^{1/2}(\epsilon - \mu)}\,.$$

Substituting $\epsilon = u^2$,

$$\int_0^\infty \frac{d\epsilon}{\epsilon^{1/2}(\epsilon - \mu)} = 2\int_0^\infty \frac{du}{u^2 - \mu} = \frac{2}{\sqrt{-\mu}}\tan^{-1}\left(\frac{u}{\sqrt{-\mu}}\right)\bigg|_0^\infty = \frac{\pi}{\sqrt{-\mu}}\,.$$

Hence

$$n_* - n = \frac{2\pi^2 kT(2m)^{3/2}\sqrt{-\mu}}{h^3}\,.$$

Solving for μ, the middle expression in Eq. (5.4.17) is obtained.

In order to express the temperature dependence in terms of T_c, to obtain the rightmost expression of Eq. (5.4.17), we write

$$n_* = n\left(\frac{T}{T_c}\right)^{3/2}$$

(see Exercise 4.3). Hence

$$\mu \approx -\frac{h^6}{32\pi^4 m^3}\left(\frac{n}{kT}\right)^2\left[\left(\frac{T}{T_c}\right)^{3/2}-1\right]^2 .$$

We note that [see (5.4.12)]

$$\frac{h^6 n^2}{8\pi^3 m^3} = 2.612^2 (kT_c)^3 ,$$

and hence we can write

$$\mu \approx -\frac{(kT_c)^3}{4\pi}\cdot\left(\frac{2.612}{kT}\right)^2\left[\left(\frac{T}{T_c}\right)^{3/2}-1\right]^2$$

$$= -\frac{kT}{4\pi}\left[2.612\left(\frac{T_c}{T}\right)^{3/2}\right]^2\left[\left(\frac{T}{T_c}\right)^{3/2}-1\right]^2 ,$$

which leads to the the rightmost expression in (5.4.17).

Solution 4.7 *Exercise on page 496*

From Eq. (5.4.18) we obtain

$$E_* - E = \frac{2\pi V(2m)^{3/2}(e^{-\beta\mu}-1)}{h^3}\int_0^\infty \frac{e^{\beta\epsilon}\epsilon^{3/2}d\epsilon}{(e^{\beta\epsilon}-1)(e^{\beta(\epsilon-\mu)}-1)} .$$

Since μ is small we expand the exponential in front of the integral. But we cannot expand the factors in the denominator of the integrand, as was done in the solution of the previous exercise, since the integral obtained in this way does not converge. To calculate the integral we use the fact that it converges for $\mu = 0$. After the variable change $x = \beta\epsilon$ we obtain

$$E_* - E \approx -\mu\frac{2\pi V(2mkT)^{3/2}}{h^3}\int_0^\infty \frac{e^x x^{3/2}dx}{(e^x-1)^2} .$$

Denoting the integral by I and integrating by parts, we obtain

$$I = -\int_0^\infty x^{3/2}\frac{d}{dx}\left[\frac{1}{e^x-1}\right]dx = \frac{3}{2}\int_0^\infty \frac{x^{1/2}dx}{e^x-1} = \frac{3}{2}\cdot 1.306\sqrt{\pi} ,$$

where the last integral was calculated in Solution 4.2 using (5.A.14).

We arrive at

$$E_* - E \approx -\frac{3}{2}\mu\cdot\left[\frac{2.612V(2\pi mkT)^{3/2}}{h^3}\right] ,$$

where the expression in the square brackets is $N_*(= Vn_*)$ [Eq. (5.4.9)]. From here Eq. (5.4.19) is immediately obtained.

Solution 4.8 *Exercise on page 499*

(a) Substituting $\mathbf{p}' = \mathbf{p} - \hbar\mathbf{q}$ into the energy equation we obtain

$$\frac{p^2}{2M} = \frac{p^2 - 2\hbar\mathbf{q}\cdot\mathbf{p} + (\hbar q)^2}{2M} + \epsilon(q)\,.$$

The initial kinetic energy term drops out of both sides. Writing $\mathbf{p} = M\mathbf{v}$, we obtain

$$0 = -\hbar\mathbf{q}\cdot\mathbf{v} + \frac{(\hbar q)^2}{2M} + \epsilon(q)\,,$$

which leads to Eq. (5.4.23).

(b) q is the wave number of a phonon in liquid helium. It can be at most $2\pi/a$, where a is the interatomic distance, which is of the order of 1 Å. Hence, $q < 6\text{Å}^{-1}$. Next we show that indeed

$$\frac{(\hbar q)^2}{2M} \ll \epsilon(q)\,.$$

We substitute $M \sim 1$ g, to find that

$$\frac{(\hbar q)^2}{2M} \approx 10^{-67}\text{erg} \sim 10^{-55}\ \text{eV}\,.$$

This energy is smaller by many orders of magnitude than typical phonon energies, which are

$$\epsilon \approx \hbar v_s q \sim 10^{-14}\ \text{erg} \sim 10^{-2}\ \text{eV}\,,$$

where v_s, the speed of sound in liquid helium, is about 240 m s^{-1}. Hence, for a macroscopic body, even a thousand times lighter,

$$\hbar\mathbf{q}\cdot\mathbf{v} = \epsilon(q)\,.$$

Solution 4.9 *Exercise on page 501*

(a) Measuring the slope near the origin we obtain

$$\Delta q = 0.6\ \text{Å}^{-1}\,,$$

$$\frac{\Delta\epsilon}{k} = 10.77\ \text{K}\,,$$

and since the slope of the graph of ϵ versus q is $\hbar v_s$,

$$\hbar v_s = \frac{\Delta\epsilon}{\Delta q} \Rightarrow v_s = \frac{\Delta\epsilon}{\hbar\Delta q} = \frac{10.77\times 1.38\times 10^{-23}}{1.05\times 10^{-34}\times 0.6\times 10^{10}} = 236\ \text{m s}^{-1}\,,$$

and this value is very close to the speed of sound in liquid helium.

(b) The critical speed is obtained from the solution of Eq. (5.4.26), which is equivalent to the solution of (5.4.24) for the value at the minimum point of the dispersion curve:

$$v_c = \frac{8.67 \times 1.38 \times 10^{-23}}{1.05 \times 10^{-34} \times 1.94 \times 10^{10}} = 59 \text{ m s}^{-1}\,.$$

Solution 4.10 *Exercise on page 503*

The description of He II as a boson gas below its condensation temperature is not very accurate. In a description that takes into account the interatomic forces, He II behaves as a phonon gas. As we have seen in Part IV, the specific heat of a phonon gas at low temperatures is proportional to T^3 and not to $T^{3/2}$. Thus we expect the specific heat of He^4 at low temperatures to behave this way. And indeed experiments show that this is the case below a temperature of 0.6 K.

Solution 4.11 *Exercise on page 503*

Substituting the mass density of liquid He^3, d, and its atomic mass, Eqs. (5.3.23) and (5.3.24) give

$$T_F = \frac{h^2}{2mk}\left(\frac{3n}{8\pi}\right)^{2/3} = \frac{h^2}{2m^{5/3}k}\left(\frac{3d}{8\pi}\right)^{2/3} = 4.5 \text{ K}\,.$$

Solution 4.12 *Exercise on page 505*

Substituting (5.4.29a) into (5.4.30), we obtain

$$\eta \approx \frac{mv_F}{3\sigma}\left(\frac{\epsilon_F}{kT}\right)^2,$$

and using the expression for the Fermi energy, Eq. (5.3.23), we have

$$\eta \approx \frac{h^5}{12\sigma m^2 (kT)^2}\left(\frac{3n}{8\pi}\right)^{5/3}.$$

Namely,

(a) $\eta \propto 1/m^2$;
(b) At constant T, $\eta \propto n^{5/3}$.

Note the significant difference between this behavior and the behavior of the viscosity of a classical gas; see Exercise 3.14 of Part I.

Solution 4.13 *Exercise on page 506*

Since the bosons in this case are pairs of He^3 atoms, we substitute into Eq. (5.4.12) the mass of two He^3 atoms, which we shall denote by $2m$, and for the density, one half that of ordinary He^3. We thus obtain

$$T_c = \frac{h^2}{2^{5/3} \cdot 2\pi mk} \left(\frac{n}{2.612} \right)^{2/3} = \frac{h^2}{(2m)^{5/3} 2\pi k} \left(\frac{d}{2.612} \right)^{2/3} ,$$

where d is the mass density of liquid He^3 and m is the mass of a He^3 atom. Substituting the numerical values we obtain

$$T_c = 0.98 \text{ K} .$$

Solutions to self-assessment exercises

Solution 1 *Exercise on page 512*

(a) The grand canonical partition function given in Eq. (5.1.3) can be written using the general form of the canonical partition function of an ideal gas (with the Gibbs correction), i.e. Eq. (3.5.3):

$$\mathcal{Z} = \sum_{N=0}^{\infty} e^{\beta\mu N} \frac{z^N}{N!} . \qquad \text{(i)}$$

This equation can also be written in the form

$$\mathcal{Z} = \sum_{N=0}^{\infty} \frac{(ze^{\beta\mu})^N}{N!} = \exp[z \exp(\beta\mu)] . \qquad \text{(ii)}$$

The thermodynamic potential is obtained from (5.1.8):

$$\Omega = -kT \ln \mathcal{Z} = -kTze^{\beta\mu} , \qquad \text{(iii)}$$

and the average number of particles is obtained from (5.1.10):

$$\langle N \rangle = -\frac{\partial \Omega}{\partial \mu} = ze^{\beta\mu} = -\frac{\Omega}{kT} , \qquad \text{(iv)}$$

and this is the required relation.

(b) The probability for there to be exactly N particles in the container can be obtained from an inspection of (5.1.3), since the partition function is the sum of all the (unnormalized) probabilities, or from (5.1.1), and a summation over all the microscopic states with a constant number of particles, N. In both ways we obtain

$$P_N = \mathcal{Z}^{-1} e^{\beta(\mu N - F)} = \mathcal{Z}^{-1} e^{\beta\mu N} Z(T, V, N) . \qquad \text{(v)}$$

If we write

$$\begin{cases} \mathcal{Z} = e^{-\beta\Omega} , \\ Z = \dfrac{z^N}{N!} , \end{cases} \qquad \text{(vi)}$$

we obtain from (v)

$$P_N = \frac{e^{\beta\Omega}(ze^{\beta\mu})^N}{N!}\,, \tag{vi}$$

and with the help of (iv) we obtain the required result.

Solution 2 *Exercise on page 512*

(a) Indeed, the free energy depends on N. However, N is not the number of phonons in the crystal but the number of molecules in it as well as the number of its different modes of vibration. When we describe the vibrations of the crystal as a gas of phonons, the number of phonons is arbitrary and is not subject to any restriction. Hence we can calculate the free energy from the canonical partition function, Eq. (5.2.5), without the constraint (5.2.2):

$$Z = \left(\sum_{n_1=0}^{\infty} e^{-\beta\epsilon_1 n_1}\right)\cdot\ldots\cdot\left(\sum_{n_k=0}^{\infty} e^{-\beta\epsilon_k n_k}\right)\cdots \tag{i}$$

Each occupation number n_k is the number of phonons with that k and it varies between zero and infinity, since phonons are bosons. The free energy is, therefore, the sum over all the vibration modes $\{k\}$, exactly as in calculating Ω:

$$F = -kT\ln Z = kT\sum_k \ln(1 - e^{-\beta\epsilon_k})\,. \tag{ii}$$

This sum cannot depend on the number of phonons, which we will here denote by N_P, but it does depend on the number of molecules in the crystal, N. Hence the chemical potential of the phonon gas is

$$\mu_P = \frac{\partial F}{\partial N_P} = 0\,. \tag{iii}$$

(b) In order to calculate the energy from the grand canonical partition function, it is not enough to differentiate $\ln\mathcal{Z}$ with respect to β as we did in the canonical case, since

$$\frac{\partial \ln\mathcal{Z}}{\partial\beta} = \mu N - E\,. \tag{iv}$$

But, if we also express N in terms of the partition function, we obtain our goal. We do this using Eq. (5.1.7), to obtain

$$E = \mu N - \frac{\partial \ln\mathcal{Z}}{\partial\beta} = \frac{\mu}{\beta}\frac{\partial \ln\mathcal{Z}}{\partial\mu} - \frac{\partial \ln\mathcal{Z}}{\partial\beta}\,. \tag{v}$$

We now use Ω, for a boson gas:

$$\ln \mathcal{Z}^{(B)} = -\beta\Omega^{(B)} = -\sum_k \ln(1 - e^{-\beta(\mu-\epsilon_k)}) , \tag{vi}$$

and substituting this into (v) we obtain

$$\begin{aligned} E &= \frac{\mu}{\beta}\sum_k \frac{\beta e^{\beta(\mu-\epsilon_k)}}{1 - e^{\beta(\mu-\epsilon_k)}} - \sum_k \frac{(\mu-\epsilon_k)e^{\beta(\mu-\epsilon_k)}}{1 - e^{\beta(\mu-\epsilon_k)}} \\ &= \sum_k \frac{\epsilon_k e^{\beta(\mu-\epsilon_k)}}{1 - e^{-\beta(\mu-\epsilon_k)}} = \sum_k \frac{\epsilon_k}{e^{\beta(\epsilon_k-\mu)} - 1} , \end{aligned} \tag{vii}$$

which is exactly the required expression for bosons. For a fermion gas we use

$$\ln \mathcal{Z}^{(F)} = -\beta\Omega^{(F)} = \sum_k \ln(1 + e^{\beta(\mu-\epsilon_k)}) , \tag{viii}$$

and substituting into (v) we obtain this time

$$\begin{aligned} E &= \frac{\mu}{\beta}\sum_k \frac{\beta e^{\beta(\mu-\epsilon_k)}}{1 + e^{\beta(\mu-\epsilon_k)}} - \sum_k \frac{(\mu-\epsilon_k)e^{\beta(\mu-\epsilon_k)}}{1 + e^{\beta(\mu-\epsilon_k)}} \\ &= \sum_k \frac{\epsilon_k e^{\beta(\mu-\epsilon_k)}}{1 + e^{\beta(\mu-\epsilon_k)}} = \sum_k \frac{\epsilon_k}{e^{\beta(\epsilon_k-\mu)} + 1} , \end{aligned} \tag{ix}$$

which is the required result.

Solution 3 *Exercise on page 513*

First we calculate the thermodynamic potential Ω from Eq. (5.2.9a), as a sum over single particle states, which are eigenstates of momentum. For three-dimensional motion d^3p includes Vd^3p/h^3 states and for the volume element of a two-dimensional motion d^2p includes Ad^2p/h^2 states. We thus obtain the two-dimensional analog of Eq. (5.3.12):

$$\Omega = -kTA\int \ln\left\{1 + \exp\left[\beta\left(\mu - \frac{p^2}{2m}\right)\right]\right\}\frac{d^2p}{h^2} . \tag{i}$$

Integrating over all directions in momentum space using polar coordinates we obtain

$$\Omega = -\frac{2\pi AkT}{h^2}\int_0^\infty \ln\left\{1 + \exp\left[\beta\left(\mu - \frac{p^2}{2m}\right)\right]\right\} pdp . \tag{ii}$$

From (5.1.10)

$$N = -\frac{\partial\Omega}{\partial\mu} = \frac{2\pi A}{h^2}\int_0^\infty \frac{pdp}{\exp\left[\beta\left(\frac{p^2}{2m} - \mu\right)\right] + 1} , \tag{iii}$$

and with the usual change of variables,

$$x = \beta\epsilon = \frac{\beta p^2}{2m}, \quad dx = \frac{\beta p}{m} dp, \tag{iv}$$

we have, for the two-dimensional density n,

$$n = \frac{N}{A} = \frac{2\pi mkT}{h^2} \int_0^\infty \frac{dx}{e^{x-\beta\mu}+1}. \tag{v}$$

This equation is an implicit relation between n and μ. In the present case it is possible to calculate the integral as

$$\int_0^\infty \frac{dx}{e^{x-\beta\mu}+1} = \int_0^\infty \frac{e^{\beta\mu-x}}{1+e^{\beta\mu-x}} dx = -\ln(1+e^{\beta\mu-x})\Big|_0^\infty = \ln(1+e^{\beta\mu}).$$

From here

$$n = \frac{2\pi mkT}{h^2} \ln(1+e^{\beta\mu}), \tag{vi}$$

and inverting we have

$$1 + \exp(\beta\mu) = \exp\left(\frac{nh^2}{2\pi mkT}\right),$$

from which we find $\mu(T, n)$:

$$\mu = kT \ln\left[\exp\left(\frac{nh^2}{2\pi mkT}\right) - 1\right]. \tag{vii}$$

Solution 4 *Exercise on page 513*

(a) The pressure is obtained by differentiating the thermodynamic potential with respect to the volume. From Eq. (5.1.6) with Eq. (5.3.13) we obtain

$$P = \frac{8\pi kT}{h^2} \int_0^\infty \ln\left\{1 + \exp\left[\beta\left(\mu - \frac{p^2}{2m}\right)\right]\right\} p^2 dp.$$

If the gas is degenerate, there are no electrons above the Fermi energy ϵ_F, or above the Fermi momentum p_F. In order to use this and the fact that the occupation function behaves like a step function, we integrate by parts to find that

$$P = \frac{8\pi kT}{3h^3} \int_0^\infty p^3 \frac{d}{dp} \ln\left\{1 + \exp\left[\beta\left(\mu - \frac{p^2}{2m}\right)\right]\right\} dp \tag{i}$$

$$= \frac{8\pi}{3h^3 m} \int_0^\infty \frac{p^4 dp}{\exp\left[\beta\left(\frac{p^2}{2m} - \mu\right)\right] + 1}. \tag{ii}$$

Now, we use the fact that the occupation number appearing in the integrand vanishes above p_F and is equal to unity below p_F so that

$$P = \frac{8\pi}{3mh^3}\int_0^{p_F} p^4 dp = \frac{8\pi p_F^5}{15mh^3}\,. \tag{iii}$$

Using (5.3.23) we express the Fermi momentum in terms of the density and obtain

$$P = \frac{8\pi}{15mh^3}\left[h^5\left(\frac{3n}{8\pi}\right)^{5/3}\right] = \left(\frac{3}{8\pi}\right)^{2/3}\frac{h^2 n^{5/3}}{5m}\,. \tag{iv}$$

A simpler way of obtaining the same result is to use the free energy $F = E - TS$. At $T = 0$, $F = E$ and we can calculate the pressure from the energy:

$$P = -\left(\frac{\partial F}{\partial V}\right)_N = -\left(\frac{\partial E}{\partial V}\right)_N\,, \tag{v}$$

and we obtain the energy using the average energy at $T = 0$ calculated in Exercise 3.12:

$$E = \frac{3}{5}N\epsilon_F = \left(\frac{3}{8\pi}\right)^{2/3}\frac{3h^2}{10m}\frac{N^{5/3}}{V^{2/3}}\,. \tag{vi}$$

By differentiating E we obtain the pressure as in Eq. (ii).

Note that the pressure and the energy density E/V satisfy the usual nonrelativistic relation [Eq. (1.1.6)]:

$$P = \frac{2}{3}\frac{E}{V}\,. \tag{vii}$$

(b) The electron density in aluminum is obtained using the data appearing in Table 5.3.2 (see Solutions 3.1 and 3.6 as well):

$$n = \frac{vN_0 d}{A} = 1.8\times10^{23}\ \text{cm}^{-3} = 1.8\times10^{29}\ \text{m}^{-3}\,, \tag{viii}$$

and substituting into Eq. (iv) we obtain

$$P = 1.4\times10^{11}\ \text{N/m}^2 \approx 1.4\times10^{6}\ \text{atm}. \tag{ix}$$

Solution 5 *Exercise on page 513*

(a) For extremely relativistic particles, $\epsilon(p) = cp$. Thus we cannot use the energy distribution function obtained in Eq. (5.3.17a). We begin, therefore, from Eq. (5.2.11a). The number of states per unit volume in phase space does not change when the particles become relativistic.

The only change is in the relation between the momentum and the energy. Thus we obtain from (5.2.11a)

$$N = \frac{2V}{h^3}\int \frac{d^3p}{\exp\{\beta[\epsilon(p)-\mu]\}+1} = \frac{8\pi V}{h^3}\int_0^\infty \frac{p^2dp}{\exp\{\beta[\epsilon(p)-\mu]\}+1}\,.$$

This is, after integrating over all angles, the generalization of Eq. (5.3.14). Writing $cp = \epsilon$,

$$n = \frac{8\pi}{(hc)^3}\int_0^\infty \frac{\epsilon^2 d\epsilon}{e^{\beta(\epsilon-\mu)}+1}\,, \tag{i}$$

and the energy distribution function is identified as

$$f_\epsilon(\epsilon) = \frac{8\pi\epsilon^2}{n(hc)^3}\cdot\frac{1}{e^{\beta(\epsilon-\mu)}+1}\,. \tag{ii}$$

When the gas is degenerate the occupation number is a step function [like (5.3.22)], and the relativistic Fermi energy, ϵ_R, is obtained from

$$n = \frac{8\pi}{(hc)^3}\int_0^{\epsilon_R}\epsilon^2 d\epsilon = \frac{8\pi\epsilon_R^3}{3(hc)^3}\,,$$

hence

$$\epsilon_R = hc\left(\frac{3n}{8\pi}\right)^{1/3}\,. \tag{iii}$$

We can write $f_\epsilon(\epsilon)$ in terms of ϵ_R in the form

$$f_\epsilon(\epsilon) = \frac{3\epsilon^2}{\epsilon_R^3}\frac{1}{e^{\beta(\epsilon-\mu)}+1}\,. \tag{iv}$$

The average energy is, in the degenerate case,

$$\langle\epsilon\rangle = \int_0^\infty \epsilon f_\epsilon(\epsilon)d\epsilon = \frac{3}{\epsilon_R^3}\int_0^{\epsilon_R}\epsilon^3 d\epsilon = \frac{3}{4}\epsilon_R\,, \tag{v}$$

and the energy density,

$$\frac{E}{V} = n\langle\epsilon\rangle = \left(\frac{3}{8\pi}\right)^{1/3}\times\frac{3}{4}hcn^{4/3}\,. \tag{vi}$$

The energy density in the relativistic case is proportional to $n^{4/3}$, in contrast to the nonrelativistic case, in which it is proportional to $n^{5/3}$.

Such a relativistic treatment is required when the average energy per particle is much larger than its rest energy. Since the average energy is of the same order of magnitude as ϵ_R we can write

$$\epsilon_R \gg mc^2\,,$$

and from here

$$\frac{h}{mc} \gg \left(\frac{8\pi}{3n}\right)^{1/3} . \qquad \text{(vii)}$$

The left hand side is the wavelength that characterizes the quantum motion of the relativistic particle and is called the Compton wavelength. Condition (vii) requires, therefore, that the average distance between particles in the gas be much smaller than their Compton wavelength. For an electron, the Compton wavelength is 0.02 Å, and the relativistic correction appears only at very high electron densities (above 10^{30} cm^{-3}). Such densities do not exist on earth. It is possible to find such densities in dense stars, such as white dwarfs, whose mass is the same as the sun's and whose radius is a hundredth of its radius. In fact such stars do not collapse under their self-gravity even though their nuclear fuel has run out, precisely due to the pressure of the degenerate electron gas.

(b) The pressure can be calculated by differentiating the thermodynamic potential Ω with respect to V, but at $T = 0$ it is simpler to use E. From (vi)

$$E = \frac{3}{4}\left(\frac{3}{8\pi}\right)^{1/3} \frac{hcN^{4/3}}{V^{1/3}} , \qquad \text{(viii)}$$

and then

$$P = -\left(\frac{\partial E}{\partial V}\right)_N = \frac{1}{4}\left(\frac{3}{8\pi}\right)^{1/3} hcn^{4/3} . \qquad \text{(ix)}$$

Comparing this to (vi) we obtain the regular relativistic relation between the pressure and the energy density, (1.1.7):

$$P = \frac{1}{3}\frac{E}{V} .$$

Solution 6 *Exercise on page 513*

(a) To calculate the pressure we write the potential Ω of Eq. (5.4.10) using (5.4.5):

$$\Omega = \Omega_0 + \frac{2\pi VkT(2m)^{3/2}}{h^3} \int_0^\infty \epsilon^{1/2} \ln(1 - e^{\beta(\mu-\epsilon)}) d\epsilon ,$$

and then

$$P = -\left(\frac{\partial \Omega}{\partial V}\right)_{T,\mu} = -\frac{2\pi kT(2m)^{3/2}}{h^3} \int_0^\infty \epsilon^{1/2} \ln(1 - e^{\beta(\mu-\epsilon)}) d\epsilon .$$

Note that the first term, which represents the contribution of the bosons occupying the ground level, does not contribute to the pressure since these bosons have zero kinetic energy. Below the critical temperature, $\mu = 0$ and we obtain

$$P = -\frac{2\pi(2m)^{3/2}(kT)^{5/2}}{h^3}\int_0^\infty x^{1/2}\ln(1-e^{-x})dx\,,$$

where $x = \beta\epsilon$. We integrate by parts and use Eq. (5.A.14) to find that

$$\int_0^\infty x^{1/2}\ln(1-e^{-x})dx = -\frac{2}{3}\int_0^\infty \frac{x^{3/2}dx}{e^x-1}$$
$$= -\frac{2}{3}\Gamma\left(\frac{5}{2}\right)\zeta\left(\frac{5}{2}\right) = -0.67\sqrt{\pi}\,, \qquad \text{(i)}$$

and

$$P = \frac{1.341(2\pi m)^{3/2}(kT)^{5/2}}{h^3}\,. \qquad \text{(ii)}$$

The fact that the pressure is independent of the volume is because the system is below the critical temperature. In this case decreasing the volume does not increase the pressure but reduces the density of bosons in the excited states, n_* [see (5.4.9)], transferring them to the ground state.

Comparison with the expression for the energy (5.4.14) gives once more

$$P = \frac{2}{3}\cdot\frac{E}{V}\,. \qquad \text{(iii)}$$

(b) Even when the fermion gas is fully degenerate the occupation of the ground level cannot become macroscopic as in the case of the boson gas, since the Pauli principle forbids this. Hence the relative error due to the fact that the contribution of the ground level is not included in Eq. (5.3.14), for instance, is negligible: of relative order $1/N$.

Solution 7 *Exercise on page 513*

To determine if the phenomenon of Bose–Einstein condensation occurs, we have to check the relation between the density of particles (per unit area) in the excited states n_e and the chemical potential μ. If n_e is bounded from above and hence also $N_e = n_e A$, particles added above the maximum of N_e will accumulate in the ground state, and Bose–Einstein condensation will occur.

From Eq. (5.4.3) or (5.4.2), we obtain in the two-dimensional case

$$n_e = \frac{N_e}{A} = \frac{1}{h^2}\int_0^\infty \frac{d^2p}{\exp\left[\beta\left(\frac{p^2}{2m}-\mu\right)\right]-1}\,,$$

and after integrating over the angles and changing to the energy variable, we obtain

$$n_e = \frac{2\pi m}{\hbar^2} \int_0^\infty \frac{d\epsilon}{e^{\beta(\epsilon-\mu)} - 1} \,. \tag{ii}$$

The right hand side is again a monotonic increasing function of μ.

When $\mu \to 0$ the denominator vanishes in the low energy region and the integral diverges. To see this we note that at $\mu = 0$, the integrand behaves as

$$\frac{1}{e^{\beta\epsilon} - 1} \sim \frac{1}{\beta\epsilon} \,, \tag{iii}$$

for small ϵ. The integral of such a function diverges. The reason why the corresponding integral in three dimensions does not diverge is that there is an additional factor of $\epsilon^{1/2}$ in the numerator, which causes the integrand to behave like $\epsilon^{-1/2}$ near $\epsilon \to 0$. The integral of $\epsilon^{-1/2}$ converges.

This implies that n_e is unbounded, and hence additional particles can be accommodated in the excited levels, and the ground level will not be occupied by a macroscopic number of bosons.

It is true that as the energy ϵ decreases, the occupation number increases as implied by (ii), but this is not Bose–Einstein condensation, since for any finite number of particles $\mu < 0$.

In order to see the explicit dependence of μ upon n_e, which is actually n, we calculate the integral as in Self-Assessment Exercise 3:

$$\begin{aligned} n &= \frac{2\pi mkT}{h^2} \int_0^\infty \frac{dx}{e^{x-\beta\mu} - 1} = \frac{2\pi mkT}{h^2} \int_0^\infty \frac{e^{\beta\mu - x}}{1 - e^{\beta\mu - x}} dx \\ &= -\frac{2\pi mkT}{h^2} \ln(1 - e^{\beta\mu}) \,, \end{aligned}$$

and hence

$$\mu = kT \ln\left[1 - \exp\left(-\frac{nh^2}{2\pi mkT}\right)\right] . \tag{iv}$$

Note that μ (as well as $\beta\mu$) is always negative and only vanishes in the limit $n \to \infty$ or $T \to 0$.

Solution 8 *Exercise on page 513*

Since the fermions are massless and have zero chemical potential, they can be created and destroyed freely like photons, and their number is not conserved. The *average* number of fermions at temperature T is obtained by substituting $\mu = 0$ in the general expression for n which was obtained for extremely relativistic fermion gas in the solution of Self-Assessment Exercise 5:

$$n = \frac{8\pi}{(hc)^3} \int_0^\infty \frac{\epsilon^2 d\epsilon}{e^{\beta\epsilon} + 1} \,. \tag{i}$$

The number of fermions per unit volume and per unit energy is thus

$$n_\epsilon(\epsilon) = \frac{8\pi}{(hc)^3}\frac{\epsilon^2}{e^{\beta\epsilon}+1}\,. \tag{ii}$$

We transform to a frequency variable, using $\epsilon = \hbar\omega$, and write Eq. (i) as

$$n = \frac{1}{\pi^2 c^3}\int_0^\infty \frac{\omega^2 d\omega}{e^{\beta\hbar\omega}+1}\,, \tag{iii}$$

and their density per unit frequency:

$$n_\omega(\omega) = \frac{1}{\pi^2 c^3}\frac{\omega^2}{e^{\beta\hbar\omega}+1}\,. \tag{iv}$$

It is instructive to compare with the corresponding expression for photons, Eq. (4.4.13). The energy density of the fermions per unit frequency is obtained by multiplying $n_\omega(\omega)$ by the fermion energy, $\hbar\omega$:

$$\rho(\omega) = \frac{1}{\pi^2 c^3}\frac{\hbar\omega^3}{e^{\beta\hbar\omega}+1}\,, \tag{v}$$

and we find that the total energy density is an integral over all frequencies:

$$u = \frac{1}{\pi^2 c^3}\int_0^\infty \frac{\hbar\omega^3 d\omega}{e^{\beta\hbar\omega}+1} = \frac{(kT)^4}{\pi^2(\hbar c)^3}\int_0^\infty \frac{x^3 dx}{e^x+1}\,. \tag{vi}$$

For the calculation of the integral we use Eq. (5.A.13) from the appendix. It gives the value

$$\frac{7}{8}\Gamma(4)\zeta(4) = \frac{7}{8}\cdot\frac{\pi^4}{15}\,, \tag{vii}$$

and hence

$$u = \frac{7\pi^5 k^4}{15(hc)^3}T^4\,.$$

The radiation density of a black body satisfies a similar law:

$$u_{BB} = \frac{8\pi^5 k^4}{15(hc)^3}T^4$$

[see Eqs. (4.4.16) and (4.4.17)] and their ratio is 7/8.

Solution 9 *Exercise on page 513*

We begin from the condition

$$\mu_+ + \mu_- = 0\,, \tag{i}$$

where $\mu_\pm$ are the chemical potentials of $e^\pm$, given implicitly by Eq. (i) or (ii) in the solution of Self-Assessment Exercise 5. If the gas is electrically neutral, the densities of the electrons and the positrons are equal and so are their chemical potentials:

$$\mu_+ = \mu_- \,. \tag{ii}$$

Hence,

$$\mu_\pm = 0\,. \tag{iii}$$

At extremely relativistic temperatures it is possible to neglect the mass of the particles and then we can use Eq. (i) in the solution of Self-Assessment Exercise 5 with $\mu = 0$. The result is

$$n_\pm = \frac{8\pi}{(hc)^3}\int_0^\infty \frac{\epsilon^2 d\epsilon}{e^{\beta\epsilon}+1} = \frac{8\pi}{(hc)^3}(kT)^3\int_0^\infty \frac{x^2 dx}{e^x+1}\,, \tag{iv}$$

with $x = \beta\epsilon$. The integral can be calculated using Eq. (5.A.13), and its value is $\frac{3}{2}\zeta(3)$.

The density of electrons and positrons is, therefore,

$$n_\pm = \frac{12\pi\zeta(3)}{(hc)^3}(kT)^3\,. \tag{v}$$

Index